JMP® Start Statistics

A Guide to Statistical and Data Analysis
Using JMP® and JMP IN® Software

by John Sall and Ann Lehman, SAS Institute

Duxbury Press
An Imprint of Wadsworth Publishing Company

I(T)P® An International Thomson Publishing Company

Belmont • Albany • Bonn • Boston • Cincinnati • Detroit • London • Madrid • Melbourne
Mexico City • New York • Paris • San Francisco • Singapore • Tokyo • Toronto • Washington

Software Development Editor: Stan Loll
Editorial Assistant: Martha O'Connor
Production Services Coordinator: Gary Mcdonald
Production: Ann Lehman/SAS Institute
Marketing Manager: Joanne M. Terhaar

Print Buyer: Stacey Weinberger
Permissions Editor: Peggy Meehan
Copy Editor: Robin Gold
Cover: Stuart Paterson/Image House
Compositor: SAS Institute
Printer: Globus Printing Company

Printed in the United States of America
1 2 3 4 5 6 7 8 9 10

For more information, contact Duxbury Press at Wadsworth Publishing Company:

Wadsworth Publishing Company
10 Davis Drive
Belmont, California 94002, USA

International Thomson Editores
Campos Eliseos 385, Piso 7
Col. Polanco
11560 México D.F. México

International Thomson Publishing
EuropeBerkshire House 168-173
High Holborn
London, WC1V 7AA, England

International Thomson Publishing GmbH
Königswinterer Strasse 418
53227 Bonn, Germany

Thomas Nelson Australia
102 Dodds Street
South Melbourne 3205
Victoria, Australia

International Thomson Publishing Asia
221 Henderson Road
#05-10 Henderson Building
Singapore 0315

Nelson Canada
1120 Birchmount Road
Scarborough, Ontario
Canada M1K 5G4

International Thomson Publishing Japan
Hirakawacho Kyowa Building, 3F
2-2-1 Hirakawacho
Chiyoda-ku, Tokyo 102, Japan

ISBN 0-534-26565-0 (JMP Start Statistics)
ISBN 0-534-26564-2 (JMP IN 3 Windows Version)
ISBN 0-534-26562-6 (JMP IN 3 Macintosh Version)

Contents

Part III REFERENCE DOCUMENTATION

PREFACE

JMP® is statistical discovery software. JMP helps you explore data, fit models, discover patterns, and discover points that don't fit patterns. This book is a guide to statistics using JMP.

This book is distributed in two ways:

- It is the documentation for JMP IN software, the student version of JMP distributed by Duxbury Press. This book covers most of the features of JMP and JMP IN in the context of learning statistics. JMP IN is for student use only.
- It is supplemental documentation for the full professional version of JMP. The books that come with the full version of JMP are reference manuals; this book focuses on learning statistics.

The Software

As statistical discovery software, the emphasis in JMP is to interactively work on data to find out things.

- With graphics, you are more likely to make discoveries. You are also more likely to understand the results.
- With interactivity, you are encouraged to dig deeper, for one analysis to lead to a refinement, for one discovery to lead to another discovery, and to try out more things that might improve your chances of discovering something important.
- With a progressive structure, you build context that maintains a live analysis, so you don't have to redo things, so that details come to attention at the right time.

Software's job is to create a virtual workplace. The software has facilities and platforms where the tools are located and the work is performed. JMP provides the workplace that we think best for the job of analyzing data. With the right software workplace, researchers will celebrate computers and statistics, rather than avoid them.

JMP aims to present a graph beside every statistic. You can and should always see the analysis in both ways, statistical text and graphics, without having to ask for it that way. The text and graphs stay together.

JMP is controlled largely through point and click, mouse manipulation. If you click on a point in a plot, JMP identifies and highlights the point in the plot, also highlights the point in the data table, and highlights it everywhere else the point is represented.

JMP has a progressive organization. You begin with a simple surface at the top, but as you analyze you see more and more depth. The analysis is alive, and as you dig deeper into the data, more and more options are offered according to the context of the analysis.

In JMP, completeness is not measured by the "feature" count, but by the range of applications, and the orthogonality of the tools. In JMP, you get more feeling of being in control in spite of less awareness of the control surface. In JMP you get a feeling that statistics is an orderly discipline that makes sense, rather than an unorganized collection of methods.

The statistical software package is often the point of entry into the practice of statistics. JMP endeavors to offer fulfillment rather than frustration, empowerment rather than intimidation.

If you give me a truck, I will find someone to drive it for me. But if you give me a sports car, I will learn to drive it myself. Believe that statistics can be interesting and reachable so that people will want to drive that vehicle.

SAS Institute

JMP is from SAS Institute, a large private research institution specializing in data analysis software. SAS Institute's principal commercial product is the SAS System, a large software system that performs much of the world's large scale statistical data processing. JMP is positioned as the small personal analysis tool, involving a much smaller investment than the SAS System.

This Book

Software Manual and Statistics Text

This book is a mix of software manual and statistics text. It is designed to be a complete and orderly introduction to analyzing data. It is a teaching text, but is especially useful when used in conjunction with a standard statistical textbook.

Not Just the Basics

A few of the techniques in this book are not found in most introductory statistics courses, but are accessible in basic form using JMP IN. These techniques include logistic regression, correspondence analysis, principal components with biplots, leverage plots, and density estimation. All these techniques are used in the service of understanding other more basic methods. Where appropriate, we have labeled supplemental material as "Special Topics" so that they are recognized as optional material that is not on the main track.

JMP IN also includes several advanced methods that are not covered in this book, such as nonlinear regression and multivariate analysis of variance. If you are planning to use these features, we recommend that you refer to the help system, or obtain the documentation for the professional version of JMP.

Examples Both Real and Simulated

Most examples are real-world applications. A few simulations are included too, so that difference between the true value and the estimate can be discussed, along with the variability in the estimates. Some examples are unusual, calculated to surprise you in the service of emphasizing some important concept. Everything in the book can be done by the JMP IN software included with the student package. The data for the examples are on the disks, with step-by-step instructions in the text. JMP IN can also import data from text files distributed on diskettes with other textbooks.

Acknowledgments

The software developers for JMP are John Sall, Katherine Ng, Michael Hecht, David Tilley, and Richard Potter. Ann Lehman directs documentation.

Thanks also to Jeff Perkinson, Bill Gjertsen, Annie Dudley, Duane Hayes, Aaron Walker, Ike Walker, Elizabeth Sall, Pratima Virkar, Fouad Younan, Annette Sanders, and the Statistics R&D group at SAS Institute.

A special thanks to Al Best for helping us get started with the book and for contributing several sections of chapters.

Thank you to the testers for both JMP and JMP IN. A number of features in JMP IN are due to feedback from Robert Stine. Thanks to Andy Mauromoustakos and Mike Deaton for lobbying to restore a number of features to JMP IN that were originally not in the plan. Also thanks go to Katherine Monti and Hiroshi Yamauchi.

PRELIMINARIES

JMP IN Package Contents and Installation

Your JMP IN package contains program disks and this book. Installation instructions are contained on the installation disks.

Macintosh

If you are using a Macintosh, you install JMP IN by inserting "Disk 1." There could be a README file on the first disk that contains late-breaking news (features that emerged too late to be in this book) or warnings about known problems. It is always a good idea to read this file. Double-click the file to open it. There is also the JMP IN installer on Disk 1. Begin installation by double-clicking the installer program. You will be prompted by the installer. Follow the instructions to complete a standard install.

Windows

If you are using a Windows computer, the installation instructions are on the disk labels. You install JMP IN by inserting "Disk 1." There could be a README file on the first disk that contains late-breaking news (features that emerged too late to be in this book) or warnings about known problems. It is always a good idea to read this file. Double-click the file to open it. There is also the JMP IN installer on Disk 1. Begin installation by double-clicking the installer program. You will be prompted by the installer. Follow the instructions to complete a standard install.

The Windows version of JMP runs best on Windows 95 or Windows NT. When JMP is run on the older Windows 3.1, it uses a special system library called Win32S, which is installed as part of the JMP installation process if your Windows 3.1 system does not already have it.

What You Need to Know

...about your computer

Before you begin using JMP IN, you should be familiar with standard operations and terminology such as *click*, *double-click*, *COMMAND-click* and *OPTION-click* on the Macintosh (*CONTROL-click* and *ALT-click* under Windows), *SHIFT-click*, *drag*, *select*, *copy*, and *paste*. You should also know how to use menu bars and scroll bars, move and resize windows, and manipulate files. If you are using your computer for the first time, consult the reference guides that came with it for more information.

...about Statistics

This book is designed to help you learn about statistics. Even though JMP IN has many advanced features, you need no background of formal statistical training to use it. All analysis platforms include graphical displays with options that help you review and interpret the results. Each platform also includes access to help windows that offer general help and some statistical details.

Learning About JMP

...on your own with JMP Help

If you are familiar with Macintosh or Microsoft Windows software, you may want to proceed on your own. After you install JMP, you can open any of the JMP sample data files and experiment with analysis tools. Help is available for most menus, options, and reports.

There are several ways to see JMP Help:

- Select **About JMP** from the Apple menu on the Macintosh or the **Contents** command from **Help** menu of Microsoft Windows. Your cursor becomes a question mark as it passes over the list of buttons on the right side of the About JMP window. When you click on a button, JMP leads you through help about that topic.
- Click the **Statistical Guide** button on the Macintosh About JMP screen, or select **Statistical Guide** from the **Help** menu under Windows. The JMP Statistical Guide is a scrolling alphabetical reference that tells you how to generate specific analyses using JMP and accesses further help for that topic.
- You can choose **Help** from popup menus in JMP report windows.
- After you generate a report, select the help tool (**?**) from the **Tools** menu and click the report surface. Context-sensitive help tells about the items in that report window.
- If you are using Microsoft Windows, help in typical Windows format is available under the **Help** menu on the main menu bar.
- If you use a Macintosh with Version 7 system software, balloon help is available.

...hands-on examples

This book, *JMP Start Statistics*, documents JMP by describing its features, reinforced with hands-on examples. By following along with these step-by-step examples, you can quickly become familiar with JMP menus, options, and report windows. Follow-along steps for example analyses begin with the § symbol in the margin.

...reading about JMP

The full version of JMP version 3 is accompanied by three books called *The JMP Introductory Guide*, *The JMP User's Guide*, and *The JMP Statistics and Graphics Guide*. These references cover all the commands and options in JMP and have extensive examples of the Analyze and Graph platforms. These books may be available in your department, computer lab, or library.

Chapter Organization

This book, *JMP Start Statistics*, contains chapters of documentation supported by guided actions you can take to become handy with the JMP IN product. It is divided into three parts:

Part I is called **JMP In with Both Feet**. Its purpose is to get you started quickly. It has three chapters that tell you how to use JMP tables, how to use the JMP calculator, and give and overview of how to obtain results from the Analyze and Graph menus.

Chapter 1, "JMP Right In," starts at the beginning. It tells you how to start and stop JMP IN, get at JMP tables, and takes you on a short guided tour. You are introduced to the general *personality* of JMP. You will see how data are handled by JMP. There is an overview of all analysis and graph commands, information about how to navigate a platform of results, and a description of the tools and options available for all analyses. The Help system is covered in detail.

Chapter 2 "JMP Data Tables" focuses on using the JMP data table. It shows how to create tables, subset, sort, and manipulate them with built in menu commands, and how to get data and results out of JMP and into a report.

Chapter 3, "Calculator Adventures," covers the calculator available to each column in every table. There is a description of the calculator components and overview of the extensive functions available for calculating column values.

Part II is called **Statistical Sleuthing**. It covers the array of analysis techniques offered by JMP IN. Chapters begin with simple-to-use techniques and work gradually to more complex methods. Emphasis is on learning to think about these techniques and on how to visualize data analysis at work. JMP IN offers a graph for almost every statistic and supporting tables for every graph. Using highly interactive methods you can learn more quickly and discover what your data have to say.

Chapter 4, "What is Statistics," gives you some things to ponder about the nature and use of statistics. It also attempts to dispel statistical fears and phobias that are prevalent among students and professionals alike.

Chapter 5, "Univariate Distributions: One Variable, One Sample," covers distributions of continuous and categorical variables and statistics to test univariate distributions.

Chapter 6, "The Difference Between Two Means," covers
t tests of independent groups and tells how to handle paired data. The nonparametric approach to testing related pairs is shown.

Chapter 7, "Comparing Many Means: One-Way Analysis of Variance," covers one-way analysis of variance, with standard statistics and a variety of graphical techniques.

Chapter 8, " Fitting Curves through Points: Regression," shows how to fit a regression model for one factor.

Chapter 9, "Categorical Distributions," discusses how to think about the variability in single batches of categorical data. It covers estimating and testing probabilities in categorical distributions, shows Monte Carlo methods, and introduces the Pearson and Likelihood ratio chi-square statistics.

Chapter 10, "Categorical Models," covers fitting categorical responses to a model, starting with the usual tests of independence in a two-way table, and continuing with graphical techniques, and logistic regression.

Chapter 11, "Multiple Regression," describes the parts of a linear model, talks about fitting models with multiple numeric effects, and shows a variety of examples including the use of stepwise regression to find active effects.

Chapter 12, "Fitting Linear Models," is an advanced chapter that continues the fitting discussion, moving on to categorical effects, complex effects such as interactions and nested effects.

Chapter 13, "Bivariate and Multivariate Relationships," looks at ways to examine two or more response variables using correlations, scatterplot matrices, three-dimensional plots, principal components, and other techniques. Outliers are discussed.

Chapter 14, "Design of Experiments," looks at JMP IN's built-in commands that generate specified experimental designs. Also, examples of how to analyze common screening and response level designs are covered.

Chapter 15, "Statistical Quality Control," is a survey of types of control charts (Shewhart charts) available in JMP IN and an explanation of when to use them.

Chapter 16, "Elementary Time Series," discusses some beginning methods for looking at data with correlations over time.

Chapter 17, "Machines of Fit" is an essay about statistical fitting that may prove enlightening to those who have a mind for mechanics.

Part III, "Reference Documentation," is a set of appendices that document all commands, options, and calculator functions in the JMP IN product.

Appendix A lists and describes all the main menu commands.

Appendix B lists and describes all options with the Analyze and Graph platforms.

Appendix C lists and defines all calculator functions.

Appendix D lists differences between JMP IN and JMP.

Typographical Conventions

The following manual conventions help you relate written material to information you see on your screen:

- Most open data table names used in examples are capitalized (ANIMALS or ANIMALS.JMP) in this document. In the Macintosh SAMPLE DATA folder ,the file names in the Macintosh SAMPLE DATA folder are capitalized (ANIMALS.JMP). Files in the Windows data directory are in lower case (animals.jmp), but show in uppercase when the table is opened.

- Reference to menu names (**File** menu) or menu items (**Save** command) appear in the **Helvetica bold** font, similar to the way they appear on your screen.
- When you are asked to choose a command from the main menu, this is written as **Rows→Add Rows** or Choose **File→Save As**, meaning go to the **File** menu and choose the **Save As** command.
- Likewise, items on popup menus in reports are shown in the **Helvetica bold** font, but you are given a more detailed instruction about where to find the command or option. For example, you might be asked to select the **Show Points** option from the check-mark menu, or select the **Save Predicted** command from the Fitting popup menu. The popup menus will always be visible on the platform or on its window border.
- References to variable names in data tables and items in reports show in either Helvetica but can appear in illustrations in either a plain or boldface font. These items show on your screen as you have specified them in your JMP Preferences.
- Words or phrases that are important or have definitions specific to JMP are in *italics* the first time you see them.
- When there is an action statement, you can follow along with the example by following the instruction. These statements are preceded with an action symbol (§) in the margin. An example of an action statement is:

§ Highlight the Month column by clicking the area above the column name, and then choose **Cols→Column Info**.

- Occasionally side-comments or special paragraphs are included and shaded in gray.

Chapter 1
Jump Right In

Hello

JMP software is so easy to use that after reading this chapter you'll find yourself confident enough to learn everything on your own. So let's cover the essentials fast before you escape this book. This chapter offers you the opportunity to make a small investment in time for a large return later on.

If you are already familiar with JMP IN and want to dive right into statistics, you can skip ahead to Chapters 5–17 in Part II, "STATISTICAL SLEUTHING."

You can always return later for more details about using JMP or for more details about statistics.

> *By the way, JMP is pronounced "jump." We use JMP and JMP IN interchangeably. JMP IN is the student version of JMP.*

Chapter 1 Contents

You can mouse along with the examples whenever you see an action symbol (§)

First Session

This first section just gets your toes wet. In most of the chapters of this book, you can follow along in a hands-on fashion. Watch for the action symbol (§) and perform the action it describes. Try it now:

§ To Start a JMP IN session, double-click the JMP IN application icon or select its icon and choose **File→Open** from your system's main menu.

When the application is active, you see the following JMP menu bar:

JMP Main Menu Bar

File Edit Tables Rows Cols Analyze Graph Tools Window

File Edit Tables Rows Cols Analyze Graph Tools Window Help

As with other applications, the **File** menu has all the strategic commands, like opening data tables or saving them. At the end of this menu is the most drastic command of all, **Exit** (on Windows) or **Quit** (on Macintosh).

Let's start by opening a JMP data table and doing a simple analysis.

Open a JMP Data Table

When you first start JMP, it makes an empty untitled data table, which you can fill with data as shown in Chapter 2, "JMP Data Tables." Rather than type data now, or import data from text files, let's open a JMP data table from the collection of tables that comes with JMP IN, the *sample data.*

§ Choose the **Open** command in the **File** menu (choose **File→Open**). When the Open File dialog appears, as shown in **Figure 1.1**, select BIGCLASS.JMP from the list of sample data files. You may have to navigate through your directories to find it. Look for a directory (folder) with the name DATABOOK, where all the tables that are used in this book will be. Then click the **Open** button on the open-file dialog.

Figure 1.1 Open File Dialogs

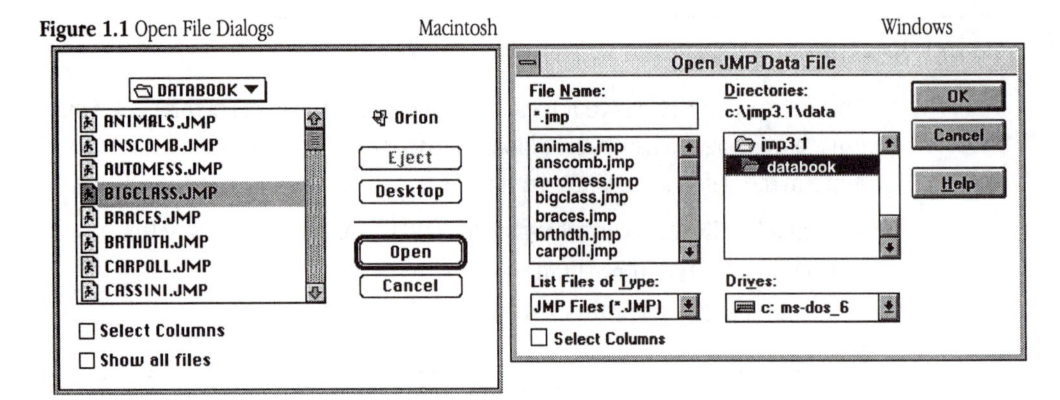

You should now see the spreadsheet presentation of the table with columns called Name, Age, Sex, Height, and Weight (shown below in **Figure 1.2**).

In Chapter 2, "JMP Data Tables," you will learn the facilities of the data table, but for now let's try an analysis.

Figure 1.2 Partial Listing of the BIGCLASS Data Table

BIGCLASS.JMP

5 Cols / 40 Rows	Name	Age	Sex	Height	Weight
1	KATIE	12	F	59	95
2	LOUISE	12	F	61	123
3	JANE	12	F	55	74
4	JACLYN	12	F	66	145
5	LILLIE	12	F	52	64
6	TIM	12	M	60	84

Launch an Analysis Platform

What is the distribution of the Weight and Age columns in the table?

§ Click on the **Analyze** menu and choose the command **Distribution of Y** (choose **Analyze→Distribution of Y**). This is called *launching the Distribution platform.*

§ A dialog (**Figure 1.3**) now appears, prompting you to choose the variables you want to analyze. Click on Weight to highlight it in the variable list on the left of the dialog. Click **Add** to carry it into the selection list. Similarly highlight and **Add** the Age variable. The term *variable* is often used to mean a column in the data table. Picking variables to fill roles is sometimes called *role assignment*.

You should now see the completed role assignment dialog shown in **Figure 1.3.**

Figure 1.3 The Role Selection Dialog

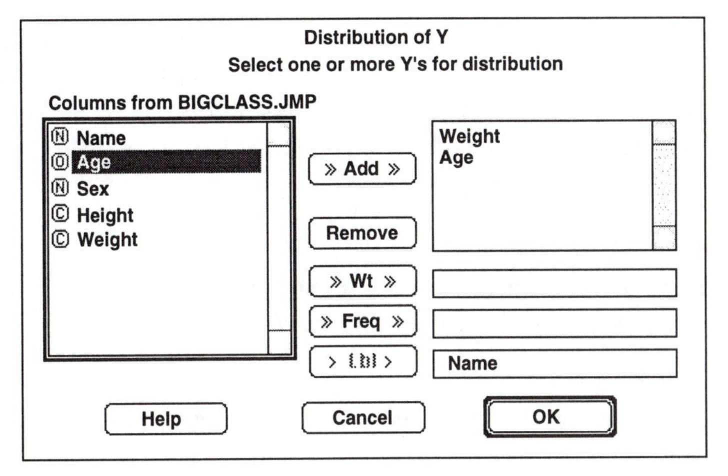

§ Click **OK** to close the dialog and open the window for the Distribution platform.

The resulting window shows the distribution of the two variables, Weight and Age as in **Figure 1.4**.

Figure 1.4 Histograms in the Distribution Platform

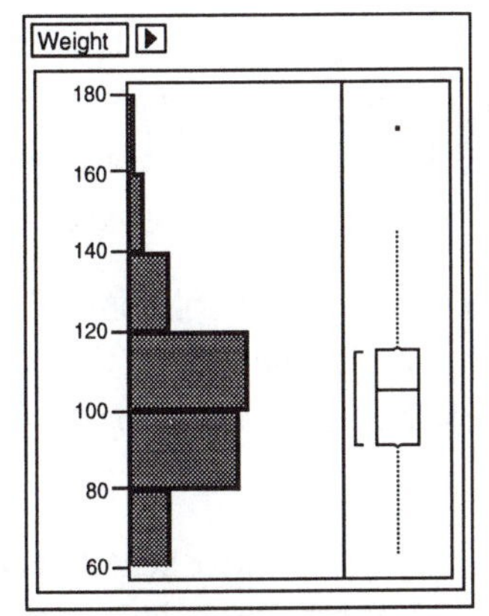

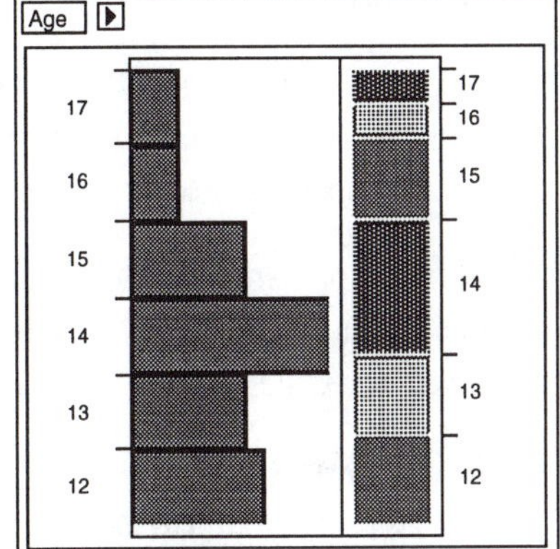

Interact with the Surface of the Platform

The idea of a platform is that you start with a basic analysis and then you work with it interactively to dig into a more detailed analysis or customize the presentation. It's a live object, not a dead report.

Row Highlighting

Click on one of the histogram bars, say the Age bar for 12. The bar highlights. Also portions of the bars in the other histogram highlight. And certain rows in the data table corresponding to that histogram bar highlight. This is the dynamic linking of the rows in the plots and data tables. Later you will see other ways of highlighting and working with attributes of the rows of a table. On the right of the **Weight** histogram is a box plot with a single point near the top. Click on that point, holding down the mouse button. The label, LAWRENCE, becomes highlighted as well as being highlighted in the data table.

Reveal Buttons

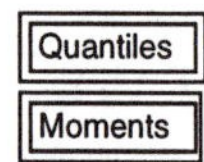

Each report title is enclosed in a frame that serves as a button. Click on a button to alternately conceal and reveal the contents.

Border Popup Menus

At the bottom of the window, level with the horizontal scroll bar, you will find three small icons, each of which offers a popup menu.

A check-mark popup menu is for commands specific to the platform, but applies to all the analyses in the window. For example, you can change the orientation of the graphs in the Distribution platform by checking or unchecking **Horizontal Layout** in check-mark menu. Try it.

The dollar (**$**) popup menu is for saving results into table columns. Try the command **Save Standardized** and you will find that it makes a new column in the data table containing **Weight** standardized to mean zero and standard deviation 1.

The star (*) popup menu lists commands common to all platforms. Try selecting the **Conceal All** command. The reports shrink into title buttons. Click on the title buttons to reveal them again, or use the star menu **Reveal All** command.

Interior Popup Menus

This icon is usually found to the right of the report name or beneath a graphical display. It contains commands to do something specific for the context in which it is located. For example, at the top of the **Weight** report, try the popup command **Stem and Leaf**, which results in a stem-and-leaf plot below the other reports.

Graph Resizing

When you click inside a graph, a small box appears in the lower-right corner of the inside frame of the graph. If you click inside this box, you can drag an outline of the graph to resize it.

Special Tools

When you need to do something special, to turn the mouse and its arrow cursor into something more specialized, pick a tool in the tools menu (or in a tool palette), and click or drag inside the analysis.

The hand is for grabbing things. Try getting this tool and then start clicking and dragging in a histogram.

The brush is for highlighting all the data in an rectangular area. Try getting this tool and dragging in the histogram. Hold down the OPTION or ALT key to change the size of the rectangle.

The lasso is for roping points to select them. Try this later for scatterplots.

? The question mark is for getting help on the analysis platform surface. Get the question mark tool and click on different areas in the Distribution platform.

The magnifier is for zooming in to certain areas in a plot. This tool is for scatterplots. Hold down the OPTION or ALT key and click to restore the original scaling.

The scissors tool is for picking out an area to copy so that you can paste its picture into another application. Hold down the OPTION or ALT key to select in a different way. See Chapter 2, "JMP Data Tables," for details.

+ The crosshairs are for sighting along lines in a graph.

In JMP, the surface of an Analysis platform bristles with interactivity. Launching an analysis is just the starting point. Then you explore, evaluate, follow clues, dig deeper, get more details, fuss with the presentation, and so on.

Modeling Type

Notice in the previous example that there are different kinds of graphs and reports for Weight and Age. This is because the variables have different modeling types. The Weight column has a modeling type of *continuous*, so JMP treated the numbers as values from a continuous scale. The Age column is categorical, having a modeling type of *ordinal* (*nominal* is similar), so JMP treated its values as labels of discrete categories.

Look at the top of the Age column in the BIGCLASS table. There is an "O" in a small box on the column heading, as shown in **Figure 1.5**. The small box is a popup menu. If you click on this box you see a menu for choosing the *modeling type* for a column, with the choices **Continuous**, **Ordinal**, and **Nominal**.

Figure 1.5 Modeling Type Popup Menu

Why does JMP make a fuss over this modeling type? For one thing, it's a convenience feature. You are telling JMP ahead of time how you want the column treated so that you don't have to say it again every time you do an analysis. It also helps reduce the number of commands you need to learn. Instead of two distribution platforms, one for continuous variables and a different one for categorical variables, a single command performs the anticipated analysis based on the modeling type you have assigned.

Here is a brief description of the three modeling types:

- *Continuous* numeric values are used in a model directly.
- *Ordinal* values are category labels, but their order is meaningful.
- *Nominal* values are treated as unordered categories, names of levels. The ordinal and nominal modeling types are treated the same in most analyses, and are often referred to collectively as categorical.

You can change the modeling type any time you want the variable treated differently. For example, if you wanted to find the mean of Age instead of the category frequency counts, then you simply change its modeling type from **Ordinal** to **Continuous** and reanalyze.

In the following section you will see that the modeling type affects the kind of analysis done in several of the platforms.

Analyze and Graph

The **Analyze** and **Graph** menus shown in **Figure 1.6** launch interactive platforms to analyze the data. The **Analyze** menu is for statistics and data analysis. The **Graph** menu is for specialized plots. That distinction, however, doesn't prevent analysis platforms from being full of graphs, or the graph platforms from computing statistics. Each platform provides a context for sets of related statistical methods and graphs. It won't take long to learn this short list of platforms.

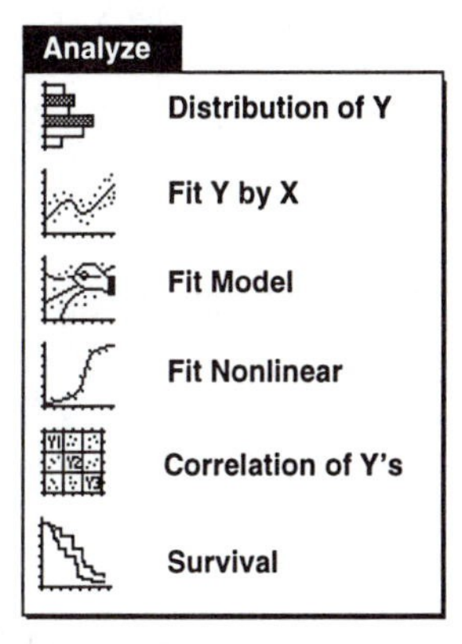

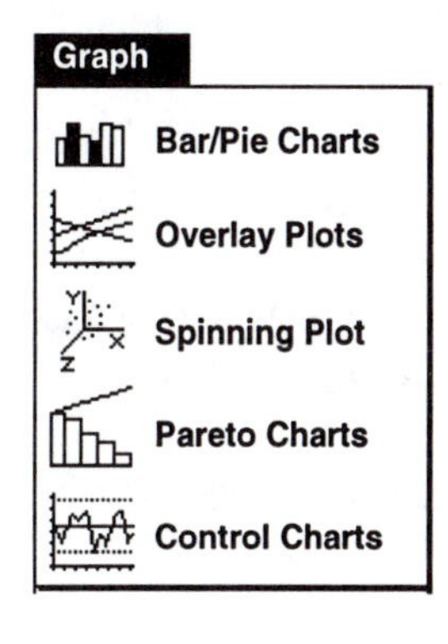

Figure 1.6 The Analyze and Graph menus launch interactive platforms

The Analyze Menu

Distribution of Y
is for univariate statistics, which describe the distribution of values for each variable, one at a time, using histograms, box plots, and other statistics.

Fit Y by X
is for bivariate analysis. A bivariate analysis describes the distribution of the Y variable as it depends on the value of the X variable. The continuous or categorical modeling type of the Y and X variables leads to the four analyses: scatterplot with regression curve fitting, one-way analysis of variance, contingency table analysis, and logistic regression.

Fit Model
launches a *general fitting platform* for linear models such as multiple regression, analysis of variance models, and others.

Fit Nonlinear
fits models that are nonlinear in their parameters, using iterative methods.

Correlation of Y's
describes relationships among variables, focusing on the correlation structure: correlations and other measures of association, scatterplot matrices, multivariate outliers, and principal components.

Survival
models the time until an event, allowing censored data. This kind of analysis is used in both reliability engineering and survival analysis.

The Graph Menu

Bar/Pie Charts
gives many forms of charts such as bar, pie, line, and needle charts.

Overlay Plots
overlays several numeric Y variables, with options to connect points, show a step plot, needle plot, and others.

Spinning Plot
shows a three-dimensional rotating scatterplot with options to see principal components and biplots.

Pareto Charts
plots frequency counts in descending order of count, with a line showing the cumulative count.

Control Charts
monitors a process through time to watch for it going out-of-control.

Navigating the Platforms, Building the Context

The first few times you use JMP, you will have navigational questions: How do I get XXX? For example, how do I do a histogram? How do I get a t test?

The strategy for approaching JMP analyses is that you build the context. Once you build that context, then the graphs and statistics become easily available–often they happen automatically, without having to ask for them specifically.

The are three keys for establishing the context:

Modeling Type
identifies a variable as either continuous or categorical (nominal and ordinal).

X or Y Role
identifies whether the variable is a response (Y) or a factor (X).

Analysis Platform
is the general approach and character of the analysis.

Once you settle your context, you can find everything.

Contexts for a Histogram

Suppose you want a *histogram*. In other packages you might find a histogram command in a **Graph** menu. But in JMP you need to think of the context. You want a histogram so that you can see the distribution of values. So you launch the

Distribution platform in the **Analyze** menu. Once you are there, then there are many graphs and reports available for focusing on the distribution of values.

Occasionally, you may want the histogram as a business graph. Then instead of using the Distribution platform, use the Bar/Pie Chart platform in the **Graph** menu.

Contexts for the t Test

Suppose you want a *t test*? Other packages might have a t test command. JMP has many t test commands because there are many contexts in which this statistic is used. So first you have to build the context of your situation.

If you want the t test to test a single variable's mean against a hypothesized value, then you are focusing on the distribution–so you launch the distribution platform, where you find lots of graphs and reports. At the top of the distribution report is a popup menu with the command **Test Mean=value**. That gives you a t test and the option to see a nonparametric test also.

If you want the t test to compare the means of two independent groups, then you have two variables to set the context, a continuous Y response and a categorical X factor. You want to find the mean Y response for each group identified by the X factor. If you launch the Fit Y by X platform, you'll see the side-by-side comparison of the two distributions, and you can use the command **Means, Anova/t-Test** from the platform popup menu.

If you want to compare the means of two continuous responses that form matched pairs, then there are several ways to build the appropriate context. You can make a third data table column to form the difference of the responses, and use the Distribution platform to do a t test that the mean of the differences is zero. Or you can use the **Fit Y by X** command to launch the bivariate platform for the two variables, where you will find the **Paired t-Test** command. In Chapter 6, "The Difference Between Two Means," you will learn more ways to do a t test.

Contexts for a Scatterplot

Suppose you want a scatterplot of two variables. Think of it more generally as a bivariate analysis. That will suggest using the Fit Y by X platform. When you launch the Fit Y by X platform for two continuous variables, the first thing you will see is a scatterplot. You can also fit regression lines inside this scatterplot because that is the part of the charter of the Fit Y by X platform.

You might also consider the **Overlay Plot** command in the **Graph** menu. It won't do regression, but will overlay multiple Y's in the same graph and connect the points. But if you have a whole series of scatterplots for many variables in mind, you are looking for many bivariate associations, which is the charter of the Correlations platform. Here you find the **Scatterplot Matrix** command.

Contexts for Nonparametric Statistics

There is no separate platform for nonparametric statistics. But there are many standard nonparametric statistics in JMP, positioned by context. When you test a mean in the Distribution of Y platform, you are also offered the option to do a (nonparametric) Wilcoxon signed-rank test. When you do a t test or one-way Anova in the Fit Y by X platform, you also have three optional nonparametric tests. The Wilcoxon rank sum test given there is equivalent to the Mann-Whitney U test. If you want a nonparametric measure of association, like Kendall's tau or Spearman's correlation, the context is the Correlation of Y's platform.

The Personality of JMP

Here are some good reasons why JMP is different from other packages:

Graphs Are in the Service of Statistics (and vice versa)

The goal of JMP is to provide a graph for every statistic, beside the statistic. The graphs shouldn't appear in separate windows. The graphs and statistics should work together. In the Analysis platforms, the graphs tend to follow the statistical context. In the Graph platforms, the statistics tend to follow the graphical context.

JMP Encourages Good Data Analysis

You didn't have to ask for a histogram, it appeared when you launched the Distribution platform. The Distribution platform was designed that way because in good data analysis you will always want to see a graph of a distribution before you start doing statistical tests on it. This is to encourage responsible data analysis.

Make Discoveries

JMP was developed with the charter to be "Statistical Discovery Software." After all, you want to find out what you didn't know, as well as try to prove what you already know. Graphs attract your attention to some outlier or other unusual feature of the data that might prove valuable to discovery. Imagine Madame Curie using a computer for her pitchblend experiment. If software had given her only the end results, rather than showing her the data and the graphs, she might not have noticed the discrepancy that lead to the discovery of radium.

Interactivity

In some products, you have to specify exactly what you want ahead of time because often that is your last chance. JMP is interactive, so everything is open to change and customization at any point in the analysis. It is easier to remove a histogram when you don't want it than decide ahead of time that you want one.

Multiple Perspectives

Did you know that a t test for 2 groups is a special case of an F test for several groups? With JMP, you tend to get general methods that are good for many situations, rather than specialty methods for special cases. You also tend to get several ways to test the same thing. For two groups there is a t test and equivalent F test. When you are ready for more, there are three nonparametric tests to use in the same situation. You also can test for and adjust for different variances across the groups. And there are two graphs to show you the separation of the means. Even after you perform statistical tests, there are multiple ways of looking at the results; in terms of the p value, the confidence intervals, least significant differences, the sample size, and least significant number. With this much statistical breadth, it is good that commands appear as you qualify the context, rather than you having to select multiple commands from the top menu bar. JMP unfolds the details progressively, as they become relevant.

Getting Help: The JMP Help System

You can access help windows almost anywhere in JMP:

- The question mark tool shows automatically when you drag the cursor over the buttons in the About JMP window. When you click these buttons, you have access to help for the whole JMP application.
- If you click the **Guide** button in the About JMP window, you have access to a guide of statistical methods.
- Many dialogs have their own help buttons.
- Help for each platform window is accessed by the asterisk popup menu located to the left of its horizontal scroll bar.

- Help for specific items in a JMP window shows when you select the question mark from the **Tools** menu and click the item.

Help from the About JMP Screen

If you have general questions or want to see a list of broad help topics,

§ On the Macintosh, choose **About JMP** from the Apple menu.

§ Under Windows, choose the **Contents** command in the **Help** menu.

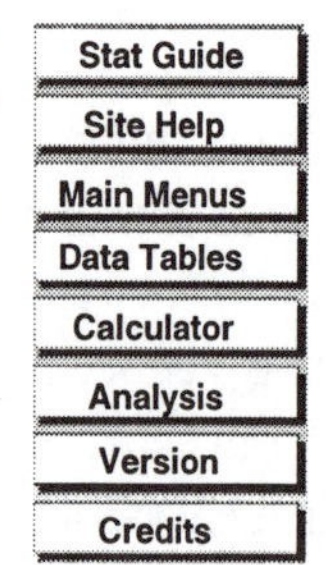

Figure 1.7 The About JMP Help Panel

This displays the About JMP window, which has the panel of buttons shown in **Figure 1.7**. The About JMP window is the topmost window in the Help tree. All the help documents are arranged hierarchically with buttons for accessing more detailed information.

Help From the JMP Statistical Guide

Figure 1.8 shows your online statistical navigation guide. To see a list of statistical and graphical procedures available in JMP,

§ On the Macintosh, click the **Stat Guide** button in the About JMP window.

§ Under Windows, you can access the Statistical Guide by choosing **Help→Contents.**

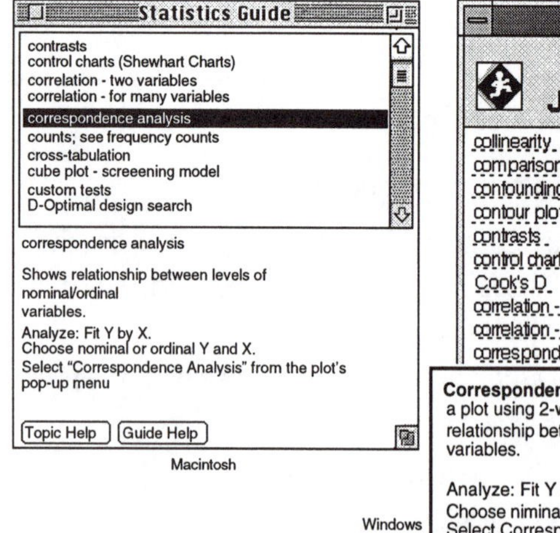

Figure 1.8 Statistical Guide

The Statistical Guide is a scrolling statistical index. When you click an analysis in the guide, the directions on how to do the analysis appear beneath the index, as shown in **Figure 1.8**. The directions tell you the menu, command, and options to use for the analysis or topic you selected.

Help from Buttons in Dialogs

Most dialog windows in JMP include a **Help** button. For example, there are help buttons on the Specify Model dialog, the calculator window, the Preferences dialog, and on data manipulation dialogs.

Help from a Platform Window

If you want help after you launch a platform, select **Help** from the asterisk popup menu located to the left of the horizontal scroll bar. For example:

§ Open any existing JMP data table from the sample data.

§ Choose **Analyze→Distribution of Y** and assign any variable as the Y variable.

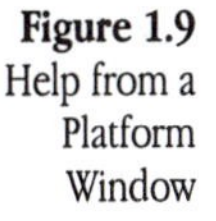

Figure 1.9 Help from a Platform Window

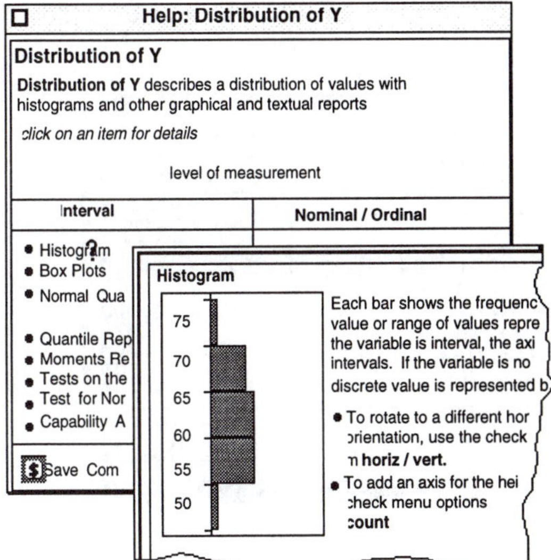

When the platform appears, select **Help** from the asterisk menu and click on a topic in the Help screen.

Figure 1.9 shows help for the **Distribution of Y** platform with a second layer of help for the histogram component of the platform.

Help by Clicking an Item

When you need help concerning a specific item in a JMP report window, select the question mark from the **Tools** menu and click the item. For example, when you click **Std Error** (see **Figure 1.10**), a help screen displays. The help screen persists as long as the help tool is on top of it. Also, you can SHIFT-click to open the help screen as a window that stays open until you close it. If a help window has buttons for further help, it automatically persists on the screen until you close it.

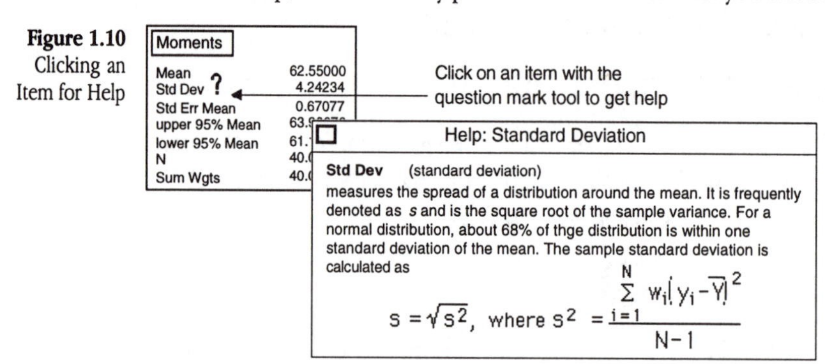

Figure 1.10 Clicking an Item for Help

Additional Help Commands under Windows

The Microsoft Windows **Help** menu has additional commands called **Search for Help On...** and **How to Use Help**. The **Search for Help On...** displays the scrolling list of topics like that shown in **Figure 1.11**. When you select a topic and click **Go To**, you immediately see the topic-specific help.

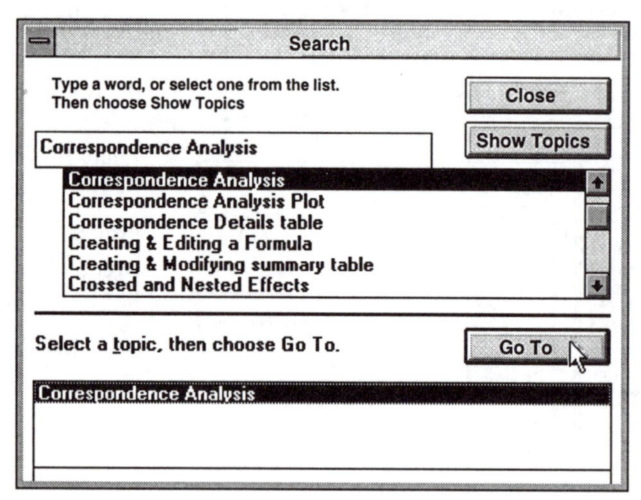

Figure 1.11 Search Index for Statistical Help

Chapter 2 JMP Data Tables

Overview

JMP data are organized in memory as rows and columns of a table referred to as the *data table*. The columns have names and the rows are numbered. An open data table is kept in memory, and you communicate with it through an active spreadsheet window. You can open as many data tables in a JMP session as memory allows. When you close a JMP data table, it is stored in a *file* on disk, sometimes called a *document*.

Commands in the **File**, **Edit**, **Tables**, **Rows**, and **Cols** menus give you a broad range of data handling operations and file management tasks such as data entry, data validation, text editing, and extensive table manipulation.

In particular, the **Tables** menu has commands that perform a wide variety of data management tasks on JMP data tables. These commands let you sort, subset, stack, or split table columns, join two tables side by side, concatenate multiple tables end to end, and transpose tables. You can also create summary tables for group processing and summary statistics.

The purpose of this chapter is to tell you about JMP data tables and give a variety of hands-on examples to help you get comfortable handling table operations.

Chapter 2 Contents

You can mouse along with the examples whenever you see an action symbol (§)

The Ins and Outs of a JMP Data Table

A JMP data table opens in a window as a spreadsheet. Using this spreadsheet view, you can do a variety of data table management tasks such as editing cells, creating, rearranging or deleting rows and columns, taking a subset, sorting, or combining tables. **Figure 2.1** identifies active areas of a spreadsheet.

There are a few basic things to keep in mind:

- Column names can have as many as 31 characters and can use any keyboard character including spaces. The size and font for names and values is a preference setting you control.
- Character fields can be as large as 255 characters long. You can drag column boundaries and enlarge the column to view long values.
- There is no limit to the number of rows or columns in a data table. However, the table must fit in memory.

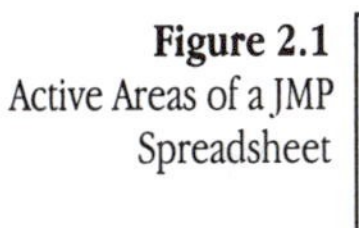

Figure 2.1
Active Areas of a JMP Spreadsheet

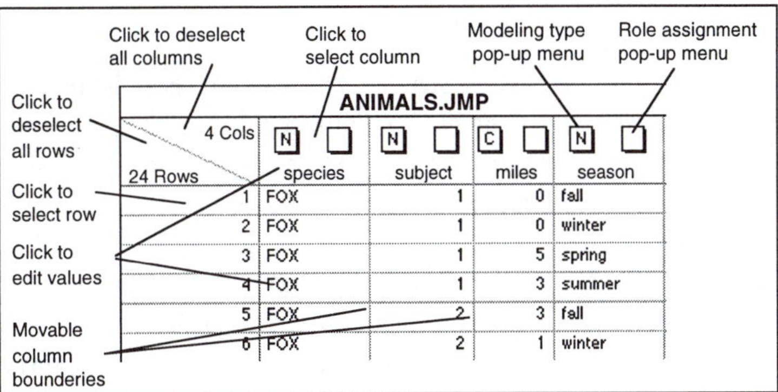

Selecting and Deselecting Rows and Columns

Many actions, including commands from the **Rows** and **Columns** menu, operate only on *selected* rows and columns. The selection of rows and columns is done by highlighting them. To highlight a row, click the space that contains the row number. To highlight a column, click the background area above the column name. These areas are shown in **Figure 2.1**.

To extend a selection of rows or columns, drag (in the selection area) across the range of rows or columns you want, or SHIFT-click the first and last row or column of the range. Use COMMAND-click or CONTROL-click to make a discontiguous selection. To select rows and columns at the same time, drag across table cells in the spreadsheet.

To deselect a row or column, use COMMAND-click or CONTROL-click on the row or column selection area. Or, click the triangular Rows or Cols area in the upper-left corner of the spreadsheet to deselect all rows or columns at once.

Mousing Around a Spreadsheet: Cursor Forms

To navigate in the spreadsheet, you need to understand how the cursor works in each part of the spreadsheet.

§ To experiment with the different cursor forms, open the ANIMALS.JMP sample table, move the mouse around on the surface areas shown in **Figure 2.1**, and see how the cursor changes form:

Arrow Cursor

The cursor is the standard arrow when it is in the modeling type box or the role assignment box at the top of a column, in the Cols or Rows triangular areas in the upper-left corner of the spreadsheet, or in the empty portion of the table.

Cross Cursor +

When the cursor is within a column heading or a data cell, it becomes a large cross indicating it can select (highlight) text. When you click the cross cursor, that cell is highlighted to show that its text is editable. The cursor then becomes an I–beam.

If you drag the cross cursor, it forms a stretch rectangle and selects the rows and columns of cells within the rectangle.

Click twice on a column name to select it for editing.

I–Beam Cursor

To edit text, position the I–beam within highlighted text. Click the I-beam to mark an insertion point, or drag it to select text for replacement. The I–beam deposits a blinking vertical bar to indicate a text insertion point or a highlighted area of text to be replaced. The default selection is the entire cell. Use the keyboard to make changes.

Open Cross Cursor

The cursor is a large open cross when you move it into a row or column selection area or a noneditable area of the spreadsheet such as a locked column cell. Click the open cross cursor to select a single row or column. SHIFT-click a beginning and ending range of rows or columns to select an entire array. Use COMMAND-click on the Macintosh (CONTROL-click under Windows) to make a discontiguous selection.

Double Arrow Cursor

The cursor changes to a filled double arrow when on a column boundary. Drag this cursor left or right to change the width of a spreadsheet column.

List Check and Range Check Cursors

The cursor changes form when you move it over values in columns that have data validation in effect (automatic checking for specific values). It becomes a small, downward-pointing arrow on a column with list checking and a large double I–beam on a column with range checking. When you click, the value highlights and the cursor becomes the standard I–beam; you enter or edit data as usual. However, you are not allowed to enter invalid data values.

§ After you finish exploring, choose **File→Close** to close the ANIMALS.JMP table.

Creating a New JMP Table

Hopefully, most data will reach you already in electronic form. But if you have to key in data, JMP provides a spreadsheet with familiar data entry features. A short example shows you how to start from scratch.

Figure 2.2 Notebook of Raw Study Data

Blood Pressure Study

Month	Control	Placebo	300 mg	450 mg
March	165	163	166	168
April	162	159	165	163
May	164	158	161	153
June	162	161	158	151
July	166	158	160	148
August	163	158	157	150

Suppose data values are blood pressure readings collected over six months and recorded in a notebook page as shown in **Figure 2.2**.

Define Rows and Columns

JMP data tables have rows and columns, sometimes called observations and variables in statistical terms. The raw data in **Figure 2.2** are arranged as five columns (treatment groups) and six rows (months March through August). The first line in the notebook names each column of values that can be used as column names in a JMP table. But first, you need a new table to work with:

§ Choose **File→New** to see a new empty data table. A new untitled table appears with one column and no rows.

The Add Columns Command

§ Choose **Cols→Add Columns** and respond to the Add Columns dialog by giving the number of columns to add, where to add them, and what type of columns to add. Four new numeric columns will be fine for our current example.

The default column names are Column 1, Column 2, and so on, but you can change them in the Add Columns dialog. You can also change them by typing in the editable column fields. Do that now:

§ Type the names from the data journal (Month, Control, Placebo, 300 mg, and 450 mg) into the columns headers of the new table.

To edit a column name, first click the column selection area to highlight the column. Then click a second time and position the I–beam within the highlighted characters. Click again to mark an insertion point for typing. The I–beam then becomes a blinking vertical bar. Alternatively, drag the I–beam to select a portion of the text for replacement.

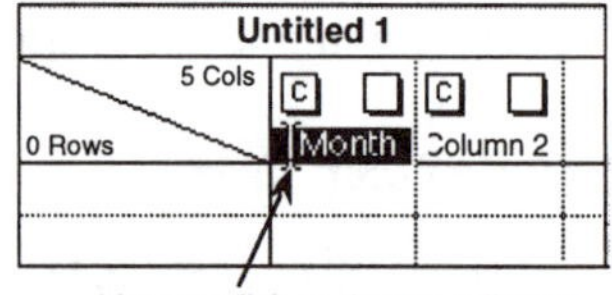

Set Column Characteristics

Columns can have different characteristics. By default, they contain numeric data, but in this example the month-names column will hold a nonnumeric character variable.

§ Highlight the Month column by clicking the area above the column name and choose **Cols→Column Info**. In the Column Info dialog, use the **Data Type** popup menu to change Month to a character variable.

You can also use the Column Info dialog to change the other column characteristics and to access the JMP calculator for computing column values.

Add Rows

§ Adding new rows is very easy: Choose **Rows→Add Rows** and ask for six new rows. Alternatively, if you double-click anywhere in the body of the table, the table automatically fills with new rows through the position of the cursor.

§ The last step is to give the table a name and save it. Choose **File→Save As** to name the table BPSTUDY.JMP and give it a disk location.

The data table is now ready to hold data values. **Figure 2.3** summarizes the table evolution so far.

Figure 2.3 JMP Data Table with New Rows, Columns, and Names

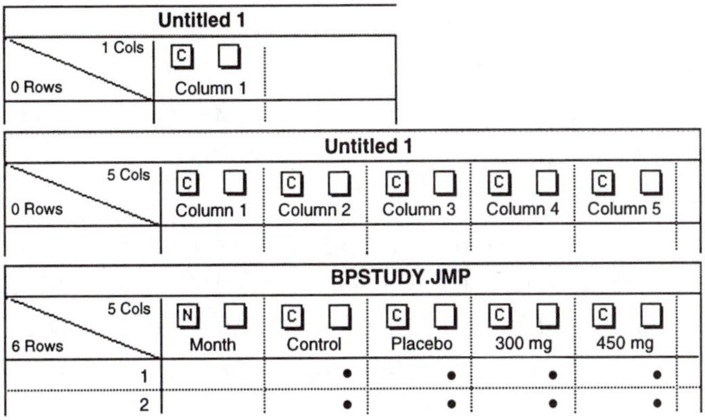

Untitled 1	
1 Cols / 0 Rows	[C] [ ] Column 1

Untitled 1					
5 Cols / 0 Rows	[C] [ ] Column 1	[C] [ ] Column 2	[C] [ ] Column 3	[C] [ ] Column 4	[C] [ ] Column 5

BPSTUDY.JMP					
5 Cols / 6 Rows	[N] [ ] Month	[C] [ ] Control	[C] [ ] Placebo	[C] [ ] 300 mg	[C] [ ] 450 mg
1		•	•	•	•
2		•	•	•	•

Enter Data

Entering data into the data table requires typing values into the appropriate table cells. To enter data into the data table, do the following:

§ Move the cursor into a data cell and click to highlight the cell. The cursor becomes an I–beam in the highlighted area.

§ Click again and the I–beam becomes a blinking vertical bar.

§ Key in the data from the notebook (**Figure 2.2**). If you make a mistake, drag the I–beam across the incorrect entry to highlight it and type the correction over it. Your result should look like the table in **Figure 2.4**.

The TAB and RETURN keys are useful keyboard tools for data entry:

- TAB moves the cursor one cell to the right. SHIFT-TAB moves the cursor one cell to the left. Moving the cursor with the TAB key automatically wraps it to the beginning of the next (or previous) row. Tabbing past the last table cell creates a new row.

- RETURN moves the cursor down one cell; SHIFT-RETURN moves the cursor up one cell.

Figure 2.4 Finished Blood Pressure Study Table

BPSTUDY.JMP

5 Cols / 6 Rows	Month (N)	Control (C)	Placebo (C)	300 mg (C)	450 mg (C)
1	March	165	163	166	168
2	April	162	159	155	163
3	May	164	158	161	153
4	June	162	161	158	151
5	July	166	158	160	148
6	August	163	158	157	150

The New Column Command

In the first part of this example, you used the **Add Cols** command from the **Cols** menu to create a group of new columns in a data table. Often you need to add a single new column with specific characteristics.

Continuing with the current example, suppose you learn that the blood pressure readings were taken at one lab, called "Accurate Readings Inc.," during March and April but at another location called "Most Reliable Measurements Ltd." for the remaining months of the study. You want to include this information in the data table:

§ Begin by choosing **Cols→New Column**, which displays a New Column dialog like the one shown in **Figure 2.5**. The New Column dialog lets you set the new column's characteristics:

1) Type a new name, Location, in the **Col Name** area.
2) Because the actual names of the location are characters, select **Character** from the **Data Type** popup menu as shown in **Figure 2.5**. Notice that the Modeling Type then automatically changes to Nominal.
3) The default field length is 8 characters, but you need to enter 31 characters for one of the locations. Type **31** into the **Field Width** box.
4) Use the Notes area to document the new column. Documentation of data tables and their columns cannot be overemphasized.

When you click **OK**, the new column appears in the table, and you can enter data as previously described.

Figure 2.5
The New Column Dialog

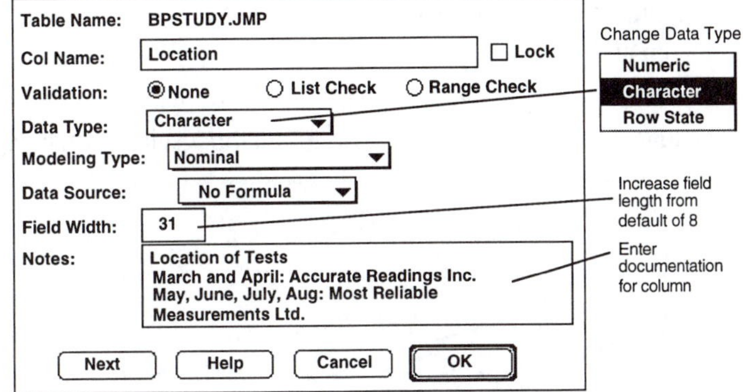

Plot the Data

There are many ways to check the data for errors. One way is to generate a plot and see that there are no obvious anomalous values. Let's experiment with the **Bar/Pie Charts** command in the **Graph** menu.

We will plot the months along the horizontal (X) axis and the columns of blood pressure statistics for each treatment group up the vertical (Y) axis.

§ Choose **Graph→Bar/Pie Charts**. Then select Month as the X variable and the 4 remaining columns as Y variables. When you Click **OK** you first see a series of bar charts.

§ Use Options! Click the check mark on the lower left of the window to see a variety of options. Select the Overlay option and then select Line to see the chart shown in **Figure 2.6** (the plot doesn't appear to have much to say yet).

Figure 2.6
Line Chart for Blood Pressure Values over Month

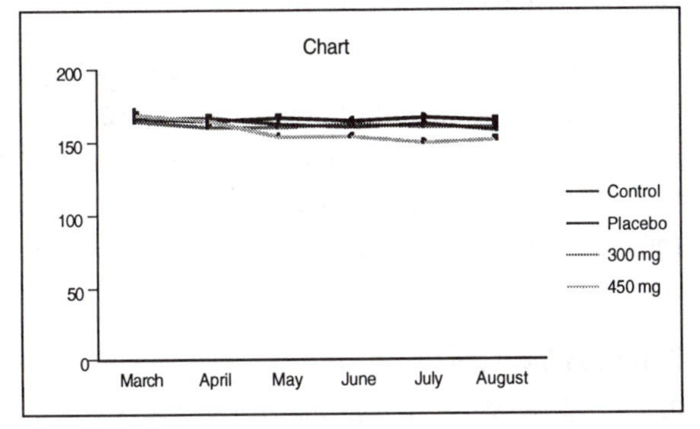

By default, Y-axis scaling begins at zero. To present easy-to-read information, the Y axis need to be rescaled.

§ Double-click anywhere in the Y axis area and use the Rescale dialog that appears. Based on what you can see in **Figure 2.6**, the plotted values range from about 145 to 175. Type these values into the Axis Rescale dialog as the minimum and maximum.

§ Also, change the increment for the tick marks from 50 to 5, which divides the range into six intervals. Then click **OK** to see the result shown in **Figure 2.7**.

Figure 2.7
Line Chart with Modified Y Axis

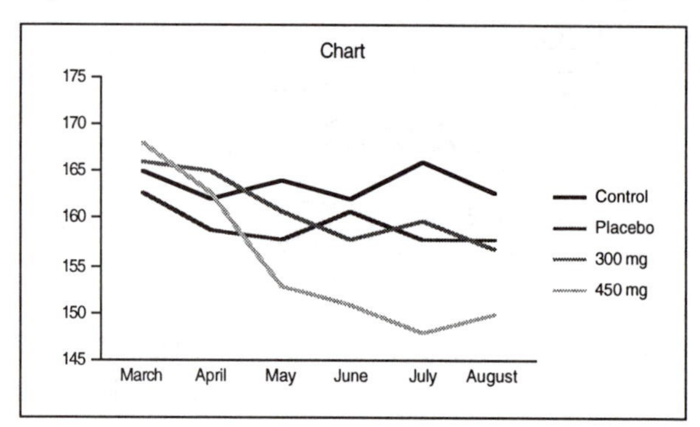

Importing Data

The **Import** command in the **File** menu lets you import raw data and create a new JMP data table by reading data from a text-format file or a SAS transport file. The **Import** command popup selections display appropriate dialogs for creating the new table:

The Import Text Command

The Text option creates a new JMP data table by reading data from a text-format file. When you import a file in text format, you can use options to identify line and field delimiters in the incoming file.

By default, the **Import→Text** command reads a text file of rows and columns and names the JMP data table columns Column 1, Column 2, and so on. If the file has header information, JMP header information, or special characters to denote ends of fields or ends of lines, or if data values are enclosed in quotes, you can use the radio buttons and check boxes in the Import dialog to correctly read the data.

The Import SAS Transport Command

The SAS Transport option displays a special dialog to specify the location of a SAS transport file. A *SAS transport file* is a portable file created by the SAS system.

Cut, Copy, and Paste Spreadsheet Data

These commands in the **Edit** menu let you move data around:

Cut

copies selected fields from the active spreadsheet to the clipboard and replaces them with missing values. It is equivalent to **Copy**, then **Clear**.

Copy

copies the values of selected data cells from the active data table to the clipboard. If you do not select columns, **Copy** copies entire rows. Likewise, you can copy values from whole columns if no rows are selected. If you select both rows and columns, **Copy** copies the subset of cells defined by their intersection. Data you cut or copy to the clipboard can be pasted into JMP tables or into other applications.

Paste

copies data from the clipboard into a JMP data table. **Paste** can be used with the **Copy** command to duplicate rows, columns, or any subset of cells defined by selected rows and columns.

To duplicate an entire row or column:

1) Select and **Copy** the row or column to be duplicated.
2) Select an existing row or column to receive the values.
3) Use the **Paste** command.

To duplicate a subset of values defined by selecting both rows and columns, follow the previous steps, but select the same arrangement of rows and columns to receive the copied values as originally contained in them. If you paste data with fewer rows into a destination with more rows, the source values recycle until all receiving rows are filled.

Paste at End

extends a JMP data table by adding rows and columns to a data table as needed to accept values from the clipboard.

To transfer data from another application into a JMP data table, first copy the data to the clipboard from within the other application. Then use the **Paste at End** command to copy the values to a JMP data table. Rows and columns are automatically created as needed when you **Paste at End**.

If you choose **Paste at End** while holding down the OPTION or ALT key, the first line of information on the clipboard is used as column names in the new JMP data table.

Moving Data and Results Out of JMP

Two questions that come up early with JMP (or with any application) are "Can I get my data back out of JMP?" and "How do I get results out of JMP?"

The Save As Command

The **Save As** command writes the active data table to a file after prompting you for a name and disk location with the dialog shown in **Figure 2.8**.

Figure 2.8 Save As Dialog and Text Formatting Options

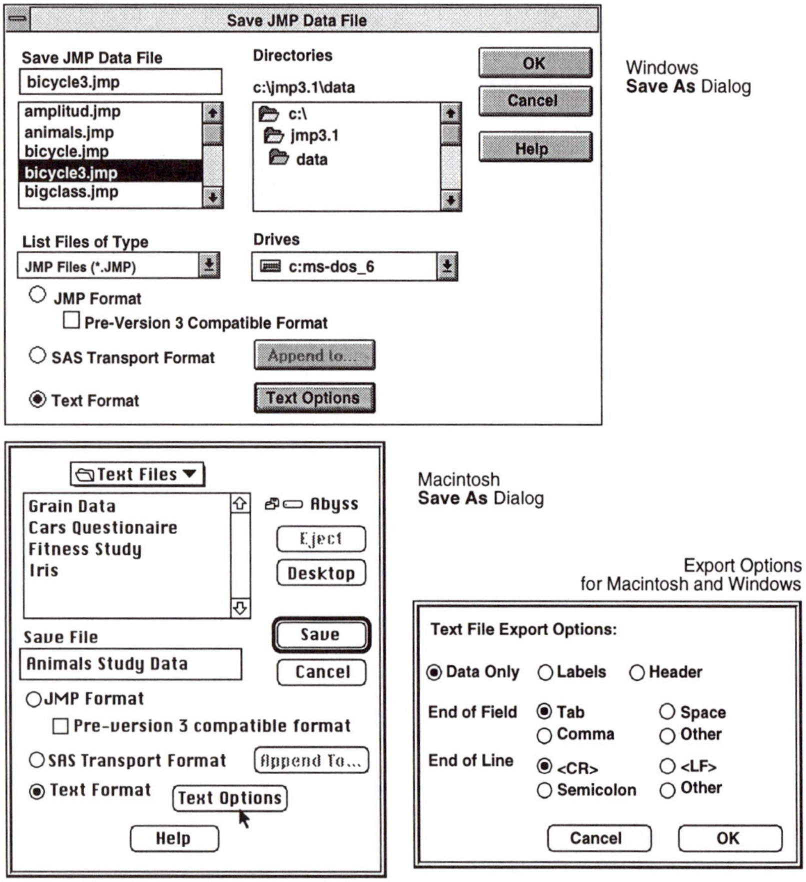

Windows **Save As** Dialog

Macintosh **Save As** Dialog

Export Options for Macintosh and Windows

Use the options and radio buttons for exporting JMP files as needed. In particular, note that you can export JMP files to be compatible with previous releases of JMP, as SAS transport files for use with the SAS system, or as text files for importing to use in other applications. When you export JMP files as text there are options to include column names and other header information with the data and to specify end of field and end of line delimiters.

Cut, Copy, and Paste Graphs and Reports

You can use standard copy and paste operations to move graphical displays and statistical reports from JMP to other applications. Although the **Edit** menu includes both **Cut** and **Copy** commands, they both perform the same tasks in report windows. **Cut** copies all or selected parts of the active report window into the clipboard, but does not delete the image from the platform. There is also a special command called **Copy as Text** (discussed later) that copies only the text from an active report window. JMP copies reports in PICT data format with hierarchical grouping information used by some drawing applications.

See the section **Print Data, Reports, and Journals** later in this chapter for information about copying JMP PICT data into other applications and printing.

Copy and Paste Reports as Pictures

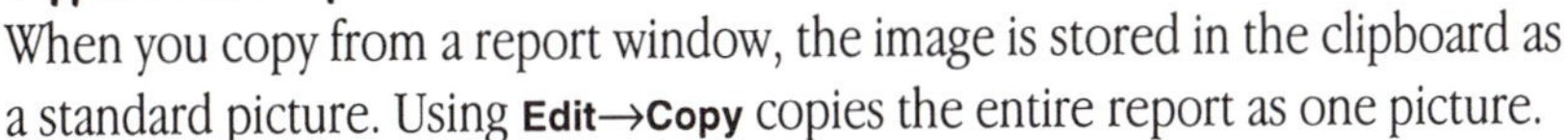

When you copy from a report window, the image is stored in the clipboard as a standard picture. Using **Edit→Copy** copies the entire report as one picture.

If you want to copy part of a report window, select the scissors tool from the **Tools** menu. Drag the scissors diagonally to draw a rectangle over the area you want to copy. The tip of the upper blade marks the starting and ending points of the selection. As you drag, a rectangle appears and selects the area to copy. Then use the **Copy** command to copy the selected area to the clipboard. Use the **Paste** command to paste the picture into a JMP journal or another application.

If you want the scissors tool to select specific components or blocks of a report, hold down the OPTION or ALT key and click or drag. Each component of the report highlights as a separate unit. Hold down both the OPTION and SHIFT keys to extend a selection of discrete report components.

You can change the selection. Drag the scissors over a new area or extend the current selection by SHIFT–dragging the scissors across an additional area. Extensions do not have to be contiguous, and you can have as many extensions as you want. To deselect all areas, click the report window (without dragging).

Copy Reports as Text Only

To copy text only from a report window, choose **Copy as Text** from the **Edit** menu. The **Copy as Text** command copies all text from the active report window (no graphical displays) and stores the text in the clipboard. Each field within a text report is separated by tabs so that you can edit the text easily in a word processing spreadsheet or application.

If you select a statistical text report with the scissors tool, **Copy as Text** copies just the contents of that report.

If you select **Copy as Text** while holding down the OPTION or ALT key, the text reports are copied to the clipboard and also show in a simple text editing window within JMP. You can do preliminary editing and again use **Cut**, **Copy**, and **Paste** commands on those results.

It is sometimes handy to capture the results from an analysis table and transfer them to a new JMP data table. To do this you use both the **Copy as Text** and **Paste at End** commands:

1) First use the scissors to select the report text you want.
2) Hold down the OPTION or ALT key and choose **Edit→Copy as Text**.
3) Choose **File→New** to create a new empty data table.
4) Choose **Edit→Paste at End** to paste the clipboard contents into the table, which will automatically create the rows and columns as needed.

Journal Data and Results

Journaling is a special feature of JMP that helps you save analysis results in a word-processing document. When you select the **Journal** command from the **Edit** menu, all text and pictures from the active JMP analysis window are written to the journal document (as text and pictures). However, if an area of an analysis window is selected with the scissors tool, the **Journal** command saves just that selected area as a picture.

The JMP journal window opens after you first select the **Journal** command. The journal window is an on-screen representation of the document you will see when you open it with a word processor. Each time you again select the **Journal** command, the text and graphs from the active analysis window are appended to the end of the journal window.

The journaled text and graphs cannot be changed until you close the file and open it with a word processor. However, when a JMP journal is open you can take these actions:

- Append text and graphs from the active analysis window to the end of the journal window each time you again select the **Journal** command.
- Type text notes at the bottom of the journal window.
- Scroll through the journal window.
- Copy individual graphs and text reports from a JMP analysis window and paste them at the end of a journal window.
- Copy text and pictures from other Macintosh applications and paste them at the end of a journal window.
- Print the contents of the journal window at any time using the **Print** command in the **File** menu.
- Hide the journal window by selecting **Hide** from the **Window** menu.

When you close the journal window, a save dialog prompts you for a disk location, file format, and name for the JMP journal. You select the file format you want from the popup menu showing beneath the **Save Journal as:** prompt. After the journal is saved, you open it with a word processor of your choice for preparing reports.

Note: You cannot open an existing journal document from within JMP. After you close your journal it has the file format you specified. It is not a JMP document. If you want continued access to a journal from within JMP, but want it out of the way, leave it open and use the **Hide** command in the **Window** menu. To again see the journal, use the **Unhide** command.

Print Data, Reports, and Journals

Select **Print** from the **File** menu to print data tables, open text reports and plots from the active report window, or an open journal window. Reports print best on PostScript output devices such as a LaserWriter, but they also print on QuickDraw laser printers and ImageWriters.

When you print a JMP text report or graph, the text reports and graphical displays automatically rearrange on each printed page for best fit. Some controls and icons do not appear on the printouts. You can use the **Include Pop-up Controls in Copy/Print** preference (see the **Preferences** command in the **File** menu) to print popup control icons. The **Print Preview** selection in the Star (*) border menu draws page-size rectangles on the report window to show you how printing will occur. The journal prints as it appears on your screen.

The PICT files that store graphical displays generated by JMP contain special PostScript (Adobe Systems Corporation) information to make them print better. With this special information, lines and curves are thin, and dashed lines have uniformly sized segments. When you copy a JMP picture into most word processing applications, this special information is retained. However, some drawing programs discard extra PostScript commands.

Juggling Data Tables

Each of the following examples uses commands from the **Tables**, **Rows**, or **Cols** menus. If you want to do these examples yourself as you read, watch for the action (§) symbol.

Correct a Sort Problem: Subset, Sort, and Join

Suppose you have the following situation. Person A began a data entry task and entered state names in order of descending auto theft rates. Then Person B took over the data entry but mistakenly entered the auto theft rates in alphabetical order by state.

§ To see the result, open the AUTOMESS.JMP sample table. **Figure 2.9** shows a partial listing of this table.

Figure 2.9
Date Entered Incorrectly

AUTOMESS.JMP

51 Rows / 2 Cols	State	Auto theft
1	SOUTH DAKOTA	348
2	NORTH DAKOTA	565
3	WYOMING	863
4	WEST VIRGINIA	289
5	IDAHO	1016
48	RHODE ISLAND	447
49	CALIFORNIA	154
50	NEW YORK	416
51	DISTRICT OF COLUMBIA	149

Could this ever really happen? Yes! Never underestimate the diabolical convolution of data that can appear in an electronic table, and hence a circulated report. Always check your data with common sense.

To put the data into its correct order, you need make a copy of the AUTOMESS table, sort it in ascending order, and join the sorted result with the original table:

§ With AUTOMESS.JMP active, choose **Tables→Subset**, which automatically creates a duplicate table when no rows or columns are selected in the original table. The table name is Untitled 1, but you don't need to give this table an official name because it is only temporary.

§ With the Untitled 1 table active, choose **Tables→Sort**. When the Sort dialog appears, choose Auto theft as the sort variable and click **Sort**.

You now have a second untitled table (Untitled 2) that is sorted by auto theft rates in ascending order, which is the way you want it. Before finishing, close the Untitled 1 data table to avoid forgetting which Untitled table is which. Now you want to join the incorrectly sorted AUTOMESS table with the correctly sorted Untitled 2 table. You do this as follows:

§ Choose **Tables→Join**. When the Join dialog appears, note which table is listed next to the word **Join** (either Automess or Untitled 2), click the other table in the list of tables and then click **With** on the join dialog.

§ Because you don't want all the columns from both tables in the final result, click the **Select Columns** box. When the variables from both tables appear, select State from the AUTOMESS table, Auto theft from the Untitled 2 table, and click **OK**. When the Join dialog again appears, click the **Join** button.

Check the new joined data table called Untitled 3: the first row is South Dakota with a theft rate of 110, and the last row is the District of Columbia with a rate of 1336. To keep this table use **Save As** and specify a name and disk location for it.

Give Your Table a New Shape: Stack Columns

One typical situation occurs when response data are recorded in two columns and you need them to be stacked into a single column. For example, suppose you collect three months of data and enter it in three columns. If you then want to look at quarterly figures, you need to change the data arrangement so that the three columns stack into a single column. You can do this with the **Stack Columns** command in the **Cols** menu.

Here is an example of stacking columns:

§ Open the CHEZSPLT.JMP sample data to see the table on the left in **Figure 2.9**. This sample data has columns for four kinds of cheese, labeled A, B, C, and D. In a taste test judges ranked the cheeses on an ordinal scale from 1 to 9 (1-awful, 9-wonderful). The **Response** column shows these ratings. The counts for each cheese for each ranking of taste are the body of the table. Its form looks like a two-way table, but to analyze this contingency table JMP needs to see the cheese categories in a single column. To rearrange the data:

§ Choose **Tables→Stack Columns**. In the dialog that appear, select the cheeses (A, B, C, and D) from the **Columns** list and add them to the **Stack** list. Leave everything else as defaults. Click **OK** to see the right-hand table in **Figure 2.10**. The _ID_ column shows the cheeses, and the **_Stacked_** column is now the count variable for the **Response** categories.

§ Use the role assignment popup to set the role of the **_Stacked_** variable to **Freq**. This causes the values **_Stacked_** in the Untitled table to be interpreted by analyses as the number of times that row's response value occurred.

Figure 2.10 Stack Columns Example

CHEZSPLT.JMP

9 Rows / 5 Cols	Response	A	B	C	D
1	1	0	6	1	0
2	2	0	9	1	0
3	3	1	12	6	0
4	4	7	11	8	1
5	5	8	7	23	3
6	6	8	6	7	7

Untitled

36 Rows / 3 Cols	Response	_Stacked_	_ID_
1	1	0	A
2	1	6	B
3	1	1	C
4	1	0	D
5	2	0	A
6	2	9	B

To see how response relates to type of cheese:

§ Choose **Analyze→Fit Y by X**. Notice that the _Stacked_ column appears as the Freq variable. Select Response as Y and _ID_ as X.

Figure 2.11 Stack Columns Example

Crosstabs

_ID_ (columns) / Response (rows)

Count	A	3	C	D	
1	0	6	1	0	7
2	0	9	1	0	10
3	1	12	6	0	19
4	7	11	8	1	27
5	8	7	23	3	41
6	8	6	7	7	28
7	19	1	5	14	39
8	8	0	1	16	25
9	1	0	0	11	12
	52	52	52	52	208

When you click **OK** the contingency table platform appears with a Mosaic plot, Crosstabs table (**Figure 2.11**), Tests table, and popup menu options. You can use the Question Mark tool in the **Tools** menu and click on the platform surface to find more information about the platform components. Only the Crosstabs table is shown here, but the Cheese data is used again later for further analysis.

§ Extra Credit: For practice, see if you can use the **Split Cols** command on the stacked data table to reproduce a copy of the CHEZSPLT.JMP table.

The Group/Summary Command

One of the most powerful and useful commands in the **Tables** menu is the **Group/Summary** command.

Group/Summary creates a JMP window that contains a *summary table*. This table summarizes columns from the active data table, called its *source table*. It has a single row for each level of a grouping variable you specify. A grouping variable divides a data table into groups according to each of its values. For example, a gender variable can be used to group a table into males and females.

When there are several grouping variables, the summary table has a row for each combination of levels of all variables. Each row in the summary table identifies its corresponding subset of rows in the source table. The columns of the summary table are summary statistics you request.

Create a Table of Summary Statistics

The example data used to illustrate the **Group/Summary** command is the JMP table called COMPANYS.JMP (see **Figure 2.12**).

§ Open the COMPANYS.JMP sample table. It is a collection of financial information for 32 companies. The first column (**Type**) identifies the type of company with values "Computer" or "Pharmaceut." The second column (**Size Co**) categorizes each company by size with values "small," " medium," and "big." These two columns are typical examples of grouping information.

Figure 2.12 JMP Table to Summarize

COMPANYS.JMP

7 Cols / 32 Cols	Type	Size Co	Sales($M)	Profilts($M)	# Employ	profit/emp	Assets
1	Computer	small	855.1	31.0	7523	4120.70	615.2
2	Pharmaceut	big	5453.5	859.8	40929	21007.11	4851.6
3	Computer	small	2153.7	153.0	8200	18658.54	2233.7
4	Pharmaceut	big	6747.0	1102.2	50816	21690.02	5681.5
5	Computer	small	5284.0	454.0	12068	37620.15	2743.9
6	Pharmaceut	big	9422.0	747.0	54100	13807.76	8497.0

§ Choose **Tables→Group/Summary**. When the Group/Summary dialog appears (**Figure 2.13**) select the variable **Type** in the Columns list of the dialog and click **Group** to see it in the grouping variables list. You can select as many grouping variables as you want. Click **OK** to see the summary table.

Figure 2.13 Grouping Dialog and Summary Table

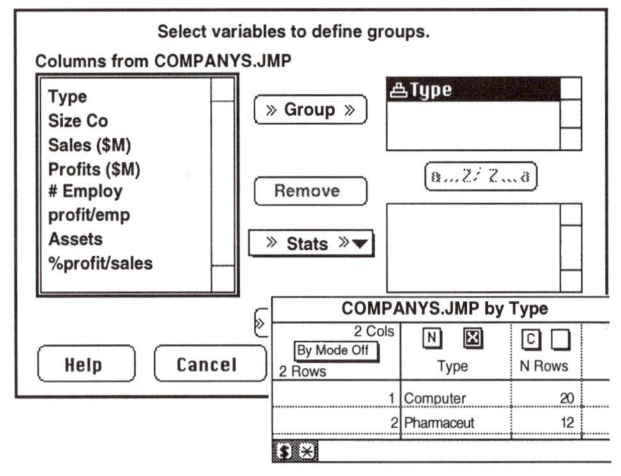

The new summary table appears in an active window. This table is linked to its source table. When you highlight rows in the summary table, the corresponding rows highlight in its source table.

Initially, a summary displays frequency counts (**N Rows**) for each level of the grouping variables. This example shows 20 computer companies and 12 pharmaceutical companies. However, you can add columns of descriptive statistics to the table. The **Stats** popup menu in the Group/Summary dialog lists standard univariate descriptive statistics.

§ To add summary statistics to the table, follow these steps:

1) Use the **Add Summary Cols** command in the dollar-sign ($) popup menu to again display the Group/Summary dialog (**Figure 2.13**).
2) Select a numeric column from the source table columns list.
3) Select the statistic you want from the **Stats** popup menu.
4) If necessary, repeat steps 2 and 3 to add more statistics to the summary table.
5) Click **OK** to add the columns of statistics to the summary table.

The righthand table in **Figure 2.14** shows the mean of **profit/emp** (profit per employee) in the summary table grouped by **Type**. The righthand table shows the mean when the source table is grouped by **Type** and **Size**.

Figure 2.14 Expanded Summary tables

Type	N	Mean(profit/emp)
Computer	20	6159.015
Pharmaceut	12	23546.12

Type	Size Co	N	Mean(profit/emp)
Computer	small	14	7998.815
Computer	medium	2	-3462.51
Computer	big	4	4530.478
Pharmaceut	small	2	38337.19
Pharmaceut	medium	5	24035.11
Pharmaceut	big	5	17140.7

Another way to add summary statistics to a summary table is with the **Subgroup** button on the Group/Summary dialog. This method creates a new column in the summary table for each level of the variable you specify with **Subgroup**. The subgroup variable is usually *nested* within all the grouping variables. The summary table now becomes a table of summary statistics.

Analyze Subsets of Data

When you highlight rows in the summary table, the corresponding rows highlight in its source table. The **Analyze** and **Graph** menu commands then process the source table differently, depending on its *mode*. The summary table has two modes, accessed by the **By Mode** popup menu (see **Figure 2.13**) in the upper-left corner of the table:

- When By-Mode is off, **Analyze** and **Graph** menu commands and the **Transpose** command act on the summary table itself.
- When By-Mode is on, the menu commands recognize highlighted subsets in the source table identified by selected rows in its summary table. **Analyze** menu commands produce a separate report window for each source table subset selected in the summary table.

Let's look at automatic subset analysis in action:

§ First, summarize the COMPANYS data table by Size Co. To do this, choose **Tables→Group/Summary** and select Size Co as the **Group** variable in the Group/Summary dialog.

§ When the summary table appears, click the By-Mode popup in the upper-left corner of the data table to turn By-Mode on.

§ Then highlight the rows for "small" and "big" companies in the summary table. Use OPTION-click or ALT-click to select these noncontiguous rows in the summary table. Note that the corresponding rows in the source table highlight at the same time. These highlighted rows in the summary table identify the two subsets in the source table with those values of Size Co. Highlighting the same rows in the source table without using a summary table does not give any subsetting information to JMP.

§ Now, with the Summary table active, choose **Analyze→Distribution of Y**.

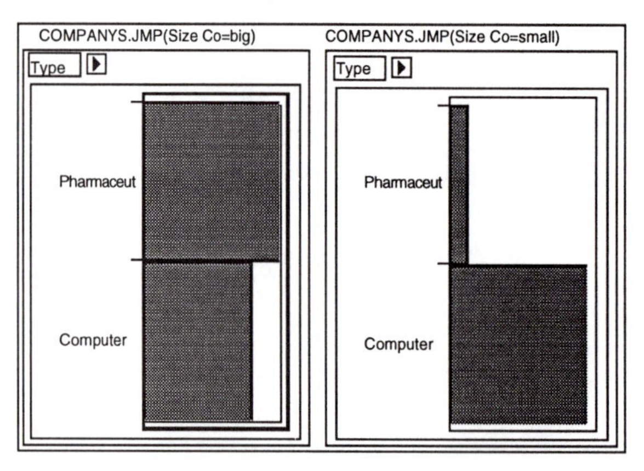

Figure 2.15 Distributions by Size of Company

With the By-Mode in effect, two separate analysis windows for the selected subsets of company type will appear as shown in **Figure 2.15**.

Plot and Chart Summary Data

The **Bar/Pie Charts** selection in the **Graph** menu lets you display summarized data easily. However, **Bar/Pie Charts** requires unique values of any column given an X role. You may need to preprocess raw data tables with the **Group/Summary** command and produce summary tables to use the **Bar/Pie Charts** command effectively. To see an example of this:

§ Choose **Graph→Bar/Pie Charts** when the source table COMPANYS.JMP is active. When the role assignment dialog appears, select Type and Size Co as X variables and profit/emp as Y.

You immediately see a message that tells you there is more than one row with the same X values, and that you must summarize your data before charting it.

§ Click **OK** on the alert box dialog.

The Group/Summary dialog then appears with the X variables you selected automatically assigned as Group variables. The Y variable you selected appears in the Stats box with the default **Sum** statistic requested. (You can change the summary statistics if you want.)

§ Now, when you click **OK** on the Group/Summary dialog the **Bar/Pie Charts** command produces the chart you see in **Figure 2.16**.

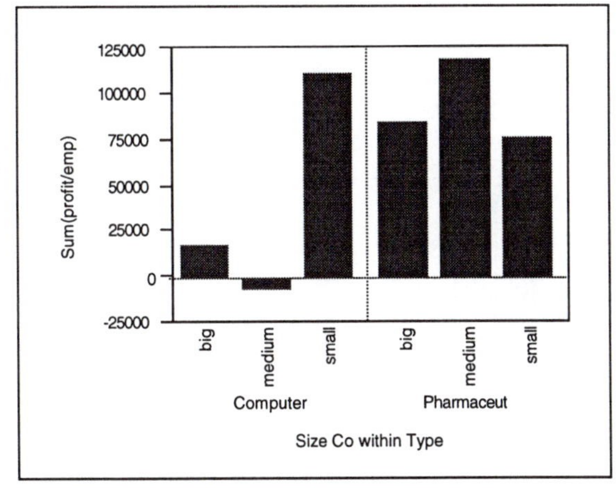

Figure 2.16 Chart of Mean Profit Per Employee for Size and Type of Company

Chapter 3
Calculator Adventures

Overview

What is *The Calculator*?

Actually, each column has its own calculator. The JMP calculator is a powerful tool for building formulas that calculate values for each cell in a column. A calculator window operates like a pocket calculator with buttons, displays, and an extensive list of easy-to-use features for building formulas.

JMP formulas can use information from other columns in the data table, built-in functions, and constants. Formulas can be simple assignments of numeric, character, or row state constants or can contain complex evaluations based on conditional clauses.

When you create a formula, that formula becomes an integral part of the data table. The formula is stored as part of a column's information when you save the data table, and it is retrieved when you reopen the data table. You can examine or change a column's formula at any time by opening its calculator window.

A column whose values are computed is both *linked* and *locked*. It is linked to (or *dependent on*) all other columns that are part of its formula. Its values are automatically recomputed whenever you edit the values in these columns. It is also locked so that its data values cannot be edited, which would invalidate its formula.

This chapter describes calculator features and gives a variety of examples. See Appendix C, "Calculator Functions," for a complete list of calculator functions.

Chapter 3 Contents

The Calculator Window

The JMP calculator is a window that computes values for a column. You can open a column's calculator window three ways:

- Select **Formula** from the **Data Source** popup menu in a New Column dialog, and click **OK**.
- Click the formula display area in the Column Info dialog of a column that has a formula.
- OPTION-click (or CNTR-click) in the white space at the top of a column that has a formula. This opens the column's calculator without first opening the Column Info dialog.

The two main areas of the calculator window are called the *calculator work panel* and the *formula display,* as illustrated by **Figure 3.1**. The calculator work panel is composed of buttons (**Help**, **Constant**, add and delete icons, and **Evaluate**), selector lists, and a constant entry field. The formula display is an editing area you use to construct and modify formulas.

Figure 3.1 The Calculator Window

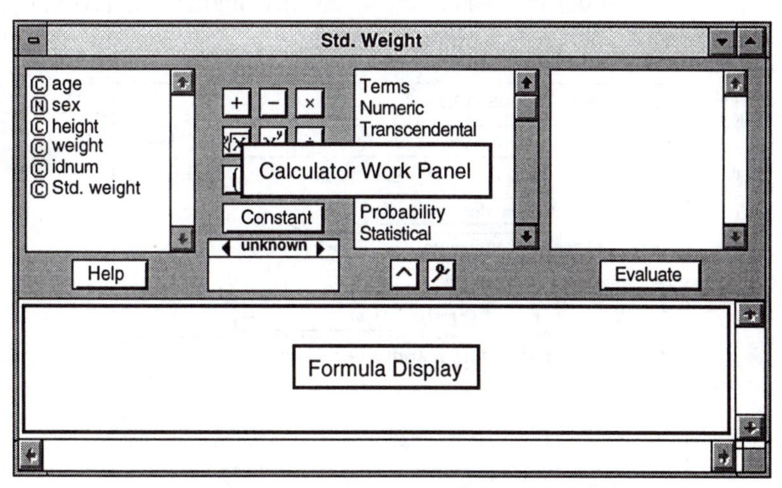

The sections that follow show you a simple example, define calculator terminology, and give the details you need to use the work panel and the formula display.

For a hands-on approach, watch for the action symbol (§).

A Quick Example

The following example gives you a quick look at the basic features of the calculator. Suppose you want to compute a standardized value. That is, for a numeric variable X you want to compute

$$\frac{\text{X} - \text{Mean of X}}{\text{standard deviation of X}}$$

for each row in a data table.

§ For this example, open the STUDENTS.JMP data table. It has a column called **weight** and you want a new column that uses the above formula to generate standardized weight values.

§ Begin by choosing **Cols→New Column**, which displays a New Column dialog like the one shown in **Figure 3.2**. The New Column dialog lets you set the new column's characteristics.

§ Type the new name, **Std. Weight**, in the **Col Name** area.

The other default column characteristics define a numeric continuous variable and are correct for this example.

Figure 3.2 The New Column Dialog

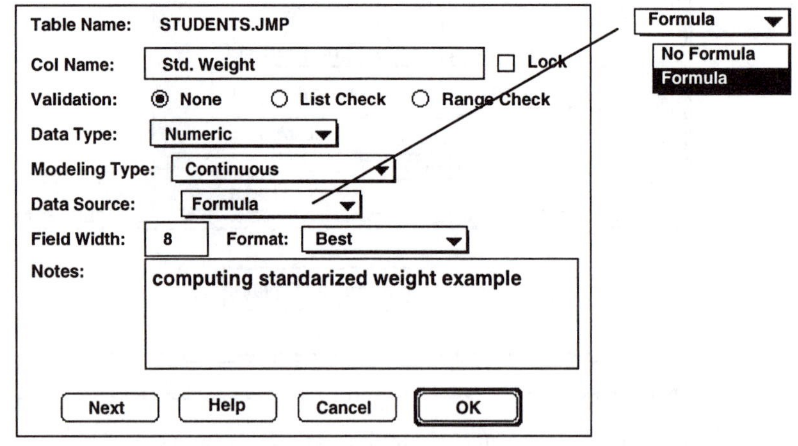

§ Select **Formula** from the **Data Source** popup menu and click **OK** to see the calculator window shown in **Figure 3.3.**

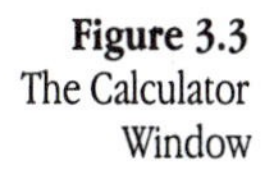

Figure 3.3
The Calculator Window

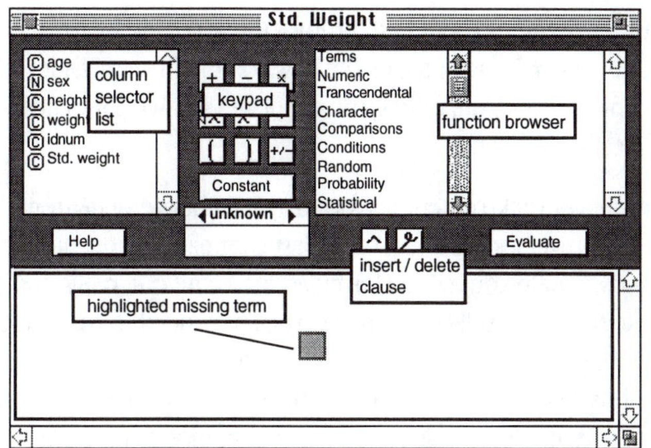

§ Next, enter the formula that standardizes the weight values by following these steps:

1) While the initial missing term is highlighted, click on the column named **weight** in the column selector list.
2) Click on the minus button in the keypad.
3) While the new missing term is highlighted, click on **weight** again.

1) *weight*   2) *weight* – ■   3) *weight* – *weight*

4) Click in the function browser topics list and scroll down the topics to locate **Statistical** functions. Click on this topic to see a list of statistical functions on the right half of the function browser. Then click **Mean** in this list.
5) Click the right parenthesis in the keypad twice to highlight the entire expression.

4) $weight - \overline{weight}$ 5) $weight - \overline{weight}$

6) While the entire expression is highlighted, click on the division button in the keypad.
7) Choose **weight** again from the column selector list.
8) While **weight** is still highlighted in the denominator, choose **Std. Deviation** from the **Statistical** functions in the function browser.

6) $\dfrac{weight - \overline{weight}}{■}$ 7) $\dfrac{weight - \overline{weight}}{weight}$ 8) $\dfrac{weight - \overline{weight}}{\text{std } weight}$

You have now entered your first formula. Close the calculator window (or click the **Evaluate** button) to see the new column fill with values. If you change any of the **weight** values, the calculated **std. weight** values automatically recompute.

If you make a mistake entering a formula, first choose **Undo** from the **Edit** menu. **Undo** reverses the effect of the last command. There are other editing commands to help you modify formulas, including **Cut**, **Copy**, and **Paste**. The DELETE key removes selected expressions. If you need to rearrange terms or expressions, you can use the hand tool to move or swap formula pieces. See the sections **Selecting Expressions** and **Dragging Expressions** later in this chapter for examples of using the hand tool to modify formulas.

This example may be all you need to proceed. However the rest of the chapter covers details about the calculator and gives a variety of other examples. Also, Appendix C, "Calculator Functions," lists all of the available functions and briefly describes them.

Calculator Pieces and Parts

This section begins with calculator terminology used when discussing examples. The calculator has distinct areas, so we also describe its geography. This section a brief description of all the function categories. Later sections give examples of specific functions.

Terminology

The following list is a glossary of terminology used in discussions about the calculator.

term

A term is an indivisible part of an expression. Constants and variables are terms.

std *weight* — term

argument

An argument is a constant, a column, or temporary variable *expression* (including mathematical operands) that is operated on by a *function*.

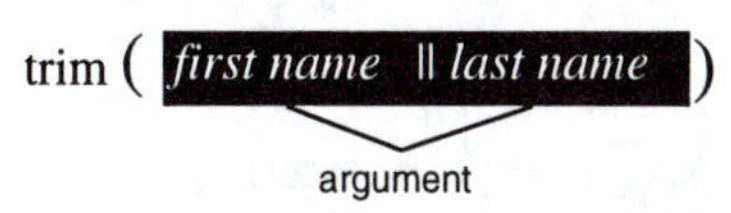

function

A mathematical or logical operation that performs a specific action on one or more *arguments*. Functions include most items in the function browser and all keypad operators.

function — std *weight*

expression

A formula or any part of it that can be highlighted as a single unit, including *terms, missing terms,* and *functions* grouped with their *arguments*.

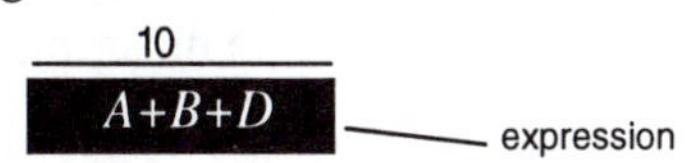

missing term

Any empty place holder for an *expression,* represented by a small empty box.

POPCORN.JMP

5 Cols / 16 Rows	batch	yield
1	large	8.2
2	large	8.6
3		?
4	large	9.2

missing values

missing value

Missing values are excluded or null data consisting of the missing value mark (a period or a ?) for numeric data, or null character strings for character data.

The Calculator Work Panel

The top part of the calculator is called the work panel. It is composed of buttons and selection lists as illustrated in **Figure 3.4**.

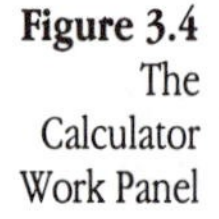

Figure 3.4 The Calculator Work Panel

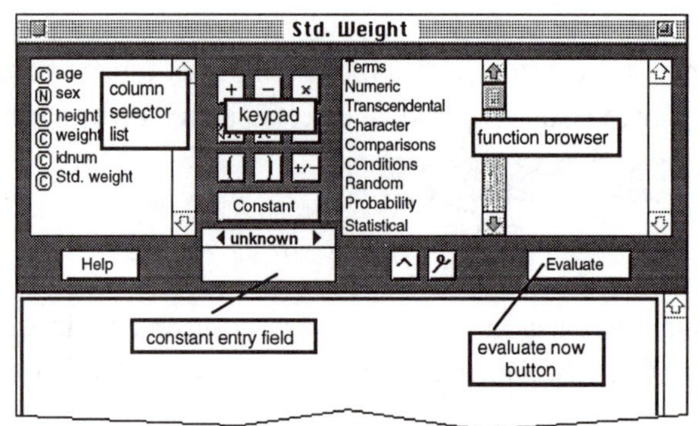

Some of the calculator features such as those in the keypad are like a hand calculator. Other features are unique to the JMP calculator. Here is a brief description of the calculator work panel areas shown in **Figure 3.4**.

Column Selector List

The column selector list displays all columns from the current data table. To choose a column, highlight an expression in the formula and click on the column name you want in the column selector list.

Keypad

The keypad is a set of buttons for commonly used calculator functions.

Function Browser

The function browser groups calculator functions and features in lists organized by topic. To enter a function in a formula, highlight a formula expression and click any item in one of the function browser topics.

Constant Entry Field

The constant entry field is an editable field used to enter any type of constant. The bar above this field identifies the current data type, and the triangles on either side of the bar scroll to other data types.

Constant Button

The **Constant** button, when clicked, enters the contents of the constant entry field into the formula, replacing the highlighted expression.

Insert and Delete Clause Buttons

The insert and delete clause buttons are located to the right of the constant entry field.

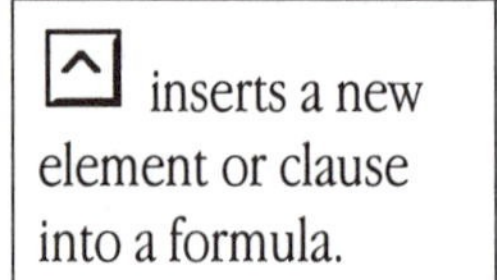

^ inserts a new element or clause into a formula.	deletes an existing element or clause from a formula.

A clause is a set of arguments to a function. To insert a clause into a formula, first select the existing clause that you want the new clause to follow. Then select the function in the function browser whose clause you want to add (such as an additional **if, otherwise** clause). When you click the insert button, the new highlighted clause appears. To delete a clause from a formula, select it and click the delete button.

Help Button

The **Help** button displays a help window with information to help you use the calculator.

Evaluate Button

The spreadsheet columns automatically fill with calculated values whenever you change a formula and then close the calculator window or make it inactive. Use the **Evaluate** button to calculate a column's values if you want the calculator window to remain open.

The Formula Display

The formula display is the area where you build and view a formula. To compose a formula, you highlight expressions in the formula display and apply functions and terms from the calculator panel. Functions always *operate* upon highlighted expressions, terms always *replace* highlighted expressions, and arguments are always *grouped* with functions. To find which expressions serve as a function's arguments, highlight that function. These groupings also show how order of precedence rules apply and show which arguments will be deleted if you delete a function.

Function Browser Definitions

The *function browser* groups the calculator functions by the topic. To enter a function, highlight an expression and click any item in the function browser topics. Examples of some commonly used functions are included later in this chapter. Also, see Appendix C, "Calculator Functions," for a description of all the functions available in the JMP calculator.

The function categories are briefly described in the following list. They are presented in the order you find them in the function browser.

Terms

are a list of commonly needed constants or functions such as e, π, current row number, or total number of rows.

Numeric
lists functions such as **Round**, **Absolute Value**, **Maximum**, and **Minimum**.

Transcendental
lists the standard transcendental functions such as natural and common log, sine, cosine, tangent, inverse functions, and hyperbolic functions.

Character
lists functions that operate on character arguments for trimming, finding the length of a string, and changing numbers to characters or characters to numbers.

Comparisons
are the standard logical comparisons such as less than, less than or equal to, not equal to, and so forth.

Conditions
are the logical functions **Not**, **And**, and **Or**. They also include programming-like functions such as **Assign**, **If/Otherwise**, **Match**, and **Choose**.

Random
is a collection of functions that generate random numbers from a variety of distributions.

Probability
lists functions that compute probabilities and quantiles for the normal, Student's t, chi-square, and F distributions.

Statistical
lists a variety of functions that calculate standard statistical quantities such as the mean or standard deviation.

Dates
are functions that require arguments with the *date* data type, which is interpreted as the number of seconds since January 1, 1904. You assign **Date** to **Data Type** in the New Column or Column Info dialog. **Date** functions return values such as day, week, or month of the year, compute dates, and can find date intervals.

Row State
lists functions that assign or detect special row characteristics called row states. Row states include color, marker, label, hidden (in plots), excluded (from analyses), and selected or not selected.

Parameters
are named constants that you create and can use in any formula.

Variables

are named, temporary variables that you assign an expression and use in other expressions for a given column.

Editing functions

let you alter the size of the font used in the formula display, replace a formula with its derivative, or discard any changes made to a formula.

Terms Functions

To do the next examples, first create an empty data table and give it rows and columns:

§ Choose **File→New**.

§ When the new table appears, choose **Rows→Add Rows** and ask for 5 rows (or any number of rows you want).

§ Choose **Cols→Add Columns** and ask for 9 new columns.

The first category in the function browser is called Terms. When you click Terms in the function browser, you see the list of functions shown to the right. (It will help you to store these functions in your own (organic) memory as they are necessary in many formula structures.) You often need to use these Terms:

π (or pi)
e
i(row#)
n(# of rows)
Subscript

i (row #)

is the current row number when an expression is evaluated for that row. You can incorporate i in any expression, including those used as column name subscripts (discussed in the next section). The default subscript of a column name is i unless otherwise specified.

n (# of rows)

is the total number of rows in the data table.

Subscript

enables you to use a column's value from a row other than the current row. Highlight a column name in the formula display and click **Subscript** to display the column's default subscript i. The i highlights and can be changed to any numeric expression. Subscripts that evaluate to nonexistent row numbers produce missing values. A column name without a subscript refers to the current row. To remove a subscript from a column, select the subscript and delete it. Then delete the missing box. The formula

$RowID_{i-1}$

uses subscript of i–1 to calculate the lag of the column named **Row ID**. The formula

$$\begin{cases} 1, & \text{if } i \le 2 \\ (Fib_{i-1})+(Fib_{i-2}), & \text{otherwise} \end{cases}$$

calculates a column called **Fib**, which contains the terms of the *Fibonacci* series (each value is the sum of the two preceding values in the calculated column). It shows the use of subscripts to do recursive calculations. A recursive formula includes the name of the calculated column, subscripted such that it references previously evaluated rows (rows 1 through i–1).

Using a Subscript

A simple use for subscripts is to create a *lag variable*. A lag variable is a column whose rows have values that are previous values of another column. Follow these steps to create a lag variable:

§ Give the first column in the empty data table the name **RowID** and open its calculator. To open the calculator choose **Cols→Column Info** and select **Formula** from the **Data Source** popup menu as shown previously in **Figure 3.2**.

§ Click **Terms** in the Function Browser, select **i(row#)** from its list of functions, and close the calculator window for **RowID**.

§ Name the second column **TotRows**, open its calculator, select **n(# of Rows)** from the list of **Terms** functions and close the calculator for **TotRows**.

§ Name the third column **Lag**, open its calculator and build the lag formula as follows:

1) Click **Row ID** in the list of columns.
2) Select **Subscript** from the list of **Term**.
3) Click the minus sign on the calculator work panel.
4 Type "1" in the constant area and click the **Constant** button (or type "1" on your keyboard and press RETURN). You should now see the lag formula, $RowID_{i-1}$

5) Click the Evaluate button on the calculator (or close the calculator) to see the results shown in **Figure 3.5**.

Note that the values in RowID and Lag are a functions of individual rows, but the constant value in the TotRows column is a function of the data table (look ahead at **Figure 3.5**).

Conditional Expressions and Comparison Operators

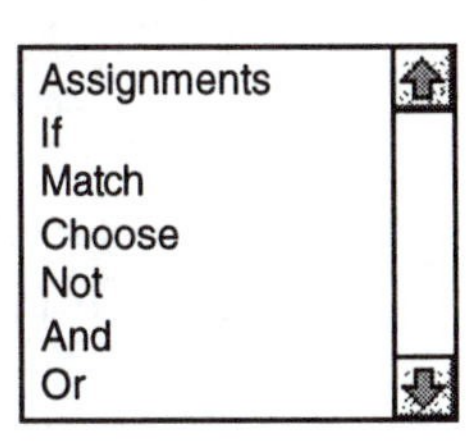

The formula to calculate the Fibonacci series is shown in the previous section. It includes a conditional expression and a comparison expression. These function categories have most of the familiar programming functions and a few you might not have seen before. This section shows examples of *conditionals* used with comparison operators.

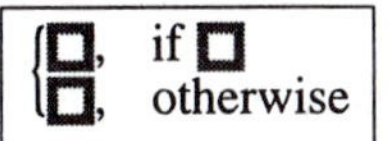

One of the most familiar conditional functions uses If, Otherwise clauses. When you highlight an expression and click If, the calculator creates a new conditional expression like the one shown here to the left. It has one If clause and one otherwise clause. A conditional expression is usually a comparison. However, any expression that evaluates as a numeric value can be a conditional expression. By definition, expressions that evaluate as zero or missing are *false*. All other numeric expressions are *true*.

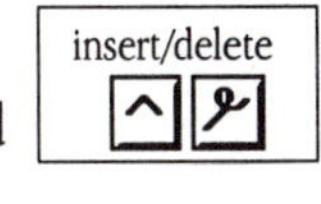

To enter a conditional expression, fill the three missing terms with expressions. Click the insert icon to add a new clause and the delete icon to remove unwanted clauses.

Using the If, Otherwise Condition Function

To create your own Fibonacci sequence:

§ Give the name Fib to the next column in your table, open its calculator, enter the Fibonacci formula as follows, and then close the calculator.

1) Click Conditions in the Function Browser, and select if from its list of functions.

2) Type "1" and click the **Constant** button for the highlighted missing term showing first in the **if** clause.
3) Highlight the **if** clause to make it active and select **x ≤ y** from the **Comparisons** list.
4) Select **i** from the **Terms** function list for the left-hand comparison, then click the righthand comparison, type "2," and press RETURN.
5) Highlight the **otherwise** clause and click the plus sign.

So far, your actions should have progressed like this:

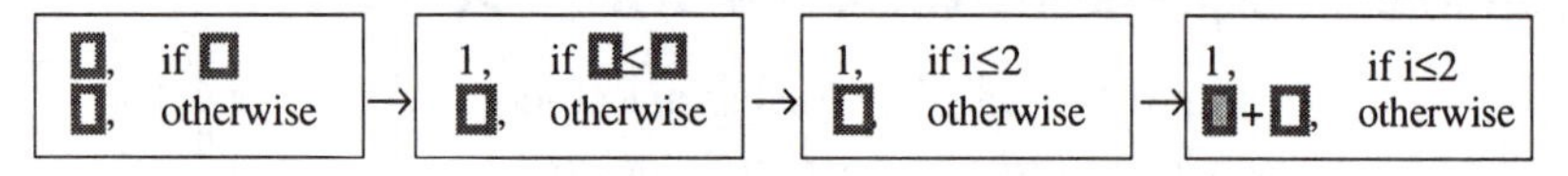

To complete the formula, construct the **otherwise** clause as follows:

6) Highlight the first missing term in the addition expression and select **Fib** (the column you are constructing) from the column selector list.
7) Create a subscript for **Fib** as before—select **Subscript** from the **Terms** functions, click the minus sign and then type a constant "1," which creates the subscript **i–1**.
8) Highlight the second missing term in the addition expression to make it active and again select **Fib** from the column selector list. This time create **i–2** as its subscript.

Figure 3.5 Results of Calculator Example

Untitled 1

10 Cols / 10 Rows	Row ID	TotRows	Lag	Fib	Group
1	1	10	•	1	1
2	2	10	1	1	0
3	3	10	2	2	1
4	4	10	3	3	0
5	5	10	4	5	1
6	6	10	5	8	0
7	7	10	6	13	1

You are finished! When you close the calculator, the formula, you entered generates the values shown here.

FYI—the Fibonacci sequence has many interesting and easy to understand properties that you can find discussed in number theory textbooks.

§ For practice, create the values in the **Group** column shown in **Figure 3.5**. Use **Modulo** from the **Numeric** functions with **RowID** as its argument, as follows:

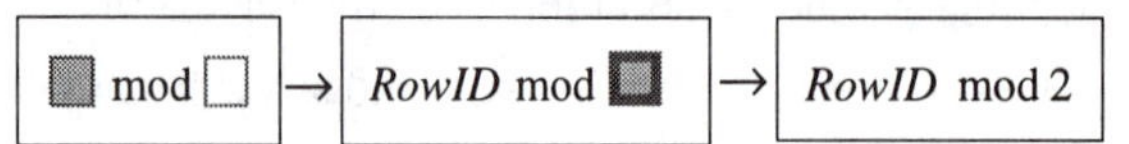

Summarize Down a Column or Summarize Across Rows

The calculator evaluates statistical functions differently from other functions. Most functions evaluate data for the current row only. However, all statistical functions require a set of values upon which to operate.

Note: The **Sum** and **Product** functions always evaluate for an explicit range of column values. All other statistical functions *always* evaluate for $i = 1$ to n values *on every row*.

Except for the **# of non-missing** function, statistical functions apply only to numeric data. The calculator excludes missing numeric values from its statistical calculations.

The Nuts and Bolts of the Quantile Function

$\text{quantile}_{0.5}\square$ The calculator computes a quantile for a column of N nonmissing values. It computes the value at which a specific percentage of the values is less than or equal. The **Quantile** function's subscript, *p*, represents the quantile percentage divided by 100. The 25% quantile, also called the lower quartile, corresponds to $p = .25$, and the 75% quantile, called the upper quartile, corresponds to $p = .75$. The default value of p is 0.5 (the median).

The pth quantile value is calculated using the formula $I = p \bullet (N+1)$ where p is the quantile and **N** is the total number of nonmissing values. If I is an integer, then the quantile value is $y_p = y_i$. If I is not an integer, then the value is interpolated by assigning the integer part of the result to i, and the fractional part to *f*, and by applying the formula

$$q_p = (1-f)y_i + (f)y_{i+1}$$

The following are example **Quantile** formulas for a column named **age**:

- $\text{quantile}_1\, age$ calculates the maximum *age*.
- $\text{quantile}_{0.75}\, age$ calculates the upper quartile *age*.
- $\text{quantile}_{0.5}\, age$ calculates the median *age*.
- $\text{quantile}_{0.25}\, age$ calculates the lower quartile *age*.
- $\text{quantile}_0\, age$ calculates the minimum *age*.

A Quantile Function Challenge Problem

In Chapter 2, you saw a data table called AUTOMESS.JMP that had auto theft rates entered as though the states were in alphabetical order. But the states were entered to match auto theft rates as though they were in ascending order. You can use the **Subset**, **Sort**, and **Join** commands in the **Tables** menu to correct this problem, but suppose the **Tables** menu is broken.

The challenge is to create a new column that has the auto theft rates sorted in ascending order, but use only a calculator formula to do it. In the following example you will enter Formula (1) shown below and verify that it solves the sort problem. What does the second formula do?

(1) $\text{quantile}_{\left(\frac{i-1}{n-1}\right)}$ *Auto theft*   (2) $\text{quantile}_{\left(\frac{n-i}{n-1}\right)}$ *Auto theft*

Hint: For entering complicated quantile subscripts, use parentheses judiciously at each stage of the formula construction, and highlight the terms within the parentheses to work on them.

§ Open the AUTOMESS.JMP table. Choose **Cols→New Column** to create a new column, and specify **Formula** as the **Data Source** in its New Column dialog.

Here is how you build the first formula shown above:

§ Create an empty subscript in parentheses. To do this, select **Quantile** from the **Statistical** functions and click **Auto theft** in the variables list to be its argument. Then click the quantile subscript to highlight it and press DELETE on your keyboard to delete the default subscript, i.

§ To create the new subscript, click the left parenthesis on the calculator to enclose the empty subscript in a set of parentheses. The subscript remains highlighted.

§ To create the numerator of the subscript, enter i–1 as shown previously (i is the row number function from the **Terms** functions). Click the right parenthesis on the calculator once to highlight i–1. So far your actions should have progressed like this:

$\text{quantile}_{(\blacksquare)}$ *Auto theft* ⟶ $\text{quantile}_{(i-1)}$ *Auto theft* ⟶ $\text{quantile}_{(i-1)}$ *Auto theft*

§ To create the denominator of the subscript click the divide symbol on the calculator. Then click the left parenthesis on the work panel to enclose the whole denominator in a set of parentheses. Enter **n–1** where **n** is the **Terms** function that gives the total number of rows in the table.

$$\text{quantile}_{\left(\frac{i-1}{(\blacksquare)}\right)} \textit{Auto theft} \rightarrow \text{quantile}_{\left(\frac{i-1}{n-1}\right)} \textit{Auto theft}$$

When you click the evaluate button or close the calculator, the new column has the auto rates in ascending order.

For practice, create another new column and enter formula (2) shown at the beginning of this section and see what it does.

Using the Summation Function

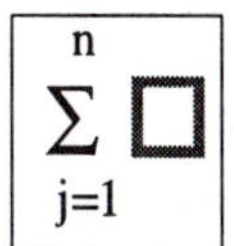

The **Sum(Σ)** function uses the summation notation shown here to the left. To calculate a sum, select **Sum(Σ)** from the **Statistics** function list and choose a variable or create an expression as its argument. The expression automatically appears with the index variable j. **Sum** repeatedly evaluates the expression for j = 1, j = 2, through j = n and then adds the nonmissing results together to determine the final result.

You can replace the index **n** (the number of rows in the active spreadsheet) and the index constant 1 with any expressions appropriate for your formula.

For example, the summation shown to the right computes the total of all **revenue** values for row 1 through the current row number, filling the calculated column with the cumulative totals of the **revenue** column.

$$\sum_{j=1}^{i} \textit{revenue}_{j}$$

Let's see how to compute a moving average using the summation function:

§ Open the XYZSTOCK.JMP sample table. Create a new column called **Moving Avg** and select **Formula** from the **Data Source** popup menu in the New Column dialog. Also, use the dialog to change the format from **Best** to **Fixed Dec** to make more sense of the table

When you specify **Use Formula**, and click **OK**, the new column appears in the table with missing values, and its calculator window opens. You should see a table like the one shown in **Figure 3.6.**

Figure 3.6 Example Table for Building a Moving Average

XYZSTOCK.JMP

6 Cols / 66 Rows	Date	DJI High	DJI Low	DJI Close	XYZ	Moving Avg
1	4/15/91	2957.18	2933.17	2896.29	62.250	•
2	4/16/91	2995.79	2986.88	2912.13	64.250	•
3	4/17/91	3030.45	3004.46	2963.12	63.250	•
4	4/18/91	3027.72	2999.26	2976.24	61.000	•
5	4/19/91	3000.25	2965.59	2943.56	59.625	•

A moving average is the average of a fixed number of consecutive values in a column, updated for each row. The following example shows you how to compute a 10-day moving average for the XYZ stock. This means that for each row the calculator computes of the sum of the current XYZ value with the 9 preceding values, then divides that sum by 10.

§ Begin the formula by selecting the Sum(Σ) function from the Statistics function category. Now tailor the summation indices to sum just the 10 values you want:

1) Highlight the 1 in the lower index (j=1) and click DELETE to change it to a missing term, giving j = ■.
2) Click the left parenthesis on the calculator keypad to enclose the missing term in a set of parentheses.
3) Enter the expression i–9 inside the parentheses, where i is selected from the Terms functions.
4) Click the upper index n to highlight it and select i from the Terms functions.
5) Click the summation argument to highlight it and select XYZ from the variables list. The summation expression should now look like the one shown to the right.

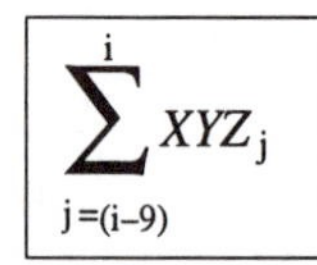

§ To finish the moving average formula, you want to divide the sum by 10 but not start the averaging process until you actually have 10 values to work with.

6) Click on the right parenthesis in the control panel to highlight the whole summation expression. (Clicking on the expression itself will also highlight it.)
7) Click the divide operator on the control pane, and then enter the constant 10 into the highlighted denominator that appears. The result should look like the one shown here to the right.

$$\frac{\sum_{j=(i-9)}^{i} XYZ_j}{10}$$

All that is left to do is use a conditional so that you don't compute anything for the first 9 values in the table:

8) Click on the entire expression to highlight it and select **if** from the Conditions function list. Then complete the **if** expression with the comparison **i > 9** (don't forget that i is the term, not the alphabetical character.)

$$\begin{cases} \dfrac{\left(\sum_{j=(1-9)}^{i} XYZ_j\right)}{10}, & \text{if } i>9 \\ \square, & \text{otherwise} \end{cases}$$

When you click **Evaluate** or close the calculator, the **Moving Avg** column fills with values.

Now generate a plot to see the result of your efforts:

§ Choose **Graph→Overlay Plots**, and select **Date** as **X** and both **XYZ** and **Moving Avg** as **Y**. When you click **OK** you see the plot in **Figure 3.7**, which compares the XYZ stock price with its 10-day moving average.

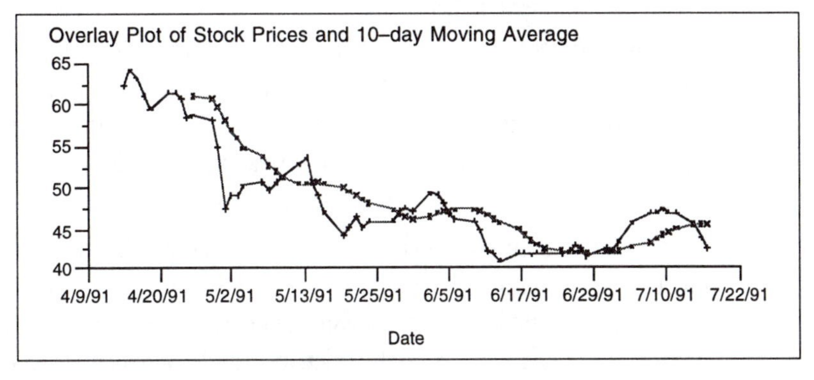

Figure 3.7 Plot of Stock Prices and Their Moving Average

The Of Function Computes Statistics Across Rows

of(□) The **Of** function is very helpful to know about because it lets you examine values across columns for each row. The **Of** function is usually used with any of the statistical functions. For example:

std dev of (*hist0*, *hist1*, *hist3*, *hist5*)

computes the standard deviation of the variables in the argument list for each row. When the arguments are column names, you compute statistics across columns.

However, the Of function accepts any list of values or expressions, including references down a column. Another way to generate the moving average in the previous example is to explicitly list the range of XYZ prices in an Of function and use the Mean function on the result as shown below. (The line above the Of sequence results when you select the Mean function.)

$$of(XYZ_{i}, XYZ_{i-1}, XYZ_{i-2}, XYZ_{i-3}, XYZ_{i-4}, XYZ_{i-5}, XYZ_{i-6}, XYZ_{i-7}, XYZ_{i-8}, XYZ_{i-9})$$

insert/delete

To create the list of arguments for the Of function, use either the comma key or the insert clause button on the calculator panel. Use the DELETE key or the delete clause button to remove unwanted arguments.

Note: When you use Of and the summation functions together, the sum index n becomes the number of items in the Of list instead of the number of rows in the data table.

Random Number Functions

Random number functions generate real numbers by effectively "rolling the dice" within the constraints of the specified distribution. You can use the random number functions with a default 'seed', that provides a pseudo-random starting point for the random series, or use the Random Number Seed function and give a specific starting seed.

Each time you click **Evaluate** in the calculator window, the random number function produces a new set of numbers. This section shows examples of three random functions called Uniform, Normal, and Shuffle. Other random number functions available in the calculator are Exponential, Cauchy, Gamma, Triangular, Poisson, Binomial, Geometric, and Negative Binomial.

The Uniform Distribution

The Uniform function (denoted ?Uniform in a calculator formula) generates random numbers uniformly between 0 and 1. This means that any number between 0 and 1 is as likely to occur as any other. The result is an approximately even distribution. You can use the Uniform function to generate any set of numbers by modifying the function with the appropriate constants.

You can see simulated distributions using the **Uniform** function and the Fit Y by X platform:

§ Choose **File→New**, to create a new data table.

§ Select (highlight) Column 1 and choose **Cols→Column Info**. When the Column Info dialog appears, select **Formula** from the **Data Source** popup menu and click **OK**. (You can use a better name for the column if you want.)

§ When the calculator window opens, select **Uniform**, from the **Random** function list in the function browser, and then close the calculator.

§ Choose **Cols→New Column** to create a second column. Follow the same steps as before except modify the Uniform function to generate the integers from 1 to 10 as follows:

1) Click **Random** in the function browser and select **Uniform**, from its list.
2) Click the multiply sign on the calculator key pad
3) Enter 10 in the constant entry box and click **Constant** (or enter 10 on your keyboard and press enter).
4) Click the addition sign on the calculator key pad
5) Enter 1 in the constant entry box and click **Constant** (or enter 1 on your keyboard and press enter).

Your actions should have gone like this so far

?uniform →?uniform • ▢→?uniform • 10 →?uniform • 10 +▢→?uniform • 10 + 1

The next steps are the key to generating a uniform distribution of integers (as opposed to real numbers):

6) Click the right parenthesis on the key pad to highlight the entire formula.
7) Click **Numeric** in the function browser and select the **Floor** function from its list. The final formula is

$\lfloor$(?uniform•10)+1$\rfloor$

§ Close the calculator. You now have a table template for creating two uniform distributions. All you have to do is add as many columns as you want. So choose **Rows→Add Rows** and add 500 rows.

§ The table will fill with values. Change the modeling type of the column of integers to nominal so it will be treated as a discrete distribution.

§ Choose **Analyze→Distribution of Y**, use both columns as **Y** variables, and then click **OK**.

You will see two histograms similar to those shown in **Figure 3.8**. The left-hand histogram represents simple uniform random numbers and the right-hand histogram shows random integers from 1 to 10.

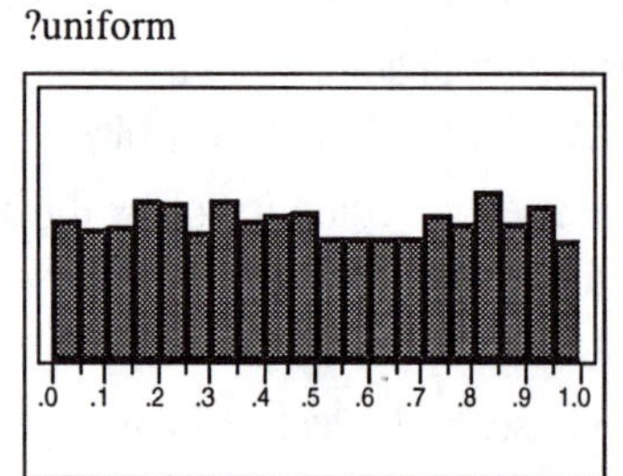

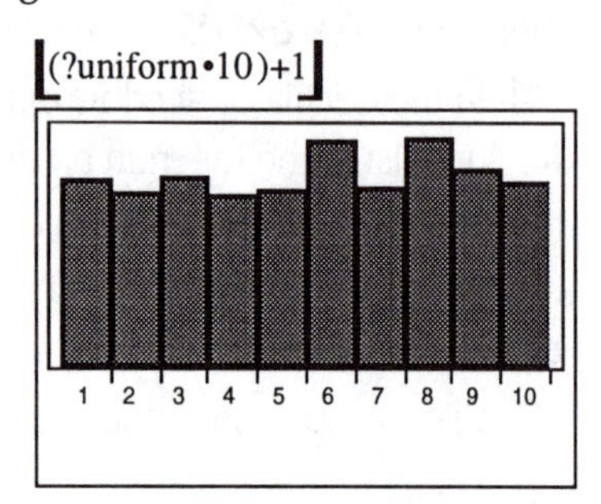

Figure 3.8 Example of Two Uniform Distribution Simulations

The Normal Distribution

?Normal generates random numbers that approximate a normal distribution with a mean of 0 and variance of 1. The normal distribution is bell shaped and symmetrical. You can modify the **Normal** function with constants to specify a normal distribution with a different mean and standard deviation.

§ As an exercise, follow the same instructions described previously for the **Uniform** random number function. Create a table with columns for a standard normal distribution and a normal distribution with mean 30 and standard deviation 10. The modified normal formula is ?normal •5+30 .

Figure 3.9 shows the Distribution of Y platform for these normal simulations.

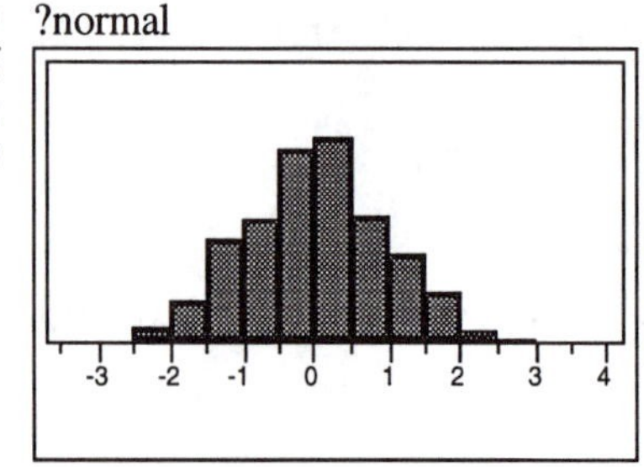

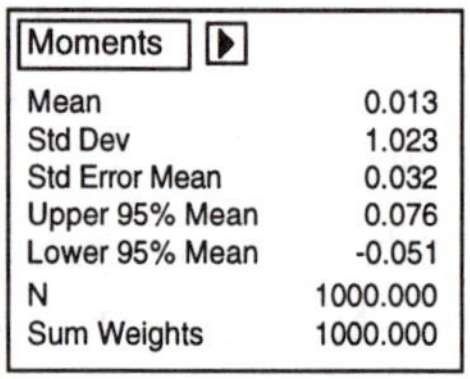

Moments	
Mean	0.013
Std Dev	1.023
Std Error Mean	0.032
Upper 95% Mean	0.076
Lower 95% Mean	-0.051
N	1000.000
Sum Weights	1000.000

Figure 3.9 Illustration of Selecting Expressions

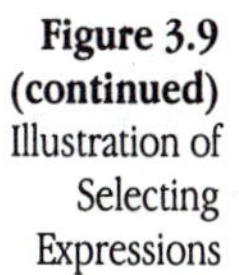

Figure 3.9 (continued) Illustration of Selecting Expressions

?normal•5+30

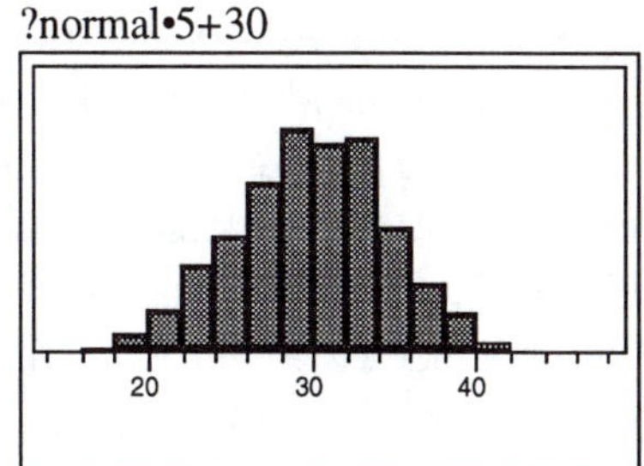

Moments	
Mean	29.934
Std Dev	4.921
Std Error Mean	0.156
Upper 95% Mean	30.239
Lower 95% Mean	29.629
N	1000.000
Sum Weights	1000.000

The Shuffle Function

?Shuffle selects a row number at random from the current data table. Each row number is selected only once. When Shuffle is used as a subscript, it returns a value selected at random from the column that serves as its argument. Each value from the original column is assigned only once. For example, to identify a 50% random sample without replacement, use the following formula:

$$\begin{cases} X_{\text{?shuffle}}, & \text{if } i \le \frac{n}{2} \\ \blacksquare, & \text{otherwise} \end{cases}$$

This formula chooses half the values (n/2) from the column *X* and assigns them to the first half of the rows in the computed column. The remaining rows of the computed column fill with missing values.

Parameters

Parameters are named constants created in the calculator that can be used in any formula. Numeric parameters are most useful in formulas created for nonlinear fitting.

Clicking the New Parameter item in the function browser brings up a New Parameter dialog. You use the dialog to assign a name, value, and data type to the new parameter. The default names for parameters are p1, p2, and so on. You can create as many parameters as you need for a column. After a parameter is created, it shows at the bottom of the function browser parameter list for that column.

You assign a data type (**numeric**, **character**, or **row state**) with the **Data Type** popup menu in the parameter dialog. Click **Done** when the dialog is complete or **Cancel** to exit the dialog without creating a new parameter. To change parameter settings or remove a parameter, OPTION-click or ALT-click on the parameter name. This displays the New Parameter dialog again with the **Remove** button activated. Parameters are added to formulas in much the same way that variables are added. To insert parameters into your formula, click the parameter name in the function browser list. Parameters are easy to recognize in formulas because they are displayed in **bold** type. For example, in the formula

$$\mathbf{b0}\bullet\left(1-e^{-\mathbf{b1}\bullet X}\right)$$

b0 and **b1** are parameters.

Parameter data types must be valid in the context of the expressions where they are used. If a parameter data type is invalid, an error message appears. When a parameter is used in a model for the nonlinear platform, the initial value or starting value should be given as the parameter value. After completing a nonlinear fit or after using the **Reset** button in the nonlinear control panel, the parameter's value is the most recent value computed by the nonlinear platform.

When you paste a formula with parameters into a column, the parameters are automatically created for that column unless it has existing parameters with the same names.

Tips on Building Formulas

This section describes some techniques for building calculator formulas. First, you need to keep the computational order of precedence in mind. All of the functions have an order of precedence defined by levels 1 through 6 as described in **Table 3.1**, where 1 is the highest order of precedence. Expressions with a high order of precedence are evaluated before those at lower levels. When an expression has operators of equal precedence, it is evaluated from left to right. You can use parentheses to override other precedence rules when necessary because any expression within parentheses is always evaluated first. Terms have no order of precedence because they cannot be evaluated further.

Table 3.1 Precedence Order of Functions

Level 1:	Parentheses
Level 2:	Functions: |x|†, Ceiling†, Char to Num, Color, ColorOf, Conditions†, Exclude, Floor†, Hide, Hidden, Label, Labeled, Length, Log, ln, Marker, MarkerOf, Max†, Mean†, Min†, Munger, N, Num to Char, Power, Probability†, Product, Quantile, Root, Row State Combinations†, Select, Selected, Std. Deviation, Summation, Transcendental†, Trim
Level 3:	*, ÷, Modulo
Level 4:	+, –
Level 5:	Comparisons: <, ≤, =, ≠, >, ≥, ≤ x ≤, ≤ x <, < x ≤, < x <
Level 6:	Logical Functions: And, Not, Or
†	When one of these functions has an expression as its argument, the argument has a higher order of precedence (level 1) as it would if enclosed in parentheses.

It is best to build a formula starting with any expression that serves as an argument. This is because functions have a high order of precedence and are always grouped with their corresponding arguments. It is a good idea to create expressions working from highest to lowest order of precedence when possible. If you need parentheses, be sure to click the left parenthesis in the keypad before entering the expression to be enclosed. The left parenthesis key always places a set of parentheses around whatever you highlight.

Because order of precedence determines which arguments are affected by each function, order of precedence also affects the grouping of expressions. Highlight functions in the formula to verify how the order of precedence rules have been applied.

Constant Expressions

Once JMP has evaluated a formula, you can highlight a constant expression to see its value in the constant entry field. This is true for both parameters and expressions that evaluate to a constant value.

Focused Work Areas

Functions, column names, and constants can only be entered into formulas when their area of the calculator is *focused*. Clicking the mouse in the column selector list, function browser, constant entry field, or formula display area, causes a keyboard focus shift to that item. For example, clicking a column name in the column selector list makes the scroll bars visible and the column name highlight. Likewise, clicking any function group in the function browser makes the scroll bars of both function browser panels appear and the first function of the selected group highlight.

When the formula display area is focused, the gray border around it changes to black. Keystrokes and editing commands now affect the formula.

There are several methods of changing focus. One way is to use the mouse to make all selections. You can also repeatedly press the TAB key on the keyboard to change focus by cycling through the column selector list, Constant entry field, function browser, and formula display area. Also, typing the name of a column changes focus to the column selector list and selects the first column beginning with the typed letters. Typing a digit changes the Constant entry field focus to a numeric data type. Typing a quote changes the Constant entry field focus to a character data type.

Cutting and Pasting Formulas

You can cut or copy an entire formula or any expression and paste it into another formula display. If you paste a formula in the scrapbook, it is saved both as PICT and Expr (expression format) data. You can copy expression format data into any formula display, and you can paste PICT format information into other applications.

When you copy an expression from one data table to another, it expects to find the same column names. If a formula column name does not appear in the table, a dialog asks you for a substitute from the list of available column names.

Selecting Expressions

You can use the keyboard arrow keys to select expressions for editing or to view the order of precedence within a formula when parentheses are not present. Clicking an operand in an associative expression, such as addition or multiplication, selects the operand. Clicking any associative operator, such as a + or • sign, selects the operator and its operands.

Once an operand is selected, the left and right arrow keys move the selection across other associative operands having equal precedence within the expression. The up arrow extends the current selection by adding the operand and operator of higher precedence to the selection. The down arrow reduces the current selection by removing an operand and operator from the selection.

The example in **Figure 3.10** demonstrates how the arrow keys navigate through expressions.

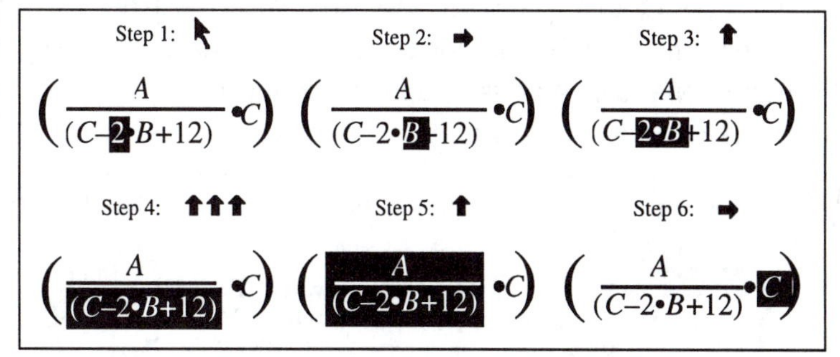

Figure 3.10 Illustration of Selecting Expressions

Dragging Expressions

When you place the arrow cursor inside a selected expression, it changes to a hand cursor. This enables you to drag the selected expression to a new location. As you drag the expression, possible destinations highlight. The originally selected expression replaces the highlighted expression when the mouse button is released.

The following steps show how to move the text "Male" from the female to the male result clause in the sex match statement in **Figure 3.11**:

1) Press the mouse button over the selected term "Male". The cursor turns into the hand tool.
2) Without releasing the mouse, move the hand and drag until the male result clause highlights.
3) Release the mouse button and the text "Male" moves from the female to the male result clause.

Figure 3.11 Example of Dragging Expressions

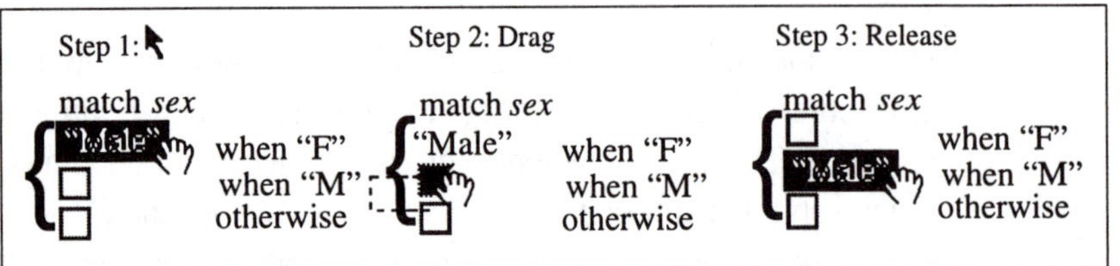

Using the OPTION key or ALT key with the hand tool enables you to drag a copy of an expression to a new destination. An example of OPTION-dragging (or ALT-dragging) is shown in **Figure 3.12**. Using the COMMAND key on the Macintosh or the CONTROL key under Windows allows you to swap the selected expression with the destination expression. To swap or drag a copy of an expression you must press the appropriate key (OPTION or ALT, COMMAND or CONTROL) *before* dragging it

Figure 3.12 Example of Option-Dragging (Copy)

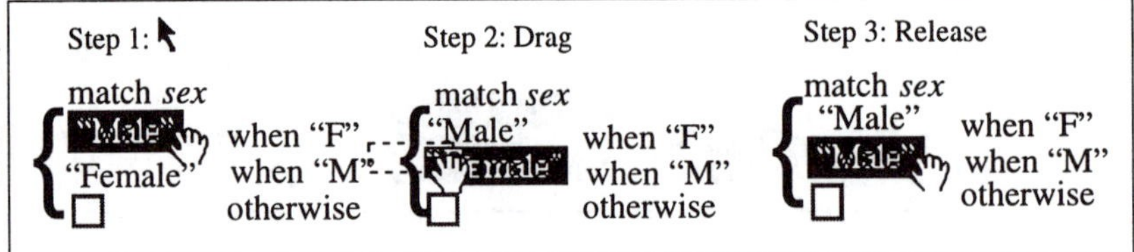

Changing a Formula

If you need to change a formula, double-click the corresponding column's selection area to bring up its Column Info dialog. To reopen the calculator window, click the picture of the formula in its dialog. Note: You can take a shortcut to a column's formula window by OPTION-clicking the selection area at the top of the column. This bypasses the Column Info dialog.

Deleting a function also deletes its arguments. Deleting a required argument or missing term from a function deletes the function as well. You can save complicated expressions that serve as arguments and paste them where needed. Use the **Copy** command to copy the arguments to the clipboard or scrapbook, or delete all but one function and argument and *peel* the function from the remaining argument as shown in **Figure 3.13**.

To peel a function from a single argument, first highlight the function with the mouse and then hold down the OPTION (or ALT) key and press the DELETE key. If you prefer to use the menus, you can choose **Clear** from the **Edit** menu in place of pressing the DELETE key. For example, to remove the Trim function (shown in **Figure 3.13**) leaving only n, highlight the Trim(n) and press OPTION-DELETE on the Macintosh or ALT-DELETE under Windows

Figure 3.13 Peeling an Argument

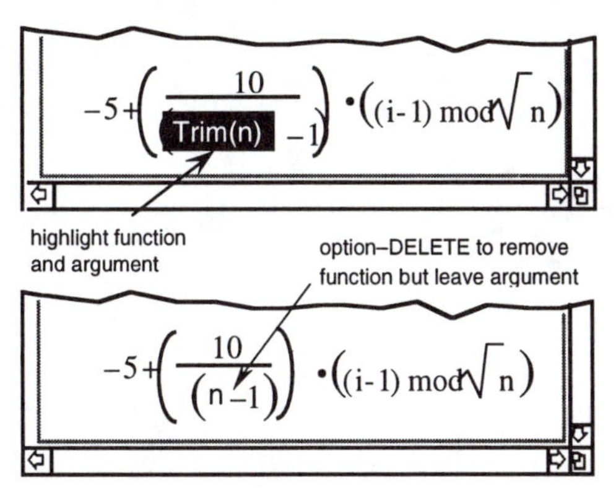

Or, you can use the hand tool to drag the argument on top of its function. Simply highlight an argument and drag it to the left until it is over the left parenthesis of its function. When you release the mouse the function disappears.

Once you have created a formula, you can change values in columns that are referenced by your formula. The calculator automatically recalculates all affected values in the formula's column.

Caution and Error Messages

Error and warning messages alert you that there is an error in your formula and warn you whenever you choose a command or use a JMP feature that can have unforeseen effects on calculated data. A beep often serves as the first warning of a problem, followed by a message if the problem persists.

Calculator messages include both *Caution Alerts* and *Stop Alerts*.

Caution Alerts allow you to continue a specific course of action but warn you of possible problems. They typically give you a choice of response buttons.

Stop Alerts detect errors that must be corrected before processing can continue. They often inform you that an invalid action has been ignored. Stop Alerts also appear each time you attempt to close the calculator window or click in another window while an invalid formula is in the formula display. When this happens, click **OK** and correct or delete the invalid formula.

You might get a lot of messages when you first start using the calculator, but with practice and perseverance, you will find that almost nothing is beyond the capability of the JMP calculator.

Chapter 4
What are Statistics?

Overview

Statistics are numbers, but the practice of statistics is the craft of measuring imperfect knowledge. That's one definition and there are many more. This chapter is collection of short essays to get you started on the many ways of statistical thinking and to get you used to the terminology of the field.

Chapter 4 Contents

Ponderings

The Business of Statistics

The discipline of statistics provides the framework of balance sheets and income statements for scientific knowledge. Statistics is an accounting discipline, but instead of accounting for money, it is accounting for scientific credibility. It is the craft of weighing and balancing observational evidence. Scientific conclusions are based on experimental data in the presence of uncertainty, and statistics is the mechanism to judge the merit of those conclusions—the statistical tests are like credibility audits. Of course you can juggle the books and make poor science sometimes look better than it is. But there are important phenomena that you just can't uncover without statistics.

A special joy in statistics is when it can be used as a discovery tool to find out new phenomena. There are many views of your data—the more perspectives you have on your data, the more likely you are to find out something new. Statistics as a discovery tool is the auditing process that unveils phenomena that are not anticipated by a scientific model and are unseen with a straightforward analysis. These anomalies lead to better scientific models.

Statistics fits models, weights evidence, helps identify patterns in data and then helps find data points that don't fit the patterns. Statistics is induction from experience; it is there to keep score on the evidence that supports scientific models.

Statistics is the science of uncertainty, credibility accounting, measurement science, truth- craft, the stain you apply to your data to reveal the hidden structure, the sleuthing tool of a scientific detective.

Statistics is a necessary bureaucracy of science.

Two Sides of Statistics

The Yin and Yang of Statistics

There are two sides to Statistics.

First there is the Yang of statistics, a shining sun. The Yang is always illuminating, condensing, evaporating uncertainty, and leaving behind the precipitate of knowledge. It pushes phenomena into forms. The Yang is out to prove things in the presence of uncertainty and ultimately compel the data to confess its form, conquering ignorance headlong. The Yang demolishes hypotheses by ridiculing their improbability. The Yang mechanically cranks through the ore of knowledge and distills it to the answer.

On the other side, we find the contrapositive Yin, the moon, reflecting the light. The Yin catches the shadow of the world, feeling the shape of truth under the umbra. The Yin is forever looking and listening for clues from the creator, nurturing seeds of pattern and anomaly into maturing discoveries. The Yin whispers its secrets to our left hemisphere. It unlocks doors for us in the middle of the night, planting dream seeds, making connections. The Yin draws out the puzzle pieces to tantalize our curiosity. It teases our awareness and tickles our sense of mystery until the climax of revelation—Eureka!

The Yin and Yang are forever interacting, catalyzed by Random, the agent of uncertainty. As we see the world reflected in the pool of experience, the waters are stirred, and in the agitated surface we can't see exactly how things are. Emerging from this, we find that the world is knowable only by degree, that we have knowledge in measure, not in absolute.

The Faces of Statistics

Everyone has a different view of statistics.

Match the definition on this side....	with someone likely to have said it on this side.
1. The literature of numerical facts.	a. Engineer
2. An applied branch of mathematics.	b. Original meaning
3. The science of evidence in the face of uncertainty.	c. Social scientist
4. A digestive process that condenses a mass of raw data into a few high-nutrient pellets of knowledge.	d. Philosopher
5. A cooking discipline with data as the ingredients and methods as the recipes.	e. Economist
6. The calculus of empiricism.	f. Computer scientist
7. The lubricant for models of the world.	g. Mathematician
8. A calibration aid.	h. Physicist
9. Adjustment for imperfect measurement.	i. Baseball fan
10. An application of information theory.	j. Lawyer
11. Involves a measurable space, a sigma algebra, and Lebesgue integration.	k. Joe College
12. The nation's state.	l. Politician
13. The proof of the pudding.	m. Businessman
14. The craft of separating signal from noise.	n. Statistician
15. A way to predict the future.	

An interesting way to think of statistics is as a toy for grownups; remember that toys are proxies that children use to model the world. Children use toys to learn behaviours and develop explanations and strategies, as aids for internalizing the external. This is the case with statistical models. You model the world with a mathematical equation, and then see how the model stacks up to the observed world.

Statistics lives in the interface of the real world data and mathematical models, between induction and deduction, empiricism and idealism, thought and experience. It seeks to balance real data and a mathematical model. The model addresses the data and stretches to fit. The model changes and the change of fit is measured. When the model doesn't fit, the data suspends from the model and leaves clues. You see patterns in the data that don't fit—this leads to a better model, and points that don't fit into patterns can lead to important discoveries.

Don't Panic

Some university students have a panic reaction to the subject of statistics. Yet most science, engineering, business, and social science majors usually have to take at least one statistics course. What are some of the sources of our phobias about statistics?

Abstract Mathematics

Though statistics can become very mathematical to those so inclined, applied statistics can be used effectively with only a very basic level of mathematics. You can talk about statistical properties and procedures without plunging into abstract mathematical depths. In this book, we are interested in looking at applied statistics.

Lingo

Statisticians often don't bother to translate terms like 'heteroschedasticity' into 'varying variances' or 'multicollinearity' into 'closely related variables.' Or, for that matter, further translate 'varying variances' into 'difference in the spread of values between samples,' and 'losely related variables' into 'variables that give similar information.' We will tame some of the common statistical terms in the discussions that follow.

Awkward Phrasing

There is a lot of subtlety in statistical statements that can sound awkward, but the phrasing is very precise and means exactly what it says. Sometimes statistical statements include multiple negatives. For example, "The statistical test failed to reject the null hypothesis of no effect at the specified alpha level." That is a quadruple negative statement—count the negatives: 'fail,' 'reject,' 'null,' and 'no effect.' You can reduce the negatives by saying "the statistical results are not significant" as long as you are careful not to confuse that with the statement "there is no effect." Failing to prove something does not prove the opposite!

A bad reputation

The tendency to assume the proof of an effect because you cannot statistically prove the absence of the effect is the origin of the saying, "Statistics can prove anything." This is what happens when you twist a term like 'nonsignificant' into 'no effect.' This idea is common in a court room; you can't twist the phrase "there is not enough evidence to prove beyond reasonable doubt that the accused committed the crime" with "the accused is innocent." What nonsignificant really means is that there is not enough data to show a significant effect—it does not mean that you are certain there is no effect at all.

Uncertainty

Although we are comfortable with uncertainty in ordinary daily life, we are not used to embracing it in our knowledge of the world. We think of knowledge in terms of hard facts and solid logic, though much of our most valuable real knowledge is less than solid. We can say when we know something for sure (yesterday it rained), and we can say when we don't know (don't know whether it will rain tomorrow). But when we describe knowing something with incomplete certainty, it sounds apologetic or uncommitted. For example it sounds like a form of equivocation to say that there is a .9 chance that it will rain tomorrow. Yet much of what we think we know is really just that kind of uncertainty.

Preparations

A few fundamental concepts will prepare you for absorbing details in upcoming chapters.

Three Levels of Uncertainty

Statistics is about uncertainty, but there are several different levels of uncertainty that you have to keep in separate accounts:

The Randomness of Events:

Even if you know everything possible about the world, you still have events you can't predict. You can see an obvious example of this in any gambling casino. You can be an expert at playing blackjack, but the randomness of the card deck renders the outcome of any game indeterminate. We make models with random error terms to account for uncertainty due to randomness. Some of the error term may be due to ignoring details; some may be measurement error; but much of it is attributed to inherent randomness.

The Unknownness of Parameters:

Not only are you uncertain how an event is going to turn out, you often don't even know what the numbers (parameters) are in the model that generates the events. You have to estimate the parameters and test if hypothesized values of them are plausible, given the data. This is the chief responsibility of the field of statistics.

The Not-knowing About Whether a Model Is Correct

Sometimes you not only don't know how an event is going to turn out, and you don't know what the numbers are in the model, but you don't even know if the form of the model is right.

Statistics is very limited in its help for certifying that the model is correct. Most statistical conclusions assume that the model is correct. The correctness of the model is the responsibility of the subject- matter science. Statistics might give you clues if the model is not carrying the data very well. Statistical analyses can give diagnostic plots to help you see patterns that could lead to new insights, to better models.

Probability and Randomness

In the old days, statistics texts all began with chapters on probability. Today the many popular statistics books discuss probability in later chapters. We will skip the balls in urns entirely, though probability is the essence of our topic.

Randomness makes the world interesting and *probability* is needed as the measuring stick. Probability is the aspect of uncertainty that allows the information content of events to be measured. If the world were deterministic, then the information value of an event would be zero because it would already be known to occur; the probability of the event occurring in 1. The sun rising tomorrow is a nearly deterministic event and doesn't make the front page of the newspaper when it happens. The event that happens but has been attributed to have probability near zero would be big news. For example, the event of extraterrestrial intelligent life forms landing on earth would make the headlines.

Statistical language uses the term probability on several levels:

- When we make observations or collect measurements, our responses are said to have a *probability distribution*. For whatever reason, we assume that something in the world adds randomness to our observed responses, which makes for all the fun in analyzing data that has uncertainty in it.
- We calculate statistics using probability distributions, seeking the safe position of *maximum likelihood*, which is the position of least improbability.
- The significance of an event is reported in terms of probability. We demolish statistical null hypotheses by making their consequences look incredibly improbable.

Assumptions

Statisticians are naturally conservative professionals. Like the boilerplate of official financial audits, statisticians' opinions are full of provisos such as "assuming that the model is correct, and assuming that the error is normally distributed, and assuming that the observations are independent and identically distributed, and assuming that there is no measurement error, and assuming...." Even then the conclusions are hypothetical, with phrases like "if you say the hypothesis is false, then the probability of being wrong is less than .05."

Statisticians are just being precise, though they sound like they are combining the skills of equivocation like a politician, technobabble like a technocrat, and trick-prediction like the Oracle at Delphi.

Ceteris Paribus

A crucial assumption is the *ceteris paribus* clause, which is Latin for *other things being equal*. This means we assume that the response we observed was really only affected by the model and random error; all other factors that might affect the response were maintained at the same controlled value across all observations or experimental units. This is, of course, often not the case, especially in observational studies, and the researcher must try to make whatever adjustments, appeals, or apologies to atone for this. When statistical evidence is admitted in court cases, there are endless ways to challenge it, based on all the assumptions that may have been violated.

Is the Model Correct?

The most important assumption is that your model is right. There are no easy tests for this assumption. Statistics almost always measure one model against a submodel, and these have no validity if neither model is appropriate in the first place.

Is the Sample Valid?

The other supremely important issue is that the data relate to your model; that is, you have collected your data in a way that is fair to the questions that you will be asking it. If your sample is ill-chosen, or if you have skewed your data by rejecting data in a process that relates to its applicability to the questions, then your judgments will be flawed. If you have not taken careful consideration of the direction of causation, you may be in trouble. If taking a response affects the value another response, then they are not independent of each other, which can affect the study conclusions.

In brief, are your samples fairly taken and are your experimental units independent?

Data-Mining?

One issue that most researchers are guilty of to a certain extent is stringing together a whole series of conclusions and assuming that the joint conclusion has the same confidence as the individual ones. An example of this is *data mining*, in which hundreds of models are tried until one is found with the hoped-for results. Just think about the fact that if you collect purely random data, you will find a given test significant at the .05 level about 5% of the time. So you could just repeat the experiment until you get what you want, discarding the rest. That's obviously bad science, but something similar often happens in studies. This multiple-testing problem remains largely unaddressed by statistical literature and software except for some special cases such as means comparisons, a few general methods that may be inefficient (Bonferroni's adjustment), and expensive, brute-force approaches (resampling methods).

Another problem with this issue is that the same kind of bias is present across unrelated researchers because nonsignificant results are often not published. Suppose that 20 unrelated researchers do the same experiment, and by random chance one researcher got a .05- level significant result. That's the result that gets published.

In light of all the assumptions and pitfalls, it is appropriate that statisticians are cautious in the way they phrase results. Our trust in our results has limits.

Statistical Terms

Statisticians are often unaware that they use certain words completely differently from other professionals. In the following list, some definitions will be the same as you are used to, and some will be the opposite:

Model

The statistical model is the mathematical equation that predicts the response variable as a function of other variables, together with some distributional statements about the random terms that allow it to not fit exactly. Sometimes this model is taken very casually in order to look at trends and tease out phenomena, and sometimes the model is taken very seriously.

Parameters

To a statistician, parameters are the unknown coefficients in a model, to be estimated and to test hypotheses about. They are the indices to distributions; the mean and standard deviation are the location and scale parameters in the normal distribution family.

Unfortunately, engineers use the same word (parameters) to describe the factors themselves.

Statisticians usually name their parameters after Greek letters, like mu, sigma, beta, and theta. You can tell where statisticians went to school by which Greek and Roman letters they use in various situations. For example, in multivariate models, the L-Beta-M fraternity is distinguished from C-Eta-M.

Hypotheses

In science, the hypothesis is the bright idea that you want to confirm. In statistics, this is turned upside down because it is using logic analogous to a proof-by-contradiction. The so-called *null hypothesis* is usually the statement that you want to demolish. The usual null hypothesis is that some factor has no effect on the response. You are of course trying to support the opposite, which is called the alternative hypothesis. You support the alternative hypothesis by statistically rejecting the null hypothesis.

Two-Sided versus One-Sided, Two-Tailed versus One-Tailed

Most often, the null hypothesis can be stated as some parameter in a model being zero. The alternative is that it is not zero, which is called a two-sided alternative. In some cases, you may be willing to state the hypothesis with a one-sided alternative, that the parameter is greater than zero, for example. The one-sided test has greater power at the cost of less generality. These terms have only this narrow technical meaning, and it has nothing to do with common English phrases like presenting a one-sided argument (prejudiced, biased in the everyday sense) or being two-faced (hypocrisy or equivocation). You can also substitute the word "tailed" for "sided." The idea is to get a big statistic that is way out in the tails of the distribution where it is highly improbable. You measure how improbable by calculating the area of one of the tails, or the other, or both.

Statistical Significance

Statistical significance is a precise statistical term that has no relation to whether an effect is of practical significance in the real world. Statistical significance usually means that the data gives you the evidence to believe that some parameter is not the null value. If you have a ton of data, you can get a statistically significant test when the values of the estimates are practically zero. If you have very little data, you may get an estimate of an effect that would indicate enormous practical significance, but it is supported by so little data that it is not statistically significant. A nonsignificant result is one that might be the result of random variation rather than a real effect.

Significance Level, p value, Alpha Level

You want small *p values*, because the p value is the probability of being wrong if you declare an effect to be non-null. The p value is sometimes labeled the significance probability, or sometimes labeled more precisely in terms of the distribution that is doing the measuring. The p value labeled "Prob>|t|" is read as "the probability of getting a greater t (in absolute value)." The alpha-level is your standard of the p value that you claim, so that p values below this reject the null hypothesis (that is, they show that there is an effect).

Power, Beta Level

Power is how likely you are to detect an effect if it is there. The more data you have, the more statistical power. The greater the real effect, the more power. The less random variation in your world, the more power. The more sensitive your statistical method, the more power. It you had a method that always declared an effect significant, regardless of the data, it would have a perfect power of 1 (but it would have an alpha-level of 1, too, the probability of declaring significance when there was no effect). The goal in experimental design is usually to get the most power you can afford, given a certain alpha-level. It is not a mistake to connect the statistical term *power* with the common sense of power as persuasive ability. It has nothing to do with work or energy, though.

Confidence Intervals

A *confidence interval* is an interval around a parameter estimate that has a given probability of containing the true value of a parameter. Most often the probability is 95%. Confidence intervals are now considered one of the best ways to report results. It is expressed as a percentage of 1–alpha, so a .05 alpha level for a two-tailed t quantile can be used for a 95% confidence interval. (For linear estimates, it is constructed by multiplying the standard error by a t statistic and adding and subtracting that to the estimate. If the model involves nonlinearities, then the linear estimates are just approximations, and there are better confidence intervals called "profile likelihood confidence intervals." If you want to form confidence regions involving several parameters, it is not valid to just combine confidence limits on individual parameters.)

Biased, Unbiased

Unbiased means that the expected value of an estimator is the parameter being estimated. It is considered a desirable trait, but not an overwhelming one. There are cases when statisticians recommend biased estimators. The maximum likelihood estimator of the variance has a small but nonzero bias, for example.

Sample Mean versus True Mean

The *sample mean* is the one you calculate from your data–the sum divided by the number. It is a statistic, that is, a function of your data. The true mean is the expected value of the probability distribution that generated your data. You usually don't know the true mean, and that is why you collect data, so you can estimate the true mean with the sample mean.

Variance and Standard Deviation, Standard Error

Variance is the expected squared deviation of a random variable from its expected value. It is estimated by the *sample variance. Standard deviation* is the square root of the variance, and we prefer to report it because it is in the same units as the original random variable (or response values). The *sample standard deviation* is the square root of the sample variance. The term *standard error* describes an estimate of the standard deviation of another (unbiased) estimate.

Degrees of Freedom

The specific name for a value indexing some popular distributions of test statistics. It is called *degrees of freedom* because it relates to differences in numbers of parameters that are or could be in the model. The more parameters a model has, the more *freedom* it has to fit the data better. The *DF* (degrees of freedom) for a test statistic is usually the difference in the number of parameters between two models.

Chapter 5 Univariate Distributions: One Variable, One Sample

Overview

This chapter introduces statistics in the simplest possible setting—the distribution of a single batch of values in one variable. The **Distribution of Y** command in the **Analyze** menu launches JMP's *Distribution platform*, which is used to describe the distribution of a column of values from a table using a number of graphs, and by calculating various summary statistics.

This chapter also introduces the concept of the distribution of a statistic, and how confidence intervals and hypothesis tests can be obtained.

Chapter 5 Contents

You can mouse along with the examples whenever you see an action symbol (§)

Looking at Distributions

Let's take a look at some actual data and start noticing aspects of their distribution.

§ Begin by opening the data table called BRTHDTH.JMP, which contains the 1976 birth and death rates of 74 nations (**Figure 5.1**).

Figure 5.1 Partial Listing of the BRTHDTH Data Table

BRTHDTH.JMP

4 Cols / 74 Rows	country	birth	death	Region
1	AFGHANISTAN	52	30	Asia
2	ALGERIA	50	16	Africa
3	ANGOLA	47	23	Africa
4	ARGENTINA	22	10	S America
5	AUSTRALIA	16	8	Pacific

§ Choose **Analyze→Distribution of Y**.

§ Select the birth, death, and Region columns as the Y variables and click **OK**.

When you see the report, be adventuresome: scroll around and click in various places on the surface of the report. Notice that a histogram or statistical table can be opened or closed by clicking its report title button.

§ Open and close tables, and click on bars until you have the configuration shown in **Figure 5.2**.

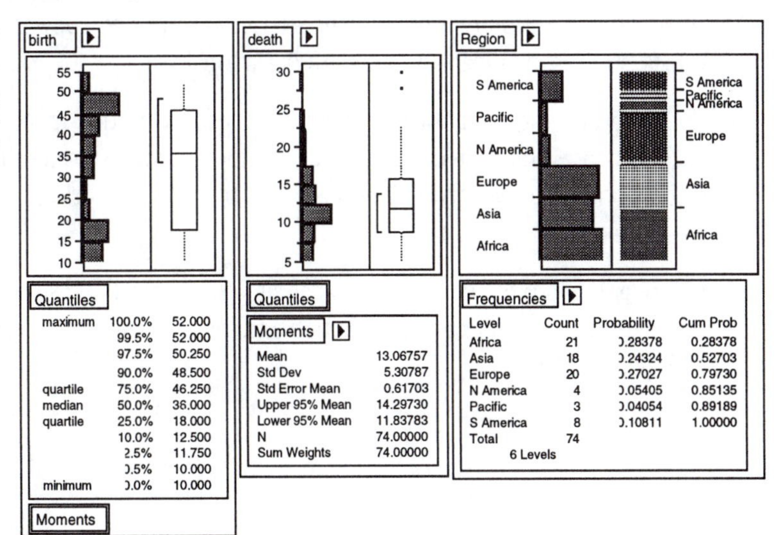

Quantiles (birth)

maximum	100.0%	52.000
	99.5%	52.000
	97.5%	50.250
	90.0%	48.500
quartile	75.0%	46.250
median	50.0%	36.000
quartile	25.0%	18.000
	10.0%	12.500
	2.5%	11.750
	).5%	10.000
minimum	).0%	10.000

Moments (death)

Mean	13.06757
Std Dev	5.30787
Std Error Mean	0.61703
Upper 95% Mean	14.29730
Lower 95% Mean	11.83783
N	74.00000
Sum Weights	74.00000

Frequencies (Region)

Level	Count	Probability	Cum Prob
Africa	21	).28378	0.28378
Asia	18	).24324	0.52703
Europe	20	).27027	0.79730
N America	4	).05405	0.85135
Pacific	3	).04054	0.89189
S America	8	).10811	1.00000
Total	74		
6 Levels			

Figure 5.2 Histograms, Quantiles, Moments, and Frequencies

Note that there are two kinds of analyses:

- The analyses for **birth** and **death** are appropriate for continuous distributions. These are the kind of reports you get when the column in the data table has the continuous modeling type. The **C** in the column heading indicates continuous.
- The analysis for **Region** is appropriate for categorical distributions. These are the kind of graphs and reports you get when the column in the data table has the modeling type of nominal or ordinal, **N** or **O** in the column heading.

You can change the modeling type in the data table to control which kind of report you get.

For the continuous distributions, the graphs give a general idea of the shape of the distribution. The **death** data have a natural center with most values near the center. Distributions with one peak are called unimodal. The **birth** data has two peaks and is therefore bimodal. There are many countries with high birth rates, and many with low birth rates, but few with middle rates.

The text reports for **birth** and **death** show a number of measurements concerning the distributions. There are two broad families of measures. *Quantiles* are the points at which various percentages of the total sample are above or below. *Moments* are measures based on sums of a power of the values, such as the mean and standard deviation.

The report for the categorical distribution focuses on frequency counts. This chapter concentrates on continuous distributions and postpones the discussion of categorical distributions to Chapter 9, "Categorical Distributions."

Before going into the details of the analysis, let's review the distinctions between the properties of a distribution and the estimates that can be obtained. from it.

Review: Probability Distributions

A probability distribution is the mathematical description of how an abstract random process distributes its values. Continuous distributions are described by a density function, such that the probability that a random value falls in any interval is described by the area under the density curve in this interval, as illustrated in **Figure 5.3**. The area under the whole density curve is 1.

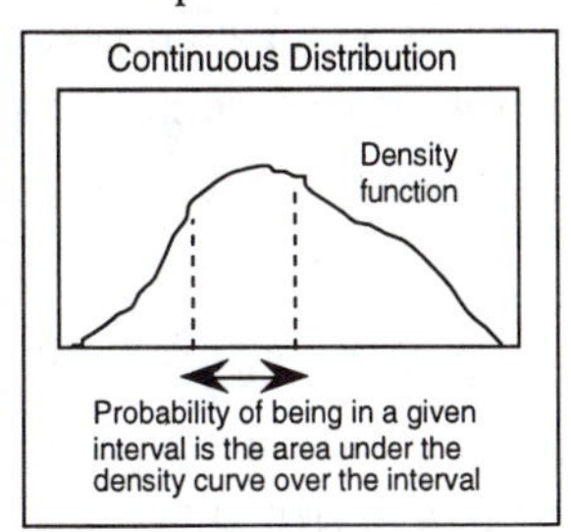

Figure 5.3 Continuous Distribution

These mathematical probability distributions are useful because they can model distributions of values in the real world. This book won't show the formulas for any distributions, but you will need to learn their names and what they are useful for.

True Distribution Function versus Real-World Sample Distribution

Sometimes it is hard to keep straight when you are talking about the real data sample and when you are talking about the abstract mathematical distribution.

Table 5.1 Continuous Distribution

Concept	Abstract mathematical form, probability distribution	Numbers from the real world, data, sample
Average	Expected value or true mean, the point that balances each side of the density	Sample mean, the sum of values divided by the count
Median	Median, the mid-value of the density area, where 50% of the density is on either side	Sample median, the middle value where 50% of the data are on either side
Quantile	The value where some percent of the density is below it	Sample quantile, the value for which some percent of the data are below it
Spread	Variance, the expected squared deviation from the expected value	Sample variance, the sum of squared deviations from the sample mean divided by n–1
General Properties	Any function of the distribution: parameter, property	Any function of the data: estimate, statistic

This distinction of the property from its estimate is crucial in avoiding misunderstanding. Consider the following problem:

When you talk about variability in a sample of values, you can see the variability because you have many different values. But after you compute a mean, you have only one number, so how can you talk about its variance or standard error of the mean itself? There is only one number, not a sample of values to measure the spread of.

To get the idea of variance, you have to separate the abstract quality from the estimate of it. When you do statistics, you are assuming that the response data comes from some process that has a random element to it. Even if you have a single response value, there is variance associated with it, whether you know its value or not. Its the same process whether you have one observation or millions. But you can't get an estimate of the variance when you have only one observation.

Now, if you take a collection of values from realizations of this random process, sum them, and divide by the number of them, then you can calculate the variance associated with this single number. There is a simple algebraic relationship between the variance of the responses and the variance of the sum of the responses divided by n. This is how you can think of the variance of the mean. The square root of the estimated variance of a mean is called the standard error of the mean. Its one number, but you can still estimate its standard deviation.

The Normal Distribution

The most notable continuous probability distribution is the *Normal distribution*, also known as the Gaussian distribution, or the bell curve, like the one shown in **Figure 5.4**. It is an amazing distribution. Its greatest distinction is that it is the most random distribution for a given mean and variance. (It is most random in a very technical sense, having maximum expected unexpectedness or entropy.) Its values are as if they had been realized by adding up billions of little random events.

Much of real world data are normally distributed. The normal distribution is so basic that it is the benchmark used to compare with the shape of

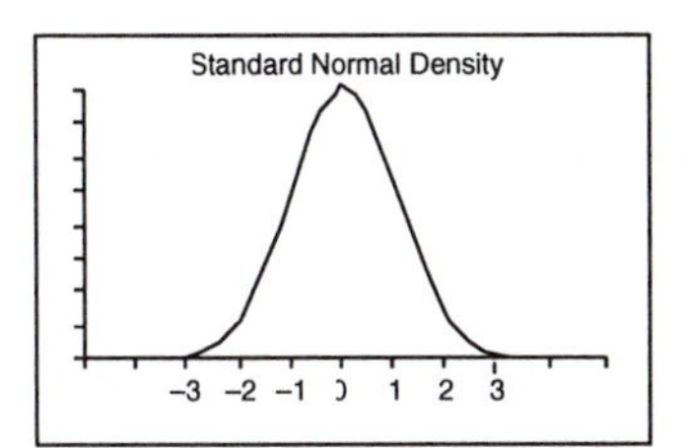

Figure 5.4 Standard Normal Density Curve

distributions of real values. We talk about sample distributions by saying how they differ from the normal. Many of the methods in JMP serve mainly to highlight how a distribution of values differs from a normal distribution. But the usefulness of the Normal doesn't end there. The Normal distribution is also the standard from which to derive the distribution of estimates and test statistics.

The famous *Central Limit Theorem* says that under various fairly general conditions, the sum of a large number of independent and identically distributed random variables is approximately normally distributed. Because most statistics can be written as sums like this, they are normally distributed if you have enough data. Many other useful distributions can be derived as simple functions of normal random distributions.

Later in this chapter, you will meet the distribution of the mean and learn how to test hypotheses about it. The next sections introduce the four most useful distributions of test statistics: the Normal, Student's t, chi-square, and the F distributions.

Describing Distributions of Values

"Honey, I shrunk the data!"

The following sections take you on a tour of the graphs and statistics in the Distribution platform. These statistics try to show the properties of the distribution of a sample, especially these four focus areas:

Location
  is the center and range of the distribution.

Spread
  is how concentrated or spread out the distribution is.

Shape
: refers to symmetry, unimodal (one peak) or not, and especially how it compares to a normal distribution.

Extremes
: are outlying values far away from the rest of the distribution.

Generating Random Data

Before getting into more real data, let's make some random data with known distributions, and then see what an analysis does to it. This is an important exercise because there is no other way to get experience on the distinction between the true distribution of a random process and the distribution of the values you get in a sample.

Remember that in Plato's mode of thinking, the "true" world is some ideal form, and what you are looking at as real data is only a shadow from which you can get hints at what the true world is like. Most of the time, you don't know the true state, so an experience where you do know it will be valuable.

In the following example the true world will be a distribution, and you can use a random number generator in JMP to obtain realizations of the random process to make a sample of values. Then you will see that the sample mean of those values is not exactly the same as the true mean used to generate the numbers. This distinction is fundamental to what statistics is all about.

To create your own random data, do this:

§ Open RANDDIST.JMP. This data table has four columns, but no rows. The columns contain formulas to generate random data having the distributions **Uniform, Normal, Exponential,** and **Dbl Exponential** (double exponential).

Figure 5.5 Partial Listing of the RANDDIST Table

RANDDIST.JMP

4 Cols / 1000 Rows	Uniform	Normal	Exponential	Dbl Expon
1	0.411453	0.264441	1.486959	-1.73878
2	0.943497	-2.00857	0.781836	0.626366
3	0.411179	0.617756	0.57025	-1.76179
4	0.573288	-0.20338	0.23546	-1.48534
5	0.894821	2.239007	0.373961	2.081764
6	0.530533	1.836579	0.018278	0.94255

§ Choose **Rows→Add Rows** and enter 1000 to see a table like the one in **Figure 5.5**. Adding rows generates the random data using column formulas. When you do this example, your random results will be a little different from those shown in **Figure 5.5** because random number generations produce a slightly different set of numbers each time a table is created.

§ To look at the distributions of the columns in the RANDDIST table, choose **Analyze→Distribution of Y**, and select the four columns as Y variables.

The resulting analysis automatically shows a number of graphs and statistical reports. Further graphs and reports can be obtained by using the popup menus located at the bottom of the window (the check-mark icon), and also with icons that show inside the reports. The following sections review the graphs and the text reports available in the Distribution of Y platform.

Histograms

A histogram defines a set of intervals and shows how many values in a sample fall into each interval. It shows the shape of the density of a batch of values.

Try out the following histogram features:

§ Choose Normal Curve in the check-mark menu at the lower left of the window. This command superimposes over each histogram the normal density corresponding to the mean and standard deviation in your sample. **Figure 5.6** shows the histogram portion of what your analysis will have.

Figure 5.6 Histograms of Various Continuous Distributions

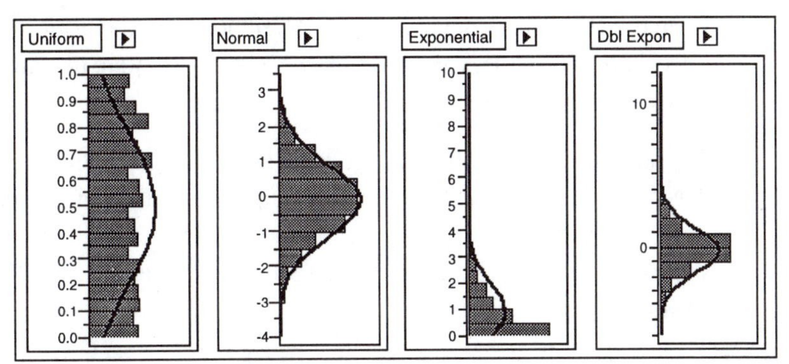

In these histograms, it is clear which one fits the normal curve. The uniform does not have the bell shape. The exponential is skewed to one side. The double exponential has a higher central peak and is somewhat skewed.

§ Click in a histogram bar. As the bar highlights, the corresponding portions of bars in other histograms also highlight, and so do the corresponding data table rows. When you do this with real data, you can see conditional distributions, the distributions in other variables corresponding to a subset of a given variable's distribution.

§ Get the grabber (hand) tool from the **Tools** menu. Now click on the histogram and drag to the right, then back to the left. The histogram bars get narrower and wider as shown in **Figure 5.7**. Make them wider and then drag up and down to change the position of the bars.

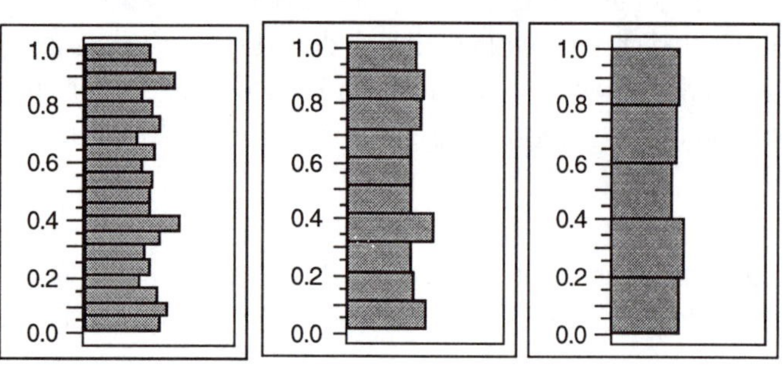

Figure 5.7 Histograms Before and After Grabbing with the Hand Tool to Adjust Bar Widths

Outlier and Quantile Box Plots

Box plots are schematics that help show how your data are distributed. JMP's Distribution platform offers two varieties of box plot that you can turn on or off with options in the check-mark menu.

§ From the check-mark menu at the lower-left border of the platform window, select both **Quantile Box Plot** and **Outlier Box Plot**, if they are not already checked (see **Figure 5.8**).

The box part within each plot surrounds the middle half of the data, from the lower quartile to the median, then to the upper quartile, which defines a distance called the interquartile range. The lines extending from the box show the tails of the distribution, where the rest of the data are. (These lines are sometimes called whiskers.)

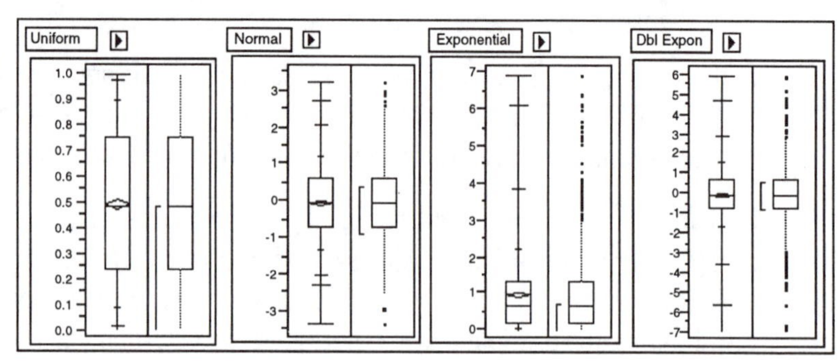

Figure 5.8 Quantile and Outlier Box Plots for Various Continuous Distributions

In the outlier box plots, shown on the right of each panel in **Figure 5.8**, the tail extends to the farthest point that is still within 1.5 interquartile ranges from the quartiles. Points farther away are shown individually as outliers.

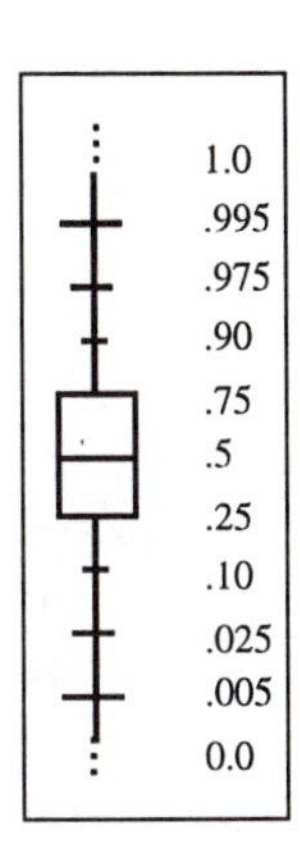

In the quantile box plots (the left-hand plot in each panel) the tails are marked at certain quantiles. The quantiles are chosen so that if the distribution is normal, the marks will be about equidistant with the same spacing as the quartiles from the median. So you look at the spacing of the marks to see how it differs from a normal distribution.

Looking at the boxes in the four distributions in **Figure 5.8**, the middle half of the data are wide in the uniform, thin in the double exponential, and very one-sided in the exponential distribution.

In the outlier box plot, the shortest half (the shortest interval containing half the data) is shown by a red bracket on the side of the box plot. The shortest half is at the center for the symmetric distributions, but at the short tail of the exponential.

In the quantile boxplot, the mean and its 95% confidence interval are shown by a diamond. With a thousand observations, the mean is estimated with great precision giving a very short confidence interval. Confidence intervals are talked about in the following sections.

Normal Quantile Plots

Normal quantile plots show all the values of the data as points in a plot, such that if the data are normal, the points tend to follow a straight line.

The Y vertical coordinate is the actual value. The X horizontal coordinate is the normal quantile associated with the rank of the value after sorting the data.

If you are interested in the details, the precise formula used for the normal quantile values is

$$\Phi^{-1}\left(\frac{r_i}{N+1}\right)$$

where r_i is the rank of the observation being scored, N is the number of observations, and Φ^{-1} is the function that returns the normal quantile associated with a probability argument ($r_i/(N+1)$). The normal quantile is the value on the x-axis of the normal density that has the portion p of the area below it, where p is the probability argument. For example, the quantile for .5 (the probability of being less than the median) is zero, because half (.5 or 50%) of the density of the standard normal is below zero. The technical name for the quantiles we use is the van der Waerden normal scores, and they are cheap but good approximations to the very expensive exact expected normal order statistics.

§ From the check-mark popup menu, select **Normal Quantile Plot**. The histograms and normal quantile plots for the four simulated distributions are shown in **Figures 5.9** to **5.12**.

- A red straight line with confidence limits shows where the points would tend to lie if the data were normal. This line is purely a function of the mean and standard deviation of the sample. The line crosses the mean of the data at the normal quantile of 0. The slope of the line is the standard deviation of the data.
- Dashed lines surrounding the straight line form a confidence interval for the normal distribution. When the points fall outside these dashed lines, you are seeing a significant departure from normality. For more details see the section **Special Topic: Testing Normality** later in this chapter.
- Where the slope across the points is small (relative to the normal) then you are crossing a lot of ranked data with very little variation in the real values, and therefore you are encountering a dense cluster. When the slope across the points is large, then you are crossing a lot of space that has only a few ranked points. Dense clusters make flat sections, and thinly populated regions make steep sections. The overall slope is the standard deviation.

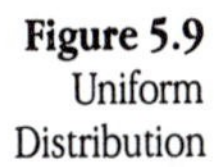

Figure 5.9
Uniform
Distribution

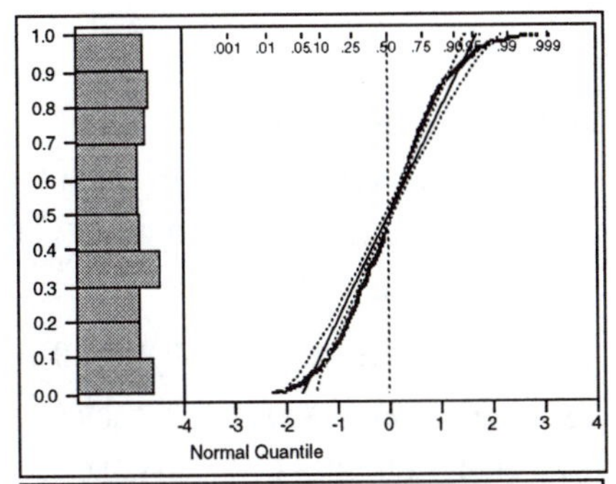

In the middle, the uniform distribution is steeper (less dense) than the normal. In the tails, the uniform is flatter (more dense) than the normal. In fact the tails are truncated at the end of the range, where the normal tails extend (infinitely).

Figure 5.10
Normal
Distribution

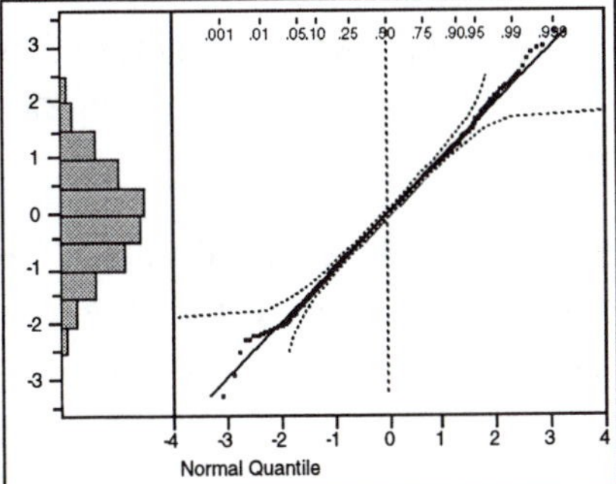

The normal distribution has a normal quantile plot that tends to follow a straight line. The points at the end have the highest variance and are most likely not to fall near the line. This is reflected by the flair in the confidence limits near the ends.

Figure 5.11
Exponential
Distribution

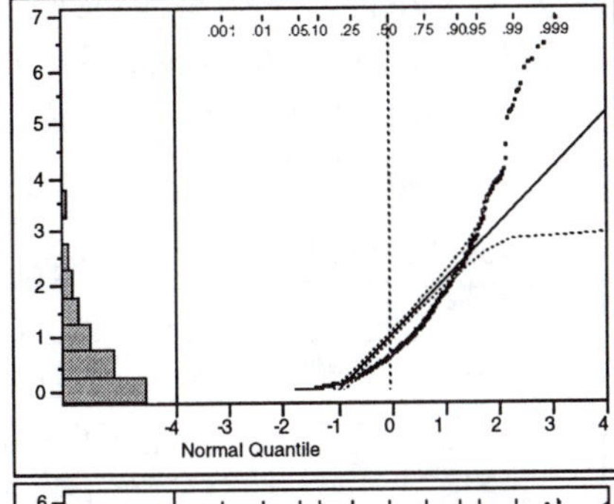

The exponential distribution is skewed, that is, one-sided. The top tail runs steeply past the normal line; it is spread out more than the normal. The bottom tail is shallow, and much denser than the normal.

Figure 5.12
Double
Exponential
Distribution

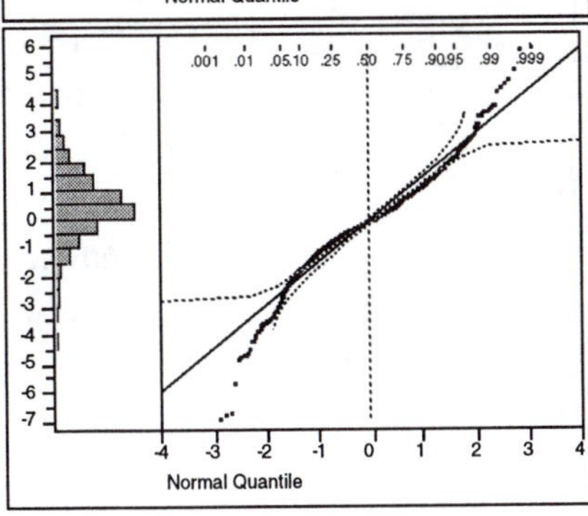

The middle portion of the double exponential is denser (more shallow) than the normal. In the tails, it spreads out more (is steeper) than the normal.

Stem-and-Leaf Plots

A stem-and-leaf plot is a variation on the histogram. It was developed for tallying data in the days when computers were rare. Each line of the plot has a **Stem** value that is the leading digits of a range of column values. The **Leaf** values are made from the next-in-line digits of the values.

§ To see two examples, open the BIGCLASS.JMP and the AUTOMESS.JMP tables.

§ For each table choose **Analyze→Distribution of Y**. The **Y** variables are **Weight** from the BIGCLASS table and **Auto theft** from the AUTOMESS table.

§ When the histograms appear, select **Stem and Leaf** from the options popup menu next to the histogram name. This option appends stem-and-leaf plots to the end of the text reports.

Figure 5.13 shows the plot for **Weight** on the left and the plot for **Auto theft** on the right. The values in the stem column of the plot are chosen as a function of the range of values to be plotted, with the scale factor, or multiplier, at the bottom of the plot.

You can reconstruct the data values by joining the stem and leaf and checking the scale factor. For example, on the first line of the **Weight** plot, you can read the values 64 and 67 (6 from the stem, 4 and 7 from the leaf). At the bottom, the weight is 172 (17 from the stem, 2 from the leaf).

Figure 5.13 Examples of Stem-and-Leaf Plots

Plot of **Weight** variable

Stem and Leaf

Stem	Leaf	Count
6	47	2
7	499	3
8	1445	4
9	122355899	9
10	45567	5
11	122223569	9
12	3888	4
13	4	1
14	25	2
15		
16		
17	2	1

Multiply Stem.Leaf by 10

Plot of **Auto theft** variable

Stem and Leaf

Stem	Leaf	Count
0		
1	13557788	8
2	01144489	8
3	3445789	7
4	2344569	7
5	14779	5
6	0047	4
7	113	3
8	36	2
9	1245	4
10	24	2
11		
12		
13	4	1

Multiply Stem.Leaf by 10**2

The leaves respond to mouse clicks. Click on the two 5s on the first stem of the **Auto theft** plot. This highlights the corresponding rows in the data table, which are California with the value 154 and the District of Columbia with 149.

Mean and Standard Deviation

The *mean* of a collection of values is the average value, which is the sum of the values divided by the number of values in the sum.

The sample *variance* is the average squared deviation from the sample mean, which is shown as the expression:

$$s^2 = \sum (x_i - \bar{x})^2/(n-1).$$

The sample *standard deviation* is the square root of the sample variance. The standard deviation is preferred in reports because it is in the same units as the original data (rather than squares of units).

If you assume a distribution is normal, the only things you need to know to completely characterize the distribution are the mean and standard deviation.

When you say "mean" and "standard deviation," you are allowed to be ambiguous as to whether you are referring to the true (and usually unknown) parameters of the distribution, or the sample statistics you use to estimate the true values.

The sample mean has these properties:

- It is the balance point. The sum of deviations of each sample value from the sample mean is zero.
- It is the least squares estimate; that is, the sum of squared deviations of the values from the estimate is minimized. That sum is less than would be computed from any estimate other than the sample mean.
- It is also the maximum likelihood estimator of the true mean when the distribution is normal. It is the estimate that makes the data you collected more likely than any other estimate of the true mean would.

Median and Other Quantiles

The sample median is a value where half the data are above and half below. It estimates the 50% quantile of the distribution. A sample quantile can be defined for any probability between 0 and 1; the 100% quantile is the maximum value, 100% of the data values are at or below it. The 75% quantile is the upper quartile, the value for which 75% of the data values are at or below it.

There is an interesting indeterminacy about how to report the median and other quantiles. If you have an even number of observations there may be an interval where half the data are above, half below. There are about a dozen different ways for reporting medians in the statistical literature, many of which are only different if you have tied points on either of both sides of the middle. You can take one side, the other, the midpoint, or a weighted average of the middle values, with a number of weighting options. For example, if the sample values are {1,2,3,4,4,5,5,5,7,8}, the median can be defined anywhere between 4 and 5, including one side or the other, or half way, or two-thirds of the way into the interval.

Another interesting property of the median is that it is the least-absolute-values estimator. That is, it is the number that minimizes the sum of the absolute differences between itself and each value in the sample. Least-absolute-values estimators are also called L1 estimators, or MAD estimators (minimum absolute deviation).

Mean versus Median

If the distribution is symmetric, the mean and median are estimates of both the expected value of the underlying distribution and its 50% quantile. If the distribution is normal, the mean is a better estimate than the median, by a ratio of 2 to 3.1416 (pi) in terms of variance. In other words, the mean has only 63% of the variance of the median.

If the data are contaminated by an outlier, the median will be little affected, but the mean could be greatly influenced, especially if the outlier is far out. The median is said to be outlier-resistant, or robust.

Suppose that you have a skewed distribution, like income, which has lots of extreme points on the high end, but is limited to zero on the low end. If you want to know how much income a typical person made, then the median makes more sense to report than the mean. But if you want to track per-capita income as an aggregating measure, then the mean income might be better to report.

Higher Moments: Skewness and Kurtosis

Moment statistics are those that are formed from sums of powers of the values. The first four moments are defined as follows:

- The first moment is the mean, which is calculated from a sum of values to the power 1. The mean measures the center of the distribution.
- The second moment is the variance (and standard deviation), which are calculated from sums of the values to the power 2. They measure the spread of the distribution.
- The third moment is skewness, which is calculated from sums of the values to the power 3. Skewness measures the asymmetry of the distribution.
- The fourth moment is kurtosis, which is calculated from sums of the values to the power 4. Kurtosis measures the relative shape of the middle and tails of the distribution.

One use of skewness and kurtosis is to help determine if a distribution is normal and, if not, what the distribution might be. A problem with the higher order moments is that the statistics have higher variance and are more sensitive to outliers.

§ To get the skewness and kurtosis, use the Distribution of Y platform and select them from the **More Moments** popup menu beside the Moments report.

Extremes, Tail Detail

The extremes (the minimum and maximum) are the 0% and 100% quantiles.

At first glance the most interesting aspect of a distribution appears to be where its center lies. But statisticians often look first at the outlying points, because they can carry useful information. That's where the unusual values are, the possible contaminants, the rogues, the potential discoveries.

In the normal distribution (with infinite tails) the extremes tend to reach farther as you collect more data. However, this is not necessarily the case with other distributions. For data that is uniformly distributed across an interval, the extremes tend to change less and less as you collect more data, and the extremes are the most informative statistics on the distribution.

Statistical Inference on the Mean

The previous sections talked about descriptive graphs and statistics. This section moves on to the real business of statistics, inference. We want to form confidence intervals for the mean and test hypotheses about the mean.

Standard Error of the Mean

Suppose there exists some true but unknown population mean and you estimate that true mean with the sample mean. The sample mean comes from realizations of a random process, so there is variability associated with it.

The mean is the arithmetic average–the sum of n values divided by n. The variance of the sample mean is 1/n of the variance of the original data. Taking the square root of this, the standard deviation of the sample mean is 1/sqrt(n) of the standard deviation of the original data.

Substituting in the estimate of the standard deviation of the data, we now define the *standard error of lhe mean*, which estimates thc standard deviation of the sample mean. It is the standard deviation of the data divided by the square root of n. The mean and its standard error are the key quantities involved in statistical inference concerning the mean.

std error mean = (std dev estimate of data)/ (square root of (n))

Confidence Intervals for the Mean

The sample mean is sometimes called a point estimate, because it's only one number. You know that the true mean is not this point. It would be more useful to have an interval between two numbers that you are pretty sure contains the true mean, say 95% sure. This interval is called a 95% *confidence interval* for the true mean.

To construct this interval, first make some assumptions. Assume that the data are normal and that its true standard deviation is the sample standard deviation. (The second assumption will be revised later.) Then, the exact distribution of the mean estimate is known, except for its location (because you don't know the true mean).

If you knew the true mean and had to forecast a sample mean, then you could construct an interval around the true mean that would contain the sample mean with probability .95. To do this, you first obtain the quantiles of the standard normal distribution that have 5% of the density in the tails. These quantiles are –1.96 and +1.96. Then you scale this interval between them by the standard deviation and add in the true mean.

But instead of a forecast, you already have a realization of the sample mean; instead of an interval for a realization of the sample mean, you need an interval to capture the true mean. If the sample mean is 95% likely to be within this distance of the true mean, then the true mean is 95% likely to be within this distance of the sample mean. So the interval is centered at the sample mean. The formula for the approximate 95% confidence interval is

95% C.I. for mean = (sample mean) $\pm$ 1.96 • (standard error of the mean).

Figure 5.14 illustrates the construction of confidence intervals.

Figure 5.14 Illustration of Confidence Intervals

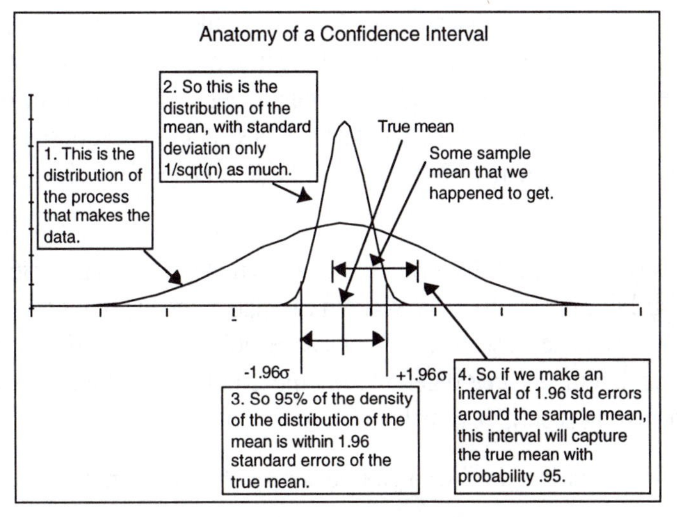

This is not exactly the confidence interval that JMP calculates. Instead of using the normal quantile of 1.96, it uses a Student's t quantile, discussed later, which takes care of the extra uncertainty that results from estimating the standard error of the mean. So the formula for the confidence interval is

($1-\alpha$) C.I. for mean = (sample mean) $\pm$ $t_{1-\alpha}$ • (standard error of the mean).

The alpha (α) in the formula is the probability that the interval does not capture the true mean. That probability is .05 for a 95% interval. The confidence interval is reported on the Distribution of Y platform in the Moments report as the Upper 95% Mean and Lower 95% Mean. It is represented in the quantile box plot by the ends of a diamond (see **Figure 5.15**).

Figure 5.15 Moments Report and Quantile Box Plot on the Distribution of Y Platform

Moments	
Mean	23.49878
Std Dev	0.01961
Std Error Mean	0.00877
Upper 95% Mean	23.52313
Lower 95% Mean	23.47443
N	5.00000
Sum Weights	5.00000

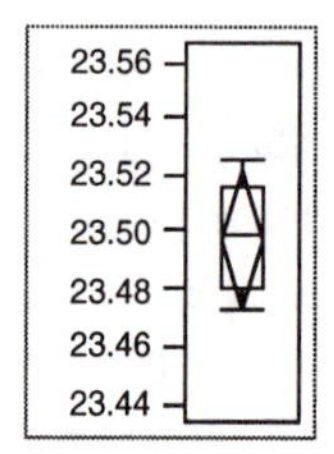

Testing Hypotheses: Terminology

Suppose that you want to test whether the mean of a collection of sample values is significantly different from some hypothesized value. The strategy is to calculate a statistic, and note that if the true mean was that hypothesized value, it would be an extremely rare event to get so large a statistic value. You would rather believe the hypothesis to be false than to believe such a rare coincidence happened. This is a probabilistic version of *proof-by-contradiction*.

The way that you see that event to be rare is that it is past a point in the tail of a probability distribution that you can calculate if the hypothesis were true. Often, researchers use .05 as a significance indicator, which means that you believe that the mean is different from the hypothesized value if the chance of being wrong is only 5% (one in twenty).

Statisticians have a precise and formal terminology for hypothesis testing:

- The possibility of the true mean being the hypothesized value is called the *null hypothesis*, which you want to reject (refute). The alternative hypothesis is that the mean is different from the hypothesized value, which can be phrased as greater than, less than, or unequal. The latter is called a two-sided alternative.
- The situation where you reject the null hypothesis when it happens to be true is called a *Type I (one) error.* This declare that some difference is

nonzero when it is really zero. Not detecting a difference when there is a difference is called a *Type II (two) error*.

- The probability of getting a Type I error in a test is called the *alpha-level* of the test. This is the probability that you are wrong if you say that there is a difference. The *beta-level* or *power* of the test is the probability of being right when you say that there is a difference. One minus this beta-level probability is the probability of a Type II error.
- Statistics and tests are constructed so that the power is maximized subject to the alpha-level being maintained.

In the past, people obtained critical values for alpha-levels and ended with an accept/reject decision based on whether the statistic was bigger or smaller than the critical value.

But what happens with computers now is you don't specify the alpha-level, but rather you end up with it, and in that context, it is called a p value. The definition of a p value can be said in many ways:

The *p value* or *significance level* is the alpha level at which the statistic would be significant. It is how unlikely getting so big a statistic would be if the true mean were the hypothesized value. It is the probability of being wrong if you rejected the null hypothesis. It is the probability of a Type 1 error. It is the area in the tail of the distribution of the test statistic under the null hypothesis.

The p value is the number you want to be very small, certainly below .05, so that you can say that the mean is significantly different from the hypothesized value. The p values in JMP are labeled according the test statistic's distribution. The label "Prob > | t | " is read as the "probability of getting an even greater absolute t statistic, given that the null hypothesis is true."

The Normal Z Test for the Mean

If the original response data are normally distributed, then when many samples are drawn, the means of the samples are normally distributed. Even if the original response data are not normally distributed, the sample mean still has an approximate normal distribution if the sample size is large enough. So the normal distribution provides a reference to use to compare the sample mean to an hypothesized value.

The Standard Normal distribution has a mean of zero and a standard deviation of 1. If the hypothesis were true, the test statistic you construct should have this standard distribution. You can center any variable to mean zero by subtracting the mean (at least the hypothesized mean). You can standardize any variable to have standard deviation 1 by dividing by the true standard deviation, assuming for now that you know what it is. Tests using the normal distribution constructed like this (hypothesized mean and known standard deviation) are called *z tests*. The formula for a z statistic is

$$\text{z test} = \frac{(\text{ the estimate} - \text{hypothesized value})}{\text{its standard deviation}}$$

You want to find out how unusual your computed z value is from the point of view of believing the hypothesis. If the value is too improbable then you doubt the null hypothesis.

To get a significance probability, you take the computed z value and find the probability of getting an even greater absolute value. This involves finding the area in the tails of the normal distribution that are greater than absolute z and less than negative absolute z. **Figure 5.16** illustrates a two-tailed z test.

Figure 5.16 Illustration of Two-Tailed z Test

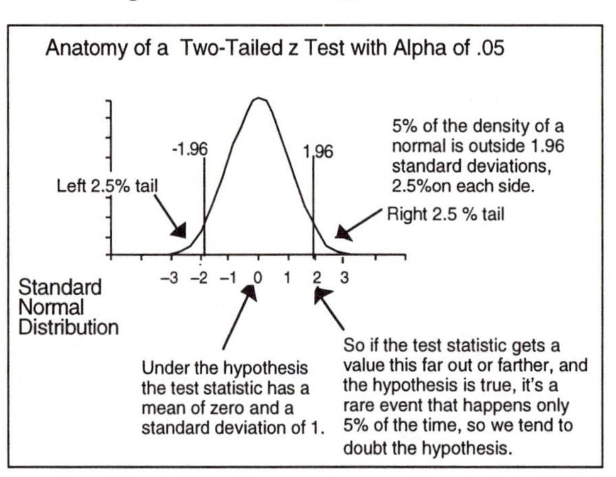

For example, in 1738 the Paris observatory determined with high accuracy that the angle of the earth's spin was 23.472. But someone suggested that the angle changes over time. Historical accounts were examined and 5 measurements were found dating from 1460 to 1570. These measurements were somewhat different than the Paris measurement, but they were done using much less precise methods. The question is whether the differences in the measurements can be attributed to the errors in measurement of the earlier observations or whether the angle of the earth's rotation actually changed. How do you test the hypothesis that the angle changed, that the difference was real?

§ Open CASSUB.JMP (Stigler, 1986).

§ Choose **Analyze→Distribution of Y** and select Obliquity as the Y variable. When you click **OK**, the Distribution platform shows a histogram of the 5 values.

§ To see the histogram in **Figure 5.17**, get the hand tool from the **Tools** menu and stretch the bars to the right.

§ Select **Test Mean=value** from the popup menu at the top of the histogram. In the dialog that appears, enter the hypothesized value of 23.47222 (the value measured by the Paris observatory), and enter the standard deviation of .0196 found in the Moments table. The resulting z test report is shown to the right in **Figure 5.17**.

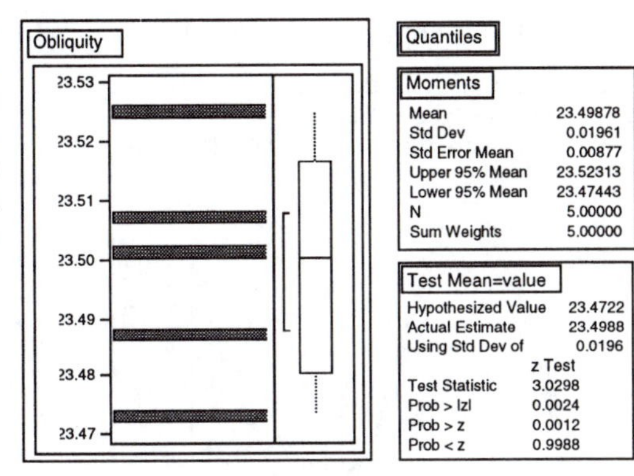

Moments

Mean	23.49878
Std Dev	0.01961
Std Error Mean	0.00877
Upper 95% Mean	23.52313
Lower 95% Mean	23.47443
N	5.00000
Sum Weights	5.00000

Test Mean=value

Hypothesized Value	23.4722
Actual Estimate	23.4988
Using Std Dev of	0.0196
	z Test
Test Statistic	3.0298
Prob > lzl	0.0024
Prob > z	0.0012
Prob < z	0.9988

Figure 5.17 Histogram and Normal Quantile Plot for Double Exponential Distribution

The test statistic z has the value 3.0298. The area under the normal curve to the right of this value is reported as Prob >z, which is the probability (p value) of getting an even greater z value if there was no difference. In this case, the p value is .001. This is the right significance value for testing that the mean is the value with the alternative that it is greater than the hypothesized value.

To test that the mean is the value with a two-sided alternative, that it is different in either direction, you need the area in both tails, listed as Prob >lzl, which is .002. The other one-sided test has a p value of .999, indicating that you are not going to prove that the mean is less than the hypothesized value. The two-sided p value is always twice the smaller of the one-sided p values.

Student's t Test

The z test requires a major assumption. It depends on knowing the standard deviation of the response, and thus the standard deviation of the mean estimate. Usually this true standard deviation value is unknown and you have to use an estimate of the standard deviation. Using this estimate in the denominator of the statistical test computation is the right thing to do, but you have to adjust the distribution that you use for the test.

Now instead of having a normal distribution, you have a *Student's t* distribution. The statistic is called the Student's t statistic and is computed

$$\text{t statistic} = \frac{\text{(sample mean)} - \text{(hypothesized value)}}{\text{(standard error of mean, s/sqrt(n))}}.$$

A large sample estimates the standard deviation very well, and the Student's t distribution is not much different than the normal distribution, as illustrated in **Figure 5.18**. However, in this example there were only 5 observations.

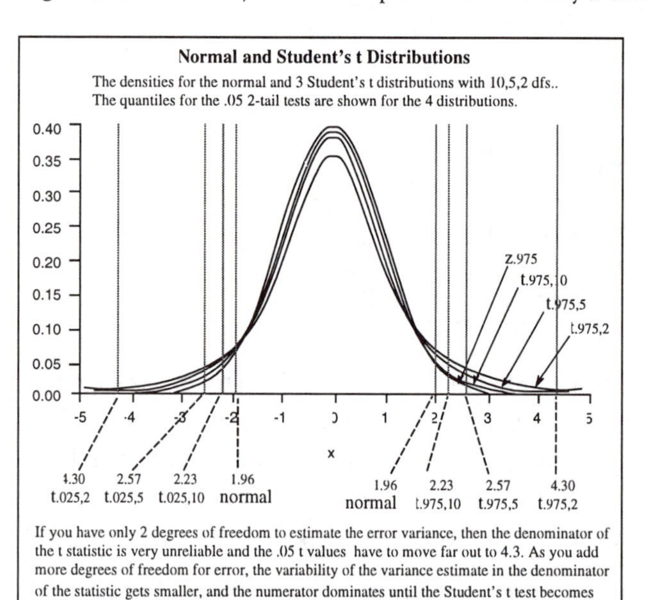

Figure 5.18 Comparison of Normal And Student's t Distributions

There is a different t distribution for each number of observations, indexed by a value called *degrees of freedom*, which is the number of observations minus the number of parameters estimated in fitting the model. In this case, five observations minus one parameter (the mean) yields 5–1=4 degrees of freedom. As you can see in **Figure 5.18**, the quantiles for the t distribution spread out farther than the normal when there are few degrees of freedom. It turns out that for a 5% two-tailed test, which uses a p value of a .975, the t-quantile for 4 degrees of freedom is 2.7764, which is far greater than the corresponding z quantile of 1.96.

So, let's do this test again:

§ Select **Test Mean=value** and again enter 23.47222 for the hypothesized mean value. But this time do not fill in the standard deviation. Click **OK.** The Test Mean=value table (**Figure 5.19**) now shows a t test instead of a z test.

Figure 5.19 Test Mean When the Standard Deviation is Unknown

Test Mean=value	
Hypothesized Value	23.4722
Actual Estimate	23.4988
t Test	
Test Statistic	3.028
Prob > |t|	0.039
Prob > t	0.019
Prob < t	0.981

When you don't specify a standard deviation, JMP uses the sample estimate of the standard deviation, but this is only an estimate because you didn't enter it yourself. The significance looks marginal, but the p value of .039 still looks somewhat convincing so you can conclude that the angle has changed.

When you have a significant result, the idea is that under the null hypothesis, the expected value of the t statistic is zero. It is highly unlikely (probability less than alpha) for the t statistic to be so far out in the tails. Therefore, you don't put much belief in the null hypothesis.

Note: You may have noticed that the test dialog offers the options of a Wilcoxon signed-rank nonparametric test. Some statisticians favor nonparametric tests because the results don't depend on the response having a normal distribution. Nonparametric tests are covered in more detail in Chapter 7, "Comparing Many Means: One-Way Analysis of Variance."

Curiosity: A Significant Difference?

Here are some data that demonstrate a point, that a statistically significant difference is quite different than a practically significant difference.

Dr. Quick and Dr. Quack are both in the business of selling diets, and they have claims that appear contradictory:

Dr. Quack studied 500 dieters and claims

> "A statistical analysis of my dieters shows a statistically significant weight loss for my Quack diet. The Quick diet, by contrast, shows no significant weight loss by its dieters."

Dr. Quick followed the progress of 20 dieters and claims

> "A statistical study shows that on average my dieters lose over three times as much weight on the Quick diet as on the Quack diet."

Figure 5.20
Diet Data

DIETS.JMP

2 Cols / 500 Rows	C Y Quack's Weight Change	C Y Quick's Weight Change
18	4.3	-7.9
19	15.7	-14.9
20	11.8	-2.2
21	12.4	•
22	-1.4	•

So which diet is better, statistically?

§ To compare the Quick and Quack diets, open the DIETS.JMP sample data table and choose **Analyze→Distribution of Y**. **Figure 5.20** shows a partial listing of the DIETS data table.

§ Select **Test Mean=value** from the popup menu at the top of each plot to compare the mean weight loss to zero. You should use the one-sided t test because you are only interested in significant weight loss (not gain).

If you look closely at the t test results in **Figure 5.21**, you can verify both claims! Quick's average weight loss of 2.73 is over three times the .91 weight loss reported by Quack. But Quick's larger mean weight loss was not significantly different from zero, and Quack's small weight loss was. Quack may not have a better diet, but he has a more evidence—500 cases compared with 20 cases. So even though the diet produced a weight loss of less than a pound, it is statistically significant. Significance is about evidence, and having a large sample size can make up for having a small effect.

Dr. Quick needs to collect more cases, and then he can easily dominate the Quack diet, though it seems like even a 2.7 pound loss may not be enough of a practical difference to a customer.

If you have a large enough sample size, even a very small difference can be significant. If your sample size is small, even a large difference may not be significant.

Looking closer at the claims, note that Quick reports on the estimated difference between the two diets, where as Quack reports on the significance probabilities. Both are somewhat empty statements. It is not satisfying to report an estimate without a measure of variability. It is not satisfying to report a significance without an estimate of the difference.

The best report is a confidence interval for the estimate, which shows both the statistical and practical significance. The next chapter presents the tools to do a more complete analysis on data like the Quick and Quack diet data.

Figure 5.21 Histograms and Reports of the Quick and Quack Example

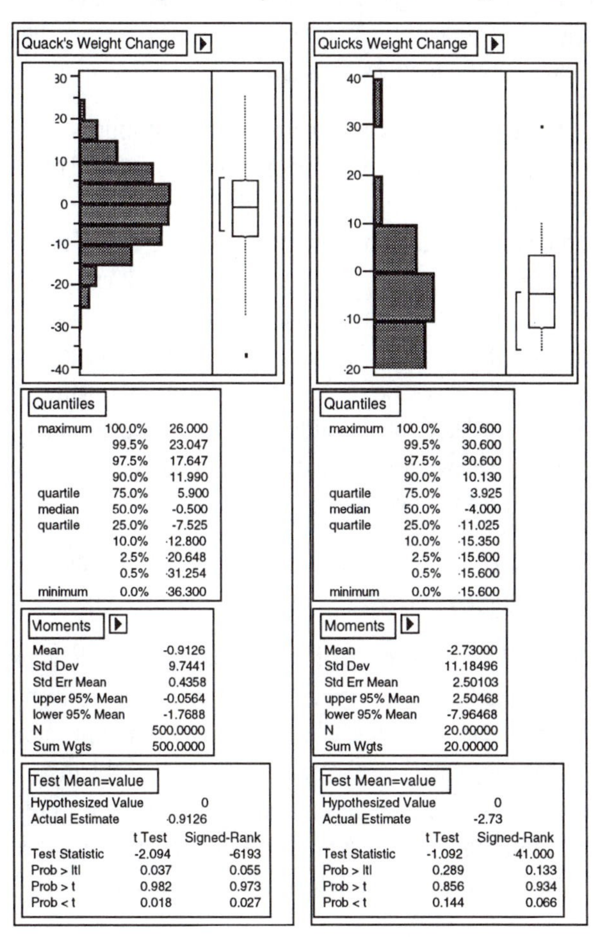

Special Topic: Testing for Normality

Sometimes you may want to test whether a set of values has a particular distribution. Perhaps you are verifying assumptions and want to test that the values have a normal distribution. Because you want the data to be normal, this is an example of an unusual situation where you hope the test fails to be significant. If you have a large number of observations, you may want to reconsider this tactic because the normality tests will be sensitive to small departures from normality, and such small departures would not jeopardize the other analyses you make (the central limit theorem), especially because they will also probably be highly significant.

A widely used test that the data are from a specific distribution is the Kolmogorov test (also called the Kolmogorov-Smirnov test). The test statistic is the greatest absolute difference between the hypothesized distribution function and the empirical distribution function of the data. The empirical distribution function goes from 0 to 1 in steps of 1/n as it crosses data values. When the Kolmogorov test is applied to the normal distribution and adapted to use estimates for the mean and standard deviation, it is called the Lilliefor's test or KSL test. In JMP, the Lilliefor's quantiles on the cumulative distribution function (cdf) are translated into confidence limits in the normal quantile plot, so that you can see where the distribution departs from normality by where it crosses the confidence curves.

Another test of normality produced by JMP is the Shapiro-Wilk test (or the W statistic), which is implemented for samples as large as 2000.

All the distributional tests assume that the data are independent and identically distributed. The most frequent use of normality tests, however, is for residuals from linear model fits, which both have different variances and are correlated. In reasonably large samples, these two problems are minimal. However, you are bound to learn more from diagnostic plots of the residuals, in which the distribution of the residuals and their relation to the factors will be more useful information than a significance p value for a normality test.

§ Look again at the BRTHDTH.JMP data table. Select **Normal Curve** and **Normal Quantile plot** from the check-mark popup menu.

§ Select **Test for Normality** from the popup menu next to the variable name at the top to get the Shapiro-Wilk test. Its results are shown in **Figure 5.22.**

Figure 5.22 Test Distributions for Normality

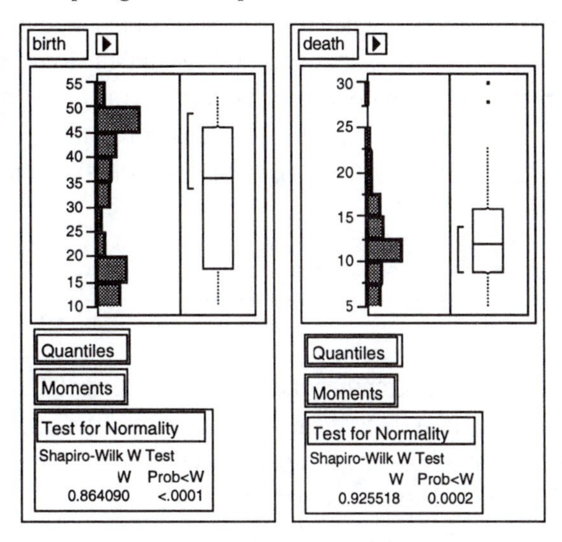

The conclusion is that neither distribution is normal, though the second is much closer than the first.

Special Topic: Simulating the Central Limit Theorem

The Central Limit Theorem says that for a very large sample size, the sample mean will be very close to normally distributed, almost regardless of the shape of the underlying distribution. That is, if you compute means from many samples of a given size, the distribution of those means approaches normality, even if the underlying population from which the samples were drawn is not.

You can see the Central Limit Theorem in action using the template called CNTRLMT.JMP.

§ Open CENTLIMT.JMP, highlight the first column and choose **Cols→Column Info**. Click on the formula in the Column Info dialog to see its calculator and formula. Do this same thing for the second column, called N=5, and display its calculator.

Looking at the formulas should give you an idea of what's going on. The uniform random number function creates a highly skewed distribution. For each row, the first column, N–1, generates a single uniform random number to the fourth power. For each row in the second column, the formula generates a sample of 5 uniform numbers, takes each to the fourth power and computes the mean. The next column does the same for a sample size of 10, and the following columns generate means for sample sizes of 50 and 100.

$$\frac{\sum_{j=1}^{1} ?\text{uniform}^4}{1} \quad \frac{\sum_{j=1}^{5} ?\text{uniform}^4}{5} \quad \frac{\sum_{j=1}^{10} ?\text{uniform}^4}{10} \quad \frac{\sum_{j=1}^{50} ?\text{uniform}^4}{50}$$

If you have a reasonably fast computer, add 500 or 1000 rows to the data table. You can add as many rows to the table as you want. (But be forewarned that if you have a slow machine, the computations might take some time—each cell in the last column requires 100 random numbers, taken to the fourth power, added together, and divided by 100.)

When the computations are complete:

§ Choose **Analyze→Distribution of Y**. Select all the variables, assign them as Y variables and click **Done**.

You should see the results in **Figure 5.23**. When the sample size is only 1, the skewed distribution s is apparent. As the sample size increases you can clearly see the distributions becoming more and more normal.

Figure 5.23 Example of the Central Limit Theorem in Action

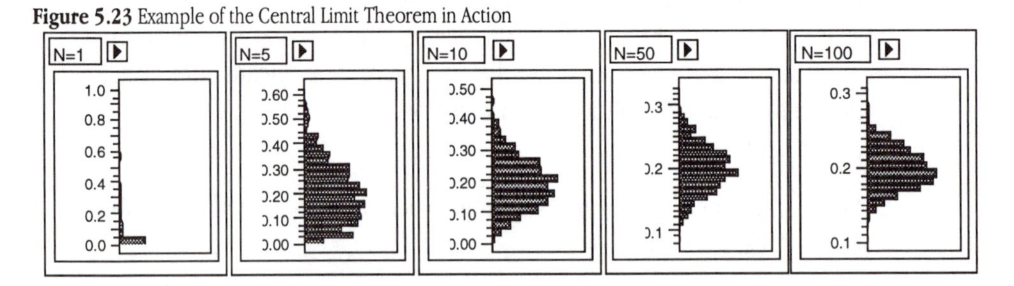

Chapter 6
The Difference Between Two Means

Overview

Are the mean responses from two groups different? What evidence would it take to convince you? This question opens the door to many of the issues that pervade statistical inference, and thus this chapter will be a long one, exploring those issues. Comparing group means also introduces an important statistical distinction regarding how the measurement or sampling process affects the way the data are analyzed. This chapter also talks about validating statistical assumptions.

There are two very different situations that lead to two different analyses:

Independent Groups—the responses from the two groups are unrelated and statistically independent. For example, the two groups might be two classrooms with two sets of students in them. The responses come from different experimental units or subjects. The responses are uncorrelated and the means from the two groups are uncorrelated.

Matched Pairs—the two responses form a pair of measurements coming from the same experimental unit or subject. For example, a matched pair might be a before and after blood pressure measurement from the same subject. The responses are correlated, and the statistical method must take that into account.

Chapter 6 Contents

Two Independent Groups

For two different groups, the goal might be to estimate the group means, and determine if they are significantly different. Along the way, you might also want to notice anything else of interest about the data.

When the Difference Isn't Significant...

A study compiled height measurements from 63 children, all age 12. It's safe to say that as they get older, the mean height for males is greater than for females, but is this the case at age 12? Let's find out:

§ Open HTWT12.JMP to see the data shown in **Figure 6.1**.

Figure 6.1 Partial Listing of the HTWT12 Data

HTWT12.JMP

63 Rows / 3 Cols	Gender (N)	Height (C)	Weight (C)
27	f	54.5	74
28	f	66	144.5
29	f	51.5	64
30	f	61	92
31	m	57.3	76.5
32	m	59.5	84
33	m	60.8	128
34	m	50.5	79

There are 63 rows and three columns. This example uses **Gender** and **Height**. **Gender** has the Nominal modeling type, with codes for the two categories, "f" and "m". **Gender** is the **X** variable for the analysis. **Height** contains the response of interest, the **Y** values.

Check the Data

Every pilot walks around the plane looking for damage or other problems before starting up. You wouldn't want to submit your analysis to the FDA without making sure that the data was not confused with your income tax, or your grocery list, or your homeowners' association phone list. Do your kids use the same computer that you do? Does your data have so many decimals of precision that it looks like it came from a random number generator? Great detectives let no clue go unnoticed. Great data analysts check their data carefully.

To check the data, first look at the distributions of both variables graphically with histograms and boxplots:

§ Choose **Analyze→Distribution of Y** from the menu bar.

§ Select **Gender** and **Height** as **Y** variables.

§ Click **OK** to see an analysis window like the one shown in **Figure 6.2**.

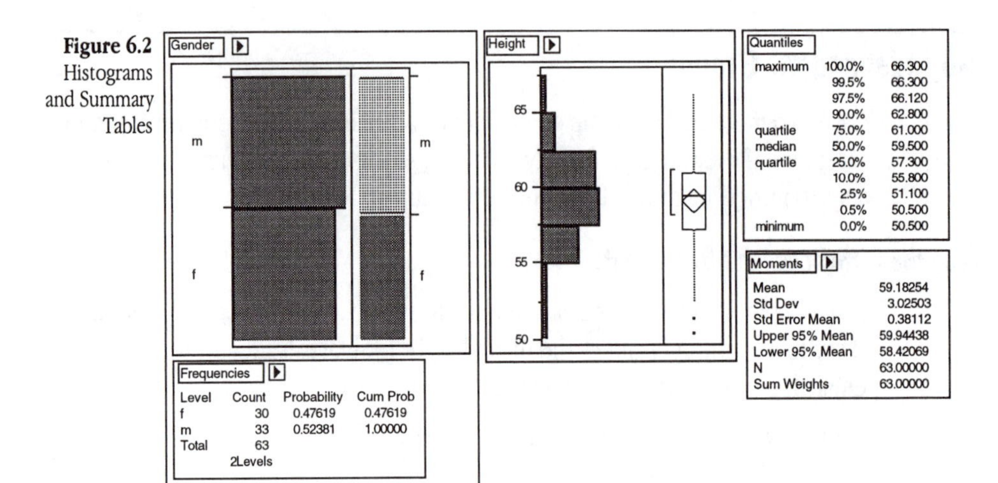

Figure 6.2 Histograms and Summary Tables

A look at the histograms for **Gender** and **Height** reveals that there are a few more males than females. The overall mean height is about 59, and there are no missing values (N is 63). The box plot indicates that two of the children seem unusually short compared to the rest of the data.

§ Move the cursor to the **Gender** histogram, and click on the bar for "f".

Clicking the bar highlights the females in the data table and also highlights that part of the distribution in the **Height** histogram. Now click on the "m" bar, which highlights the males and unhighlights the females.

By alternately clicking on the bars for males and females, as in **Figure 6.3**, you can see the distribution the subsets highlighted in the **Height** histogram. This gives a preliminary look at the height distribution within each group.

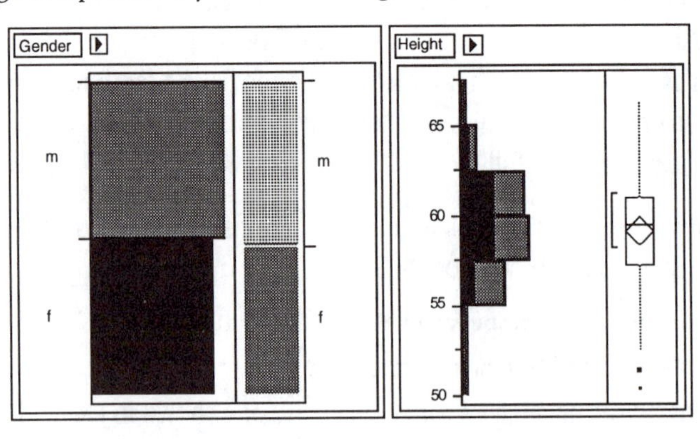

Figure 6.3 Interactive Histograms: Click on "f" to Highlight the Females

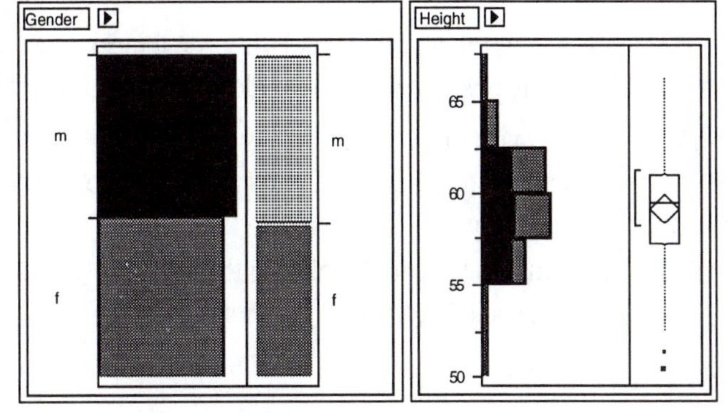

Figure 6.3 (continued) Interactive Histograms: Click on "f" to Highlight the Females

Launch the Fit Y by X Platform

You can compare group means with the Fit Y by X platform where **Height** is the continuous **Y** variable and **Gender** is the nominal (grouping) **X** variable. Begin by launching an analysis platform:

§ Choose **Analyze→Fit Y by X.**

§ In the role prompting dialog, select **Height** as **Y** and **Gender** as **X**.

Notice that the role prompting dialog indicates that you are doing a *Grouped, Means, one-way Anova* analysis. Because **Height** is continuous and **Gender** is categorical (nominal), the **Fit Y by X** command automatically gives a one-way layout for comparing distributions. When you Click **OK** you see the initial graphs, which are side-by-side vertical scatterplots for each group.

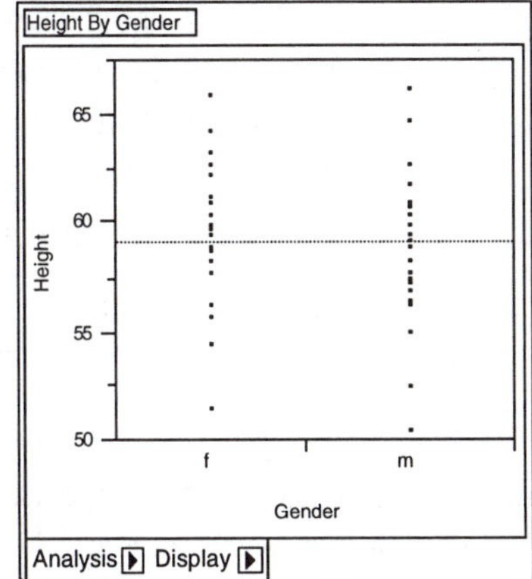

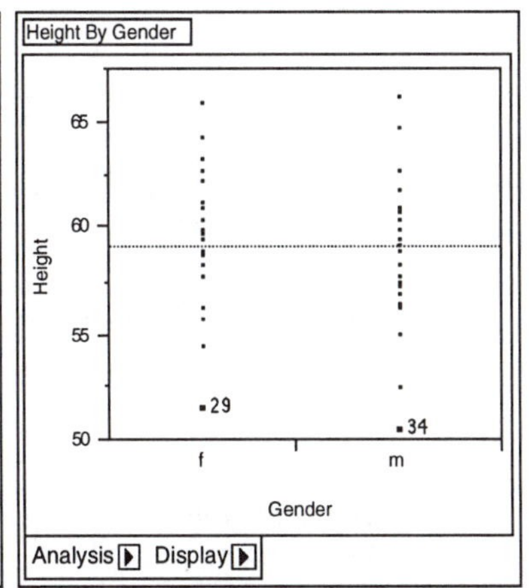

Figure 6.4 Point Plot of the Responses, Before and After Labeling the Two Points

Examine the Plot

The horizontal line across the middle shows the overall mean of all the observations. To identify possible outliers (students with unusual values):

§ Click the lowest point in the "f" vertical scatter, and SHIFT-click in the lowest point in the "m" sample. SHIFT-clicking extends a selection so that the first selection does not unhighlight.

§ Choose **Rows→Label/Unlabel** to see the plot on the right in **Figure 6.4**. Now the points are labeled 29 and 34, the row numbers.

Display and Compare the Means

The next step is to display the group means in the graph, and to obtain an analysis of them.

§ Select **Means, Anova/t Test** from the **Analysis** popup menu showing beneath the plot to add the analysis that estimates the group means and tests to see if they are different.

Figure 6.5 Diamonds to Show the Means and Their 95% Confidence Limits.

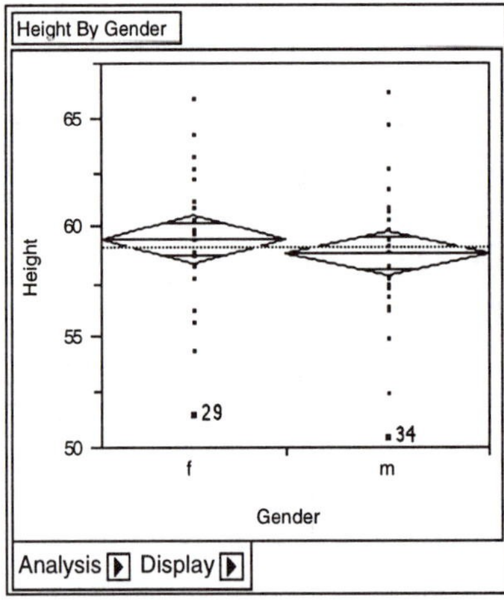

The **Means, Anova/t Test** option automatically displays the Means Diamonds as shown in **Figure 6.5**, followed by summary tables and statistical test reports.

The center lines of the means diamonds are the group means. The top and bottom of the diamonds form the 95% confidence intervals for the means. You can say the probability is .95 that this confidence interval contains the true group mean.

The confidence intervals show whether a mean is significantly different from some hypothesized value, but what can it show regarding whether two means are significantly different?

Interpretation rule for Means Diamonds:	If the confidence intervals shown by the means diamonds do not overlap, the groups are significantly different (but the reverse is not necessarily true).

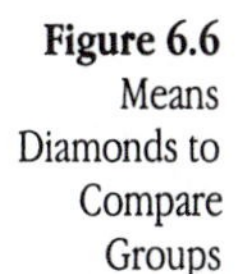

Figure 6.6 Means Diamonds to Compare Groups

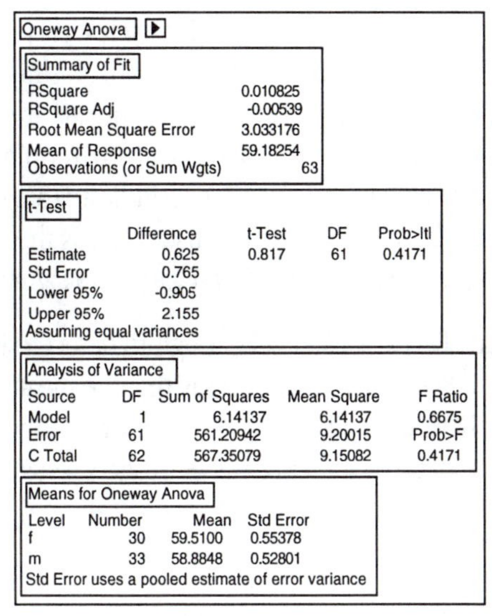

Oneway Anova

Summary of Fit

RSquare	0.010825
RSquare Adj	-0.00539
Root Mean Square Error	3.033176
Mean of Response	59.18254
Observations (or Sum Wgts)	63

t-Test

	Difference	t-Test	DF	Prob>|t|
Estimate	0.625	0.817	61	0.4171
Std Error	0.765			
Lower 95%	-0.905			
Upper 95%	2.155			

Assuming equal variances

Analysis of Variance

Source	DF	Sum of Squares	Mean Square	F Ratio
Model	1	6.14137	6.14137	0.6675
Error	61	561.20942	9.20015	Prob>F
C Total	62	567.35079	9.15082	0.4171

Means for Oneway Anova

Level	Number	Mean	Std Error
f	30	59.5100	0.55378
m	33	58.8848	0.52801

Std Error uses a pooled estimate of error variance

It is clear that the means diamonds in this example do overlap, so you need to look closer at the text report beneath the plots. The report, shown in **Figure 6.6**, includes summary statistics, a t test, an analysis of variance, and means estimates. Note that the p value in the t Test table (and in the Analysis of Variance table) is not significant.

Inside the Student's t Test:

The Student's t test was used previously to test whether a mean was significantly different from a hypothesized value. Now the situation is to test whether the difference of two means is significantly different from the hypothesized value of zero. The t ratio is formed by first finding the difference between the estimate and the hypothesized value, and then dividing that quantity by its standard error.

$$\text{t statistic} = \frac{\text{estimate} - \text{hypothesized value}}{\text{standard error of the estimate}}$$

In the current case, the estimate is the difference in the means for the two groups, and the hypothesized value is zero:

$$\text{t statistic} = \frac{(\text{mean1} - \text{mean2}) - 0}{\text{standard error of difference}}$$

It turns out that for the means of two independent groups, the standard error of the difference is the square root of the sum of squares of the standard errors of the means.

$$\text{standard error of difference} = \sqrt{s^2_{\text{mean1}} + s^2_{\text{mean2}}}$$

JMP does the se calculations and forms the tables shown in **Figure 6.6**. Roughly, you look for a t statistic greater than 2 in absolute value, to get significance (the p value) at the .05 level. The significance level is determined in part by the degrees of freedom (DF) of the t distribution. For this case DF is the number of observations (63) minus two because two means are being estimated. With the calculated t and DF, the p value can be determined. The label **Prob>|t|** is given to the p value in the test table to indicate that this is the probability of getting an even greater absolute t statistic. This is the significance level. Usually a p value less than .05 is regarded as significant.

In this example, the p value of .4171 isn't small enough to detect a significant difference in the means (see **Figure 6.6**). Is this to say that the means are the same? Not at all. You don't have enough evidence to show that they are different. If you collect more data, you might be able to show a significant, albeit small, difference.

One-Sided Version of the Test

The Student's t test in the previous example is for a two-sided alternative; the difference could go either way. If you only want to protect the test in one direction, then you can do a little arithmetic on the reported p value, forming one-sided p values by using p/2 or 1–(p/2), depending on the direction of the alternative. In this example, the mean for males was less than the mean for females, so the significance with the alternative to conclude the females higher would be the p value of 0.2085, half the 2-tailed p value. To test the other direction, the p value would be .7915.

Analysis of Variance and the All-Purpose F Test

As well as showing the t test, the report in **Figure 6.6** for comparing two groups shows an Analysis of Variance with its F test. The F test surfaces many times in the next few chapters, so an introduction here is in order, and details can unfold later as needed.

The F test compares variance estimates in two situations, one a special case of the other. It turns out that this is very useful for testing means and other things, too. Furthermore, when there are only two groups, the F test is equivalent to the t test and the F ratio is the square of the t ratio $(.81)^2 = .66$.

To begin, let's look at the different estimates of the variance as reported in the Analysis of Variance table.

First, the analysis of variance procedure pools all responses into one big population and estimates the population mean (the *grand mean*). The variance around that grand mean is estimated by taking the average sum of squared differences of each point from the grand mean.

Note: The difference between a response value and an estimate such as the mean will be called a *residual*, or sometimes the *error*.

In the Analysis of Variance table, this estimate of the variance is in the line labeled C Total (*total, corrected for the mean*). Both the Sum of Squares and the Mean Square are shown. The Mean Square shown for C Total is the estimate of the variance, and its square root is the estimate of the standard deviation, all estimated under this model of fitting only one grand mean. (Refer back to **Figure 6.6** to see that the Std Dev is 3.025, the square root of the 9.15 reported as the Mean Square for C Total.)

What happens when a separate mean is computed for each group? The variance around these individual means is calculated, and this is shown in the Error line in the Analysis of Variance table. The Mean Square for Error is the estimate of the residual variance (also called s^2), and its square root, called the *root mean squared error* (or s), is the residual standard deviation estimate.

If the true group means are different, then the separate means can give a better fit than can the one grand mean. The change in the residual sum of squares from the one-mean model to the separate-means model leads us to the F test shown in the Model line of the Analysis of Variance table. The Model Mean Square also estimates the residual variance if the hypothesis that the means are the same is true.

The F ratio (F ratio) is the Model Mean Square divided by the Error Mean Square:

$$\text{F Ratio} = \frac{\text{Mean Square for Model}}{\text{Mean Square for Error}} = \frac{6.141}{9.200} = .6675$$

The F ratio is a measure of improvement in fit when separate means are considered. It works like this: If the null hypothesis is true (if there is no difference between fitting the grand mean and individual means), then both numerator and denominator estimate the same variance (the grand mean residual variance). But if the hypothesis is not true and the separate-means model does fit better, the numerator (the model mean square) will contain more than just the grand mean residual variance. However, the denominator estimates the variance regardless of whether the hypothesis is true or not.

If the two mean squares in the F ratio are statistically independent (and they are in this kind of analysis) then you can use the F distribution associated with the F ratio to get a p value, which tells how likely you are to see the F ratio given by the analysis if there was really no difference in the means.

If the tail probability (p value) associated with the F ratio in the F distribution is smaller than .05 (or the alpha level of your choice), you can conclude that the variance estimates are different, and thus that the means are different.

In this example, the total mean square and the error mean square are not much different; in fact the F ratio is actually less than one, and the p value of .4171 (exactly the same as seen for the t test) is far from significant (much greater that .05).

The F test can be viewed as whether the variance around the group means (left-hand histogram in **Figure 6.7**) is significantly less than the variance around the grand mean (right-hand histogram). In this case, the variance isn't much different.

So by this logic, a test of variances is also a test on means. The F test turns up again and again because it is oriented to comparing the variation around two models, and most statistical tests can be constituted this way.

Figure 6.7 Histograms of Residuals for Group Means Model (left) and Grand Mean Model (right)

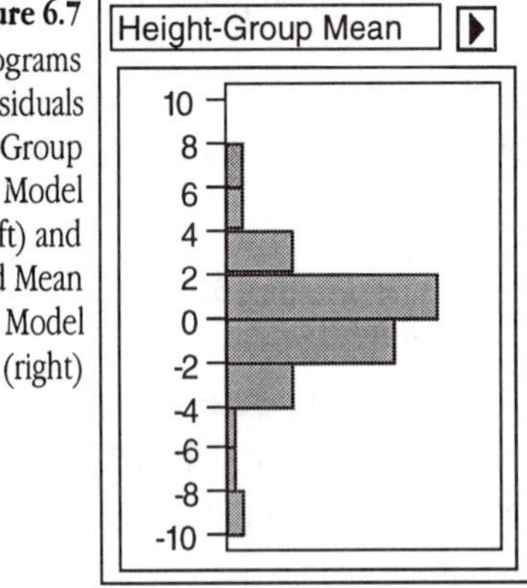

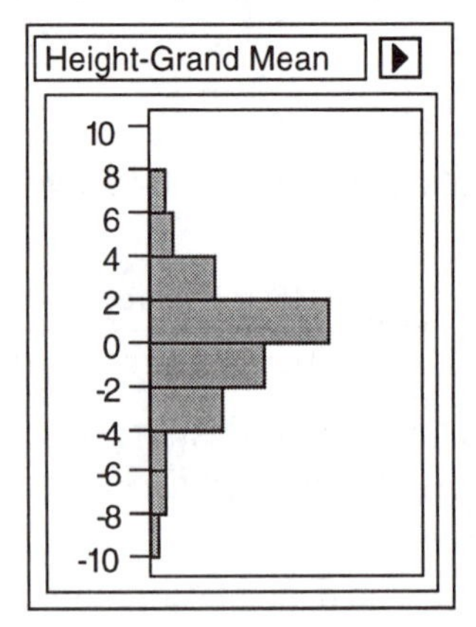

Terminology for Sums of Squares:	All disciplines that use statistics use analysis of variance in some form. However, You may find different names used for the same thing. For example, the following are different names for the same kinds of sums of squares (SS). SS(model) ≡ SS(regression) ≡ SS (between) SS(error) ≡ SS(residual) ≡ SS(within)

How Sensitive is the Test? How Many More Observations Needed?

So far in the example, there is no conclusion to report because the analysis failed to show anything. This is an uncomfortable state of affairs. It is tempting to state that we have shown *no significant difference*, but in statistics this claim is the same as saying *inconclusive.* Our conclusions can be attributed to not having enough data as easily as to there being a very small true effect.

When a test is not significant, there are additional items you can find out that might be useful:

- How small a difference could the significance test in this example detect? This is called the Least Significant Value (LSV)?
- How many more observations would make the reported difference become significant? This is called the Least Significant Number (LSN)?

Here is how to address these questions:

§ Select **Power Details** from the popup menu next to the One-way Anova table name. The analysis first displays the Power Details Dialog shown in **Figure 6.8**.

§ In the Power Details dialog check the **Solve for Least Significant Number** and **Solve for Least Significant Value**.

When you click **Done**, a report about power and sample size appears, as shown to the right in **Figure 6.8**. (The concept of *Power* is discussed in a Chapter 7, "Comparing Many Means: One-way Analysis of Variance".)

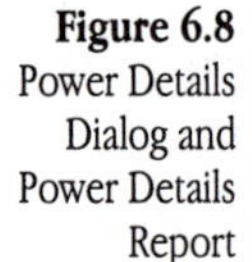

Figure 6.8 Power Details Dialog and Power Details Report

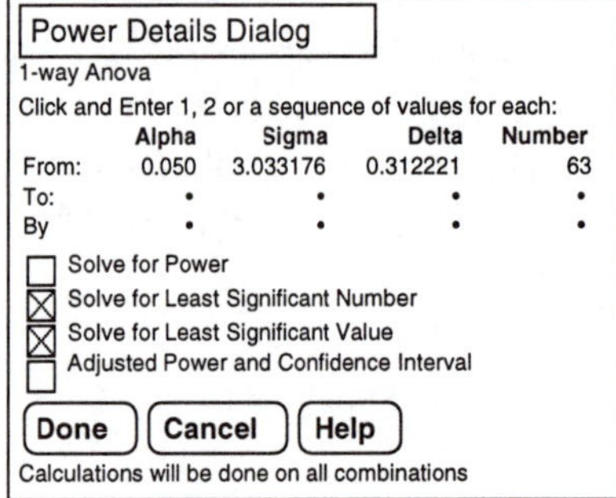

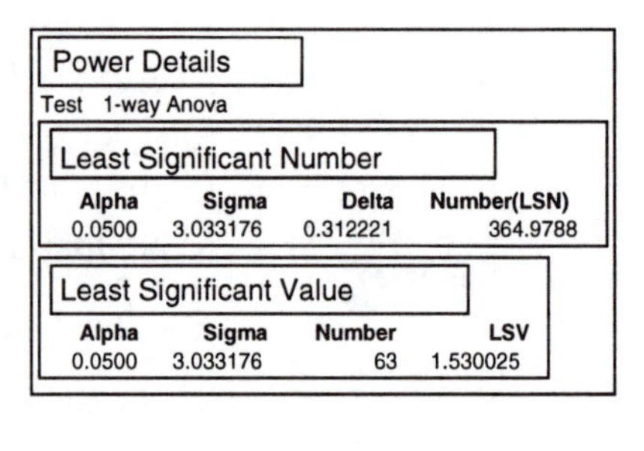

The Least Significant Value (LSV) refers to the difference in means. The actual mean height difference in this example is 59.51–58.88 = .63 inches. However, the Power Details report shows that with the existing sample sizes and variances, you need to measure differences as large as 1.53 to declare that the means are significantly different. This is the *sensitivity* of the test. The sensitivity of this test is that you would have been able to detect significant differences no less than 1.53 inches.

Knowing the sensitivity converts the inconclusive result into one that says that if there is a real difference, it is (95%) likely to be less than 1.53 inches.

The least significant number (LSN) refers to the sample size. In this case the actual sample size was 63. But with the variances and differences encountered, you need a sample size of 365 (instead of only 63) to detect a significant difference. If you are planning to collect more data, you should get even more observations than the LSN, because the probability of finding a significant difference with LSN observations can be as low as 50%.

All this converts the inconclusive result into an option to collect more data if the question is still important with this small an effect. Inconclusive results should not be regarded as noninformative. In fact a new field of statistics called meta-analysis specializes in collecting results from many studies and combining them to obtain a more conclusive result.

When the Difference Is Significant...

The 12-year-olds in the previous example don't have significantly different average heights, but let's take a look at the 15-year-olds. To start, open the sample table called HTWT15.JMP. Then proceed as before:

§ Choose **Analyze→Fit Y by X**, with Gender as X and Height as Y.

§ Select **Means, Anova/t-Test** from the Analysis popup menu showing beneath the plot to add the analysis that estimates group means and tests to see if they are different. You will see the plot and tables shown in **Figure 6.9**.

This time the difference is statistically significant.

Figure 6.9 Analysis for Mean Heights of 15-year-Olds

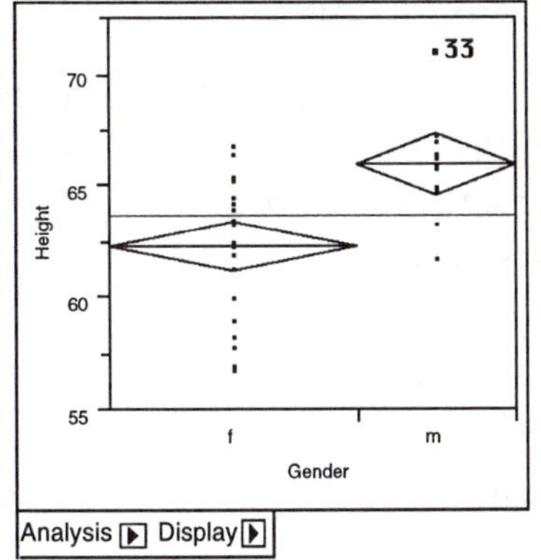

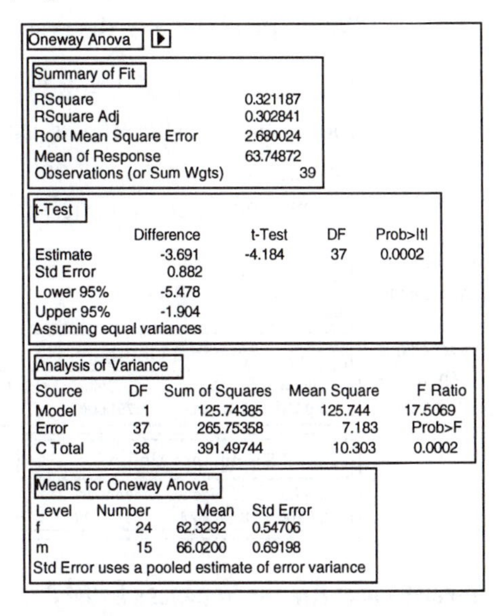

Oneway Anova

Summary of Fit

RSquare	0.321187
RSquare Adj	0.302841
Root Mean Square Error	2.680024
Mean of Response	63.74872
Observations (or Sum Wgts)	39

t-Test

	Difference	t-Test	DF	Prob>|t|
Estimate	-3.691	-4.184	37	0.0002
Std Error	0.882			
Lower 95%	-5.478			
Upper 95%	-1.904			

Assuming equal variances

Analysis of Variance

Source	DF	Sum of Squares	Mean Square	F Ratio
Model	1	125.74385	125.744	17.5069
Error	37	265.75358	7.183	Prob>F
C Total	38	391.49744	10.303	0.0002

Means for Oneway Anova

Level	Number	Mean	Std Error
f	24	62.3292	0.54706
m	15	66.0200	0.69198

Std Error uses a pooled estimate of error variance

The results for the analysis the 15-year-old heights are completely different than the 12-year-old results. Here, the males are significantly taller than the females. You can see this because the confidence intervals shown by the means diamonds do not overlap. You can see that the p values for the t test and F test are .0002, which is considered highly significant.

The F test results say that the variance around the group means (left-hand histogram in **Figure 6.10**) is significantly less than the variance around the grand mean (right-hand histogram).

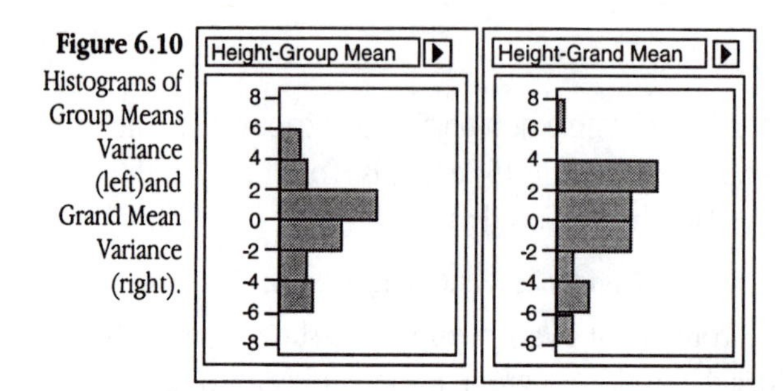

Figure 6.10 Histograms of Group Means Variance (left)and Grand Mean Variance (right).

§ Select **Power Details** from the popup menu next to the One- way Anova table name and request the LSV and LSN.

When you click **Done**, the power analysis in **Figure 6.11** is shown.

The difference of 66.02–62.33 = 3.69 is much greater than the difference that could have been detected as significant at .05, which is the LSV of 1.78. There were 39 observations, but only 12 (the LSN) might have been needed given the differences and variances in this example.

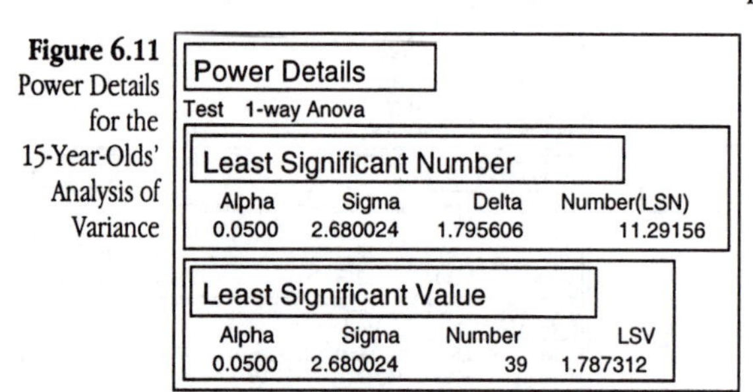

Power Details

Test 1-way Anova

Least Significant Number

Alpha	Sigma	Delta	Number(LSN)
0.0500	2.680024	1.795606	11.29156

Least Significant Value

Alpha	Sigma	Number	LSV
0.0500	2.680024	39	1.787312

Figure 6.11 Power Details for the 15-Year-Olds' Analysis of Variance

Special Topic: Are the Variances Equal Across the Groups?

The One-way Anova and related t test assume that the variance is the same within each group. The t Test table includes the note "Assuming equal variances," and the Analysis of Variance table shows the note "Std Error uses a pooled estimate of error variance," which is also the assumption of equal variances within groups. If this is not true, then the statistical tests have a problem. Let's look at this issue.

However, before you get too concerned about this, be aware that there is always a list of issues to worry about—it is not usually profitable to be overconcerned about this particular issue. Thus it is called a special topic.

For the Fit Y by X platforms given by the 12-year-old and 15-year-old analyses:

§ Select **Quantiles** from the **Analysis** popup menu showing beneath the plots.

This optional command displays quantile box plots for each group as shown in **Figure 6.12**. Note that the interquartile range is not much different for the 12 year olds, but is different for the 15-year olds, with males having a much smaller interquartile range (box height).

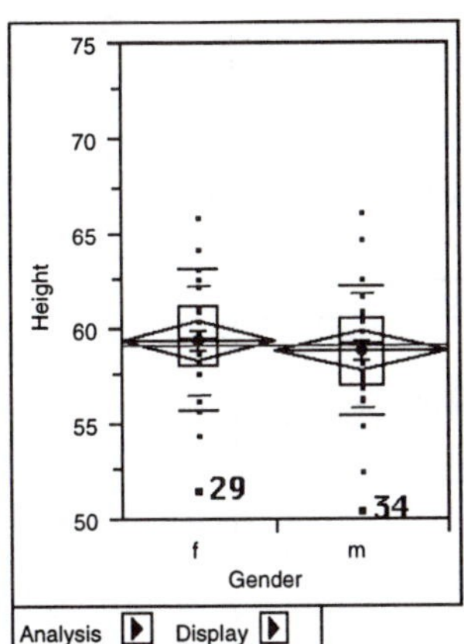

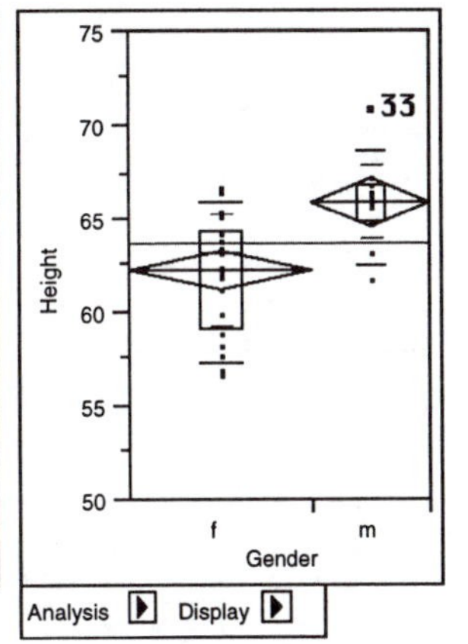

Figure 6.12 Quantile Box Plots for 12 years (left) and 15 years (right)

It's easy to get estimates of the standard deviation within each group:

§ Select **Means, StdDev, StdErr** from the **Analysis** popup menu showing beneath the plots to see the reports in **Figure 6.13**.

Figure 6.13 Means and Standard Deviations Reports

Analysis of 12-Year Olds

Means and Std Deviations

Level	Number	Mean	Std Dev	Std Err Mean
f	30	59.5100	2.98783	0.54550
m	33	58.8848	3.07369	0.53506

Analysis of 15-Year Olds

Means and Std Deviations

Level	Number	Mean	Std Dev	Std Err Mean
f	24	62.3292	3.01496	0.61543
m	15	66.0200	2.01218	0.51954

The standard deviation estimates of 2.98 and 3.07 are very close for the 12-year-olds. But the standard deviation estimates of 3.01 and 2.01 are farther apart for the 15-year-olds.

You can test these variances as follows:

§ For both HTWT12 and HTWT15 select **Unequal Variance** from the **Analysis** popup menu to see the Tests that the Variances are Equal tables in **Figure 6.14**.

Figure 6.14 Tests the Variances are Equal Report

Tests that the Variances are Equal

Level	Count	Std Dev	MeanAbsDif to Mean	MeanAbsDif to Median
f	30	2.987832	2.183333	2.183333
m	33	3.073691	2.247750	2.227273

Test	F Ratio	DF Num	DF Den	Prob>F
O'Brien[.5]	0.0160	1	61	0.8998
Brown-Forsythe	0.0071	1	61	0.9331
Levene	0.0158	1	61	0.9003
Bartlett	0.0240	1	•	0.8769

Welch Anova testing Means Equal, allowing Std's Not Equal

F Ratio	DF Num	DF Den	Prob>F
0.6694	1	60.714	0.4165

t-Test
0.8181

Tests that the Variances are Equal

Level	Count	Std Dev	MeanAbsDif to Mean	MeanAbsDif to Median
f	24	3.014960	2.482639	2.387500
m	15	2.012177	1.261333	1.260000

Test	F Ratio	DF Num	DF Den	Prob>F
O'Brien[.5]	2.7684	1	37	0.1046
Brown-Forsythe	3.3984	1	37	0.0733
Levene	5.4215	1	37	0.0255
Bartlett	2.5338	1	•	0.1114

Welch Anova testing Means Equal, allowing Std's Not Equal

F Ratio	DF Num	DF Den	Prob>F
21.0002	1	36.777	<.0001

t-Test
4.5826

First, note that the Std Dev column lists the estimates you are testing to be the same. The 12-year-olds have no evidence of different variances across Gender. However the 15-year-olds have evidence ranging from a p value of .02 (Levene's test) to .11 (Bartlett's test).

Each of the four tests in **Figure 6.12** (O'Brien, Brown-Forsythe, Levene, and Bartlett) tests whether the variances are equal, but each uses a different method for measuring variability. One way to evaluate dispersion is to take the absolute value of the difference of each response from its group mean.

- If you estimate the mean of these absolute differences for each group (shown in the table as MeanAbsDif to Mean), and then do a t test (or equivalently, an F test) on these, you have Levene's test.
- If you measure the differences from the median instead of the mean and then tested, you have the Brown-Forsythe test.

- If you trick the t test so that the means were really the variances for each group, you have O'Brien's test.
- If you derive the test mathematically, assuming that the data are normal, you obtain Bartlett's test, which, though powerful, is sensitive to departures from the Normal distribution.

Statisticians have no apologies for offering different tests with different results. Each test has its advantages and disadvantages.

Testing Means with Unequal Variances

If you think the variances are different, then you should choose a modified t test for unequal variances, which is equivalent to the F test in the Welch Anova (**Figure 6.14**). The test can be interpreted as an F test in which the observations are weighted by an amount inversely proportional to the variance estimates. This has the effect of making the variances comparable.

The p values may disagree slightly with those obtained from other programs for unequal-variance t tests. These differences arise because some methods round or truncate the denominator degrees of freedom for computational convenience. JMP uses the more accurate fractional degrees of freedom.

Let's look at the 15-year-olds. Now there are two kinds of F test to consider:

- The F test that assumes equal variances (shown previously in **Figure 6.8**) is 17.5 with a p value of .0002.
- The variance-weighted Welch F test in **Figure 6.12** is 21 with a p value of less than .0001.

With the fairly strong evidence of different variances, you can choose the Welch F to test the means. Actually, with the means being so significantly different, it makes no difference which test is used.

In practice the hope is that there are no practically conflicting results from different tests of the same hypothesis. But it could happen and there is an obligation to report the results from all reasonable perspectives.

Special Topic: Normality and Normal Quantile Plots

The t tests (and F tests) used in this chapter assume that the sampling distribution for the group means is the Normal distribution. With sample sizes of at least 30 for each group, normality is probably a safe assumption. The Central Limit Theorem says that means approach a normal distribution as the sample size increases even if the original data are not normal.

If you have small samples, or suspect that the samples are contaminated by a few bad points, or if the distribution is very nonnormal, you can consider using nonparametric methods, covered at the end of this chapter.

The normal quantile plot can be used to assess normality. This is particularly useful when overlaid for several groups, because you can see so many attributes of the distributions in one plot:

§ Return to the Fit Y by X platform showing **Height** by **Gender** for the 12-year-olds and select **Normal Quantile Plot** in the **Analysis** popup menu. Do the same for the 15-year -olds.

The normality is judged by how well the points in the Normal Quantile plot follow a straight line. The standard deviations are the slopes of the straight lines. Lines with steep slopes represent the distributions with the greater variances. The vertical separation of the lines at the middle shows the difference in the means. The separation of other quantiles shows at other points on the X axis.

The left-hand graph shown in **Figure 6.15** confirms that heights of 12-year-old males and females have the nearly the same mean and variance—the slopes (standard deviations) are the same and the positions (means) are only slightly different.

The right-hand graph in **Figure 6.15** shows 15-year-old males and females have different means and different variances—the slope (standard deviation) is higher for the females, but the position (mean) is higher for the males.

The distributions for all groups look reasonably normal.

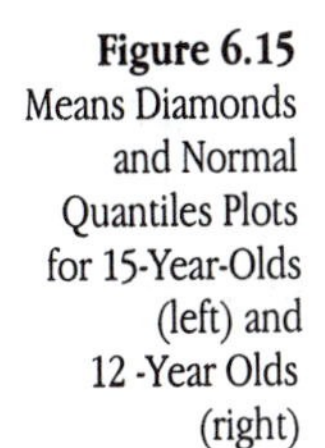

Figure 6.15 Means Diamonds and Normal Quantiles Plots for 15-Year-Olds (left) and 12 -Year Olds (right)

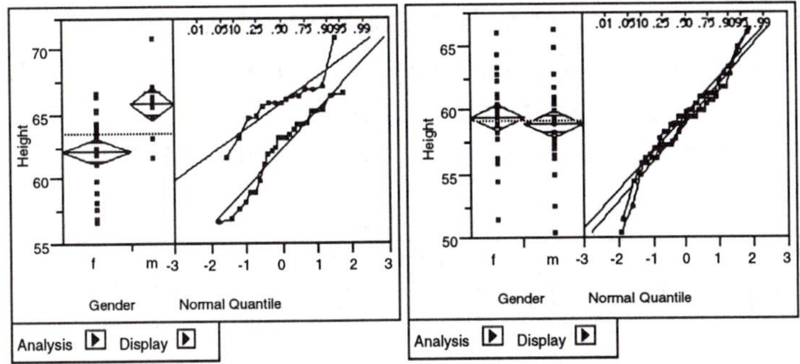

Testing Means for Matched Pairs

Consider the situation where the two responses form a pair of measurements coming from the same experimental unit or subject. A typical situation is a before-and-after measurement on the same subject. The responses are correlated, and the statistical method called the *paired t test* takes that into account.

In general, if the responses are positively correlated, the Paired t test gives a more significant p value than the t test for independent means (grouped t test) discussed in the previous sections. If responses are negatively correlated, then the Paired t test will be less significant than the grouped t test. In most cases where the pair of measurements are taken from the same individual at different times, they are positively correlated. However, if they represent competing responses the correlation can be negative.

Thermometer Tests

A health care center suspected that temperature readings from a new ear drum probe thermometer were consistently higher than readings from the standard oral mercury thermometer. To test this hypothesis temperature readings were taken twice, once with the ear-drum probe, and the other with the oral thermometer. There was variability among the readings so they were not expected to be exactly the same. However, the suspicion was that there was a systematic difference—that the ear probe was reading too high.

§ For this example, open the THERM.JMP data file.

A partial listing of the data table appears in **Figure 6.16**. The THERM.JMP data table has 20 observations and 4 variables. The two responses are the temperatures taken orally and tympanically (by ear) on the same person on the same visit.

Figure 6.16 Comparing Paired Scores

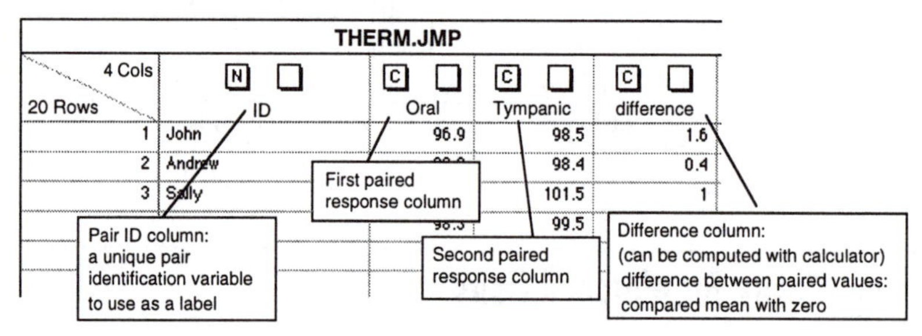

For paired comparisons, the two responses need to be arranged in two columns, each with a continuous modeling type. It is also useful to create a new column with a formula to calculate the difference between the two responses. (If your data table is arranged with the two responses in different rows, then use the **Tables→Split Columns** command to rearrange it. See the section **Managing Data** in Chapter 2, "JMP Data Tables.")

Look at the Data

Start by inspecting the distribution of the data. To do this:

§ Choose **Analyze→Distribution of Y** with Oral and Tympanic as Y variables.

§ When the results appear, select **Uniform Axes** from the check-mark options menu at the lower left of the window to force the axes of the plots to be on the same scale.

The histograms in **Figure 6.17** show the temperatures to have different distributions. You can see that the mean looks higher for the Tympanic temperatures. However, as you will see later, the picture of the difference in the means can be misleading if you try to judge the significance of the difference from this perspective.

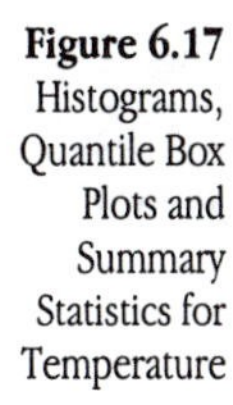

Figure 6.17 Histograms, Quantile Box Plots and Summary Statistics for Temperature

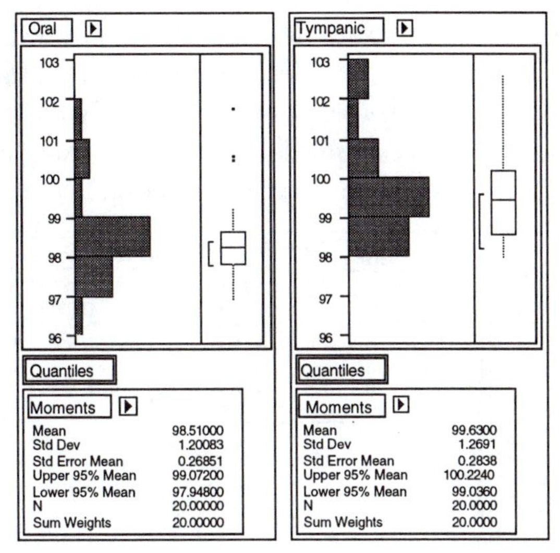

What about the four outliers at the top end of the Oral temperature distribution? (The middle point is actually two points on top of each other.) Is this of concern? Can you expect the distribution to be normal? Not really. It is not the temperatures that are of interest, but the difference in the temperatures. So there is no concern about the distribution so far. If the plots showed temperature readings of 110 or 90, there would be concern, because that would be suspicious data for human temperatures.

Look at the Distribution of the Difference

The comparison of the two means is actually done on the difference between them. Inspect the distribution of the differences:

§ Choose **Analyze→Distribution of Y** with **difference** as the Y variable.

The results in **Figure 6.18** show a distribution that seems to separate above zero. In the Moments table, the lower 95% limit for the mean is .828, above zero. The Student's t test will show this to be significant.

Student's t Test

§ Choose **Test Mean = Value** from the popup menu next to the name at the top of the histogram. When prompted for a hypothesized value, enter zero and click **OK**.

Now you have the t test for testing that the mean over the matched pairs is the same. In this case the results in the Test Mean=value table in **Figure 6.16** show a p value of less than .0001, which supports our visual guess that there is a significant difference between methods of temperature taking; the tympanic temperatures are significantly higher than the oral temperatures.

Figure 6.18 Histogram and Tests of the Difference Score

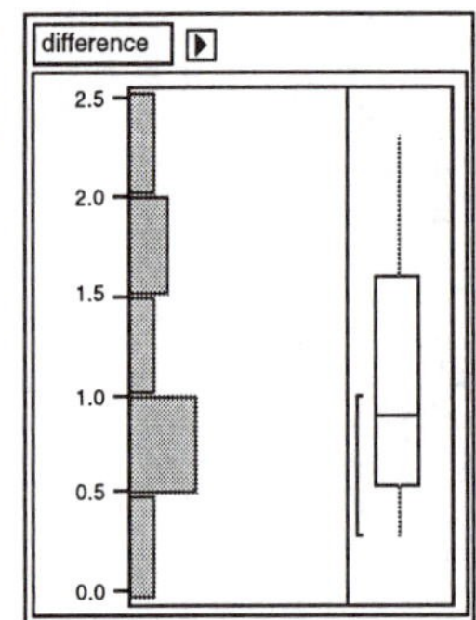

Moments	
Mean	1.12000
Std Dev	0.62374
Std Error Mean	0.13947
Upper 95% Mean	1.41192
Lower 95% Mean	0.82808
N	20.00000
Sum Weights	20.00000

Test Mean=value		
Hypothesized Value		0
Actual Estimate		1.12
	t Test	
Test Statistic	8.0302	
Prob > |t|	<.0001	
Prob > t	<.0001	
Prob < t	1.0000	

There is also a nonparametric test, the Wilcoxon signed-rank test, described at the end of this chapter, that tests the difference between two means.

An Alternative Approach for the Paired t Test

Another way to look at paired values using JMP is to examine a scatterplot that shows one score on the vertical axis and the other on the horizontal axis. Generally, if the values fall along a 1-to-1 diagonal line, the groups are not different, and if the points are farther away from the diagonal, there is a statistical difference. To use this approach:

§ Choose **Analyze→Fit Y by X** and use **Oral** as X and **Tympanic** as Y.

The role prompting dialog now indicates that you are doing a *Scatterplots, Regressions Analysis*—it turns out that there is a regression model that gives a Paired t test. Click **OK** to see the initial scatterplot of **Tympanic** by **Oral.**

To test for a difference between paired means:

§ Select **Paired t-Test** from the **Fitting** popup menu showing beneath the scatterplot.

The results window in **Figure 6.19** appears with a scatterplot showing the Y values along the vertical axis and the X values on the horizontal axis. It is best to force the vertical and horizontal axes to have the same scales, as in

Figure 6.19. To change an axis scale, double-click on the axis and respond to the dialog that appears.

The analysis first draws a gray 45-degree reference line through the origin. This is the line where the two columns are equal. If the means are equal, then the points should balance their distribution along this line. You should see about as many points above this line as below. If a point is above the gray reference line, it means that the vertical value is greater than the horizontal value. In this example, points above the line show the situation where the tympanic temperature is greater than the oral.

Parallel to the gray 45-degree line is a solid red line that is displaced from the gray line by an amount equal to the difference in means between the two responses. This red line is the line of fit for the sample. The test of the means is equivalent to asking if the red line through the points is significantly separated from the gray line through the origin.

The dashed lines around the red line of fit show the 95% confidence interval for the difference in means.

This scatterplot gives you a good idea of each variable's distribution, as well as the distribution of the difference.

Figure 6.19 Scatterplot of Paired Values Showing the Separation from the Diagonal

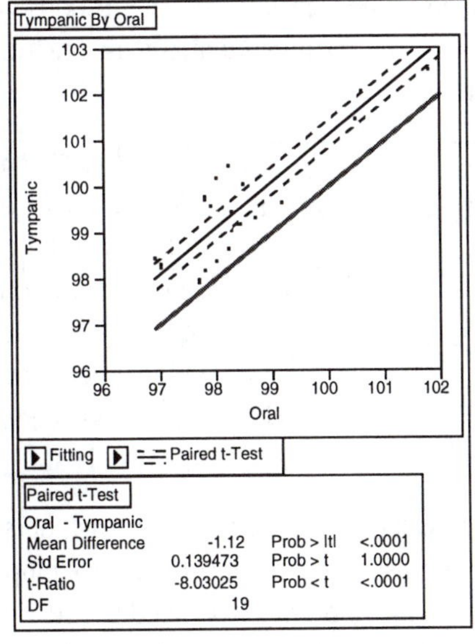

Paired t-Test

Oral - Tympanic

Mean Difference	-1.12	Prob > |t|	<.0001
Std Error	0.139473	Prob > t	1.0000
t-Ratio	-8.03025	Prob < t	<.0001
DF	19		

Interpretation rule for the Paired t test scatterplot:

If the confidence interval (represented by the dashed lines around the red line) contains the gray line through the origin, then the two means are not significantly different.

Another feature of the scatterplot is that you can see the correlation structure. If the two variables are positively correlated, they lie closer to the line of fit, and the variance of the difference is small. If the variables are negatively correlated, then most of the variation is perpendicular to the line of fit, and the variance of the difference is large. It is this variance of the difference that scales the difference in a t test and determines whether the difference is significant.

The Paired t Test table at the bottom of **Figure 6.19** gives the statistical details of the test. The results should be identical to those shown earlier with **Distribution of Y Test Mean=0**. The table shows that the observed difference of 1.12 degrees of temperature is significantly different from zero.

An Equivalent Test For Stacked Data

There is a third approach to the paired t test. Sometimes, you receive grouped data with the response values stacked into a single column instead of having a column for each group. Here is how to rearrange the THERM.JMP data table and see what a stacked table looks like:

§ Choose **Tables→Stack Columns**. When the Stack dialog appears, select **Height** and **Weight** as the columns to be stacked.

§ When you click **OK**, the table in **Figure 6.20** appears. The response values (temperatures) are in the **_Stacked_** column, identified as "oral" or "Tympanic" by the **_ID_** column.

Figure 6.20 Example of Stacked Table

Untitled 11

3 Cols / 40 Rows

	Name	difference	_Stacked_	_ID_
1	John	1.6	96.9	Oral
2	John	1.6	98.5	Tympanic
3	Andrew	0.4	98.0	Oral
4	Andrew	0.4	98.4	Tympanic
5	Sally	1	100.5	Oral
6	Sally	1	101.5	Tympanic
7	Joanie	1.2	98.3	Oral

If you do a **Fit Y by X** of **_Stacked_** (the response of both temperatures) by **_ID_** (the classification), you get the t test designed for independent groups. But this would be inappropriate for paired data.

However, fitting a model that includes an adjustment for each person fixes the independence problem because the correlation is due to temperature differences from person to person. To do this, you need to use the **Fit Model** command, which is covered in Chapter 12, " Fitting Models." The response is modeled as a function of both the category of interest (_ID_ of Oral or Tympanic) and the Name category that identifies the person.

When you get the results, the p value for the category effect is identical to the p value that you would get from the ordinary paired t test; in fact the F ratio in the effect test is exactly the square of the t test value in the paired t test. In this case the formula is

$$(\text{Paired t test statistic})^2 = 8.032^2 = 64.4848 = (\text{stacked F test statistic})$$

To get these results:

§ Choose **Analyze→Fit Model**. When the Fit Model dialog appears, use the _Stacked_ variable (temperatures) as Y, and both _ID_ and Name as **Effects in Model**. Then click **Run Model**.

§ The Fit Model platform gives you a plethora of information, but for this example you need only open the Effect Test table (**Figure 6.21**). It shows an F Ratio of 64.48, which is exactly the square of the t ratio of 8.03 found with the previous approach. It's just another way of doing the same test.

Figure 6.21 Equivalent F Test on Stacked Data

Effect Test

Source	Nparm	DF	Sum of Squares	F Ratio	Prob>F
_ID_	1	1	12.544000	64.4848	<.0001
Name	19	19	54.304000	14.6926	<.0001

t ratio from previous analysis = 8.03
F ratio = square of t ratio = 64.48

Preview: The alternative formulation for the paired means covered in the previous section is important for cases in which there are more than two related responses. Having many related responses is a *repeated-measures* or *longitudinal* situation. The generalization of the Paired t test is called the multivariate or T^2 approach, whereas the generalization of the stacked formulation is called the mixed-model or split-plot approach.

Special Topic: Examining the Normality Assumption

The paired t test assumes the differences are normally distributed. With 30 pairs or more this is probably a safe assumption–the results are reliable even if the distribution is not very normal. The temperatures example only has 20 observations, so it is a good idea to check the normality. To do this:

§ Click the Distribution of Y window previously generated to make it active, and select **Normal Curve** and **Quantile Plot** from the check-mark popup menu beneath the plot.

§ Then select the **Test Dist=Normal** from the popup menu at the top of the histogram.

The quantile plot and outlier box plot, and the S-shaped normal quantile plot all show a skewed distribution. The Test for Normality table in **Figure 6.22**, with a p value of .1326, also indicates that the distribution is not nonnormal.

If you are concerned with nonnormality, nonparametric methods or suitable transformations should be considered.

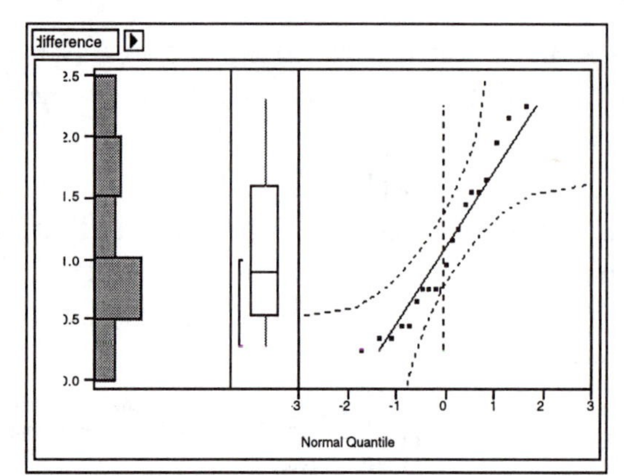

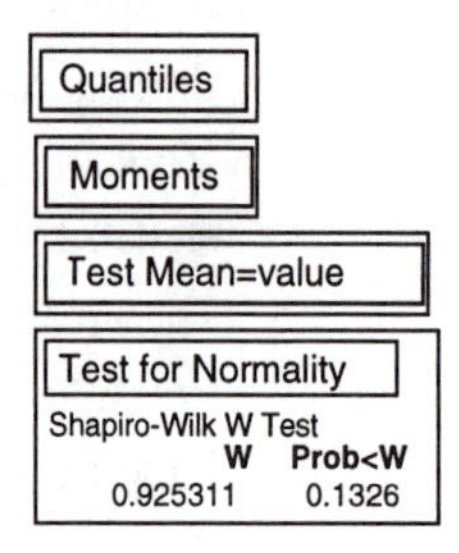

Figure 6.22 Looking at the Normality of the Difference Score

Two Extremes of Neglecting the Pairing Situation: A Dramatization

What happens if you do the wrong test? What happens if you do a t test for independent groups on highly correlated paired data?

§ Open the data table PAIRANT.JMP (**Figure 6.23**). There are two comparisons to consider. The first is to compare the **Before** and **After1** responses. The second is the compare the **Before** and **After2** responses. The column **Diff1** contains the difference for the first comparison: **After1–Before**. The column **Diff2** contains the difference for the second comparison, **After2–Before**.

Figure 6.23 The PAIRANT.JMP Data Table

PAIRANT.JMP

5 Cols / 30 Rows	Before	After 1	After 2	Diff 1	Diff 2
1	55	58	160	3	105
2	65	70	150	5	85
3	71	75	142	4	71
4	74	79	141	5	67
5	77	80	138	3	61

To begin, look at the histograms of the before and after variables:

§ Choose **Analyze→Distribution of Y** with Before, After 1, and After2 as Y.

§ When the histograms appear, choose the **Uniform Axes** option from the check-mark menu to visually compare the histograms shown in **Figure 6.24**.

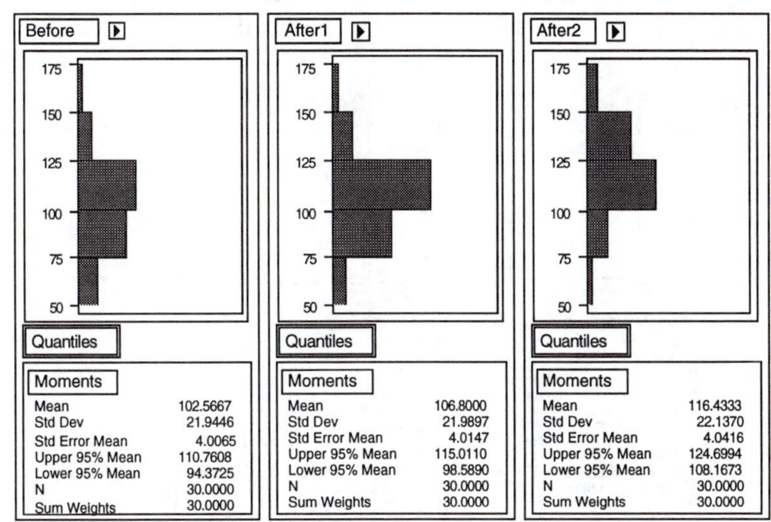

Figure 6.24 Histograms to Compare Before and After Variables

These histograms don't show much difference in the distribution of these three variables. But, as it turns out, looking at these histograms can fool your judgment on whether the means are different, for the same reason that using an independent-group t test can fool you.

Now let's do the incorrect t test–the t test for independent groups. Doing this involves reorganizing the data (using the **Stack Columns** command) twice. The results for the two comparisons are shown in **Figure 6.25**. The conclusions are that After1 is not at all significantly different from Before and that After2 is is much closer to being significantly different from Before than After1.

Figure 6.25 t Tests on Differences

After1–Before

t-Test

	Difference	t-Test	DF	Prob>|t|
Estimate	4.233333	0.746	58	0.4585
Std Error	5.671887			
Lower 95%	-7.12018			
Upper 95%	15.58685			

Assuming equal variances

After2–Before

t-Test

	Difference	t-Test	DF	Prob>|t|
Estimate	8.933333	1.575	58	0.1208
Std Error	5.672971			
Lower 95%	-2.42235			
Upper 95%	20.28902			

Assuming equal variances

Now let's do the proper test, the paired t test using the **Diff1** and **Diff2** columns. As before, do **Distribution of Y** on these differences with the **Uniform Axes** option, and then test that the mean is zero (see **Figure 6.26**).

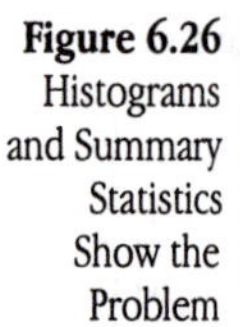

Figure 6.26 Histograms and Summary Statistics Show the Problem

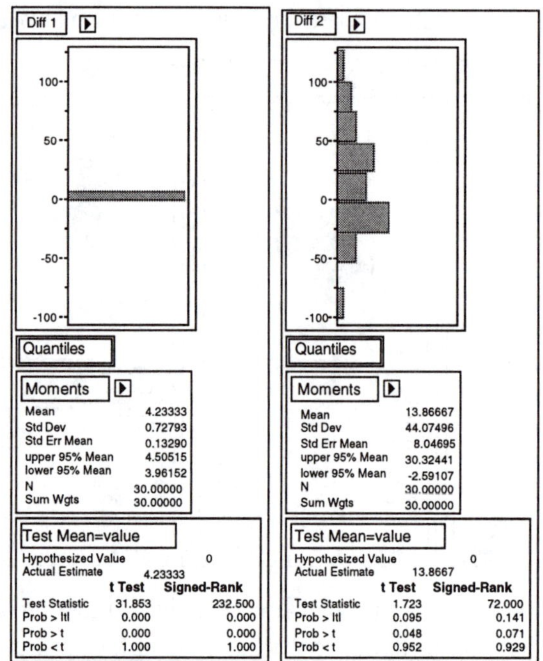

In this case the analysis of the differences leads to the opposite conclusions.

- The mean difference in the first pair (Diff1) of 4.2 is highly significant because the variance is very small.
- The mean difference in the second example (Diff2) of 13.86 is not significant because the variance of that difference is large.

So don't judge the mean of the difference by the difference in the means without noting that the variance of the difference is the measuring stick that depends on the correlation between the two responses.

The scatterplot approach in **Figure 6.27** shows what is happening. The first pair is very positively correlated, leading to a smaller variance for the difference. The second pair is highly negatively correlated, leading to a very large variance for the difference.

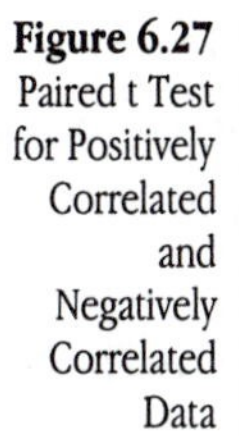

Figure 6.27 Paired t Test for Positively Correlated and Negatively Correlated Data

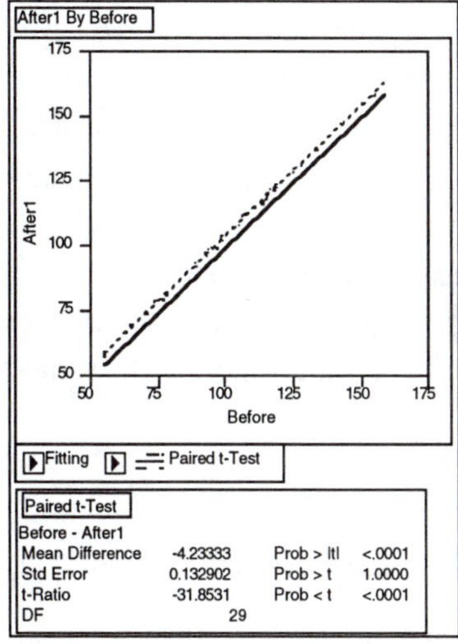

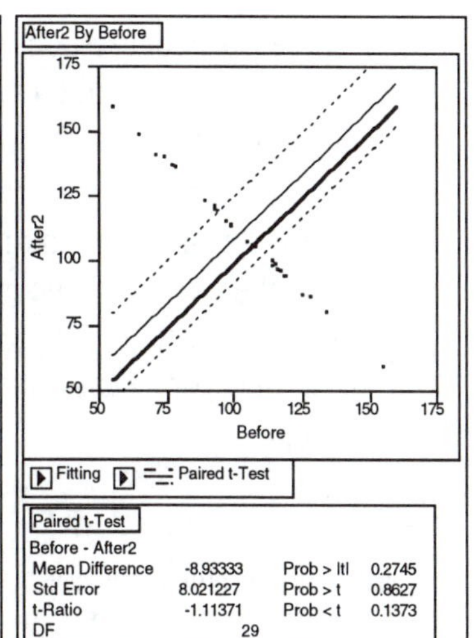

Review

Question? What is the reason that you use a different t test for matched pairs?

a. Because the statistical assumptions for the t test for groups aren't satisfied with correlated data.

b. Because you can detect the difference much better with a Paired t test—the Paired t test is much more sensitive to a given difference.

c. Because you might be overstating the significance if you used a group t test rather than a Paired t test.

d. Because you are testing a different thing.

Answer: All of the above.

a. The grouped t test assumes that the data are uncorrelated and paired data are correlated. So you would violate assumptions using the grouped t test.

b. Most of the time the data are positively correlated, so the difference has a smaller variance than you would attribute if they were independent. So the paired t test is more powerful, that is more sensitive. So you get a much more sensitive test with the paired t test.

c. There may be a situation in which the pairs are negatively correlated, and if so, the variance of the difference would be greater than you expect from independent responses, and the difference would have greater variance–the grouped t test would overstate the significance. So there exist cases where the grouped t test would overstate significance.

d. You are testing the same thing in that the mean of the difference is the same as the difference in the means. But you are testing a different thing in that the variance of the mean difference is different than the variance of the differences in the means (ignoring correlation), and the significance for means is measured with respect to the variance. So you are right.

Mouse Mystery

Comparing two means is not always straightforward. Consider this story:

A food additive showed promise as a dieting drug. An experiment was run on mice to see if it helped control their weight gain. If it proved effective, then it could be sold to millions of people trying to control their weight.

After the experiment was over, the average weight gain for the treatment group was significantly less than for the control group, as hoped for. Then someone noticed that the treatment group had fewer observations than the control group. It seems that the food additive caused the obese mice in that group to tend to die young, so the thinner mice had a better survival rate for the final weighing.

A Nonparametric Approach

Introduction to Nonparametric Methods

Nonparametric methods provide ways to analyze and test data that do not depend on assumptions about the distribution of the data. In order to ignore normality assumptions, nonparametric methods disregard some of the information in your data. Typically, instead of using actual response values, you use the rank ordering of the response.

Most of the time you don't really throw away much relevant information, but you avoid information that might be misleading. A nonparametric approach creates a statistical test that is invariant to the distribution of the response by ignoring all the spacing information between response values. This protects the test against distributions that have very nonnormal shapes and can also provide insulation from data contaminated by rogue values.

In many cases the nonparametric test has almost as much power as the corresponding parametric test and in some cases has more power. For example, even if a batch of values is normally distributed, the rank-scored test for the mean has 95% efficiency relative to the normal-theory test that is most powerful.

The most popular nonparametric techniques are based on functions (scores) of the ranks:

- The rank itself, called a Wilcoxon score.
- Whether the value is greater than the median; whether the rank is more than (n+1)/2.
- A normal quantile, computed as in normal quantile plots, called the van der Waerden score.

Nonparametric methods are not in a single platform in JMP, but are available through many platforms according to the context where that test naturally occurs.

Paired Means: The Wilcoxon Signed-Rank Test

The Wilcoxon signed-rank test is the nonparametric analog to the paired t test. You do a signed-rank test by testing the distribution of the difference of matched pairs, as discussed previously. The following example shows the advantage of using the signed-rank test when data are nonnormal.

§ Open the CHAMBER.JMP table. The data represent electrical measurements on 24 wiring boards. Each board is measured first when soldering is complete, and again after three weeks in a chamber with a controlled environment of high temperature and humidity (Iman 1995).

§ Examine the **diff** variable (difference between the outside and inside chamber measurements) with **Analyze→Distribution of Y**.

§ Select the **Test Dist=Normal** from the popup menu at the top of the histogram.

The probability of .0076 given by the Shapiro-Wilk W normality test indicates that the data are significantly nonnormal. In this situation it might be better to use signed ranks for comparing the mean of **diff** to zero.

Figure 6.28 Partial Listing of the CHAMBER Data and Test for Normality on the diff Column

CHAMBER.JMP

4 Cols / 24 Rows	board	inside	outside	diff
1	1	0.23	0.29	-0.06
2	2	1.38	1.32	0.06
3	3	0.75	0.7	0.05
4	4	1.46	2.88	-1.42
5	5	0.45	1.28	-0.83
6	6	0.48	1.36	-0.88

Test for Normality

Shapiro-Wilk W Test

W	Prob<W
0.880674	0.0076

§ Select the **Test Mean=** from the popup menu at the top of the histogram. When you respond to the dialog that appears:

1) Leave the default mean comparison at zero.
2) Leave the standard deviation blank, to be computed from the sample.
3) Click the Wilcoxon Signed-Rank check box and then click **OK**.

Figure 6.29 Comparison of t test and Shapiro-Wilk W Test

Test Mean=value

Hypothesized Value	0	
Actual Estimate	-0.4333	
	t Test	Signed-Rank
Test Statistic	<.0001	-86.500
Prob > |t|	0.1107	0.010
Prob > t	0.9446	0.995
Prob < t	0.0554	0.005

Note that the standard t test probability is insignificant (p=.1107). In this example, the signed-rank test detects a difference between the groups with a p value of .01.

Independent Means: The Wilcoxon Rank Sum Test

If you want to test the means of two independent groups, as in the t test, but nonparametrically, then you can rank the responses and analyze the ranks instead of the original data. This is the Wilcoxon Rank Sum test. It is also known as the Mann-Whitney U test because there is a different formulation of it that was not discovered to be equivalent to the Wilcoxon rank sum test until after it had become widely used.

§ Open HTWT15 again, and choose with **Analyze→Fit Y by X** for Height as Y and Gender as X. This is the same platform that gave the t test. Now select the **Nonpar-Wilcoxon** command from the **Analysis** popup menu.

The result is the report in **Figure 6.30.** This table shows the sum and mean ranks for each group, then the Wilcoxon statistic along with an approximate p value based on the large-sample distribution of the statistic. In this case, the difference in the mean heights is declared significant, with a p value of .0002. If you have small samples, you should consider also checking the tables of the Wilcoxon to obtain a more exact test because the Normal approximation is not very precise in small samples.

Figure 6.30 Wilcoxon Rank Sum Test for Independent Groups

Wilcoxon / Kruskal-Wallis Tests (Rank Sums)

Level	Count	Score Sum	Score Mean	(Mean-Mean0)/Std0
f	24	350.5	14.6042	-3.728
m	15	429.5	28.6333	3.728

2-Sample Test, Normal Approximation

S	Z	Prob>IZI
429.5	3.72806	0.0002

1-way Test, Chi-Square Approximation

ChiSquare	DF	Prob>ChiSq
14.0064	1	0.0002

Chapter 7
Comparing Many Means: One-Way Analysis of Variance

Overview

In Chapter 6, "The Difference Between Two Means" the t test was the tool needed to compare the means of two groups. But if you need to test the means of more than two groups, the t test can't handle the job. This chapter shows how to compare more than two means using the one-way analysis of variance, or *Anova* for short. The *F test* is the key element in an Anova; it is the statistical tool necessary to compare many groups, just as the t test compares 2 groups. This chapter also introduces multiple comparisons and power calculations, reviews the topic of unequal variances, and extends nonparametric methods to the one-way layout.

Chapter 7 Contents

You can mouse along with the examples whenever you see an action symbol (§)

What Is a One-Way Layout?

A *one-way layout* is the organization of data when a response is measured across a number of groups and the distribution of the response may be different across the different groups. The groups are labeled by a classification variable, which is a column with the nominal or ordinal modeling type.

Usually, one-way layouts are used to compare group means. **Figure 7.1** shows a schematic that compares two models The model on the left fits a different mean for each group, and model on the right indicates a single grand mean (a single-mean model).

Figure 7.1 Different mean for Each Group Versus a Single Overall Mean

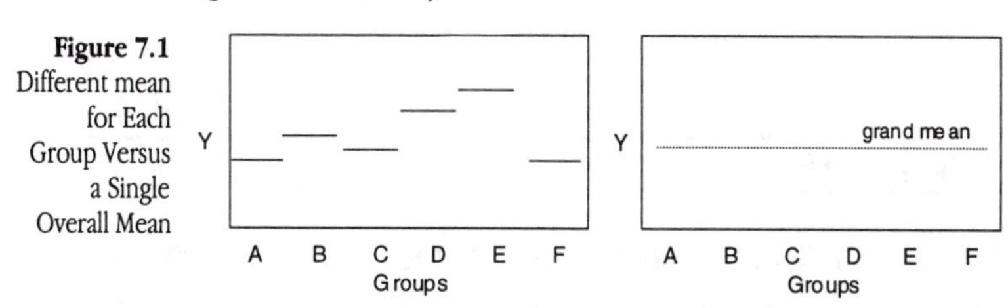

The previous chapter covered using the t test to compare two means. When there are more than two means, the t test is no longer applicable; the F test must be used.

An F test has the following features:

- An F test compares two models, one a constrained version of the other. The constrained model fits one grand mean. For the one-way layout, the unconstrained model fits a mean for each group.
- The measurement of fit is done by accumulating the sum of squared residuals, where the residual is the difference between the actual response and the mean associated with it.
- A Mean Square is calculated by dividing the sum of squares by the degrees of freedom. Mean Squares are estimates of the variance, sometimes under the assumption that certain hypotheses are true. Degrees of Freedom (DF) is the count that you divide by to get an unbiased estimate of the variance (see definitions in Chapter 4, "What are Statistics").
- An *F statistic* is constituted by ratios of Mean Squares (MS) that are independent and have the same expected value. This F statistic will have

an F distribution, at least if the hypothesis that there is no difference between the means (the null hypothesis) is true.

- If the hypothesis is not true (if there is a difference between the means), the mean square for model in the numerator of the F ratio has some effect in it besides the error variance. This numerator produces a large and significant F if you have enough data.
- When there is only one comparison, the F test is equivalent to the t test; in fact the F statistic is the square of the t statistic. This is true despite the fact that the t statistic is usually derived from the distribution of the estimates, whereas the F test is thought of in terms of the comparison of variances of residuals from two different models.

Comparing and Testing Means

The table called DRUG.JMP (**Figure 7.2**) contains the results of a study that measured the response of 30 subjects to treatment by one of three drugs (Snedecor and Cochran, 1967).

§ To begin, open DRUG.JMP. The three drug types are called “a”, “d”, and “placebo.” The **LBS** column is the response measurement. The **LBI** column is used in a more complex model, covered in Chapter 12, “Fitting Models.”

§ For a quick look at the data, choose **Analyze→Distribution of Y** and select **Drug** and **LBS** as **Y** variables.

Figure 7.2 Partial Listing of the Drug Table

DRUG.JMP

30 Rows / 3 Cols	Drug (N)	LBI (C)	LBS (C)
8	a	6	1
9	a	11	8
10	a	3	0
11	d	6	0
12	d	6	2
13	d	7	3

Note in the left-hand histogram in **Figure 7.3** that the number of observations is the same in each of the three drug groups; that is what is meant by a *balanced design*.

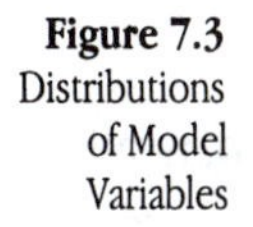

Figure 7.3 Distributions of Model Variables

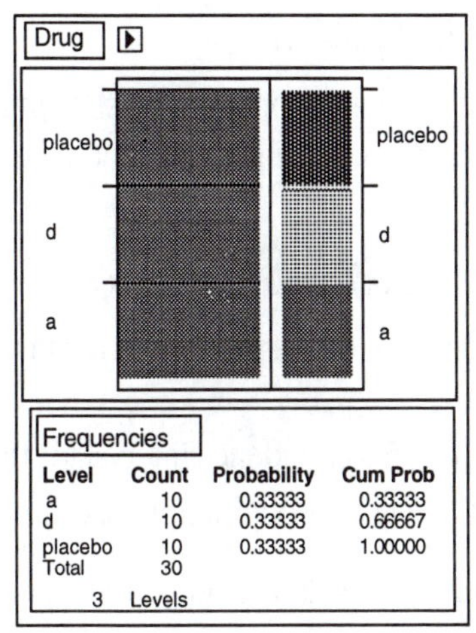

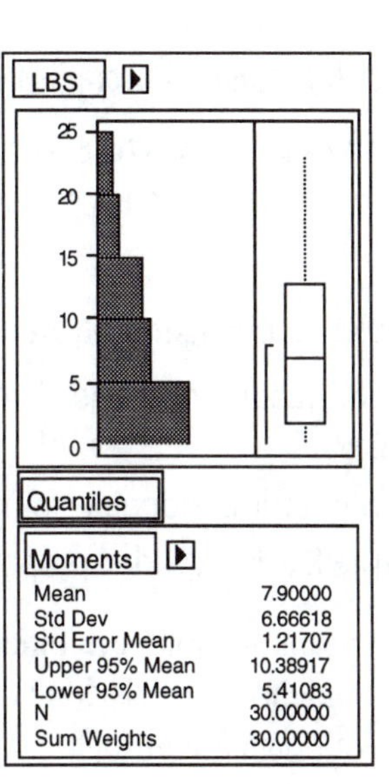

§ Next, choose **Analyze→Fit Y by X** with Drug as X and LBS as the Y variable. Notice that the role-prompting dialog displays the message that you are requesting a *Grouped, Means, One-way Anova* analysis. When you Click **OK**, the results window in **Figure 7.4** appears. The initial plot on the left shows the distribution of the response in each drug group. The dotted line across the middle is at the grand mean.

Figure 7.4 Distributions of Drug Groups

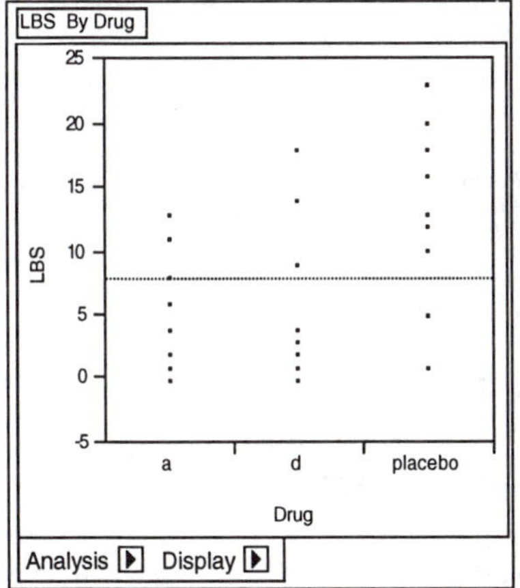

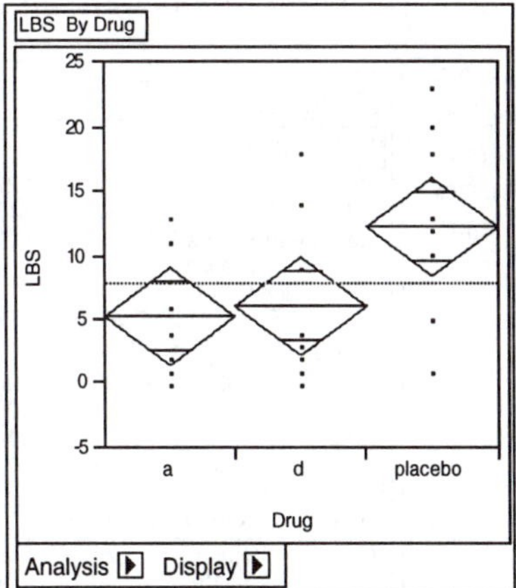

Means Diamonds: A Graphical Description of Group Means

§ Select **Means, Anova/t-Test** from the **Analysis** popup menu showing beneath the scatterplot. This adds means diamonds to the plot and also adds a set of text reports.

The plot on the right in **Figure 7.4** shows *Means Diamonds:*

- The middle line in the diamond is the response group mean for the group.
- The vertical endpoints form the 95% confidence interval for the mean.
- The X axis is divided proportionally by group sample size.

If the means are not much different, they will be close to the grand mean. If the confidence intervals don't overlap, this is enough to confirm that the means are significantly different.

Statistical Tests to Compare Means

The **Means, Anova/t-Test** command produces a report composed of the three tables shown in **Figure 7.5**:

- Summary of Fit, an overall summary of the model fit
- Analysis of Variance, which gives the F test on the means
- Means for Oneway Anova showing the means themselves.

Figure 7.5 One-Way Analysis of Variance Report

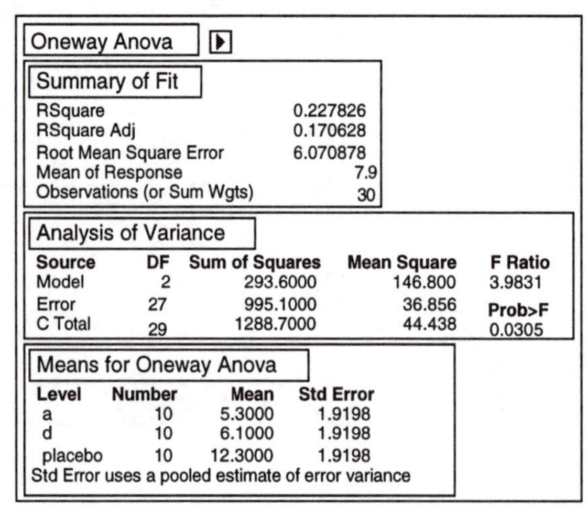

Oneway Anova

Summary of Fit

RSquare	0.227826
RSquare Adj	0.170628
Root Mean Square Error	6.070878
Mean of Response	7.9
Observations (or Sum Wgts)	30

Analysis of Variance

Source	DF	Sum of Squares	Mean Square	F Ratio
Model	2	293.6000	146.800	3.9831
Error	27	995.1000	36.856	Prob>F
C Total	29	1288.7000	44.438	0.0305

Means for Oneway Anova

Level	Number	Mean	Std Error
a	10	5.3000	1.9198
d	10	6.1000	1.9198
placebo	10	12.3000	1.9198

Std Error uses a pooled estimate of error variance

The Summary of Fit and the Analysis of Variance tables may look like a hodgepodge of numbers, but they are all derived by a few simple accounting rules. **Figure 7.6** illustrates how the statistics relate.

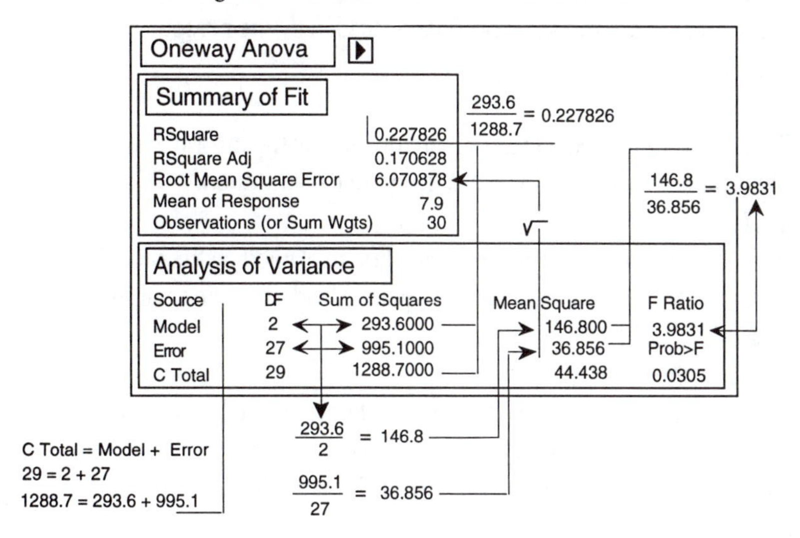

The Analysis of Variance table (**Figure 7.5** and **7.6**) describes three source components:

C Total

The **C Total Sum of Squares** (SS) is the residual SS around the grand mean. **C Total** stands for *corrected total* because it is corrected for the mean. The **C Total** degrees of freedom is the total number of observations in the sample minus 1.

Error

After you fit the group means, the remaining variation is described in the **Error** line. The **Sum of Squares** is the sum of squared residuals from the individual means. The remaining unexplained variation is **C Total** minus **Model** and is called **Error** for both the sum of squares and the degrees of freedom. The **Error Mean Square** estimates of the variance.

Model

The **Sum of Squares** for the **Model** line is the difference of **C Total** and **Error**. It is a measure of how much the Sum of Squares residuals has been reduced because of fitting the model rather than fitting only the grand mean. The degrees of freedom in the drug example is the number of parameters in the model, which is the number of groups (3) minus 1.

Everything else in the Analysis of Variance table and the Summary of Fit table is derived from these quantities:

Mean Square

Mean Squares are the sums of squares divided by their respective degrees of freedom.

F ratio

The F ratio is the model mean square divided by the error mean square. The p value for this F ratio comes from the F distribution.

RSquare

The RSquare is the proportion of variation explained by the model. In other words it is the model sum of squares divided by the total sum of squares.

Adjusted RSquare

The Adjusted RSquare is more comparable over models with different numbers of parameters (degrees of freedom). It is the error mean square divided by the total mean square, then subtracted from 1.

Root Mean Square Error

The Root Mean Square Error is the square root of the Mean Square for Error in the Analysis of Variance table. It estimates the standard deviation of the error.

So what's the verdict for the drugs? The F value of 3.98 is significant with a p value of .03, which confirms that there is a significant difference in the means. The F test does not give any specifics about which means are different, only that there is at least one pair of means that is statistically different.

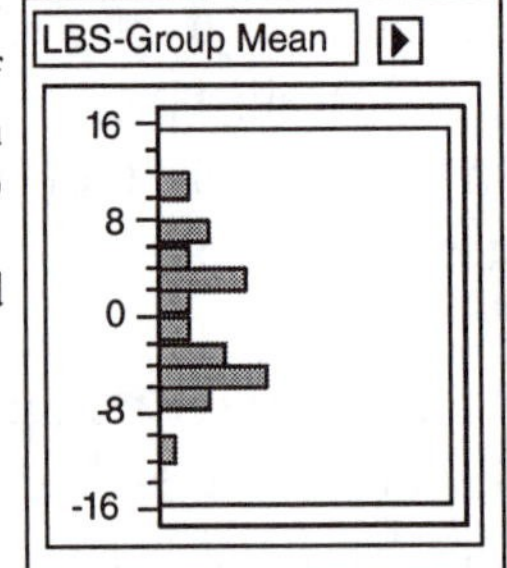

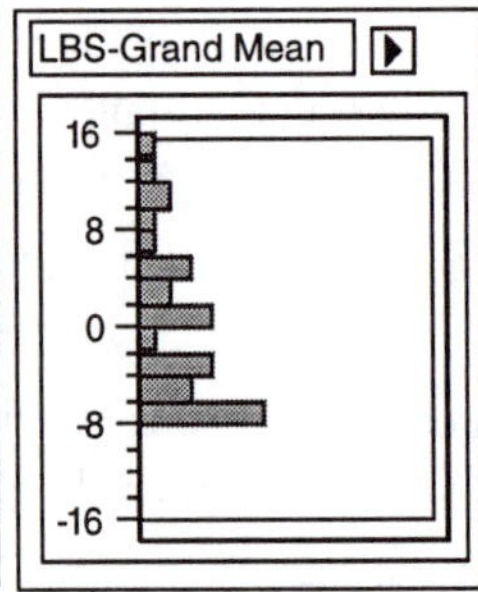

Figure 7.7 Histograms of Residuals from the Group Means (left) and the Grand Mean (right)

The F test shows whether the variance of residuals from the model is better than the variances of the residuals from only fitting a grand mean. In this case, the answer is yes (just barely).

Means Comparisons for Balanced Data

Which means are significantly different from which other means? It looks like the mean for the drug "placebo" separates from the other two. But, because all the confidence intervals for the means intersect, it takes further digging to see significance.

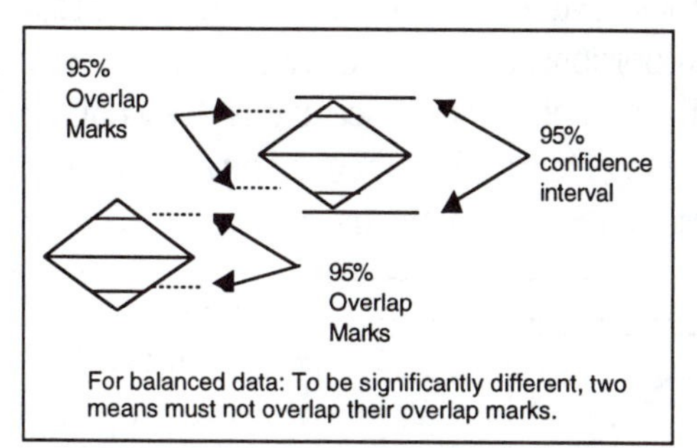

Figure 7.8 Diagram of Means Diamonds

If two means have the same number of observations behind them, then you can use the overlap marks to get a more precise graphical measure of which means are significantly different. Two means are significantly different when their overlap marks don't overlap (**Figure 7.8**). The overlap marks are placed into the confidence interval at a distance of sqrt(2)/2, which corresponds to a Student's t test of separation.

When two means do not have the same number of observations, then the design is *unbalanced* and the overlap marks no longer apply. Another technique using *comparison circles*, can be used instead. The next section describes comparison circles and shows you how to interpret them.

Special Topic: Means Comparisons for Unbalanced Data

Suppose for the sake of this example, the drug data are unbalanced. That is, there are not the same number of observations in each group. The following steps will unbalance the DRUG.JMP data in an extreme way to illustrate an apparent paradox, as well as introduce a new graphical technique.

§ Change **Drug** in row 1, rows 4 to 7, and row 9 from "a" to "placebo". Change **Drug** in rows 2 and 3 to "d". Change **LBS** in row 10 to "4." (Be careful not to save this modified table over the original copy in your sample data.)

Now drug "a" has only 2 observations, whereas "d" has 12 and "placebo" has 16. The mean for "a" will have a very high standard error because it is supported by so few observations compared with the other two levels.

Again, use the **Fit Y by X** command to look at the data:

§ Choose **Analyze→Fit Y by X** for the modified data, and as before, select the **Means, Anova/t test** option from the **Analysis** popup menu beneath the plot.

§ Select **Compare Each Pair** from the **Analysis** popup menu.

The modified data should give results like those illustrated in **Figure 7.9**. The X axis divisions are proportional to the group sample size, which causes drug "a" to be very thin because it has fewer observations. The confidence interval on its mean is large compared with the others. *Comparison circles* for Student's t tests appear to the right of the means diamonds.

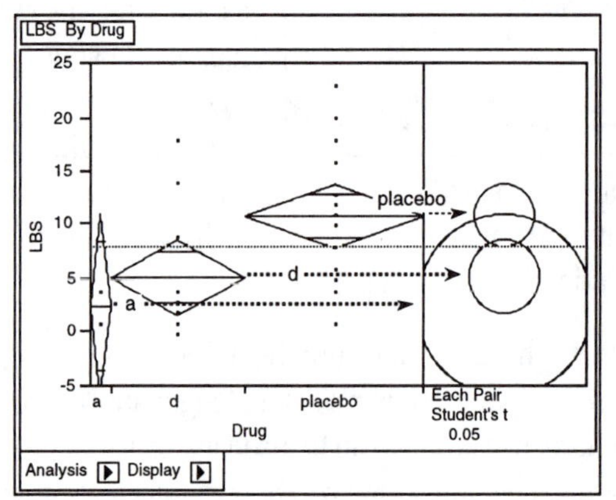

Figure 7.9 Comparison Circles to Compare Group Means

Comparison circles are a graphical technique that let you see significant separation among means in terms of how certain circles intersect. This is the only graphical technique that works in general with both equal and unequal sample sizes. The plot displays a circle for each group, with the centers lined up vertically. The center of each circle is aligned with its group mean. The radius of a circle is the 95% confidence interval for the its group mean, as you can see by comparing a circle with its corresponding means diamond. The non-overlapping confidence intervals shown by the diamonds for groups that are significantly different correspond directly to the case of non-intersecting comparison circles.

When the circles intersect, the angle of intersection is the key to seeing if the means are significantly different. If the angle of intersection is exactly a right angle, 90 degrees, then the means are on the borderline of being significantly different. This happens because then the radiuses of the circles become legs of a right triangle, of which the hypotenuse's length is the confidence limit of the difference of the two means. So the actual difference in means is the same as the confidence interval length for the difference, which is the Least Significant Difference (the LSD).

If the circles are farther apart than a right angle case, then the outside angle is more acute and the means are significantly different. If the circles are closer together, the angle is larger than a right angle, and the means are not significantly different. **Figure 7.10** illustrates these angles of intersection.

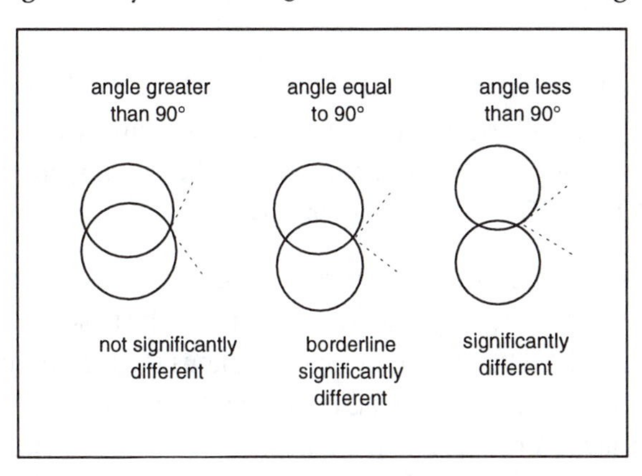

Figure 7.10 Diagram of How to Interpret Comparison Circles

So what are the conclusions for the drug example shown in **Figure 7.9**?

First, be assured that you don't need to hunt down a protractor to figure out the size of the angles of intersection. Click on a circle and see what happens—the circle highlights and becomes red. Groups that are not different from it also show in red. All groups that are significantly different remain black:

§ Click on the "placebo" circle and use the circles to compare groups means:

- The "placebo" and "d" means are represented by the smaller circles. The circles are farther separated than would occur with a right angle; the angle is acute, so these two means are significantly different.
- The circle for the "d" mean is completely nested in the circle for "a", so they are not significantly different.
- The "a" mean is much below the "d" mean, which is significantly below "placebo." So by transitivity, you expect "a" to be significantly different than "placebo." The problem with this logic is that the standard error around the "a" mean is so large that it is not significantly different from "placebo" even though it is farther away than "d." The angle of intersection is greater than a right angle, so they are not significantly different.

This complexity in relationships when the data are unbalanced is the reason a more complex graphic is needed to show relationships. The Fit Y by X platform lets you see the difference with the comparison circles and verify group differences statistically with the Means Comparisons tables as shown in **Figure 7.11**.

The Means Comparisons table uses the concept of Least Significant Difference (LSD). In the balanced case, this is the separation that any two means must have from each other to be significantly different. In the unbalanced case, there is a different LSD for each pair of means.

The bottom table in the Means Comparison report shows all the comparisons of means ordered from high to low. The elements of the table show the absolute value of the difference in two means, minus the LSD. If the means are farther apart than the LSD, then they are significantly different and the element is positive. For example the element that compares "placebo" and "d" is a positive .88, which says the means are .88 more separated than needed to be significantly different. If the means are not significantly different, then the LSD is greater than the difference, so the element in the table is negative. The elements for the other two comparisons are negative, showing no significant difference.

Figure 7.11 Statistical Text Reports to Compare Groups

Means for Oneway Anova

Level	Number	Mean	Std Error
a	2	2.5000	4.2234
d	12	5.2500	1.7242
placebo	16	10.8125	1.4932

Std Error uses a pooled estimate of error variance

Means Comparisons

Dif=Mean[i]-Mean[j]	placebo	d	a
placebo	0.00000	5.56250	8.31250
d	-5.56250	0.00000	2.75000
a	-8.31250	-2.75000	0.00000

Alpha= 0.05

Comparisons for each pair using Student's t

t 2.05181

Abs(Dif)-LSD	placebo	d	a
placebo	-4.3328	0.8826	-0.8787
d	0.8826	-5.0031	-6.6099
a	-0.8787	-6.6099	-12.2550

Positive values show pairs of means that are significantly different.

§ The last thing to do in this example is to restore your copy of the DRUG.JMP table to its original state so it can be used in other examples. To do this, choose **File→Revert** or reopen the data table.

Special Topic: Adjusting for Multiple Comparisons

Making multiple comparisons, such as comparing many pairs of means, increases the possibility of committing what is called in statistics a *Type I error*. A Type I (type one) error is the error of declaring something significant (based on statistical test results) that is not in fact significant. The more tests you do, the more likely you are to happen upon a significant difference that occurred by chance alone. If you are comparing all possible pairs of means in a large one-way layout, there are many possible tests, and a Type I error becomes very likely.

There are many methods that modify tests to control for an overall error rate. This section covers one of the most basic, the *Tukey-Kramer Honestly Significant Difference* (HSD). The Tukey-Kramer HSD uses the distribution of the maximum range among a set of random variables.

§ After reverting to the original copy of DRUG.JMP, again choose **Analyze→Fit Y by X** for the variables **LBS** as **Y Drug** as **X**.

§ Next, select three of the check-mark popup menu commands: **Means, Anova/t-Test**, **Compare Each Pair**, and **Compare All Pairs**. These commands should give you the results shown in **Figure 7.12**.

Figure 7.12 t Tests and Tukey-Kramer Adjusted t Tests for One-Way Anova

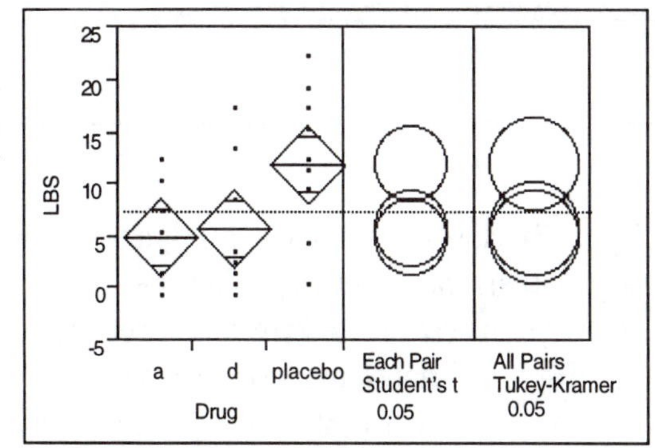

The comparison circles work as before but have different kinds of error rates.

The Tukey-Kramer comparison circles are larger than the Student's t circles. This protects more tests from falsely declaring significance, but this protection makes it harder to declare two means significantly different.

If you click on the top circle, you see that the conclusion is different between the Student's t and Tukey-Kramer's HSD for the comparison of "placebo" and "d." This comparison is significant for Student's t test but not for Tukey's test.

The difference in significance occurs because the quantile that is multiplied into the standard errors to create a Least Significant Difference has grown from 2.05 to 2.48 between Student's t test and the Tukey-Kramer test.

The only positive element in the Tukey table is the one for the "a" versus "placebo" comparison (see **Figure 7.13**).

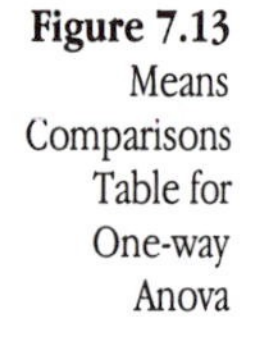

Figure 7.13 Means Comparisons Table for One-way Anova

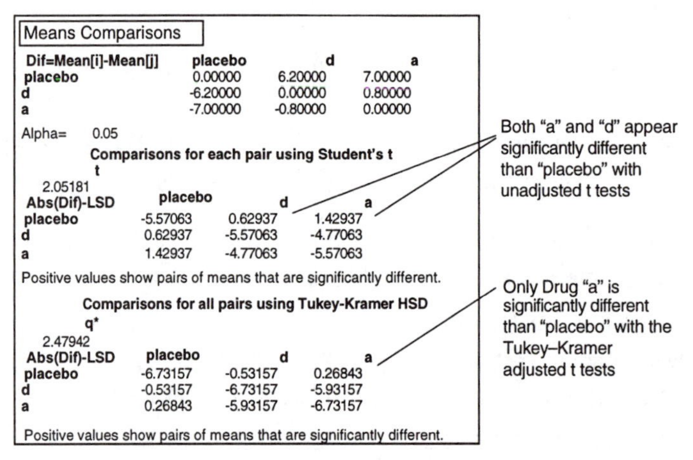

Means Comparisons

Dif=Mean[i]-Mean[j]	placebo	d	a
placebo	0.00000	6.20000	7.00000
d	-6.20000	0.00000	0.80000
a	-7.00000	-0.80000	0.00000

Alpha= 0.05

Comparisons for each pair using Student's t

t
2.05181

Abs(Dif)-LSD	placebo	d	a
placebo	-5.57063	0.62937	1.42937
d	0.62937	-5.57063	-4.77063
a	1.42937	-4.77063	-5.57063

Positive values show pairs of means that are significantly different.

Comparisons for all pairs using Tukey-Kramer HSD

q*
2.47942

Abs(Dif)-LSD	placebo	d	a
placebo	-6.73157	-0.53157	0.26843
d	-0.53157	-6.73157	-5.93157
a	0.26843	-5.93157	-6.73157

Positive values show pairs of means that are significantly different.

Both "a" and "d" appear significantly different than "placebo" with unadjusted t tests
Only Drug "a" is significantly different than "placebo" with the Tukey–Kramer adjusted t tests

Special Topic: Power

In the drug example, the F test was significant with a p value of .03. What if you wanted to be very cautious and achieve a p value of less than .01, a 1% chance of falsely declaring significance? To do this you need more data. One solution is to use the Least Significant Number (LSN) as a guide to choosing the sample size. But even with more data, it is still up to chance whether you will achieve the p value you want. *Power* is the probability of achieving a certain significance when the true means and variances are specified. You can use the power concept to help choose a sample size that is likely to give significance for certain effect sizes and variances.

Power has the following ingredients:

- the effect size—that is, the separation of the means
- the standard deviation of the error, often called *sigma*
- *Alpha*, the significance level
- the number of observations, the sample size.

Similarly, there are four ways to increase power.

Increase the Effect Size

Larger differences are easier to detect. For example, when designing an experiment to test a drug, administer as large a difference in doses as possible. Also, use balanced designs.

How do you increase the effect size? You could look for ways to determine when big differences occur. The problem with this is these bigger differences might not reflect what is going on in the process of interest. Another way to increase the observed differences is to be bold in your selection of widely separated groups. If you are studying children whose ages range from 10 to 11, the chance of finding an age-related difference is probably small. Comparing 10-year olds to 15-year olds is more likely to produce a difference. Be bold.

Decrease Residual Variance

If you have less noise it is easier to find differences. Sometimes this can be done by blocking or testing within subjects or by selecting a more homogeneous sample.

The quest of *total quality management* is to get rid of variability. Of course, this is fundamentally impossible—noise happens. One way to reduce noise is to control extraneous factors. Another way is to estimate the noise effect by including variables in the model to control for sources of nuisance variability. Either way reduces error variance. The down side of this is that it makes the study more difficult to perform and complicated to analyze. Life is not easy.

Increase the Sample Size

With larger samples the standard error of the estimate of effect size is smaller. The effect is estimated with more precision as the sample size increases. Roughly, the precision increases in proportion to the square root of the sample size.

Experimental units can be expensive, but more data are usually desirable. Another approach to increase power is to see whether it is advantageous to allocate the experimental units to a better design. Experimental design and sample size determination are the main tools to increase power. The down side is that it takes effort and expertise on your part to design good studies. Life is not simple.

Accept Less Protection

Increase alpha. There is nothing magic about alpha = .05. A larger alpha lowers the cut-off value. A statistical test with alpha = .20 declares significant differences more often (and also leads to false conclusions more often).

What if you set alpha to .5, which means that when there was no difference, you would still declare a significant difference half the time. This will reduce your experimental costs dramatically and increase your conclusiveness. Good luck in selling this to your boss or to a journal editor. You'll also need to hire a good lawyer to handle the damage claims and a psychiatrist to help you sleep at night. Protect yourself.

Power in JMP

Many JMP analyses give you the option to do a power analysis. Wherever the **Power Details** option appears in a popup menu, you can request a power analysis of the statistical test you are performing. When you select **Power Details**, the dialog shown in **Figure 7.14** appears.

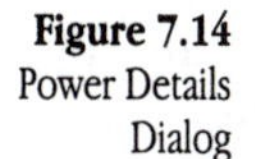

Figure 7.14 Power Details Dialog

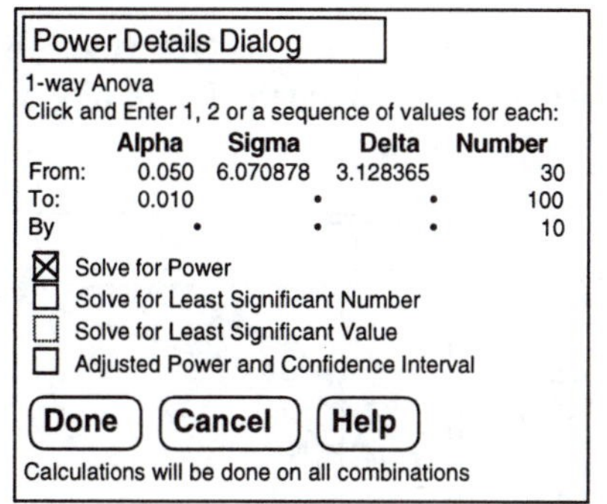

In the Power Details dialog, you set up the hypothetical scenario, specifying what you think might be true values for the effect size and the residual variance. Then the analysis tells you how likely you will detect the effect at different sample sizes and alpha levels.

You can try out this powerful tool on the analysis of variance for the **Drug** data covered in the previous sections.

§ Select the **Power Details** command from the popup menu next to the Oneway Anova table title (**Figure 7.5**) to see the Power Details dialog.

You can see the power ingredients in the dialog. Complete the editable power fields as shown in **Figure 7.14**:

Alpha

defaults to .05. Click an empty alpha field to type a second value of .01.

Sigma

shows as the Root Mean Square Error from the Summary table in the Analysis of Variance report (see **Figure 7.5**) because it estimates the standard deviation of the error. You can edit Sigma if you want to find the power for other error variances.

Delta

is a general estimate of the effect size. It is the sum of squares for the hypothesis being tested divided by the total sample size. Note that some books use a standardized effect size estimate rather than a raw effect size estimate. JMP keeps the effect size estimate separate from the error estimate so that they can be looked at separately.

Number

is the total number of observations in the table. Type a range of sample sizes from 30 to 100 in increments of 10.

§ Check the Solve for Power box in the dialog and click **Done**. Power calculations are then completed for each combination of values you entered in the dialog. The table of power calculations shown in **Figure 7.15** is appended to the analysis tables.

§ You can embellish the analysis with a plot. Select **Plot Power by N** found in the popup menu next to the Power table title to see the Power plot, which is the plot of power by sample size in **Figure 7.15**.

Figure 7.15 Power Analysis Results

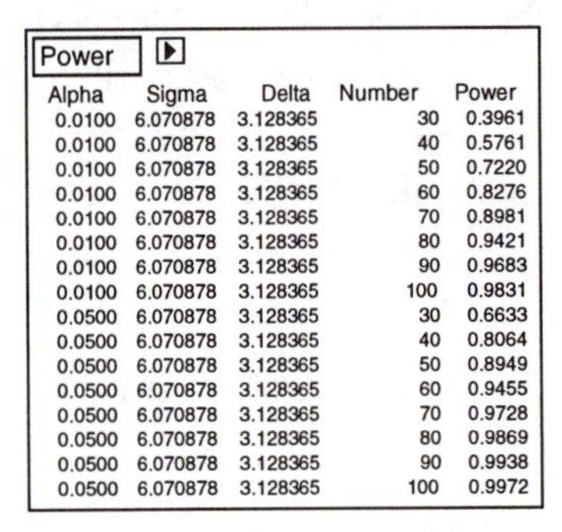

Power

Alpha	Sigma	Delta	Number	Power
0.0100	6.070878	3.128365	30	0.3961
0.0100	6.070878	3.128365	40	0.5761
0.0100	6.070878	3.128365	50	0.7220
0.0100	6.070878	3.128365	60	0.8276
0.0100	6.070878	3.128365	70	0.8981
0.0100	6.070878	3.128365	80	0.9421
0.0100	6.070878	3.128365	90	0.9683
0.0100	6.070878	3.128365	100	0.9831
0.0500	6.070878	3.128365	30	0.6633
0.0500	6.070878	3.128365	40	0.8064
0.0500	6.070878	3.128365	50	0.8949
0.0500	6.070878	3.128365	60	0.9455
0.0500	6.070878	3.128365	70	0.9728
0.0500	6.070878	3.128365	80	0.9869
0.0500	6.070878	3.128365	90	0.9938
0.0500	6.070878	3.128365	100	0.9972

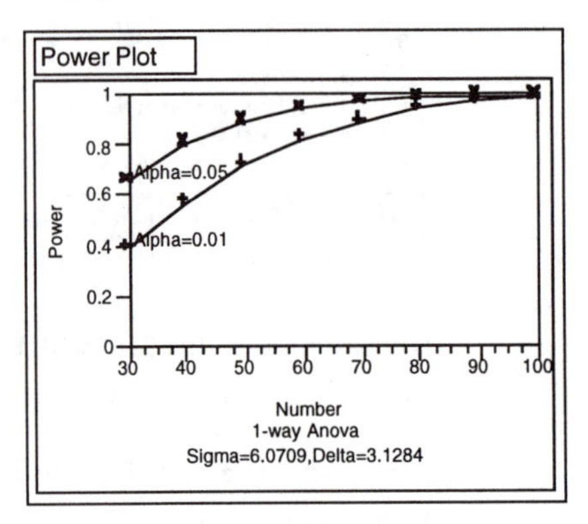

To read the table, pick values you would like to have and note the corresponding information. For example, if you want 90% probability (power) of achieving a significance of .01, then the sample size needs to be slightly above 70. For the same power at .05 level significance, the sample size only needs to be 50.

Special Topic: Unequal Variances

Analysis of Variance usually assumes the variance is the same for all samples (groups). If this is not the case, you want to know about it. To check the variance assumption:

§ Return to the Fit Y by X scatterplot of the drug groups and select **Normal Quantile Plot** in the **Analysis** popup menu showing beneath the plot. This option displays a plot next to the Means Diamonds as shown in **Figure 7.16**. The Normal Quantile plot compares mean, variance, and shape of the group distributions.

There is a line on the Normal Quantile plot for each group. The height of the line shows the location of the distribution. The slope of the line shows the standard deviation. The straightness of the line shows how close the shape of the distribution is to the Normal distribution. Note that the "placebo" group is both higher and has a greater slope, which indicates a higher mean and a higher variance, respectively.

Figure 7.16 Normal Quantiles Plot

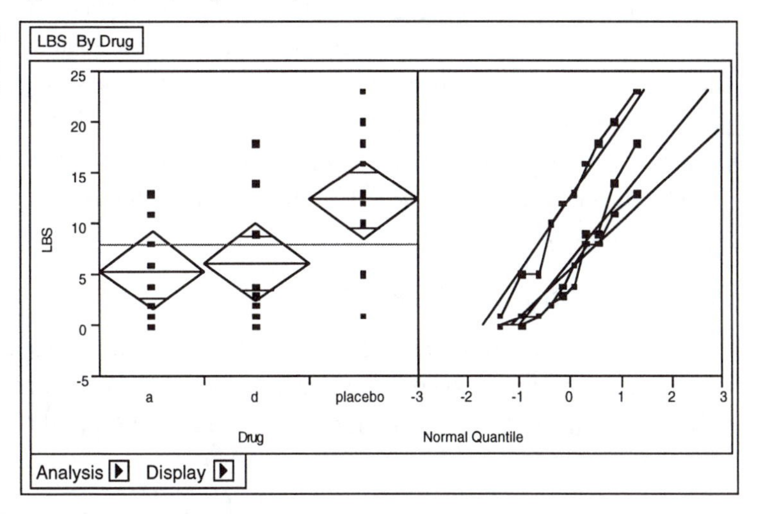

You can also check for unequal variances statistically:

§ Select **Unequal Variances** in the **Analsis** popup menu to perform homogeneity of variance statistical tests and a Welch Anova (**Figure 7.17**).

If the variances test as significantly unequal, you can use the results of the Welch Anova in the report, which is a one-way Anova F test in which the observations are weighted by the reciprocals of the estimated variances.

Figure 7.17 Tests for Unequal Variances and Welch Anova

Tests that the Variances are Equal

Level	Count	Std Dev	MeanAbsDif to Mean	MeanAbsDif to Median
a	10	4.643993	3.900000	3.900000
d	10	6.154492	5.120000	4.700000
placebo	10	7.149981	5.700000	5.700000

Test	F Ratio	DF Num	DF Den	Prob>F
O'Brien[.5]	1.1395	2	27	0.3349
Brown-Forsythe	0.5998	2	27	0.5561
Levene	0.8904	2	27	0.4222
Bartlett	0.7774	2	27	0.4596

Welch Anova testing Means Equal, allowing Std's Not Equal

F Ratio	DF Num	DF Den	Prob>F
3.3942	2	17.406	0.0569

If unequal variances are of concern, you can consider a nonparametric approach. Or, sometimes a suitable transformation of the response variable, such as the square root or the log, equalizes the variances.

Special Topic: Nonparametric Methods

JMP also offers nonparametric methods in the Fit Y by X platform in the Analysis popup menu. Nonparametric methods, introduced in the previous chapter, use only the rank order of the data and ignore the spacing information between data points. The nonparametric tests do not assume the data have a normal distribution. This section first reviews the rank-based methods, then generalizes the Wilcoxon rank sum method to the k groups of the one-way layout.

Review of Rank-Based Nonparametric Methods

The nonparametric tests are useful to test whether means or medians are the same across groups. However, the usual assumption of normality is not made. Nonparametric tests use functions of the response ranks, called rank scores (Hajek 1969).

JMP offers the following nonparametric tests for testing whether distributions across factor levels are centered at the same location. Each is the most powerful rank test for a certain distribution, as indicated in **Table 7.1**.

- Wilcoxon rank scores are the ranks of the data.
- Median rank scores are either 1 or 0 depending on whether a rank is above or below the median rank.
- Van der Waerden rank scores are the quantiles of the standard normal distribution for the probability argument formed by the rank divided by n–1. This is the same score that is used in the normal quantile plots.

Table 7.1 Guide for Using Nonparametric Tests

Fit Y by X Analysis Option	Two levels	Two or more levels	Most powerful for errors distributed as:
Nonpar-Wilcoxon	Wilcoxon rank-sum (Mann-Whitney U)	Kruskal-Wallis	Logistic
Nonpar-Median	Two sample Median	k-sample median (Brown-Mood)	Double Exponential
Nonpar-VW	Van der Waerden	k-sample Van der Waerden	Normal

The Three Rank Tests in JMP

In the DRUG.JMP example, you first check the **LBS** response for normality and then request nonparametric tests to compare the **Drug** group means of **LBS**:

§ To check the normality, choose **Analyze→Distribution of Y** to look at the distribution of **LBS**. The distribution does not look like the normal distribution.

§ Select the **Test Dist=Normal** option from the popup menu above the histogram. The p value of .0282 from the resulting Shapiro-Wilk W statistic confirms that the distribution is not normal.

Figure 7.18 Histogram and Test for Normality

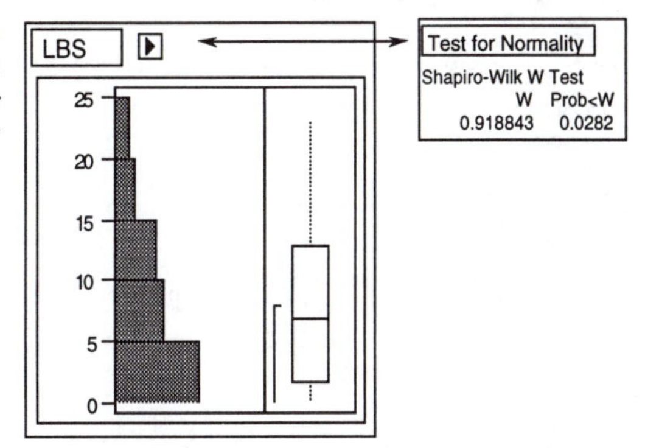

Test for Normality	
Shapiro-Wilk W Test	
W	Prob<W
0.918843	0.0282

§ Choose **Analyze→Fit Y by X** with **LBS** as **Y** and **Drug** as **X**. The one-way analysis of variance platform appears showing the distributions of the three groups, as seen previously in **Figure 7.4**.

§ In the popup menu beneath the scatterplot, select the four tests that compare groups:

- **Means, Anova /t-test**, producing the F test from the standard parametric approach
- **Nonpar-Wilcoxon**, which is known as the Kruskal-Wallis test when there are more than two groups
- **Nonpar-Median** for the median test
- **Nonpar-VW** for the Van der Waerden test.

Figure 7.19 shows the results of the four tests that compare groups. In this example the Wilcoxon and the Van der Waerden agree with the parametric F test in the Anova and show borderline significance for a .05 alpha-level. despite the lack of normality in a fairly small sample.

The median test is much less powerful than the others and doesn't detect a difference in this example.

Figure 7.19 Non-Parametric Tests for Drug Example

Analysis of Variance

Source	DF	Sum of Squares	Mean Square	F Ratio
Model	2	293.6000	146.800	3.9831
Error	27	995.1000	36.856	Prob>F
C Total	29	1288.7000	44.438	0.0305

Wilcoxon / Kruskal-Wallis Tests (Rank Sums)

Level	Count	Score Sum	Score Mean	(Mean-Mean0)/Std0
a	10	122	12.2000	-1.433
d	10	132.5	13.2500	-0.970
placebo	10	210.5	21.0500	2.425

1-way Test, Chi-Square Approximation

ChiSquare	DF	Prob>ChiSq
6.0612	2	0.0483

Median Test (Number of Points Above Median)

Level	Count	Score Sum	Score Mean	(Mean-Mean0)/Std0
a	10	4	0.400000	-0.762
d	10	4	0.400000	-0.762
placebo	10	7	0.700000	1.523

1-way Test, Chi-Square Approximation

ChiSquare	DF	Prob>ChiSq
2.3200	2	0.3135

Van der Waerden Test (Normal Quantiles)

Level	Count	Score Sum	Score Mean	(Mean-Mean0)/Std0
a	10	-3.692928	-0.36929	-1.568
d	10	-2.244516	-0.22445	-0.953
placebo	10	5.937444	0.593744	2.521

1-way Test, Chi-Square Approximation

ChiSquare	DF	Prob>ChiSq
6.4804	2	0.0392

Chapter 8
Fitting Curves through Points: Regression

Overview

Regression is a method of fitting curves through data points. Why is it called *regression*?

Sir Francis Galton, in his 1885 Presidential address before the anthropology section of the British Association for the Advancement of Science (Stigler, 1986), described a study he had made of how tall children are compared with the heights of their parents. Galton defined ranges of parents' heights and calculated the mean child height for each range. Then he drew a straight line that went through the means as best as he could. He thought he had made a discovery when he found that the child heights tended to be more moderate than the parent heights. For example, if a parent was very tall, the children tended to be tall, but shorter than the parents. If a parent was very short, the child would tend to be short, but taller than the parent. This discovery he called a *regression* to the mean, with the *regression* defined as *to come back to*.

But somehow the term *regression* became associated with the technique of fitting the line, rather than with the line itself, which was showing something about inheritance. The term has stuck for 110 years, so it's probably going to stay. At least the consequences of Galton giving a misleading term to the statistics profession was milder than the consequences from the idea he gave to anthropology (eugenics). Galton's data are covered later in this chapter.

This chapter covers the case where there is only one factor—the kind of regression situation that you can see on the scatterplot graph.

You can mouse along with the examples whenever you see an action symbol (§)

Chapter 8 Contents

Regression

Fitting one mean is easy. Fitting several means is not much harder. How do you fit a mean when it changes as a function of some other variable? You are fitting a line or a curve through the data.

Least Squares

In regression, you pick an equation and leave some of its parameters (coefficients) to be determined by the fitting mechanism. The parameters are determined by the method of least squares, which finds the parameter values that minimize the sum of squared distances from each point to the line of fit. **Figure 8.1** illustrates a least squares regression line.

Figure 8.1 Straight-Line Least-Squares Regression

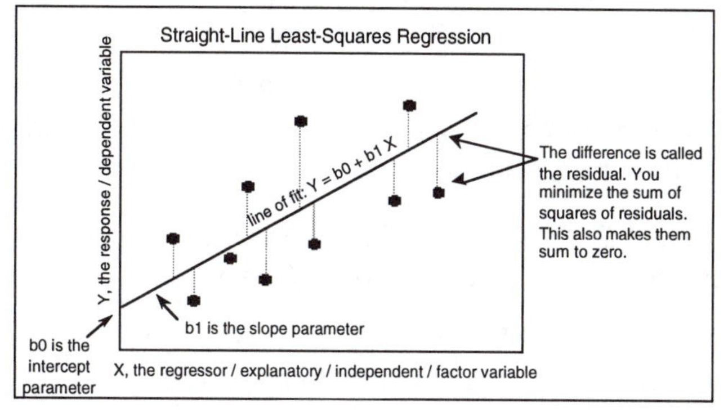

For any regression model, the term *residual* denotes the difference between the actual response value and the value predicted by the line of fit. When talking about the true (unknown) model rather than the estimated one, these differences are called the errors or disturbances, instead of the residuals.

Least squares regression is the method of fitting of a model to minimize the sum of squared residuals.

The regression line has interesting balancing properties with regard to the residuals. The sum of the residuals is always zero, which was also true for the simple mean fit. You can think of the fitted line as balancing data in the up-and-down direction. If you add the product of the residuals times the X (regressor) values, this sum will also be zero. This can be interpreted as the line balancing the data in a rotational sense. Chapter 17, "Machines of Fit," shows how these least squares properties can be visualized in terms of the forces of data acting like springs on the line of fit.

An important special case is when the line of fit is constrained to be horizontal (flat). The equation for this fit is a constant; if you constrain the slope of the line to be zero, the coefficient of the X term (regressor) is zero, and the X term drops out of the model. In this situation, the estimate for the constant is the sample mean. This special case is important because it leads to the statistical test of whether the regressor really affects the response.

Fitting the Line and Testing the Slope

Eppright *et al*. (1972) as reported in Eubank (1988) measured 72 children from birth to 70 months. You can use regression techniques to examine how the weight to height ratio changes as kids grow up:

§ Open the GROWTH.JMP sample table to see the Eppright data.
Choose **Analyze→Fit Y by X** and select ratio as Y and age as X. When you click **OK** the result is a scatterplot of ratio by age.

Look for the small icon at the lower left of the plot. Click this Fitting icon to see commands for adding regression fits to the scatterplot.

§ Select **Fit Mean** and then **Fit Line** from the Fitting popup menu.

- **Fit Mean** draws the horizontal line at the mean of ratio.
- **Fit Line** draws the regression line through the data.

These commands also add statistical tables to the regression report. You should see a scatterplot much like the one shown on the left in **Figure 8.2**. The statistical tables (on the right in **Figure 8.2**) are displayed beneath the scatterplot in the report window.

Each kind of fit you select has its own popup menu icon, next to the Fitting popup icon, which lets you request fitting details.

§ Click the popup menu for **Linear Fit**, and select **Confid Curves: Fit**. This command adds dashed lines around the regression line. The parameter estimates report shows the estimated coefficients of the regression line. The fitted regression is the equation:

$$\text{ratio} = .6656 + .005276\ \text{age} + \text{residual}$$

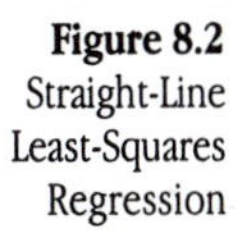

Figure 8.2 Straight-Line Least-Squares Regression

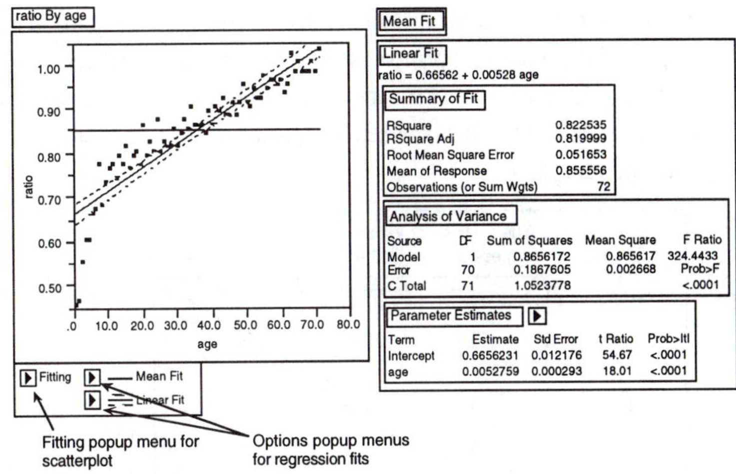

Testing the Slope by Comparing Models

Does the regressor really affect the response? Does the ratio of weight to height change as a function of age? Is the true slope of the regression line zero? Is the true value for the coefficient of age zero? Is the sloped regression line significantly different than the horizontal line at the mean? These are all the same question, assuming that the linear equation is adequate to describe the relationship of the weight to height ratio with growth.

Chapter 6, "The Differences Between Two Means," presented two analysis approaches that turned out to be equivalent. One approach used the distribution of the estimates, which resulted in the t test. The other approach compared the sum of squared residuals from two models where one model was a special case of the other. This model comparison approach resulted in an F test. In regression, there are the same two equivalent approaches: (1) distribution of estimates, and (2) model comparison.

The model comparison is between the regression line and what the line would be if the slope were constrained to be zero; that is, you compare the fitted regression line with the horizontal line at the mean (the mean fit). If the regression line is a better fit than the horizontal fit, then the slope of the regression line will test as significantly different from zero. This is often stated negatively: "If the regression line doesn't fit better than the horizontal fit, then the slope of the regression line will not test as significantly different from zero."

The F test in the Analysis of Variance table is the model comparison that tests the slope of the fitted line. It compares the sum of squared residuals from the regression fit to the sum of squared residuals from the sample mean. **Figure 8.3** diagrams the relationship between the quantities in the statistical reports and corresponding plot.

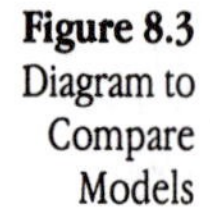

Figure 8.3 Diagram to Compare Models

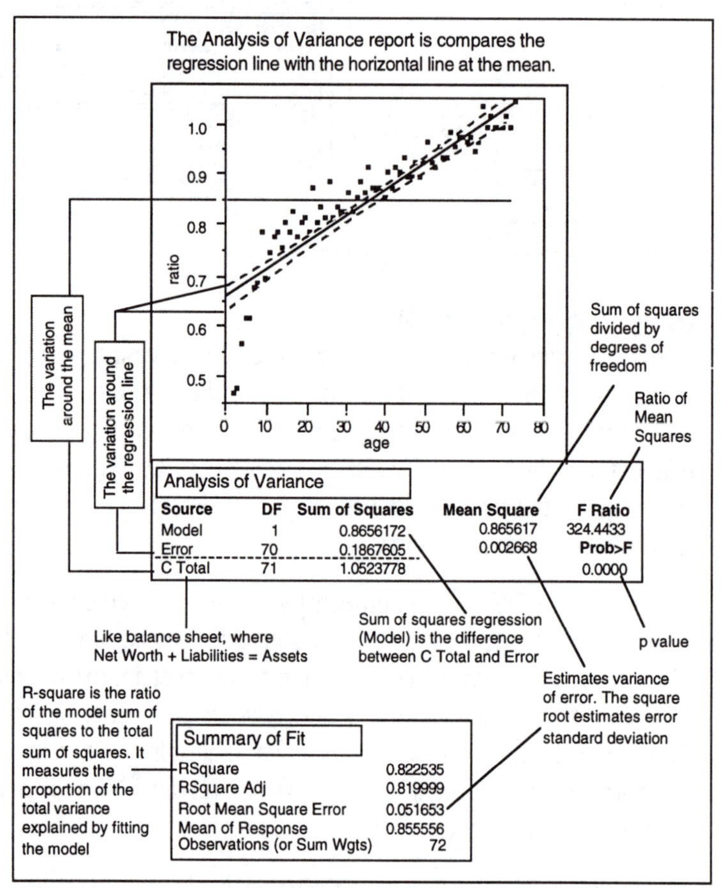

Here are descriptions of the quantities in the statistical tables:

C Total

corresponds to the sum of squares error if you only fit the mean. You can verify this by looking at the Fit Mean table from in the previous example. The C Total sum of squares (SS) is 1.0523 for both the mean fit and the line fit.

Error
: is the sum of squared residuals after fitting the line, .1867. This is sometimes casually referred to as *the residual,* or *residual error*. You can think of **Error** as left over variation—variation that didn't get explained by fitting a model.

Model
: is the difference between the sum of squares error in the two models (the mean-line model and the sloped regression line). It is the sum of squares resulting from the regression, .8656. You can think of **Model** as a measure of the variation in the data that was explained by fitting a regression line.

Mean Square
: is a sum of squares divided by its respective degrees of freedom. The Mean Square for Error is the estimate of the error variance (.002667).

Root Mean Square Error
: is found in the Summary of Fit report. It is the square root of the Mean Square for Error. It estimates the error standard deviation of the error.

If the true regression line has slope zero at the mean, then the model mean square and the error mean square both estimate the residual error variance and have the same expected value. Thus if the mean square for regression is larger than the mean square for error, you suspect that the slope is not zero. The ratio of the mean square for regression to mean square for error is the F Ratio, which has an F distribution under the hypothesis that (in this example) **age** has no effect on **ratio**.

The Distribution of the Parameter Estimates

The formula for a simple straight line only has 2 parameters (the intercept and the slope). For this example the model can be written as follows:

$$\text{ratio} = b_0 + b_1\,\text{age} + \text{residual}$$

where b_0 is the intercept and b_1 is the slope.

The Parameter Estimates Table (**Figure 8.4**) has these quantities:

Std Error
: is the estimate of the standard deviation attributed to the parameter estimates.

t-Ratio
: is a test that the true parameter is zero. The t ratio is the ratio of the estimate to its standard error. Generally, you are looking for t ratios that are greater than 2 in absolute value, which usually correspond to significance probabilities of less than .05.

Prob>|t|

is the significance probability or (p value)—you can translate this as "the probability of getting an even greater absolute t value by chance alone if the true value of the parameter is zero."

Figure 8.4 Parameter Estimates Table

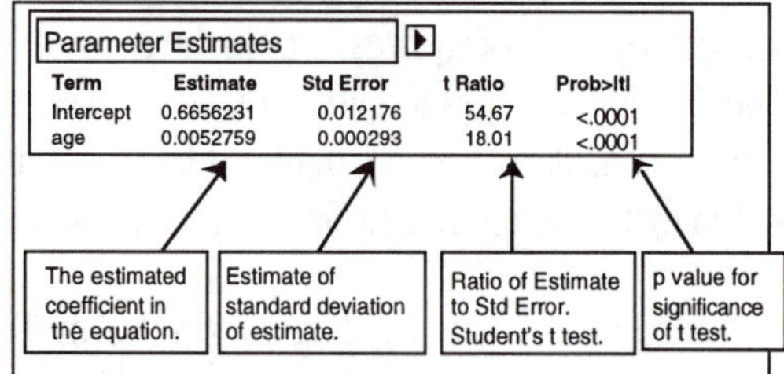

Term	Estimate	Std Error	t Ratio	Prob>|t|
Intercept	0.6656231	0.012176	54.67	<.0001
age	0.0052759	0.000293	18.01	<.0001

Note that the t ratio for the **age** parameter, 18.01, is the square root of the F Ratio in the Analysis of Variance table, 324.44. You can double-click on the p values in the tables to show them with more decimal places, and see that the p values are exactly the same. The t test for simple regression is testing the same hypothesis as the F test.

Confidence Intervals on the Estimates

There are several ways to look at the significance of the estimates. The t tests for the parameter estimates, discussed previously, test that the parameters are significantly different from zero. A more revealing way to look at the estimates is to obtain confidence limits that show the range of likely values for the true parameter values, given the estimates and their distribution.

§ In the Parameter Estimates table (**Figure 8.4**), select the **95% Confid Limits** command from the popup menu next to the table title. This command adds the confidence limits columns to the report, as shown in **Figure 8.5**.

Figure 8.5 Parameter Estimates Table

Parameter Estimates

Term	Estimate	Std Error	t Ratio	Prob>|t|	Lower 95%	Upper 95%
Intercept	0.6656231	0.012176	54.67	<.0001	0.6413397	0.6899065
age	0.0052759	0.000293	18.01	<.0001	0.0046917	0.0058601

The 95% confidence limits form the smallest interval whose range includes the true parameter values with 95% confidence. The upper and lower confidence limits are calculated by adding and subtracting respectively the standard error of the parameter times a quantile value corresponding to a (.05)/2 Student's t test.

Another way to look at this is from the point of view of the sum of squared errors. Imagine the sum of squared errors (SSE) as a function of the parameter values, so that as you vary the parameter values you can calculate the SSE at each set of parameter values. The least squares estimates are where this surface is at a minimum.

The left plot in **Figure 8.6** shows a three-dimensional view of the SSE surface for the growth data regression problem. The 3-D view shows the curvature of the SSE surface as a function of the parameters, with a minimum in the center. The X and Y axes are a grid of parameter values and the Z axis is the computed SSE for those values.

The contour plot on the right in **Figure 8.6** shows the elliptical contours corresponding to given SSE values as a function of the parameters. The least squares values for the parameters are the center of the ellipse.

Figure 8.6
Representations of Confidence Limit Regions

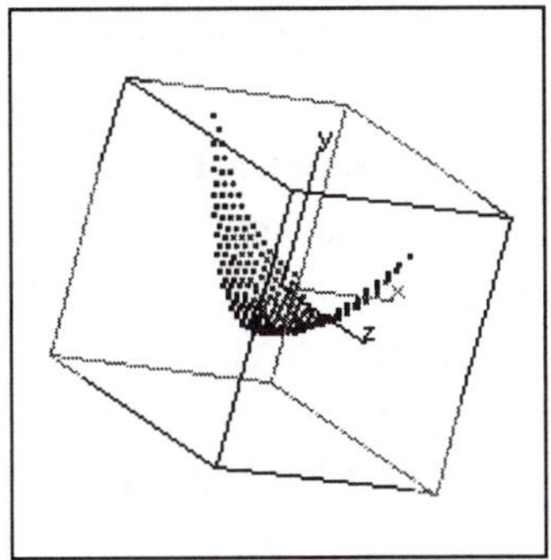

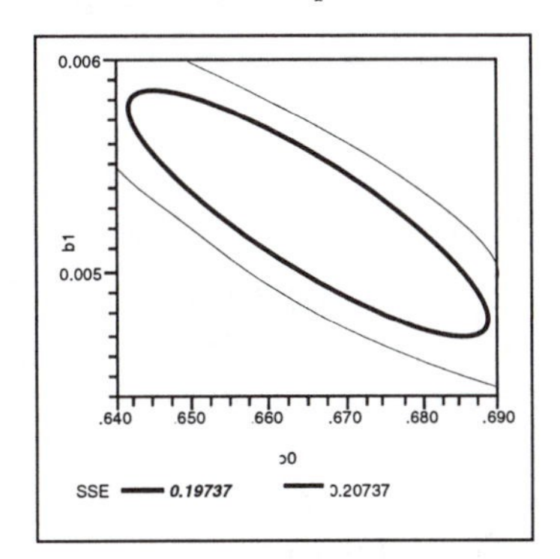

One way to form a 95% confidence interval is to turn the F test upside down. You take an F value that would be the criterion for a .05 test (3.97), multiply it by the MSE, and add that to the SSE. This gives a higher SSE of .19737 and forms a confidence region for the parameters. Anything that produces a smaller SSE is believable because it corresponds to an F test with a p value greater than .05.

The 95% confidence region is the inside elliptical shape in the plot on the right in **Figure 8.6**. The ellipse is diagonally flattened to the extent that the estimates are correlated. You can look at the plot to see what parameter values correspond to the extremes of the ellipse in each direction:

- The horizontal scale corresponds to the intercept parameter. The confidence limits are the positions of the vertical tangents to the inner contour line, indicating a low. point of .6413 and high point of .6899.
- The vertical scale corresponds to the slope parameter for **age.** The confidence limits are the positions of the vertical tangents to the inner contour line, indicating a low of .00469 and high point of .00586. These are the lower and upper 95% confidence limits for the parameters.

Examine Residuals

It is always a good idea to take a close look at the residuals from a regression (the difference between the actual values and the predicted values):

§ Select **Plot Residuals** from the **Linear Fit** popup menu beneath the scatterplot (see **Figure 8.2**). This command appends the residual plot shown in **Figure 8.7** to the bottom of the regression report.

Figure 8.7
Scatterplot to Look at Residuals

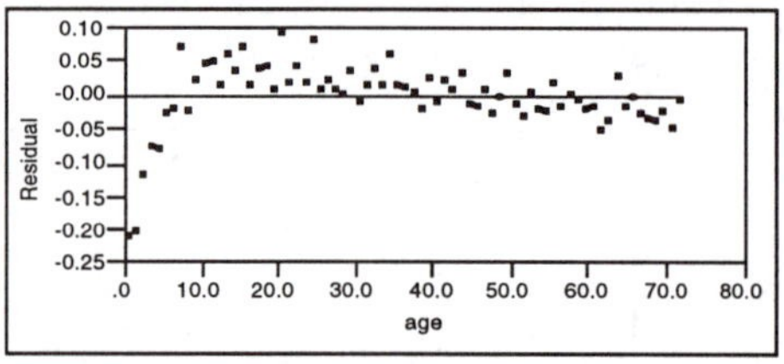

The picture you usually hope to see is the residuals scattered randomly about a mean of zero. So in residual plots like the one shown in **Figure 8.7**, you are looking for patterns and for points that violate this random scatter. This plot is suspicious because the left side has a pattern of residuals below the line. These points influence the slope of the regression line (**Figure 8.2**), pulling it down on the left. You can see what the regression would look without these points by excluding them from the analysis.

Exclusion of Rows

To exclude points (rows) from an analysis, you highlight the rows and assign them the *Exclude* row state characteristic as follows:

§ Get the Brush tool from the **Tools** menu. SHIFT-drag the brush (which changes into a rectangle) to highlight the points at the lower left of the plot.

§ Choose **Rows→Exclude/Include.** You then see a "do not use" sign on those rows in the data table window.

§ Click the original scatterplot window and select **Remove Fit** from the **Mean Fit** popup menu to clean up the plot.

§ Again select **Fit Line** from the **Fitting** popup menu to overlay a regression line with the low-age points excluded from the analysis.

The plot in **Figure 8.8** shows the two regression lines. Note that the new line of fit appears to go through the bulk of the points better, ignoring the points at the left that have been excluded.

§ To see the residuals plot for the new regression line, select **Save Residuals** from the second **Fit Line** popup (the options popup for this new fit). Click the data table to activate it and plot Residuals ratio2 by age with the **Fit Y by X** command. Note in **Figure 8.8** that the residuals no longer have a pattern to them.

Figure 8.8 Regression with Extreme Points Excluded

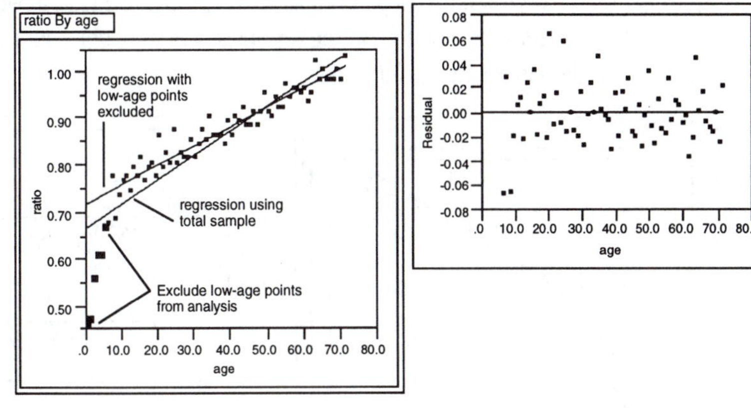

Time to Clean Up

This scatterplot is needed for the next example, so let's clean it up:

§ Choose **Rows→Select Excluded** then **Rows→Exclude/Include**. This removes the Excluded row state status from the points so that you can use them in the next steps.

§ To finish the current example, select **Remove Fit** from the second Fit Line popup menu, to remove the example regression that excluded outlying points.

Polynomial Models

Rather than excluding some points, let's try fitting a curved line through all the points. The simplest curved line is the quadratic curve (a parabola). Its equation contains a term for the squared value of the regressor, age:

$$\text{ratio} = b_0 + b_1\,\text{age} + b_2\,\text{age}^2 + \text{residual}$$

To fit a curved line to ratio by age:

§ Select the **Fit Polynomial** command on the platform Fitting popup, with **2-quadratic** from its hierarchical submenu.

The left plot in **Figure 8.9** shows the best fitting polynomial of degree 2 and line fit overlaid on the scatterplot. You can visually compare the straight line and curved line and compare them statistically with the Analysis of Variance reports for both fits that show beneath the plot.

Look at the Residuals

§ To examine the residuals, select **Plot Residuals** from the Polynomial Fit degree=2 popup menu. You should see a plot similar to the plot on the right in **Figure 8.9**. There still appears to be a pattern in the residuals, so you might want to continue and fit a model with higher order terms.

Figure 8.9 Comparison of Linear and Second Order Polynomial Fit

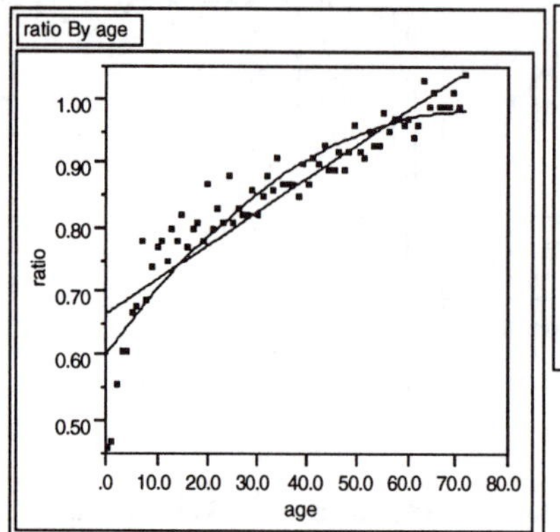

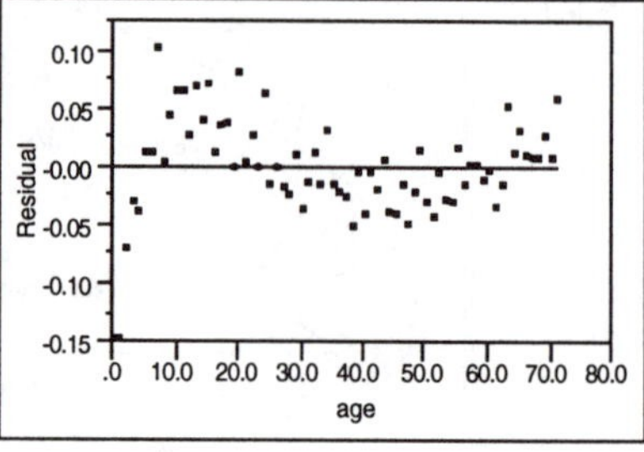

Higher Order Polynomials

If you want to give more flexibility to the curve, you can specify higher orders of polynomial, adding a term to the third power, to the fourth power, and so forth:

§ With the scatterplot active, request a polynomial of degree 4 from the Fitting popup menu. Then select **Plot Residuals** from the Polynomial Fit degree=4 options menu, which gives the plot on the right in **Figure 8.10**. Note that the residuals no longer have a pattern to them.

Figure 8.10 Comparison of Linear and Fourth Order Polynomial Fits

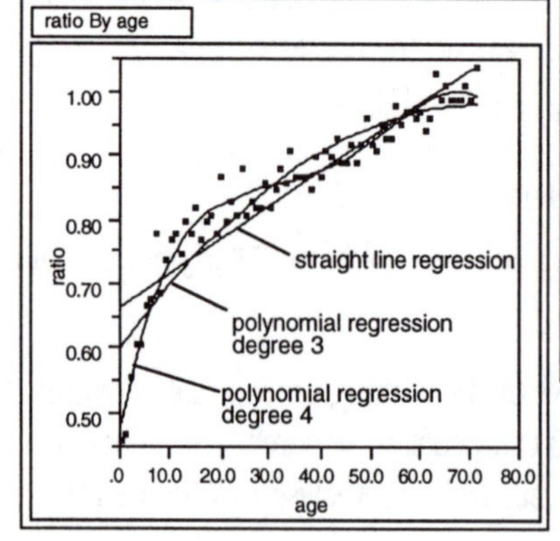

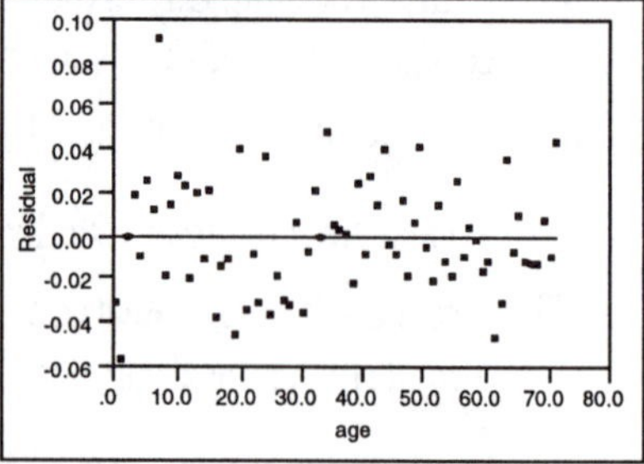

Distribution of Residuals

It is also informative to look at the shape of the distribution of the residuals. If the distribution departs dramatically from the normal, then you may be able to find further phenomena in the data.

Figure 8.11 shows histograms of residuals from the line fit, the polynomial degree-2 fit, and the polynomial degree-4 fit. You can see the distributions evolve toward normality for the better fitting models with more parameters.

§ To generate these histograms, use the **Save Residuals** command found in the popup menu for each regression fit. This forms 3 new columns in the data table. Then choose **Analyze→Distribution of Y** for the 3 columns of residual values.

Figure 8.11 Histograms to Look at Distribution of Residuals

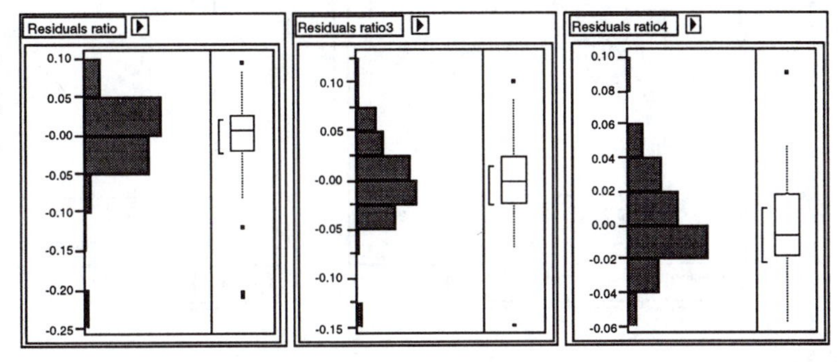

Transformed Fits

Sometimes, you can fit a curve better if you transform either the Y or X variable (or sometimes both). You can use the Fit Y by X platform to do this:

§ Again, choose **Analyze→Fit Y by X** for **ratio** by **age**. Then select **Fit Transformed** from the **Fitting** popup menu. The **Fit Transformed** command displays a dialog that lists natural log, square, square root, exponential, and reciprocal transformations as selections for both the X and Y variables. Try fitting **ratio** to the log of **age**:

§ Click the **Logarithm (ln x)** radio button for **age**. When you click **OK**, you should see the left plot in **Figure 8.12**.

Alternatively, you could create a new column in the data table and compute the log of age, then use **Fit Y by X** to do a straight line regression of ratio on this transformed age. The results are identical except you see the line of fit as a straight line in the log scale, as shown in the plot on the right in **Figure 8.12**.

If you transform the Y variable, then you are fitting a different variable and can't compare the RSquare and error sums of squares of the transformed variable fit with the untransformed variable fit.

Figure 8.12 Comparison of Fit Transform versus Transformed Fit

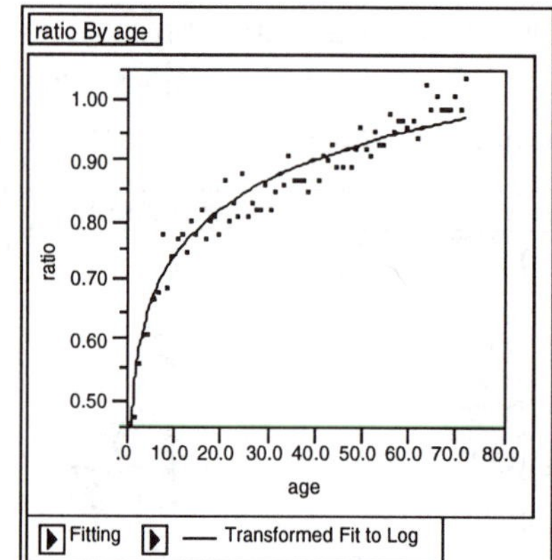

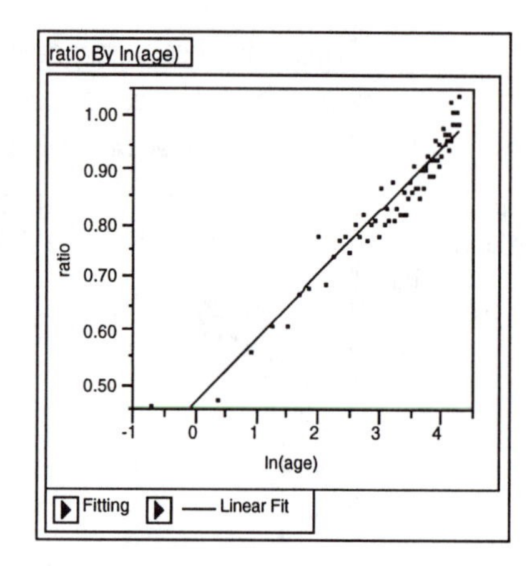

Spline Fit

It would be nice if you could fit a flexible leaf spring through the points. The leaf spring would resist bending somewhat, but would take gentle bends when it needed to fit the data better. A *smoothing spline* is that kind of fit. With smoothing splines you have to specify how stiff to make the line. If it is too rigid it looks like a straight line. But if you make the spline too flexible it curves to try to fit each point. Use these commands to see the plots in **Figure 8.13:**

§ Choose **Analyze→Fit Y by X** for ratio and age and then select **Fit Spline** from the Fitting popup menu with lambda=10.

§ Select **Fit Spline** again from the Fitting popup menu **again but** with lambda=1000.

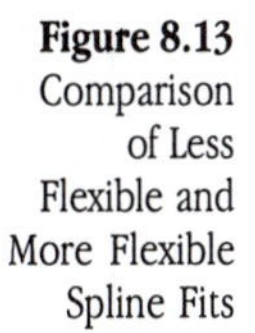

Figure 8.13 Comparison of Less Flexible and More Flexible Spline Fits

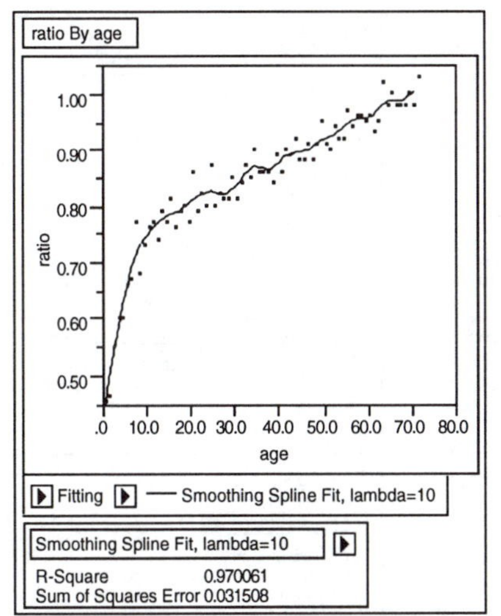

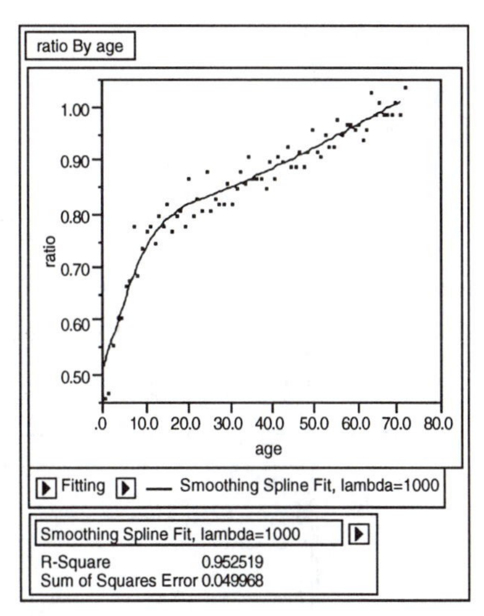

Why Graphics Are Important

Some statistical packages don't show graphs of the regression, while others require you to make an extra effort to see the graph. The following data table can help you understand the kind of phenomena that you miss if you don't look at the graph:

§ Open ANSCOMBE.JMP (Anscombe, 1973).

§ Choose **Analyze→Fit Y by X** four times, to fit Y1 by X1, Y2 by X2, Y3 by X3, and Y4 by X4. Don't look at the graphs yet. For each pair, select the **Fit Line** command. Look at the text reports (as in **Figure 8.14**) and compare them.

The thing to notice is that all the reports are nearly identical. The RSquares, the F tests, the parameter estimates and standard errors—they are all the same. So are the situations the same? Now look at the graphs of the four relations (**Figure 8.15**).

Figure 8.14 Statistical Reports for Four Analyses

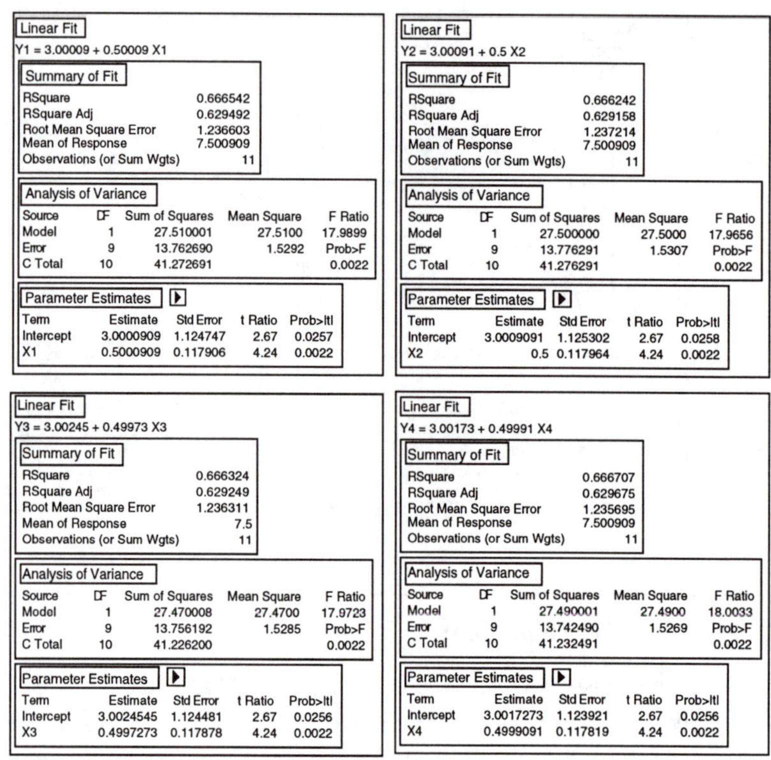

Linear Fit

Y1 = 3.00009 + 0.50009 X1

Summary of Fit

RSquare	0.666542
RSquare Adj	0.629492
Root Mean Square Error	1.236603
Mean of Response	7.500909
Observations (or Sum Wgts)	11

Analysis of Variance

Source	DF	Sum of Squares	Mean Square	F Ratio
Model	1	27.510001	27.5100	17.9899
Error	9	13.762690	1.5292	Prob>F
C Total	10	41.272691		0.0022

Parameter Estimates

Term	Estimate	Std Error	t Ratio	Prob>|t|
Intercept	3.0000909	1.124747	2.67	0.0257
X1	0.5000909	0.117906	4.24	0.0022

Linear Fit

Y2 = 3.00091 + 0.5 X2

Summary of Fit

RSquare	0.666242
RSquare Adj	0.629158
Root Mean Square Error	1.237214
Mean of Response	7.500909
Observations (or Sum Wgts)	11

Analysis of Variance

Source	DF	Sum of Squares	Mean Square	F Ratio
Model	1	27.500000	27.5000	17.9656
Error	9	13.776291	1.5307	Prob>F
C Total	10	41.276291		0.0022

Parameter Estimates

Term	Estimate	Std Error	t Ratio	Prob>|t|
Intercept	3.0009091	1.125302	2.67	0.0258
X2	0.5	0.117964	4.24	0.0022

Linear Fit

Y3 = 3.00245 + 0.49973 X3

Summary of Fit

RSquare	0.666324
RSquare Adj	0.629249
Root Mean Square Error	1.236311
Mean of Response	7.5
Observations (or Sum Wgts)	11

Analysis of Variance

Source	DF	Sum of Squares	Mean Square	F Ratio
Model	1	27.470008	27.4700	17.9723
Error	9	13.756192	1.5285	Prob>F
C Total	10	41.226200		0.0022

Parameter Estimates

Term	Estimate	Std Error	t Ratio	Prob>|t|
Intercept	3.0024545	1.124481	2.67	0.0256
X3	0.4997273	0.117878	4.24	0.0022

Linear Fit

Y4 = 3.00173 + 0.49991 X4

Summary of Fit

RSquare	0.666707
RSquare Adj	0.629675
Root Mean Square Error	1.235695
Mean of Response	7.500909
Observations (or Sum Wgts)	11

Analysis of Variance

Source	DF	Sum of Squares	Mean Square	F Ratio
Model	1	27.490001	27.4900	18.0033
Error	9	13.742490	1.5269	Prob>F
C Total	10	41.232491		0.0022

Parameter Estimates

Term	Estimate	Std Error	t Ratio	Prob>|t|
Intercept	3.0017273	1.123921	2.67	0.0256
X4	0.4999091	0.117819	4.24	0.0022

Note the characteristics of the scatterplots in **Figure 8.15**:

- Y1 by X1 show a normal regression situation.
- The points in Y2 by X2 follow a parabola, so a quadratic model would be appropriate, with the square of X2 as an additional term in the model.
- There is an extreme outlier in Y3 by X3, which brings up the slope of the line that would have been a perfect fit otherwise. It might be appropriate to exclude the outlying point and fit another line.
- In Y4 by X4 all the X values are the same except for one point, which completely determines the slope of the line. This situation is called leverage. It is not necessarily bad, but you ought to know about it.

As an exercise, use **Fit Polynomial** for Y2 and **Rows→Exclude** and **Fit Line** to refit Y3.

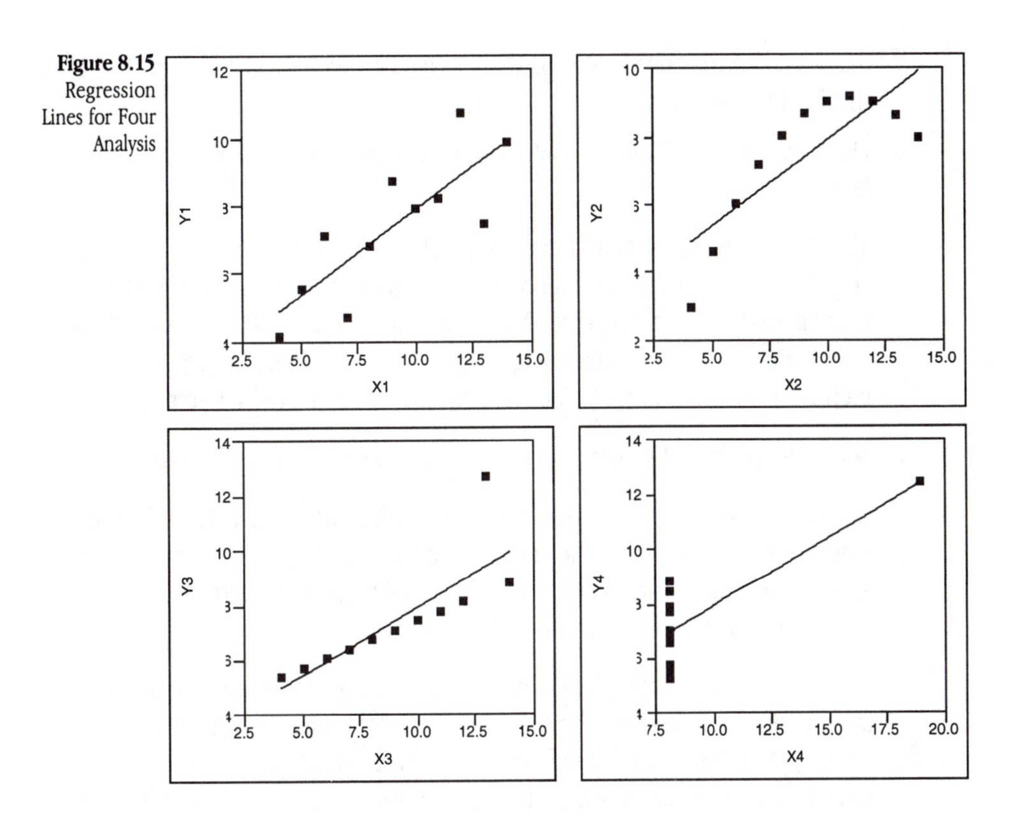

Figure 8.15 Regression Lines for Four Analysis

Why It's Called Regression: What Happens When X and Y Are Switched?

Remember the story about the study done by Sir Francis Galton mentioned at the beginning of this chapter? He examined the heights of parents and their grown children, perhaps to gain some insight into what degree height is an inherited characteristic. He concluded that the children's heights tended to be more moderate than the parent's heights, and used the term "regression to the mean" to name this phenomenon. For example, if a parent was very tall, the children would be tall, but less so than the parents. If a parent was very short, the child would tend to be short, but less so than the parent.

The case of Galton is interesting not only because it was the first use of regression, but also because Galton failed to notice some properties of regression that would have changed his mind about using regression to draw the conclusions that he drew. To investigate Galton's data:

§ Open GALTON.JMP and choose **Analyze→Fit Y by X** with **child ht** as **Y** and **parent ht** as **X**.

The data in the GALTON1 table comes from Galton's published table, but the values are *jittered* by a random amount up to .5 in either direction to each point. The jittering was done so that in plots all the points show instead of overlapping. Also Galton multiplied the women's heights by 1.08 to make them comparable to men's. The parent's height is defined as the average of the two parents.

§ Select **Fit Line** from the **Fitting** popup menu, and then select **Paired t Test.**

The **Paired t-Test** command displays a 45-degree line, which represents the t test that Galton could have hypothesized to see if the mean height of the child is the same as the parent. The comparison is done by fitting the 45-degree line and testing to see if it goes through the origin.

But that is not the test that Galton considered. He invented regression to fit an arbitrary line and then tested to see if the slope of the line was 1 (the line was vertical). If the line has a slope of one, then the predicted height of the child is the same as that of the parent, except for a generational constant. A slope of less than one indicates *regression* in the sense that the children tended to have more moderate heights (closer to the mean) than the parents.

When you examine the Parameter Estimates table and the regression line in the left plot of **Figure 8.16**, you see that the least squares regression slope is .61, which is far below 1. This confirms the *regression* toward the mean.

But is Galton's technique fair to the hypothesis? If the children's heights were more moderate than the parents, shouldn't the parent's heights be more extreme than the children's? To find out, you can reverse the model and try to predict the parent's heights from the children's heights. The analysis on the right in **Figure 8.16** shows the results when the parent's height is Y and children's height is X. Because the previous slope was less than one, if there were symmetry this analysis would give a slope greater than 1. Instead it is .28, even less than the first slope.

Figure 8.16 Child's Height as a Function of Parent. Parents Height as a Function of Child

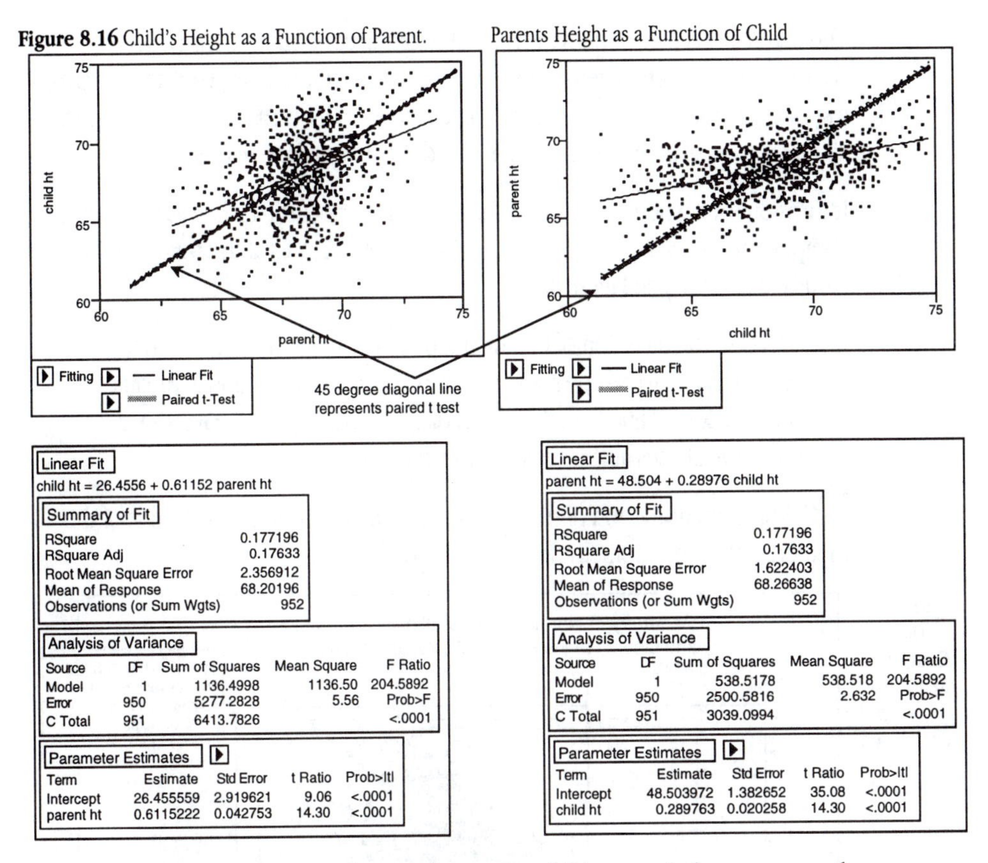

Linear Fit

child ht = 26.4556 + 0.61152 parent ht

Summary of Fit

RSquare	0.177196
RSquare Adj	0.17633
Root Mean Square Error	2.356912
Mean of Response	68.20196
Observations (or Sum Wgts)	952

Analysis of Variance

Source	DF	Sum of Squares	Mean Square	F Ratio
Model	1	1136.4998	1136.50	204.5892
Error	950	5277.2828	5.56	Prob>F
C Total	951	6413.7826		<.0001

Parameter Estimates

Term	Estimate	Std Error	t Ratio	Prob>|t|
Intercept	26.455559	2.919621	9.06	<.0001
parent ht	0.6115222	0.042753	14.30	<.0001

Linear Fit

parent ht = 48.504 + 0.28976 child ht

Summary of Fit

RSquare	0.177196
RSquare Adj	0.17633
Root Mean Square Error	1.622403
Mean of Response	68.26638
Observations (or Sum Wgts)	952

Analysis of Variance

Source	DF	Sum of Squares	Mean Square	F Ratio
Model	1	538.5178	538.518	204.5892
Error	950	2500.5816	2.632	Prob>F
C Total	951	3039.0994		<.0001

Parameter Estimates

Term	Estimate	Std Error	t Ratio	Prob>|t|
Intercept	48.503972	1.382652	35.08	<.0001
child ht	0.289763	0.020258	14.30	<.0001

Instead of phrasing the conclusion that children tended to regress to the mean, Galton could have worded his conclusion that there is a somewhat weak relationship. When you do regression, there is no symmetry between the Y and X variables. The slope of Y on X is not the reciprocal of the slope of X on Y; you cannot solve the X by Y fit by taking the Y by X fit and solving for the other variable.

The reason why regression is not symmetric is that the error that is minimized is in one direction only—that of the Y variable. So if you switch the roles, you are solving a different problem.

It always happens that the slope will be smaller than the reciprocal of inverting the variables. However, there is a way to fit the slope symmetrically, so that the role of both variables is the same. This is what is you do when you calculate a *correlation*.

The correlation characterizes the bivariate normal continuous density. The contours of the normal density form ellipses as illustrated in **Figure 8.17.** If there is a strong relationship, the ellipses become elongated along a diagonal axis. The line along this axis even has a name—it's called the *first principal component.*

It turns out that when you do least squares you are not fitting a line to the principal component. Instead, you are bisecting the contour ellipses at the points of their vertical tangents (see **Figure 8.17**).

If you reverse the direction of finding midpoints or tangents, you describe what the regression line would be if you reversed the role of the Y and X variables. If you draw the X by Y line fit in the Y by X diagram as shown in **Figure 8.17**, it intersects the ellipses at their horizontal tangents.

So Galton's phenomenon of regression to the mean was more an artifact of the method, rather than something to learn about the data.

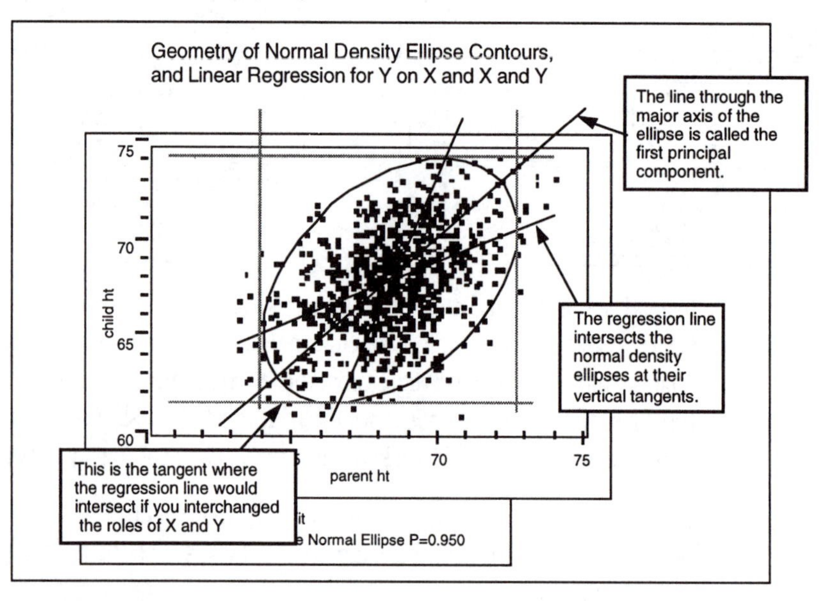

Figure 8.17 Diagram Comparing Regression and Correlation

Bivariate

Variable	Mean	Std Dev	Correlation	Signif. Prob	Number
parent ht	68.26638	1.787649	0.420947	0.0000	952
child ht	68.20196	2.59697			

Curiosities

Sometimes It's the Picture That Fools You

An experiment by a molecular biologist generated some graphs similar to the scatterplots in **Figure 8.18**. Looking quickly at the plot on the left, where would you guess the least squares regression lin lies? Now look at the graph on the right to see where the least-squares fit really appeared.

Figure 8.18 Beware of Hiding Dense Clusters

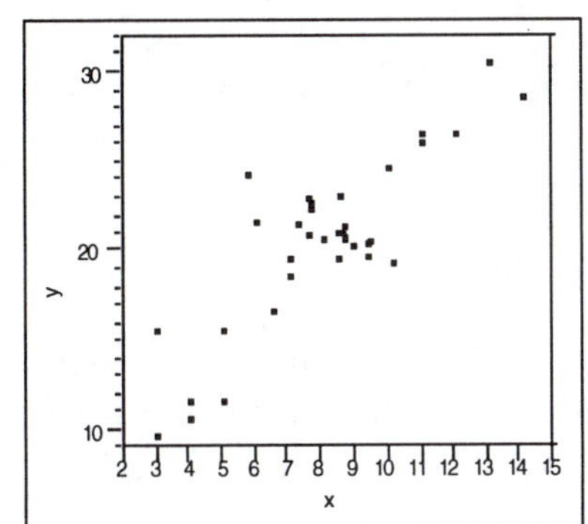

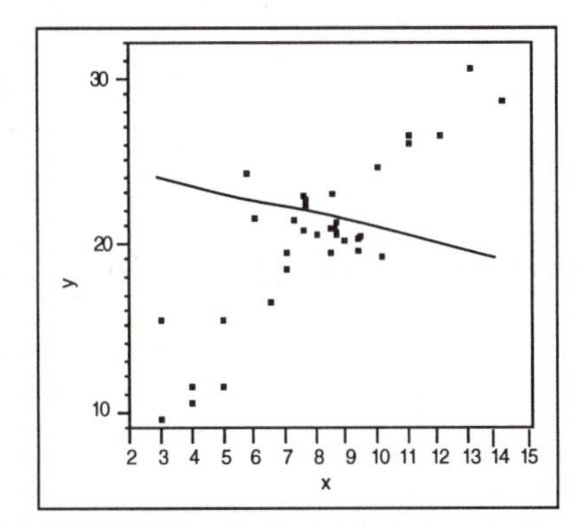

The biologist was perplexed. How did this unlikely looking regression line happen?

It turns out that there is a very dense cluster you can't see. This dense cluster of thousands of points dominated the slope estimate even though the few points farther out had more individual leverage. There was nothing wrong with the computer. Its just that your eye is fooled sometimes, especially when many points are overlaid in the same position.

High Order Polynomial Pitfall

Suppose you want to develop a prediction equation for predicting ozone based on the population of a city. The lower-order polynomials fit fine, but why not take the "more is better" approach and try a higher order one, say sixth degree. As you can see in **Figure 8.19**, the curve fit very well, too well. Would you trust the ozone prediction for a city with a population of 7500?

This overfitting phenomenon as shown in **Figure 8.19** happens in higher order polynomials when the data are unequally spaced.

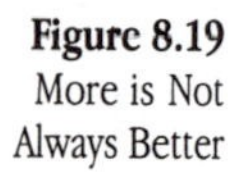

Figure 8.19 More is Not Always Better

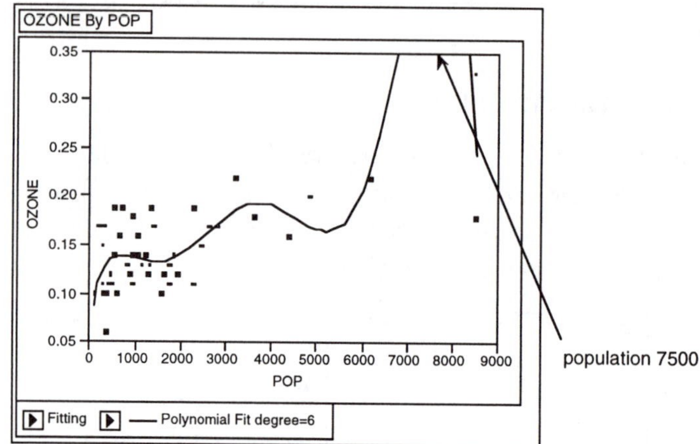

The Pappus Mystery on the Obliquity of the Ecliptic

The ancient measurements of the angle of the earth's rotation disagree with the modern measurements. Is this because the modern ones are based on different and better technology or did the angle of rotation actually change?

Chapter 6, "The Difference Between Two Means," introduced the angle-of-the-ecliptic data. The data that goes back to ancient times is in CASSINI.JMP table (Stigler 1986). **Figure 8.20** shows the regression of the obliquity (angle) by time. The regression suggests that the angle has changed over time. The mystery is that the measurement by Pappas is not consistent with the rest of the line. Was Pappas's measurement flawed or did something else happen at that time? We probably will never know.

These kinds of mysteries sometimes lead to detective work that results in great discoveries. Madame Curie discovered Radium because of a small discrepancy in measurements made with pitchblend experiments. If she hadn't noticed that discrepancy, the progress of physics might have been delayed.

Outliers are not to be dismissed casually. Moore and McCabe (1989) point out a situation in the 1980s where a satellite was measuring ozone levels over the poles. It automatically rejected a number of measurements because they were very low. Because of this, the ozone holes were not discovered until years later by experiments run from the ground that confirmed the satellite measurements.

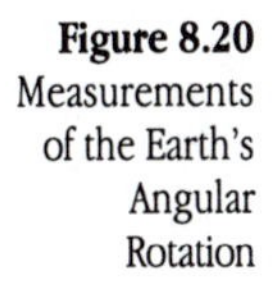

Figure 8.20 Measurements of the Earth's Angular Rotation

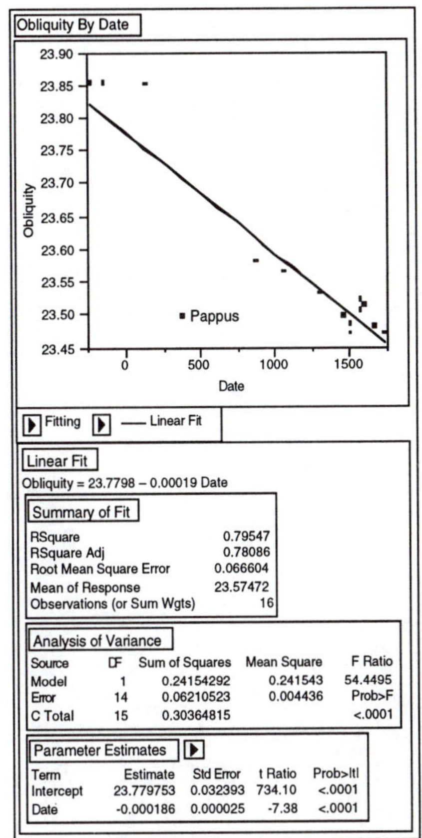

Linear Fit

Obliquity = 23.7798 – 0.00019 Date

Summary of Fit

RSquare	0.79547
RSquare Adj	0.78086
Root Mean Square Error	0.066604
Mean of Response	23.57472
Observations (or Sum Wgts)	16

Analysis of Variance

Source	DF	Sum of Squares	Mean Square	F Ratio
Model	1	0.24154292	0.241543	54.4495
Error	14	0.06210523	0.004436	Prob>F
C Total	15	0.30364815		<.0001

Parameter Estimates

Term	Estimate	Std Error	t Ratio	Prob>|t|
Intercept	23.779753	0.032393	734.10	<.0001
Date	-0.000186	0.000025	-7.38	<.0001

Chapter 9
Categorical Distributions

Overview

When the response is categorical, you need a different set of tools to analyze the data. This chapter focuses on simple categorical responses and introduces these prerequisite topics:

- There are two ways to approach categorical data. This chapter refers to them as *choosing* and *counting*. They use different tools and conventions for analysis.
- The concept of variability in categorical responses is more difficult than in continuous responses. Monte Carlo simulation helps demonstrate how categorical variability works.
- The chi-square test is the fundamental statistical tool for categorical models. There are two kinds of chi-square tests. They test the same thing in a different way and get similar results.

Fitting models to categorical response data is covered in Chapter 10, "Categorical Models."

Chapter 9 Contents

You can mouse along with the examples whenever you see an action symbol (§)

Categorical Situations

A categorical response is one in which the response is from a limited number of choices. There is a probability associated with each of the response categories, and the probabilities sum to 1.

Categorical responses are common. Consumer preferences are usually categorical: Which do you like the best—tea, coffee, juice, or soft drinks? Medical outcomes are often categorical: Did the patient live or die? Biological responses are often categorical: Did the seed germinate? Mechanical responses can be categorical: Did the fuse blow at 20 amps? Any continuous response can be converted to a categorical response: Did the temperature reach 100 degrees?

Categorical Responses and Count Data: Two Outlooks

Before starting on categorical distributions, you need to understand that there are two approaches on how to handle categorical responses. The two approaches generally get the same answers, but they use different tools and terms.

- First, imagine that each observation represents the response of a chooser. Based on conditions of the observation, the chooser is going to respond with one of the response categories. For example, the chooser might respond by choosing a dessert choice: pie, ice cream, or cake. Each response category has some probability of being chosen, and that probability varies depending on other characteristics of the observational unit.
- Now reverse the situation and think of yourself as the observation collector for one of the categories. Suppose that you sell the pies. The category *Pies* now looks like a sample category for the vendor, and the response of interest is how many pies can you sell in a day? Given total sales for the day of all desserts, you are interested in your market share.

Figure 9.1 diagrams these two ways of looking at categorical distributions.

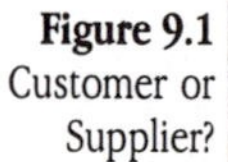

Figure 9.1 Customer or Supplier?

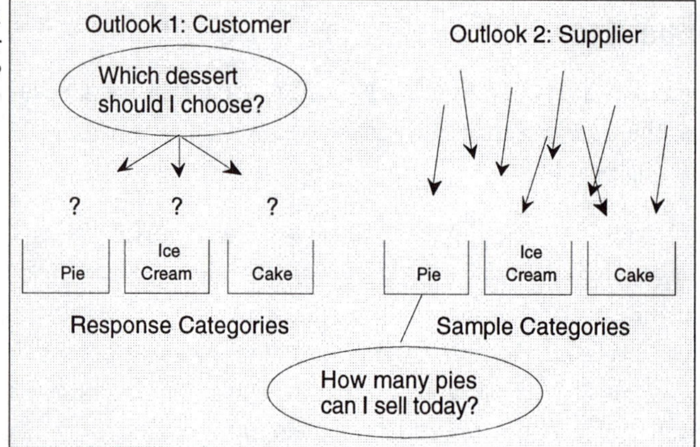

If you are the customer/ chooser, then you think in terms of logistic regression, where the Y variable is which dessert you choose and the X variables affect the probabilities associated with each dessert category. If you are the supplier/counter, then you talk about log-linear models, where the Y is the count, and the X effect is the dessert category. There can be other X's interacting with that X.

The modeling traditions are also different. Customer/chooser oriented analysts, such as the live/die medical analyst, use continuous X's (like dose, or how many years you smoked). Supplier/counter oriented analysts, typified by social scientists, are most comfortable with categorical X's (like age, race, sex, and personality type) because that keeps the data count-oriented.

The probability distributions for the two approaches are also different. This book won't go into the details of the distributions, but you can be aware of distribution names. Customer/chooser oriented analysts refer to the *Bernoulli* distribution of the choice. Supplier/counter oriented analysts refer to the *Poisson* distribution of counts in each category. However, both approaches refer to the *multinomial* distribution.

- To the customer/chooser analysts, the multinomial counts are aggregation statistics.
- To the supplier /counter analysts, the multinomial counts are the count distribution within a fixed total count.

The customer/chooser analyst thinks the basic analysis is fitting response category probabilities. The supplier/counter analyst thinks that basic analysis is a one-way analysis of variance on the counts and uses weights because the distribution is Poisson instead of normal.

Both orientations are right—they just have different outlooks on the same statistical phenomenon.

In this book, the emphasize is on the customer/chooser point of view, which is also called the logistic regression approach. With logistic regression it is important to distinguish the responses (Y's), which have the random element, from the factors (X's), which are fixed from the point of view of the model. The X's and Y's are distinguished before you do an analysis.

Let's be clear on what the X's and Y's are for the chooser point of view:

- Responses (Y's) identify a choice or an outcome. They have a random element because the choice is not determined completely by the X factors. Examples of responses are patient outcome (lived or died), or desert preference (Gobi or Sahara).
- Factors (X's) identify a sample population, an experimentally controlled condition, or an adjustment factor. They are not regarded as random even if you randomly assigned them. Examples of factors are sex (gender), age, or treatment or block.

Figure 9.2 illustrates the X and Y variables for both outlooks on categorical models.

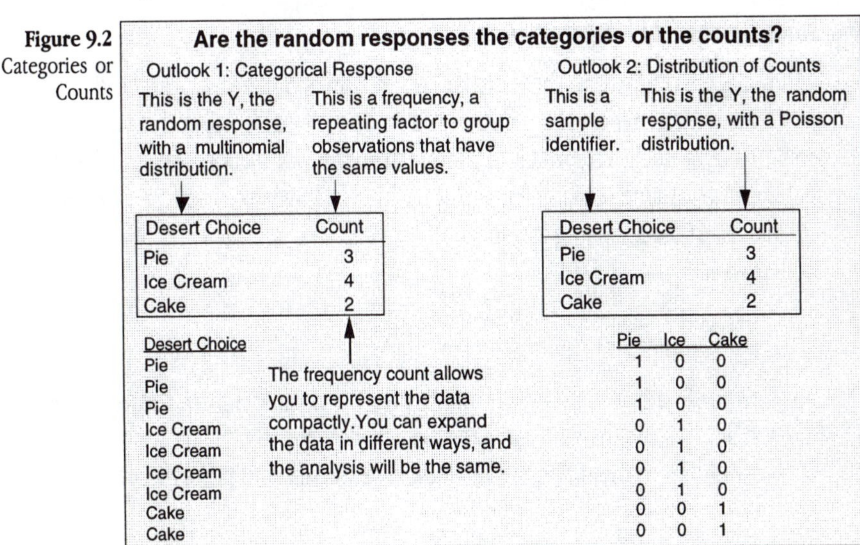

Figure 9.2 Categories or Counts

Another point of view is called the log-linear model approach. The log-linear approach regards the count as the Y variable and all the other variables as X's. After fitting the whole model, you go back and identify which effects are of interest. In particular you discard any effect that has no response category variable because that is just an artifact of the sampling design. Log-linear modeling uses a technique called iterative proportional fitting to obtain test statistics. This process is also called *raking*.

A Simulated Categorical Response

A good way to learn is to simulate data with known properties, and then analyze the simulation to see if you find the structure that you put into the simulation.

These steps describe the simulation process:

1) Simulate one batch of data, then analyze.
2) Simulate more batches, analyze them, and notice how much they vary.
3) Simulate a larger batch to notice that the estimates have less variance.
4) Do a batch of batches—simulations that for each run obtain a realization of the sample statistics over a new batch of data.
5) Use this last batch of batches to look at the distribution of the test statistics.

Simulating Some Categorical Response Data

Let's make a world where there are three soft drinks. The most popular ("Sparkle Cola") has a 50% market share and the other two ("Kool Cola" and "Lemonitz") are tied at 25% each. To simulate a sample from this population, create a data table that has one variable (call it **drink choice**), which is drawn as a random categorical variable using the following formula:

$$
\begin{array}{l}
p \Leftarrow \text{?uniform} \\
\text{results} \begin{cases} \text{"Kool Cola"}, & \text{if } p<0.25 \\ \text{"Lemonitz"}, & \text{if } p<0.5 \\ \text{"Sparkle Cola"}, & \text{otherwise} \end{cases}
\end{array}
$$

This formula first draws a random number between 0 and 1 from the **?uniform** function and assigns the result to a temporary variable p. Then it compares that random number to the **If** conditions of the 3 responses and picks the first response where p is true. Each case returns the character name of the soft drink as the response value.

Note: The calculator operations needed to construct this formula include the following operations: **Variables: New** to create a temporary variable named p, **Conditions: Assignment**, and **Conditions: If**.

This table has already been created:

§ Open the data table COLA.JMP, which contains the formula shown previously.

§ The table is stored with no rows. Choose **Rows→Add Rows** to add 50 rows to the table. A data table stored with no rows and columns with formulas is called a *table template*.

§ Choose **Analyze→Distribution of Y** with the **drink choice** variable as **Y**, which gives an analysis like that in **Figure 9.3**.

Don't expect to get the exact same numbers because the formula generates random data; each time the computations are performed gives a slightly different set of results.

Note that even though the data are based on the true probabilities of (.25, .25, and .50) the estimates came out differently (.18, .26, and .56). Your data will have random values with somewhat different probabilities.

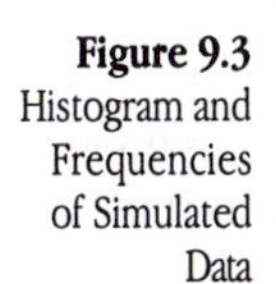

Figure 9.3 Histogram and Frequencies of Simulated Data

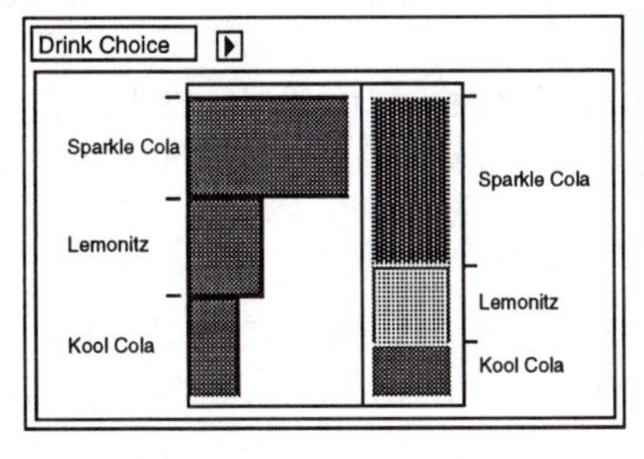

Frequencies

Level	Count	Probability	Cum Prob
Kool Cola	9	0.18000	0.18000
Lemonitz	13	0.26000	0.44000
Sparkle Cola	28	0.56000	1.00000
Total	50		
3 Levels			

Variability in the Estimates

The following sections distinguish between ρ (Greek rho), which is the true value of a probability, and its estimate, denoted p. The true value ρ is some fixed number, but the estimate p is an outcome of a random process, so it has variability associated with it.

You cannot compute a standard deviation of the original response—it has character values. But the variability in the probability estimates is well defined.

The variability of the estimate is expressed by the variance, or the standard deviation (the square root of the variance). It turns out that the variance of *p* is ρ*(1 – ρ)/n. For Sparkle Cola, having a ρ of .50, the variance of the probability estimate is .5*.5/50, or .005. The standard deviation of the estimate is the square root of the variance, 0.07071. In **Table 9.1**, first compare the difference between the true ρ and its estimate *p*. Then compare the true standard deviation of the statistic *p*, and the standard error of *p*, which estimates the standard deviation of *p*.

Remember, the term *standard error* is used to label an estimate of the standard deviation of another estimate. Only because this is a simulation with known true values (parameters) can you see both the standard errors and the true standard deviations.

Table 9.1 Simulated Probabilities and Estimates

Level	ρ, the True Probability	p, the Estimate of ρ	True Std Deviation of Estimate sqrt(ρ(1–ρ)/n)	Std Error of Estimate sqrt(*p*(1–*p*)/n) for Estimated *p*
Kool Cola	.25	.18	.06124	.05433
Lemonitz	.25	.26	.06124	.06203
Sparkle Cola	.50	.56	.07071	.07020

This simulation shows a lot of variability. As with normally distributed data, you can expect to go 2 standard deviations from the true probability about 5% of the time.

Now let's see how the estimates vary with a new set of random responses.

§ Double-click on the **Drink Choice** column heading to get the Column Info dialog (or Choose **Cols→Column Info**). Click the small formula showing at the lower left of the dialog to open the calculator window. (Alternatively, OPTION-click in the column heading area to bring up the calculator directly.)

§ Click **Evaluate** on the calculator to reevaluate the random formula.

§ Again choose **Analyze→Distribution of Y** on Drink Choice.

Repeat this evaluate/analyze cycle four times.

Each time gives a new set of random responses and a new set of estimates of the probabilities. **Table 9.2** gives the estimates from the four Monte Carlo runs.

Table 9.2 Estimates from Monte Carlo Runs

Level	Probability	Probability	Probability	Probability
Kool Cola	0.28000	0.18000	0.26000	0.40000
Lemonitz	0.26000	0.32000	0.24000	0.18000
Sparkle cola	0.46000	0.50000	0.50000	0.42000

With only 50 observations there is a lot of variability in the estimates. The "Kool Cola" probability estimate varies between .18 and .40, the "Lemonitz" estimate varies between .18 and .32, and the "Sparkle Cola" estimate varies between .42 and .50.

Larger Sample Sizes

What happens if the sample size increases from 50 to 500? With more data, the probability estimates have a much smaller variance (remember that the $\text{var}(\rho) = \rho^*(1-\rho)/n$). To see this happen:

§ Choose **Rows→Add Rows** and enter 450, giving a total of 500 rows.

§ Choose **Analyze→Distribution of Y** for the response variable **Drink Choice**.

Figure 9.4 Frequencies for Simulation Data with Increased Sample Size

Frequencies

Level	Count	Probability	StdErr Prob
Kool Cola	125	0.25000	0.01936
Lemonitz	141	0.28200	0.02012
Sparkle Cola	234	0.46800	0.02231
	500		
Total 3 Levels			

Five hundred rows give a smaller variance, .0005, with a standard deviation at about 0.02. **Figure 9.4** shows the first simulation for 500 rows.

§ Repeat the evaluate/analyze cycle four times. **Table 9.3** shows the realizations of the next 4 simulations:

Table 9.3 Estimates from Monte Carlo Runs

Level	Probability	Probability	Probability	Probability
Kool Cola	0.28000	0.25000	0.25600	0.23400
Lemonitz	0.24200	0.28200	0.23400	0.26200
Sparkle cola	0.47800	0.46800	0.51000	0.50400

Note how the probability estimates have been drawn closer to the true values and how the standard errors are smaller.

Monte Carlo Simulations for the Estimators

What do the distributions of these counts look like? Variances can be easily calculated, but what is the distribution of the estimate? Statisticians often use Monte Carlo simulations to investigate the distribution of the statistics.

To simulate estimating a probability (which has a true value of .25 in this case) over a sample size (50 in this case), construct the formula shown to the right.

$$\frac{\sum_{j=1}^{50} (?\text{uniform} < 0.25)}{50}$$

The ?uniform function generates a random value distributed uniformly between 0 and 1. The random value is checked to see if it is less than .25. The term in parentheses will be 1 or 0 depending on this comparison. It will be 1 about 25% of the time, and 0 about 75% of the time. This random number is generated 50 times, and the sum of them is divided by zero.

This formula is a complete trial of a simulation of a Bernoulli event with 50 samplings. The result estimates the probability of getting a 1, and you happen to know the true value of this probability, .25, because you generated the data.

Now, it is important to see how well these estimates behave. Theoretically, the mean (expected value) of the estimate, p, is ρ=.25 (the true value), and the standard deviation that is the square root of $(\rho(1-\rho)/n)$, or 0.061237.

Distribution of the Estimates

The sample data has a table template called SIMPROB.JMP that is a Monte Carlo simulation for the probability estimates of .25 and .5, based on 50 and 500 trials. You can add 1000 rows to the data to draw 1000 Monte Carlo trials. (Note: If you have a slow computer, be aware that the calculations can be time intensive. Adding 1000 rows requires processing 1,100,000 random numbers.)

To see how the estimates are distributed:

§ Open SIMPROB.JMP. Then choose **Rows→Add Rows** and enter 1000.

§ Next choose **Analyze→Distribution of Y** and use all the columns as Y variables. When the histograms appear, select the **Uniform** option from the check-mark popup menu at the lower left of the window.

§ Get the grabber tool (hand) from the **Tools** menu (or palette) and drag the histograms to adjust the bar widths and positions.

Figure 9.5 and **Table 9.4** show the properties as expected:

- The variance narrows as the sample size goes up.

- The distribution of the estimates is approximately normally distributed, especially as the sample size gets large.

Figure 9.5 Histograms for Simulations with Various n and p Values

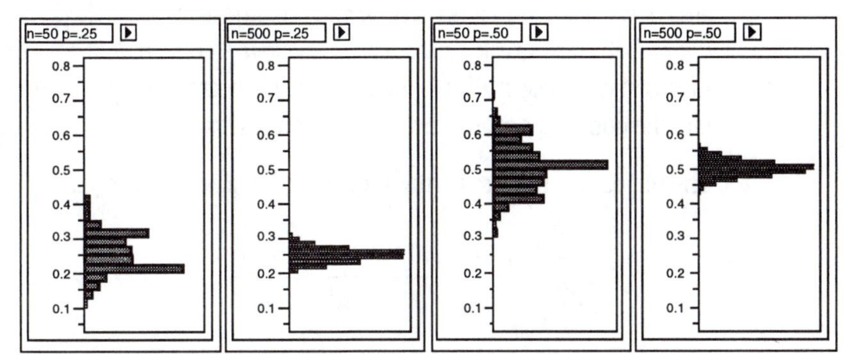

The estimates of the probability p of getting response indicator values of 0 or 1 are a kind of mean. So, as the sample gets larger the value of p gets closer and closer to .50, which is the mean of 0 and 1. Like the mean for continuous data, the standard error of the estimate relates to the sample size by the factor 1/(square root of n). The Central Limit Theorem applies here. It says that the estimates, which are a kind of mean, approach a normal distribution when there are a large number of observations.

Table 9.4 Summary of Simulation Results

True Value of p	N Used to Estimate p	Mean of the Trials of the Estimates of p	Std Dev (s) of Trials of the Estimates of p	True Mean of Estimates	True Std Dev (s) of the Estimates sqrt(p(1-p)/n)
.25	50	0.25166	0.06106	.25	0.061237
.25	500	0.25039	0.01968	.25	0.019365
.50	50	0.50074	0.07028	.50	0.070711
.50	500	0.50021	0.02247	.50	0.022361

The X^2 Pearson Chi-Square Test Statistic

Because of the normality of the estimates, it is reasonable to use normal-theory statistics on categorical response estimates. Remember that the Central Limit Theorem says that the sum of a large number of independent and identically-distributed random values will have close to a normal distribution.

However, there is a big difference between having categorical responses and continuous responses. With categorical responses the variances of the differences are known. They are a function of *n* and the probabilities. The hypothesis specifies the probabilities, so calculations can be made *under the null hypothesis*. This means that the test statistic is the Pearson chi-square test that would have gone into the numerator of an F test for a continuous response.

The standard chi-square for this model is this scaled sum of squares:

$$\chi^2 = \sum_{j=1}^{n} \left(\frac{(Observed_j - Expected_j)^2}{Expected_j} \right)$$

where observed and expected refer to cell counts rather than probabilities.

The G^2 Likelihood Ratio Chi-Square Test Statistic

Whereas the Pearson chi-square assumes normality of the estimates, another kind of chi-square test is calculated with direct reference to the probability distribution of the response and so does not require normality.

Define the *maximum likelihood* estimator to be the one that takes the joint probability function of the data and finds the values of the unknown parameters that maximize this joint probability. In other words, it finds parameters that make the data that actually occurred less improbable than they would be with any other parameter values.

There are two fortunate short cuts for finding a maximum likelihood estimator:

- Because observations are assumed to be independent, the joint probability across the observations is the product of the probability functions for each observation.
- And because addition is easier than multiplication, instead of multiplying the probabilities to get the joint probability, you add the logarithms of the probabilities, which gives the *log-likelihood*.

The term *likelihood* means the probability has been evaluated as a function of the parameters with the data fixed.

This makes computing it easy. Remember that an individual response has a multinomial distribution, so the probability is ρ_i for the i=1 to r probabilities over the r response categories.

Consider the first five responses, which happen to be

Kool Cola, Lemonitz, Sparkle Cola, Sparkle Cola, Lemonitz

For Kool Cola, Lemonitz, and Sparkle Cola, denote the probabilities as ρ_1, ρ_2, and ρ_3 respectively. The joint log-likelihood is:

$$\log(\rho_1) + \log(\rho_2) + \log(\rho_3) + \log(\rho_3) + \log(\rho_2)$$

It turns out that this likelihood is maximized by setting the probability parameter estimates to the category count divided by the total count, giving

$$p_1 = n_1/n = 1/5$$
$$p_2 = n_2/n = 2/5$$
and $$p_3 = n_3/n = 2/5$$

Substituting this into the log-likelihood gives the maximized log-likelihood of

$$\log(1/5) + \log(2/5) + \log(2/5) + \log(2/5) + \log(2/5)$$

At first it may seem that taking logarithms of probabilities is a mysterious and obscure thing to do, but it is actually very natural. You can think of the negative logarithm of p as the number of binary questions you need to ask to determine which of $1/p$ equally likely outcomes happens. The negative logarithm converts units of probability into units of information. You can think of the negative loglikelihood as the *surprise* value of the data because surprise is a good word for unlikeliness.

Likelihood Ratio Tests

So one way to measure the credibility for an hypothesis is to compare how much surprise (–log-likelihood) there would be in the actual data with the hypothesized values compared with the surprise at the maximum likelihood estimates. If there is too much surprise, then you have reason to throw out the hypothesis.

It turns out that the distribution of twice the difference in these two surprise (–log-likelihood) values approximately follows a chi-square distribution.

Here is the setup: Fit a model twice. The first time you fit by maximum likelihood with no constraints on the parameters. The second time you fit by maximum likelihood, but constrain the parameters by the null hypothesis that the outcomes are equally likely. It happens that twice the difference in log-likelihoods has an approximate chi-square distribution (under the null hypothesis). These chi-square tests are called *likelihood ratio* chi-squares, or *LR* chi-squares.

Twice the difference in the log-likelihood is a likelihood ratio chi-square test.

The likelihood ratio tests are very general. They occur not only in categorical responses, but also in a wide variety of situations.

The G^2 Likelihood Ratio Chi-Square Test

Let's focus on Bernoulli probabilities for categorical responses. The log-likelihood for a whole sample is the sum of natural logarithms of the probabilities attributed to the events that actually occurred.

log-likelihood = Σ $\log_e$ (probability the model gives to event that occurred in data)

The likelihood ratio chi-square is twice the difference in the two likelihoods, when one is constrained by the hypothesis and the other is unconstrained.

$$G^2 = 2\,(\text{log-likelihood(unconstrained)} - \text{log-likelihood(constrained)})$$

Using the words observed and hypothesized, this is formed by the sum over all observations:

$$G^2 = 2\,\Sigma\,((\log(\rho_{y_i}) - \log(p_{y_i}))$$

where ρ_{y_i} is the hypothesized probability and p_{y_i} is the estimated probability for the events y_i that actually occurred.

If you have already collected counts for each of the responses, and bring the subtraction into the log as a division, the formula becomes:

$$G^2 = 2\,\Sigma\,n_i\,(\log(\rho_{y_i} / p_{y_i}))$$

To compare with the Pearson chi-square, which is written schematically in terms of counts, the LR chi-square statistic can be written:

$$G^2 = 2 \sum \text{observed} \left(\log(\text{expected} / \text{observed})\right)$$

Univariate Categorical Chi-Square Tests

A company gave 998 of its employees the Myers-Briggs Type Indicator (MBTI) questionnaire. The test is scored to result in a 4-character personality type for each person; there are 16 possible outcomes, represented by 16 combinations of letters (see **Figure 9.6**). The company wanted to know if its employee force was statistically different in personality from the general population.

Comparing Univariate Distributions

The data table MB-DIST.JMP has a column called **TYPE** to use as a **Y** response, and a **Count** column to use as a frequency. To see the company test results:

§ Open the sample table called MB-DIST.JMP. Choose **Analyze→Distribution of Y** and use **Type** as the **Y** variable to see the report in **Figure 9.6**.

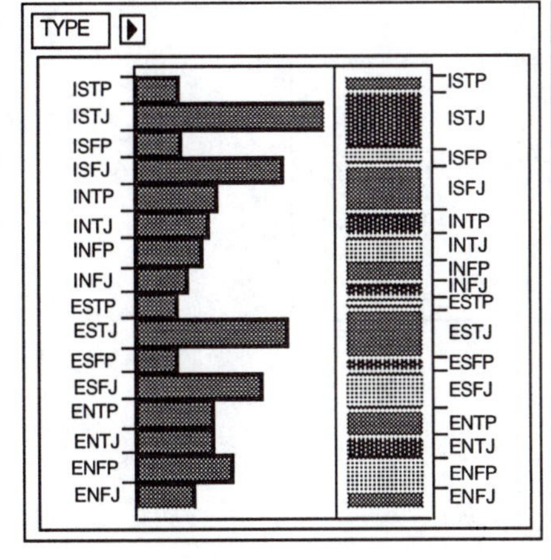

Figure 9.6 Histogram, Mosaic Plot, and Frequencies for Myers-Briggs Data

Frequencies

Level	Count	Probability	Cum Prob
ENFJ	43	0.04309	0.04309
ENFP	71	0.07114	0.11423
ENTJ	57	0.05711	0.17134
ENTP	57	0.05711	0.22846
ESFJ	92	0.09218	0.32064
ESFP	31	0.03106	0.35170
ESTJ	109	0.10922	0.46092
ESTP	31	0.03106	0.49198
INFJ	38	0.03808	0.53006
INFP	49	0.04910	0.57916
INTJ	53	0.05311	0.63226
INTP	59	0.05912	0.69138
ISFJ	106	0.10621	0.79760
ISFP	34	0.03407	0.83166
ISTJ	136	0.13627	0.96794
ISTP	32	0.03206	1.00000
Total	998		
	16 Levels		

Now, to test the hypothesis that the personalities test results at this company occur at the same rates as the general population:

§ Select the **Test Probabilities** command in the popup menu next to the histogram name at the top of the report.

§ A dialog section appears at the end of the report. Edit the **Hypoth Prob** (hypothesized probability) values by clicking and then entering the values as shown in **Figure 9.7**. These are the general population rates for each personality type.

§ When you click **Done** you see the test results appended to the Test Probabilities table, as shown at the bottom in **Figure 9.7**.

Note that the company does have a significantly different profile. Both chi-square tests are highly significant. The company appears to have more ISTJ's (introvert sensing thinking judging) and fewer ESFP's (extrovert sensing feeling perceiving) than the general population.

Figure 9.7 Test Probabilities Report for the Myers-Briggs Data

Test Probabilities

Level	Estim Prob	Hypoth Prob
ENFJ	0.04309	0.04950
ENFP	0.07114	0.04950
ENTJ	0.05711	0.04950
ENTP	0.05711	0.04950
ESFJ	0.09218	0.12871
ESFP	0.03106	0.14851
ESTJ	0.10922	0.12871
ESTP	0.03106	0.12871
INFJ	0.03808	0.00990
INFP	0.04910	0.00990
INTJ	0.05311	0.00990
INTP	0.05912	0.00990
ISFJ	0.10621	0.05941
ISFP	0.03407	0.04950
ISTJ	0.13627	0.05941
ISTP	0.03206	0.05941

Test	Chi-square	DF	Prob>Chisq
Likelihood Ratio	722.1019	15	<.0001
Pearson	1013.214	15	<.0001

By the way, some people find it upsetting that different statistical methods get different results. Actually, the G^2 (likelihood ratio) and X^2 (Pearson) chi-square statistics are usually close.

Charting to Compare Results

§ Optional: The report in **Figure 9.8** was done by using **Tables→Stack**, followed by **Graph→Bar/Pie Charts**, and adjusting chart type and dimensions.

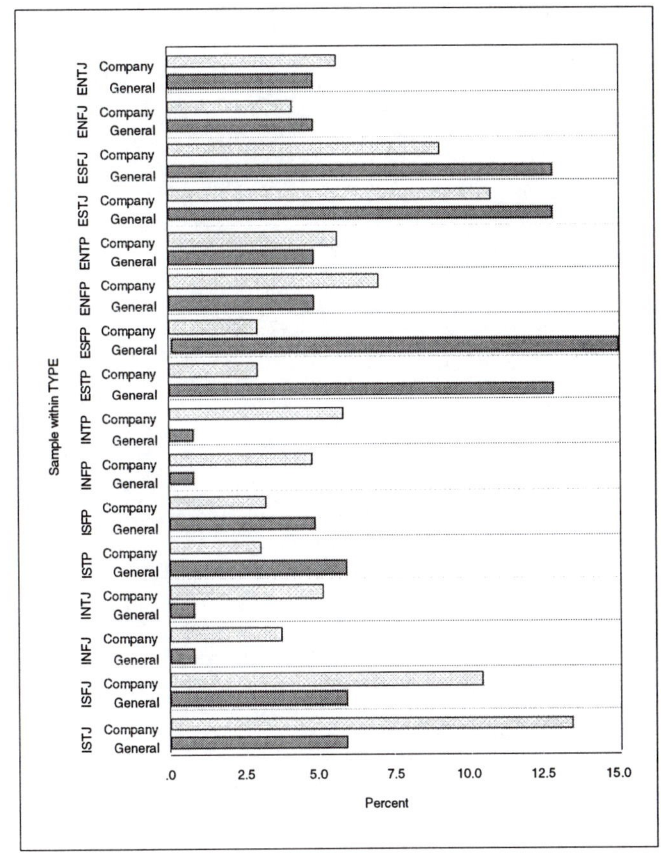

Figure 9.8 Comparison of Sample Personality Scores to Scores for General Population

Chapter 10
Categorical Models

Overview

Chapter 9, "Categorical Distributions," introduced the distribution of a single categorical response. You were introduced to the Pearson and the likelihood ratio chi-square tests and saw how to compare univariate categorical distributions.

This chapter covers the distribution of categorical responses modeled as a function of factor variables. In the simplest case the data can be presented as a two-way contingency table of frequency counts, with the expected cell probabilities and counts formed from products of marginal probabilities and counts. The chi-square test for the contingency table is same as testing multiple categorical responses for independence.

Correspondence analysis is shown as a graphical technique useful when the response and factors have many levels (values).

Also, a more general categorical response model is used to introduce nominal and ordinal logistic regression, which allows multiple continuous or categorical factors.

Chapter 10 Contents

You can mouse along with the examples whenever you see an action symbol (§)

Fitting Categorical Responses to Categorical Factors: Contingency Tables

When a categorical response is examined in relationship to a categorical factor, the question is: do the response probabilities vary across subgroups of the factor population? When the response is continuous, the usual technique is to fit means and test that they are equal with a one-way analysis of variance, as shown in Chapter 7, "Comparing Many Means: One-Way Analysis of Variance." When the response is categorical, the equivalent technique is to estimate response probabilities for each subgroup and test that they are the same across the subgroups.

The subgroups are defined by the levels of a nominal X factor. For each subgroup, you fit a set of response probabilities that add up to 1. For example, consider the following:

- The probability of whether a patient lives or dies (response probabilities) depending on whether the treatment (the X factor) was drug or placebo.
- The probability that type of car purchased (response probabilities) was a function of marital status (the X factor).

To estimate response probabilities for each subgroup, you take the count in a given response level and divide it by the total count for that subgroup.

You want to test whether the factor affects the response. The null hypothesis is that the response probabilities are the same across subgroups. The model comparison is to compare the fitted probabilities over the subgroups to the fitted probabilities combining all the groups into one population (a constant response model).

As a measure of fit for the models you want to compare, you use the negative log-likelihood to compute a likelihood ratio chi-square test. To do this you subtract the log-likelihoods for the two models and multiply by 2. For each observation, the log-likelihood is the log of the probability attributed to the response level of the observation.

Warning: When the table is sparse, neither the Pearson or likelihood ratio chi-square is a very good approximation to the true distribution. (The Cochran criterion defines sparse as when more than 20% of the cells have expected counts less than 5). The Pearson chi-square tends to be better behaved in sparse situations than the likelihood ratio chi-square.

Looking at Survey Data

Survey data often yield categorical data suitable for contingency table analysis. For example, a company did a survey to find out what factors relate to the brand of automobile people buy— what kind of people buy what kind of cars. Cars were classified into three brands: American, European, and Japanese (which includes other Asian brands).

The results of the survey are in the sample table called CARPOLL.JMP. The first step is to find probabilities that predict brand when nothing else is known about the car buyer. Looking at the distribution of car brand gives this information:

§ Open CARPOLL.JMP and choose **Analyze→Distribution of Y** to see the report on the distribution of brand shown in **Figure 10.1**. Overall, the Japanese brands lead with a 48.8% share.

Figure 10.1 Histograms and Frequencies for CARPOLL Data

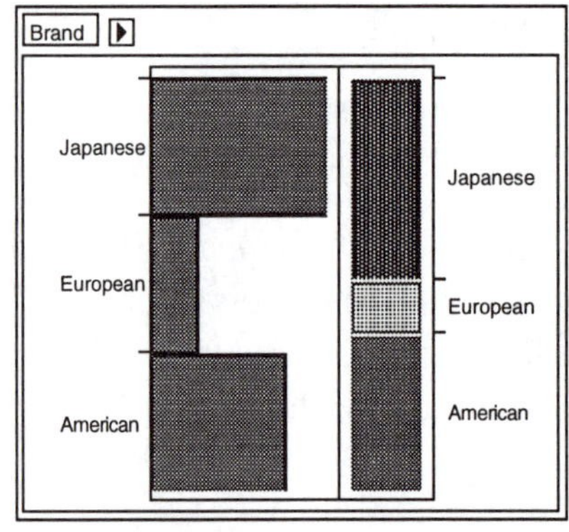

Frequencies

Level	Count	Probability	Cum Prob
American	115	0.37954	0.37954
European	40	0.13201	0.51155
Japanese	148	0.48845	1.00000
Total	303		

3 Levels

The survey questionnaire also collected demographic information (gender and marital status) and some specifics about type of car (size and type). The next step is to look at these specifics as they relate to brand of auto:

§ Choose **Analyze→Fit Y by X** with Brand as Y, and Gender, Marital Status, and Size, the X variables. When you click **OK**, the Fit Y by X platform displays Mosaic plots and Crosstabs tables for the combination of Brand with each of the X variables.

§ Select **Col%** from the popup menu at the top of the Crosstabs table to get column percent as shown in the table in **Figure 10.2**.

Contingency Table: Car Brand by Gender

Is the distribution of the response levels different over different levels of other categorical variables? In principal, this is like a one-way analysis of variance, estimating separate means for each sample, but this time they are rates over response categories rather than means.

Figure 10.2 Crosstabs (Contingency) Table for Car Brand by Gender

Crosstabs

Brand	Count Col %	Gender: Female	Gender: Male	
	American	54 39.13	61 36.97	115
	European	19 13.77	21 12.73	40
	Japanese	65 47.10	83 50.30	148
		138	165	303

In the contingency table shown here, you see the response probabilities presented as percentages in the Col% value in the bottom of each cell. The column percents are not much different between "Female" and "Male."

Mosaic Plot

The Distribution platform for a categorical variable displays table information graphically with *mosaic plots* like the one shown in **Figure 10.3**.

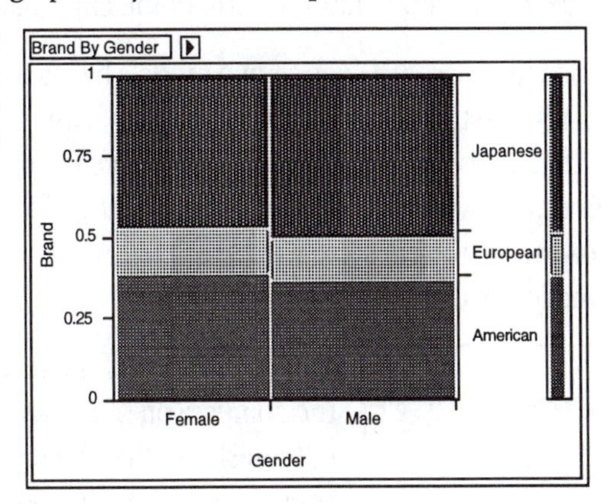

Figure 10.3 Mosaic Plot or Car Brand by Gender

A mosaic plot is a set of side-by-side divided bar plots to compare the subdivision of response probabilities for each sample. The mosaic is formed by first dividing up the horizontal axis according to the sample proportions. Then each of these cells is subdivided vertically by the estimated response probabilities. The area of each rectangle is proportional to the frequency count for that cell.

Testing Marginal Homogeneity

Now ask the question: Are the response probabilities significantly different across the samples (gender subgroups in this example)? Specifically, is the proportion of car brand sales the same for males and females? The hypothesis that the distributions are the same across the subgroup samples is sometimes referred to as the hypothesis of *marginal homogeneity*.

Instead of regarding the categorical X variable as fixed, you can consider it as another Y response variable and look at the relationship between two Y response variables. The test would be the same, but it would be known by a different name, as the test for *independence*.

When the response was continuous, there were two ways to get a test statistic:

1) Look at the distribution of the estimates, usually leading to a t test.
2) Compare the fit of a model with a submodel, leading to an F test.

The same two approaches work for categorical models. The two approaches to getting a test statistic for a contingency table both result in chi-square tests:

1) If the test is derived in terms of the distribution of the estimates, then you are led to the Pearson X^2 form of the chi-square test.
2) If the test is derived by comparing the fit of a model with a submodel, then you are led to the likelihood ratio G^2 form of the chi-square test.

For the likelihood ratio chi-square (G^2) you fit two models by maximum likelihood. One model is constrained by the hypothesis that assumes a single response population and the other is not constrained. Twice the difference of the log-likelihoods from the two models is a chi-square statistic for testing the hypothesis. **Figure 10.4** shows the chi-square tests that test whether Brand of car purchased is a function of Gender.

Figure 10.4 Table of Hypothesis Test Results

Tests

Source	DF	-LogLikelihood	RSquare (U)
Model	2	0.15594	0.0005
Error	299	298.29531	
C Total	301	298.45125	
Total Count	303		

Test	ChiSquare	Prob>ChiSq
Likelihood Ratio	0.312	0.8556
Pearson	0.312	0.8556

The top portion of the table shows the comparison of models. The line C Total shows that the model constrained by the hypothesis (fitting only one set of response probabilities) has a negative log-likelihood of 298.45. After you partition the sample by the gender factor, the negative log-likelihood is reduced to 298.30 as reported in the Error line. The difference in log-likelihoods is .1559, reported in the Model line (not much improvement). The likelihood ratio (LR) chi-square is twice this difference, that is, $G^2 = .312$, and has a nonsignificant p value of .8556.

These statistics don't support the conclusion that the car brand purchase response depends on the gender of the driver.

If you want to think about the distribution of the estimates, then in each cell you can compare the actual proportion to the proportion expected under the hypothesis, square it, and divide by something close to its variance, giving a cell chi-square. The sum of these cell chi-square values is the Pearson chi-square statistic, χ^2, which here is also .312, with a p value of .8556. In this example the Pearson chi-square is the same as the likelihood ratio chi-square.

Car Brand by Marital Status

Let's look at the relationships of brand to other categorical variables. In the case of marital status (**Figure 10.5**), there is a more significant result, with the p value of .076. Married people are more likely to buy the American brands. You can speculate that this is probably because the American brands are generally larger vehicles, which make them more comfortable for families.

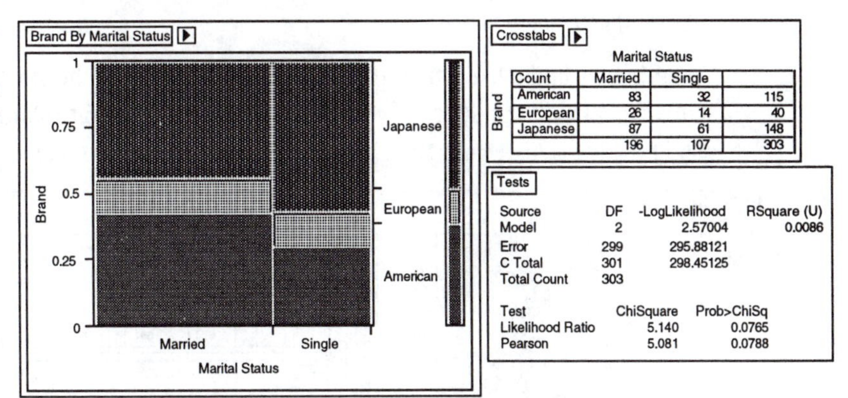

Count	Married	Single	
American	83	32	115
European	26	14	40
Japanese	87	61	148
	196	107	303

Source	DF	-LogLikelihood	RSquare (U)
Model	2	2.57004	0.0086
Error	299	295.88121	
C Total	301	298.45125	
Total Count	303		

Test	ChiSquare	Prob>ChiSq
Likelihood Ratio	5.140	0.0765
Pearson	5.081	0.0788

Figure 10.5 Mosaic Plot, Crosstabs, and Tests Table for Car Brand by Marital Status

Car Brand by Size of Vehicle

If marital status is a proxy for size of vehicle, looking at brand by size will give more direct information.

The Tests table for brand by size shows a strong relationship with a very significant chi-square. The Japanese dominate the large market for small cars, the Americans dominate the smaller market for large cars, and the European share is about the same in all three markets. The relationship is highly significant, with p values less than .0001.

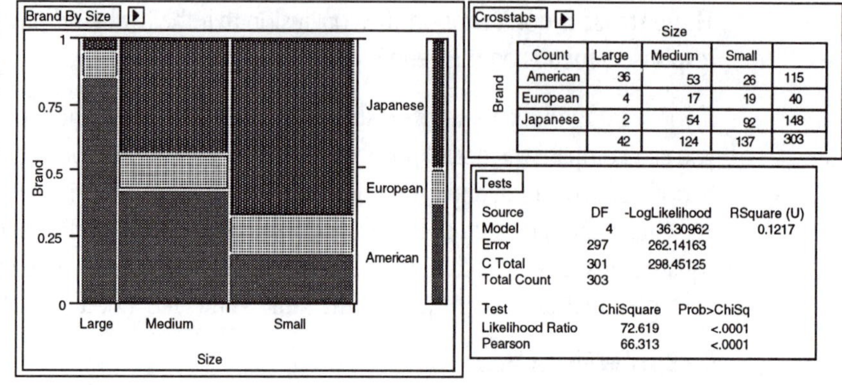

Crosstabs

Count	Large	Medium	Small	
American	36	53	26	115
European	4	17	19	40
Japanese	2	54	92	148
	42	124	137	303

Tests

Source	DF	-LogLikelihood	RSquare (U)
Model	4	36.30962	0.1217
Error	297	262.14163	
C Total	301	298.45125	
Total Count	303		

Test	ChiSquare	Prob>ChiSq
Likelihood Ratio	72.619	<.0001
Pearson	66.313	<.0001

Figure 10.6 Mosaic Plot, Crosstabs, and Tests Table for Car Brand by Size

If You Have a Perfect Fit

If a fit is perfect, every response category is predicted with probability 1. The response is completely determined by each sample (each subgroup). If the fit contributes nothing, then each distribution of the response in each sample subgroup is the same.

You probably find it likely that there is a perfect fit between the city and the state of a person's residence. If you know the city, then you are almost sure you know the state. **Figure 10.7** shows what a perfect fit looks like.

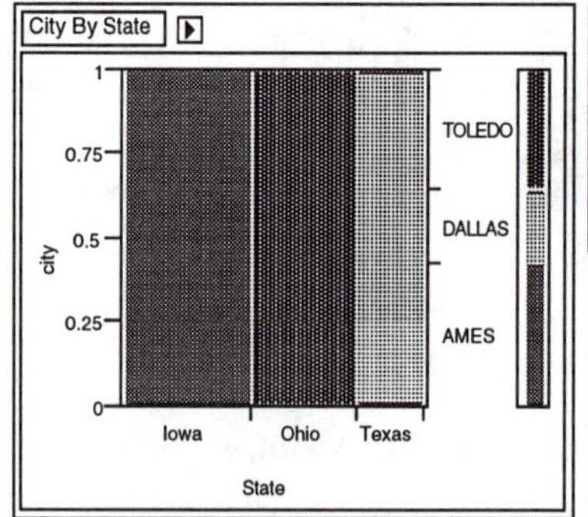

Crosstabs

Count	Iowa	Ohio	Texas	
AMES	67	0	0	67
DALLAS	0	0	34	34
TOLEDO	0	55	0	55
	67	55	34	156

Tests

Source	DF	-LogLikelihood	RSquare (U)
Model	4	165.76355	1.0000
Error	150	0.00000	
C Total	154	165.76355	
Total Count	156		

Test	ChiSquare	Prob>ChiSq
Likelihood Ratio	331.527	<.0001
Pearson	312.000	<.0001

Figure 10.7 Mosaic Plot, Crosstabs, and Tests Table for City by State

Note that when the analysis includes the people from Austin, a second city in Texas, city still predicts state perfectly, but not the other way around (state does not predict city). The chi-squares are the same because they are invariant if you switch the Y and X variables. But the mosaic and the attribution of the log-likelihood and RSquare are different.

Figure 10.8 Comparison of Plots, Tables, and Tests When X and Y are Switched

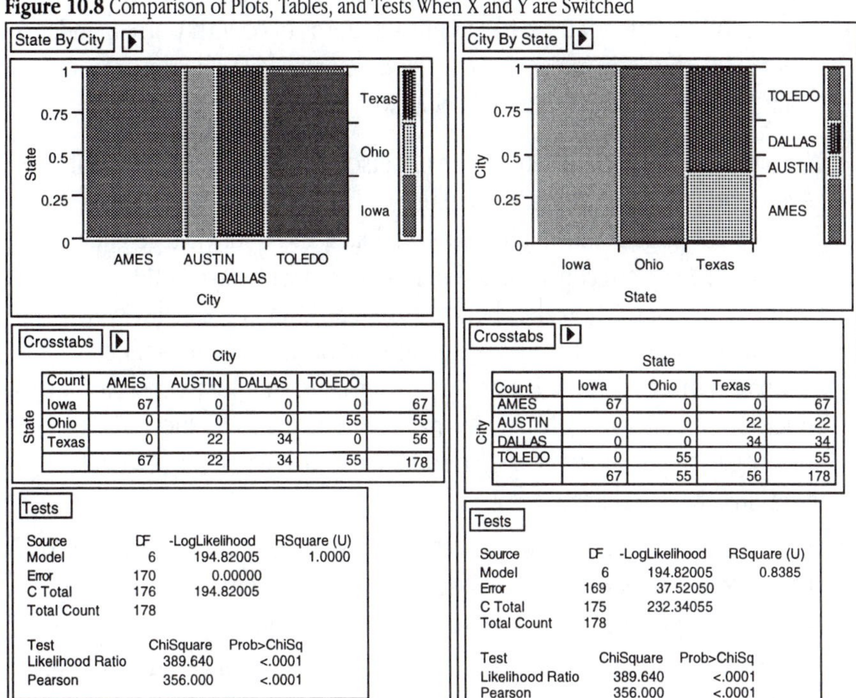

What happens if the response rates are the same in each cell as shown in the table in **Figure 10.9**? If you manufacture data for this situation, the mosaic levels line up perfectly and the chi-squares are zero.

Figure 10.9 Mosaic Plot, Frequencies, and Tests Table when Response Rates Are the Same

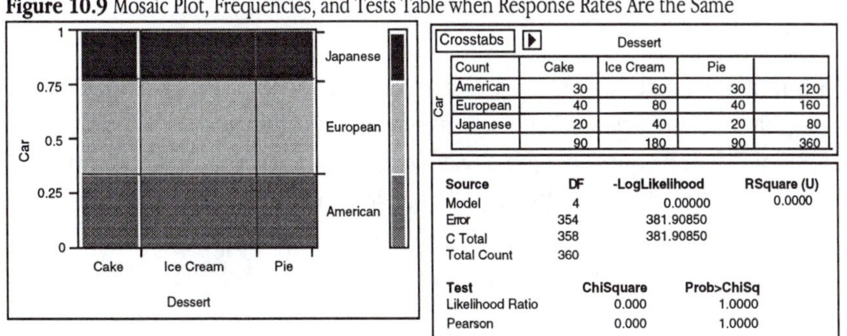

Correspondence Analysis: Looking at Data with Many Levels

Correspondence analysis is a graphical technique that shows which rows or columns of a frequency table have similar patterns of counts. Correspondence analysis is particularly valuable when you have many levels because it is difficult to find patterns in tables or mosaic plots with many levels.

The data table MBTIED.JMP has counts of personality types by educational level (**Educ**) and **Gender** (Myers and McCaulley). The values of educational level are D for Dropout, HS for high school graduate, and C for College graduate. **Gender** and **Educ** are concatenated to form the variable **GenderEd**. The goal is to determine the relationships between **GenderEd** and personality type as evidenced by the data. There is no implication of any cause-and-effect relationship because there is no way to tell from it whether personality affects education or education affects personality. But the data can show trends. The following example shows how correspondence analysis can help identify trends in categorical data:

§ Open the data table called MBTIED.JMP and choose **Analyze→Fit Y by X** with **MBTI** as **Y** and **GenderEd** as **X**. Now try to make sense out of the resulting table. It has with 96 cells—too big to understand at a glance.

§ Now try a correspondence analysis. Select **Correspondence Analysis** from the popup menu next to the Mosaic Plot title to see the plot in **Figure 10.11**.

The Correspondence Analysis plot organizes the row and column profiles in a two-dimensional space so that the X values that have similar Y profiles tend to cluster together, and the Y values that have similar X profiles cluster together. In this case, you want to see how the **GenderEd** groups are associated with the personality groups.

This plot quickly shows patterns. Gender and the Feeling(F)/Thinking(T) component form a cluster, and education clusters with the Intuition(N)/Sensing(S) personality indicator. The Extrovert(E)/Introvert(I) and Judging(J)/perceiving(P) types do not separate much. The most separation among these is the Judging(J)/Perceiving(P) separation among the Sensing(S)/Thinking (T) types (mostly noncollege men).

Figure 10.10 Mosaic Plot and Table for MBTI by GenderEd

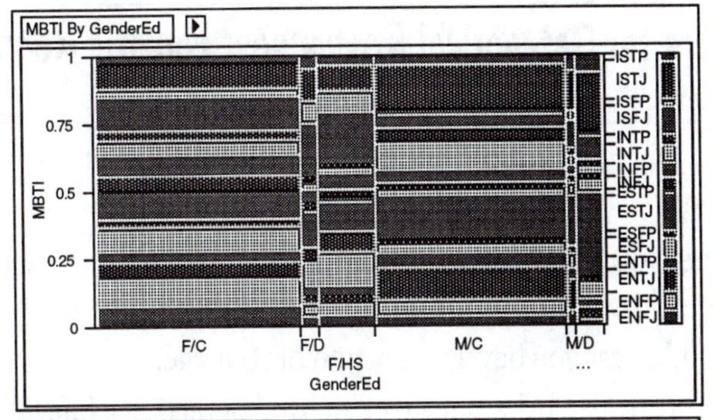

Crosstabs

GenderEd

Count	F/C	F/D	F/HS	M/C	M/D	M/HS	
ENFJ	376	15	39	185	5	8	628
ENFP	453	20	81	264	7	9	834
ENTJ	321	13	39	508	23	26	930
ENTP	193	4	19	251	8	17	492
ESFJ	423	41	189	158	11	38	860
ESFP	142	23	108	69	6	12	360
ESTJ	400	47	126	683	38	186	1480
ESTP	40	10	35	117	10	33	245
INFJ	325	10	48	145	6	7	541
INFP	351	18	71	203	7	5	655
INTJ	274	12	32	425	8	14	765
INTP	168	7	18	248	6	17	464
ISFJ	575	78	256	200	20	46	1175
ISFP	144	32	106	85	8	13	388
ISTJ	485	49	129	761	30	137	1591
ISTP	66	17	41	144	12	38	318
	4736	396	1337	4446	205	606	11726

(MBTI)

Figure 10.11 Correspondence Analysis Plot

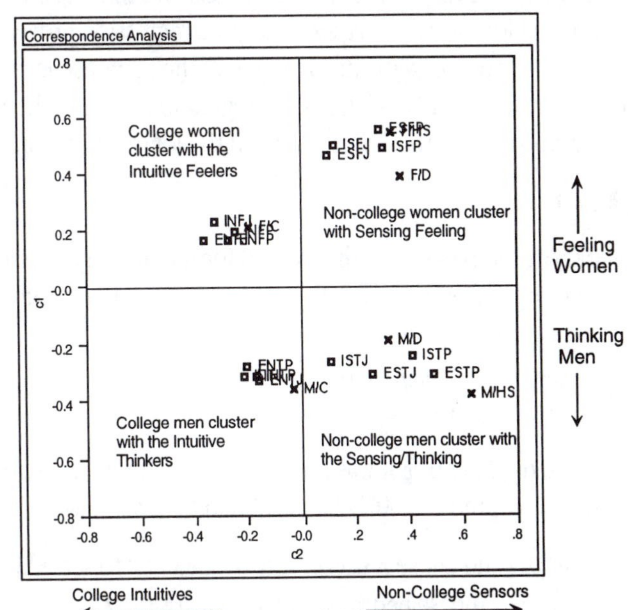

College Intuitives ← → Non-College Sensors

The Extravert/Introvert and Judging/Perceiving do not separate well for education and gender.

Continuous Factors for Categorical Responses: Logistic Regression

Suppose that a response is categorical, but the probabilities for the response change as a function of a continuous predictor. Some situations like this are the following:

- Whether you bought a car this year as a function of your disposable income. (The more you make, the more likely you are to buy a new car.)
- The kind of car you buy as a function of your age.
- The probability of whether a patient lived or died as a function of blood pressure.
- Whether favorite color is a function of height.
- Whether you avoided infection as a function of your dose of innoculation.
- Who you voted for as a function of your frustration index.
- Whether the treatment made your headache better as a function of the level of some chemical in your blood.

Problems where the response (Y) is categorical and the factor (X) is continuous call for *logistic regression*. Logistic regression provides a method to estimate the probability of choosing one of the response levels as a smooth function of the factor. It is called logistic regression because the S-shaped curve it uses to fit the probabilities is called the logistic function.

Fitting a Logistic Model

The SPRING.JMP sample is a weather record for the month of April. The variable **Precip** measures rain fall and the variable **Rained** categorizes rainfall using this calculator formula:

$$\begin{cases} \text{"Rainy"}, & \text{if } Precip > 0.02 \\ \text{"Dry"}, & \text{otherwise} \end{cases}$$

Out of the 30 days in April, it was rainy 9 days. Therefore, if there is no other information, you predict a 30% chance of rain for every day.

§ Open SPRING.JMP and chose **Analyze→Distribution of Y** to see a histogram and frequency table of the **Rained** variable as in **Figure 10.12.**

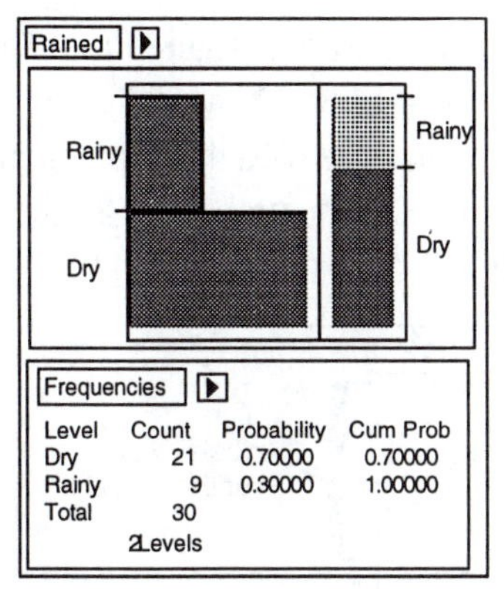

Level	Count	Probability	Cum Prob
Dry	21	0.70000	0.70000
Rainy	9	0.30000	1.00000
Total	30		

2 Levels

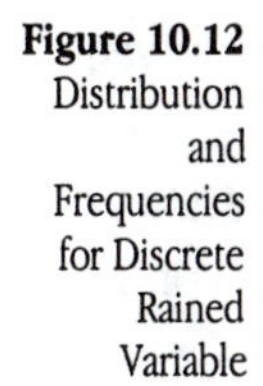

Figure 10.12 Distribution and Frequencies for Discrete Rained Variable

Now suppose you look at the temperature in the morning and then try to give a precipitation probability. Knowing the barometric pressure also might help in finding a more informative prediction.

In each case, the thing being modeled is the probability of getting one of several responses. The probabilities are constrained to add to 1. In the simplest situation, like this rain example, the binary response has two levels. Remember that statisticians like to take logs of probabilities. In this case what they fit is the difference in logs of the two probabilities as a linear function of the factor variable.

If p denotes the probability for the first response level, then $1-p$ is the probability of the second, and the linear model is written:

$$\log(p) - \log(1-p) = b_0 + b_1 {*} X \quad \text{or} \quad \log(p/(1-p)) = b_0 + b_1 {*} X$$

where $\log(p/(1-p))$ is called the *logit* of p or the *log odds-ratio*.

There is no error term here because the predicted value is not a response level. It is a probability distribution for a response level. For example, if the weather report predicts a 90% chance of rain, you don't say he erred if it didn't rain.

The accounting is done by summing the negative logarithms of the probabilities attributed by the model to the events that actually did occur. So if p is the precipitation probability from the weather model, then the score is $-\log(p)$ if it rains, and $-\log(1-p)$ if it doesn't. A weather reporter that is a perfect predictor comes up with a p of 1 when it rains ($-\log(p)$ is zero if p is 1) and a p of zero when it doesn't rain ($-\log(1-p)=0$ if $p=0$]). The perfect score is zero. No surprise ($-\log(p) = 0$) means perfect predictions.

So the inverse logit of the model $b_0+b_1{*}X$ expresses the probability for each response level, and the estimates are found so as to maximize the likelihood. That is the same as minimizing the negative sum of logs of the probabilities attributed to the response levels that actually occurred for each observation.

You can graph the probability function as shown in **Figure 10.13.** The curve is the inverse expression for $\log(p/(1-p)) = b_0+b_1*X$, which is $p = 1/(1+\exp(-(b_0+b_1*X)))$. For a given value of X, this expression evaluates the probability of getting the first response. The probability for the second response is the remaining probability, $1-p$, so that they sum to 1.

Figure 10.13 Logistic Regression Fits Probabilities of a Response Level

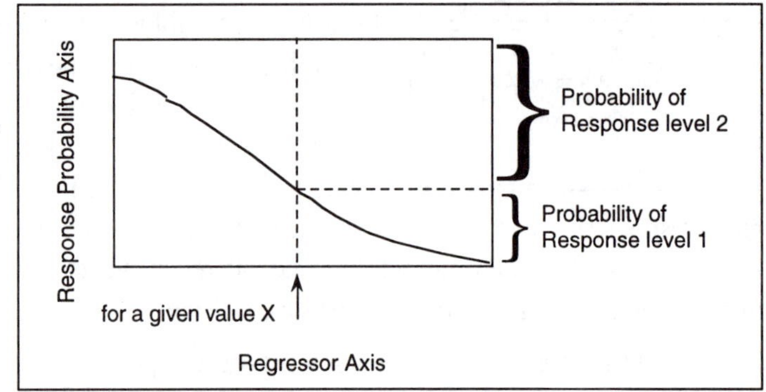

To fit the rain column by temperature and barometric pressure for the spring rain data:

§ Chose **Analyze→Fit Y by X** specifying the nominal column Rained as Y, and the continuous columns Temp and Pressure as the X's. The Fit Y by X platform does a separate logistic regression for each predictor column.

The cumulative probability plot on the left in **Figure 10.14** shows that the relationship with temperature is very weak. As the temperature ranges from 35 to 75, the probability of dry weather only changes from .73 to .66. The line of fit partitions the whole probability into the response categories. In this case, you read the probability of Dry directly on the vertical axis. The probability of Rained is the distance from the line to the top of the graph, which is 1 minus the axis reading. The weak relationship is evidenced by the very flat line of fit; the precipitation probability doesn't change much over the temperature range.

The plot on the right in **Figure 10.14** indicates a much stronger relationship with barometric pressure. When the pressure is 29 inches, the fitted probability of rain is near 100% (0 probability for Dry at the left of the graph). The curve crosses the 50% level at 29.32. At 29.8, the probability of rain drops near zero (near 1.0 for Dry).

Figure 10.14 Cumulative Probability Plot for Discrete Rain Data as a Function of Temperature and Pressure

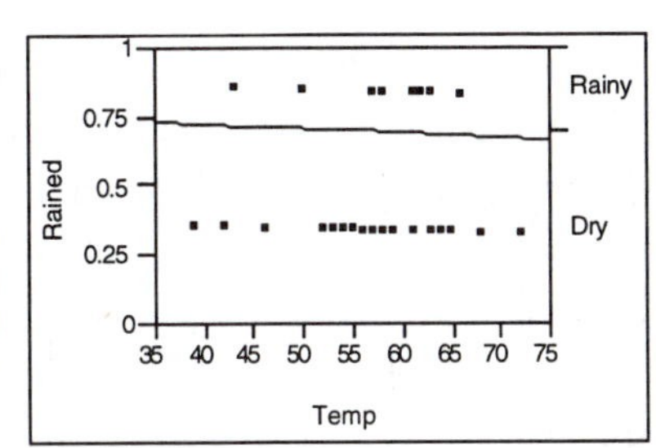

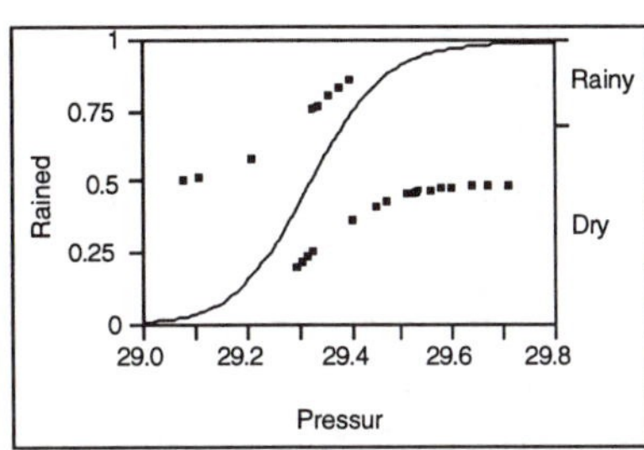

The Whole-Model Test table and the Parameter Estimates table support the plot. The RSquare measure of fit, which is on a scale of 0 to 100%, is only .07% (see **Figure 10.15**). To get 100%, the model would have to predict outcomes with certainty and be right every time. The likelihood ratio chi-square is not at all significant. The coefficient on temperature is a very small, –.008. Note that the parameter estimates are marked unstable. This is because they have high standard errors with respect to the estimates.

Figure 10.15 Logistic Regression for Discrete Rain Data as a Function of Temperature and Pressure

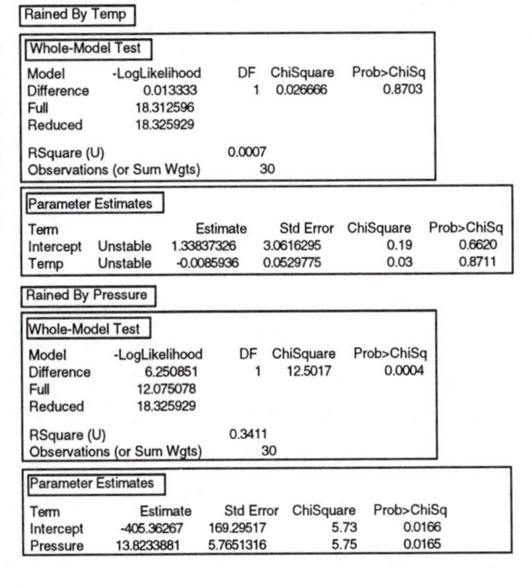

Rained By Temp

Whole-Model Test

Model	-LogLikelihood	DF	ChiSquare	Prob>ChiSq
Difference	0.013333	1	0.026666	0.8703
Full	18.312596			
Reduced	18.325929			

RSquare (U)	0.0007
Observations (or Sum Wgts)	30

Parameter Estimates

Term		Estimate	Std Error	ChiSquare	Prob>ChiSq
Intercept	Unstable	1.33837326	3.0616295	0.19	0.6620
Temp	Unstable	-0.0085936	0.0529775	0.03	0.8711

Rained By Pressure

Whole-Model Test

Model	-LogLikelihood	DF	ChiSquare	Prob>ChiSq
Difference	6.250851	1	12.5017	0.0004
Full	12.075078			
Reduced	18.325929			

RSquare (U)	0.3411
Observations (or Sum Wgts)	30

Parameter Estimates

Term	Estimate	Std Error	ChiSquare	Prob>ChiSq
Intercept	-405.36267	169.29517	5.73	0.0166
Pressure	13.8233881	5.7651316	5.75	0.0165

In contrast, the overall RSquare measure of fit with barometric pressure is 34%. The likelihood ratio chi-square is significant and the parameter coefficient for Pressureincreased to 13.8 (see **Figure 10.15**).

The conclusion is that if you want to predict whether the weather will be rainy, it doesn't help to know the temperature, but it does help to know the barometric pressure.

Degrees of Fit

The illustrations in **Figure 10.16** summarize the degree of fit as shown by the cumulative logistic probability plot.

When the fit is weak, the parameter for the slope term (X factor) in the model is small, which gives little slope to the line in the range of the data. A perfect fit means that before a certain value of X, all the responses are another level, and after that value of X, all the responses are the other level. A strong model can bet almost all of its probability on one event happening. A weak model has to bet conservatively with the background probability, less affected by the X factor's values.

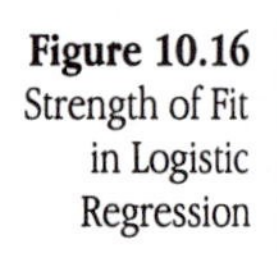

Figure 10.16 Strength of Fit in Logistic Regression

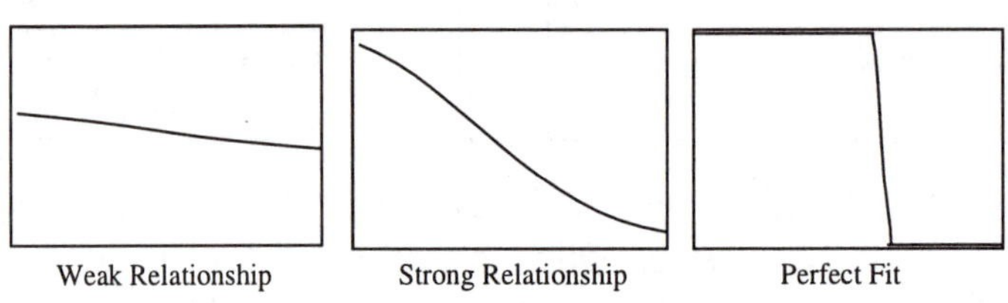

Note that when the fit is perfect, as shown on the right graph of **Figure 10.16**, the slope of the logistic line approaches infinity. This means that the parameter estimates become infinite also. In practice, the estimates are allowed to inflate only until the likelihood converges and are marked as *unstable* by the computer program. You can still test hypotheses because they are handled through the likelihood, rather than using the estimate's (theoretically infinite) values.

Special Topics

A Discriminant Alternative

There is another way to think of the situation where the response is categorical and factor is continuous. You can reverse the roles of the Y and X and treat this problem as one of finding the distribution of temperature and pressure on rainy and dry days. Then work backwards to obtain prediction probabilities. This technique is called discriminant analysis.

§ For this example, choose **Analyze→Fit Y by X** specifying the continuous columns Temp and Pressure as the Y variables and the nominal variable Rained as X.

§ Select **Means Diamonds** from the Display popup menu showing beneath the plots to see the results in **Figure 10.17**.

You can quickly see that the difference between the relationships of temperature and pressure to raininess. But the discriminant approach is a somewhat strange way to go about this example and has some problems:

- The standard analysis of variance assumes that the factor distributions are normal.
- Discriminant analysis works backwards: First, in the weather example you are trying to predict rain. But the Anova approach designates **Rained** as the independent variable, from which you can say something about the predictability of temperature and pressure. Then, you have to reverse engineer your thinking to infer raininess from temperature and pressure.

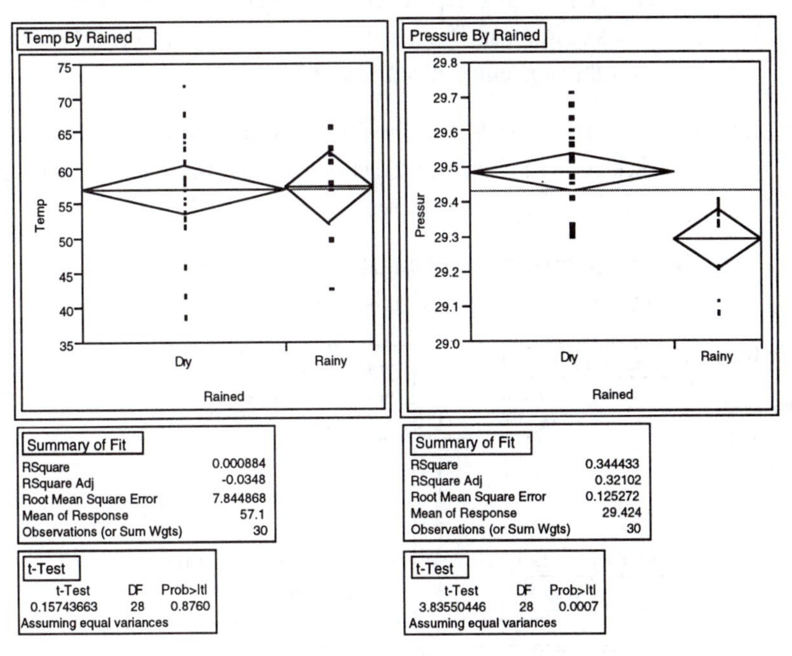

Figure 10.17 One-Way Analysis of Variance for Temperature and Pressure as a Function of Discrete Rain Variable

Inverse Prediction

If you want to know what value of the X regressor yields a certain probability, you can solve the equation, $\log(p/(1-p)) = b_0 + b_1 * X$, for X, given p. This is often done for toxicology situations, where the X value for p=50% is called an LD50 (Lethal Dose for 50%). Confidence intervals for these inverse predictions (called *fiducial confidence intervals)* can be obtained.

The Fit Model platform has an inverse prediction facility. Let's find the LD50 for pressure in the rain data—that is the value of pressure that gives a 50% chance of rain:

§ Choose **Analyze→Fit Model.** When the Fit Model dialog appears, select **Rained** in the variable selection list and assign it as **Y**. Select **Pressure** and assign it as the **Effect in Model.**

§ Click **Run Model**. When the platform appears, select **Inverse Prediction** from the check-mark popup menu at the lower left of the window. This displays the dialog at the left in **Figure 10.18**.

The Probability and 1–Alpha fields are editable. You can fill the dialog with any values of interest. The result is an inverse probability for each probability request value you entered, at the alpha level specified.

§ For this example, click .5 as the first entry in the **Probability** column. When you click **Done** the Inverse Prediction table shown in **Figure 10.18** appears appended to the platform tables.

The inverse prediction computations say that there is a 50% chance of rain when the barometric pressure is 29.32.

Figure 10.18 Inverse Prediction Dialog

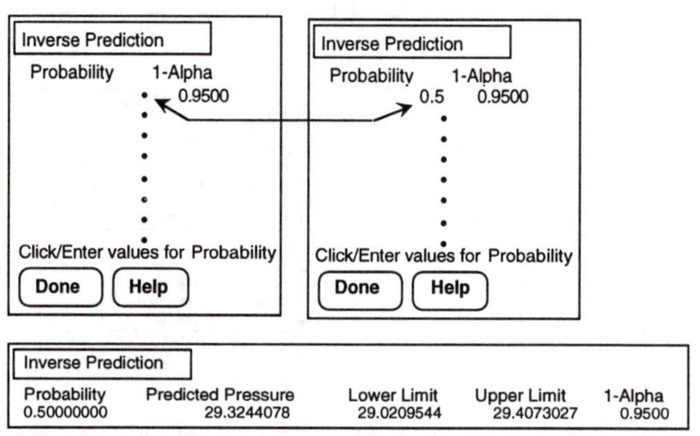

Probability	Predicted Pressure	Lower Limit	Upper Limit	1-Alpha
0.50000000	29.3244078	29.0209544	29.4073027	0.9500

You can verify these calculations visually with the Cumulative Logistic plot for **Rained** and **Pressure**. (This plot was the previous result when you did **Analyze→Fit Y by X** with **Rained** as **Y** and **Pressure** as **X**.)

§ Get the crosshair tool from the **Tools** menu. Click and drag the crosshair tool on the logistic plot until the horizontal line is at the .50 value of **Rained**.

§ Hold the crosshair at that value and drag to the logistic curve. You can then read the Pressure value, just over 29.3, on the X axis.

Figure 10.19 Inverse Prediction with Reference Lines

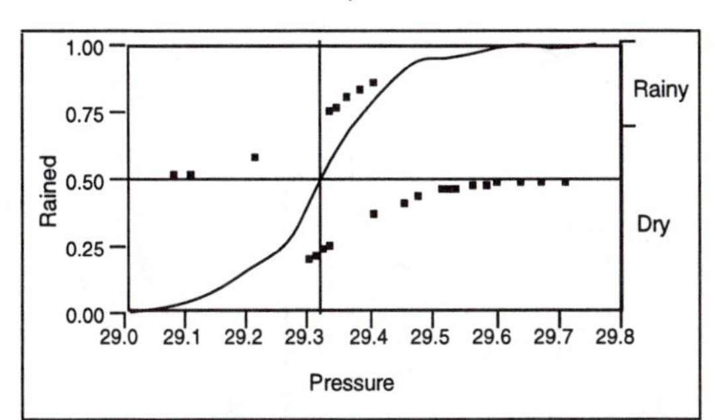

Alternatively, you can add reference lines at the known X and Y values. Double-click on the Rain (Y) axis to bring up an axis modification dialog. Enter .5 as a reference line. Double-click on the Pressure (X) axis and enter 29.32 in the axis modification dialog. When both reference lines appear, they intersect on the logistic curve as shown in **Figure 10.19**.

Polytomous: More Than 2 Response Levels

If you have more than two response probabilities, then you use a generalization of the logistic model for these *polytomous* responses. For the curves to be very flexible, you have to fit a set of linear model parameters for each of *r*–1 response levels. The logistic curves are cumulated in such a way as to form a smooth partition of the probability space as a function of the regression model. In **Figure 10.20**, the probabilities are the distances between the curves, which add up to 1.

Where the curves come close together, the model is saying the probability of a response level is very low. Where the curves separate widely, the fitted probabilities are large. See **Figures 10.20** and **10.21**.

Figure 10.20 Polytomous Logistic Regression with 5 Response Levels

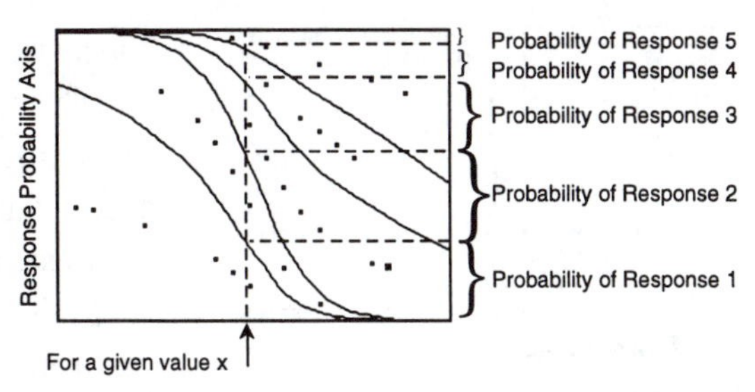

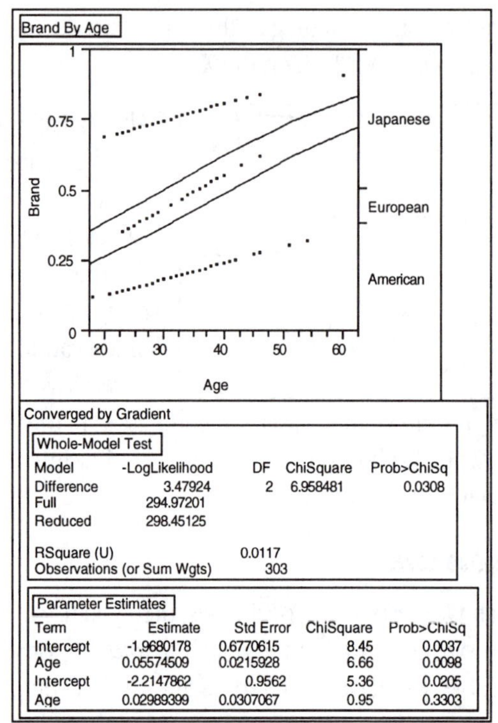

Converged by Gradient

Whole-Model Test

Model	-LogLikelihood	DF	ChiSquare	Prob>ChiSq
Difference	3.47924	2	6.958481	0.0308
Full	294.97201			
Reduced	298.45125			

RSquare (U)	0.0117
Observations (or Sum Wgts)	303

Parameter Estimates

Term	Estimate	Std Error	ChiSquare	Prob>ChiSq
Intercept	-1.9680178	0.6770615	8.45	0.0037
Age	0.05574509	0.0215928	6.66	0.0098
Intercept	-2.2147862	0.9562	5.36	0.0205
Age	0.02989399	0.0307067	0.95	0.3303

Figure 10.21 Cumulative Probability Plot and Logistic Regression for Car Poll Data, Country by Age

Ordinal Responses: Cumulative Ordinal Logistic Regression

In some cases, you don't need the full generality of multiple linear model parameter fits for the r–1 cases, but can assume that the logistic curves are the same, only shifted by a different amount over the response levels. This means that there is only one set of regression parameters on the factor, but r–1 intercepts for the r responses.

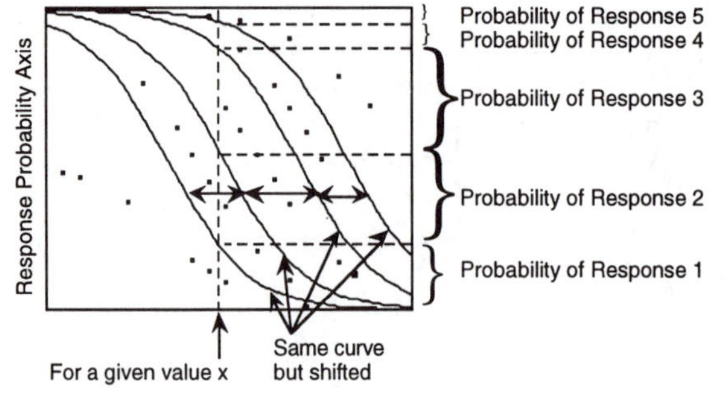

Figure 10.22 Ordinal Logistic Regression Cumulative Probability Plot

The logistic curve is actually fitting the sum of the probabilities for the responses at or below it, so it is called a cumulative ordinal logistic regression.

In the SPRING data table there is a column called **SkyCover** with values 1 to 10. First, note that you don't need to treat the response as nominal because the data have a natural order. Also, in this example, there is not enough data to support the large number of parameters needed by a 10-response level nominal model. Instead you can use a logistic model that fits **SkyCover** as an ordinal variable with the continuous variables **Temp** and **Humid1: PM.**

§ Change the modeling type of the **SkyCover** column to Ordinal using the modeling type popup menu at the top of its column in the SPRING table.

§ Chose **Analyze→Fit Y by X and** specify **SkyCover** as Y, and columns **Temp** and **Humidity** as the X's.

Figure 10.23 indicates that the relationship of **SkyCover** to **Temp** is very weak, with an RSquare of .09%, fairly flat lines, and a nonsignificant chi-square. The direction of the relation is that the higher sky covers are more likely with the higher temperatures.

Figure 10.23 Ordinal Cumulative Logistic Regression for Ordinal Sky Cover with Temperature

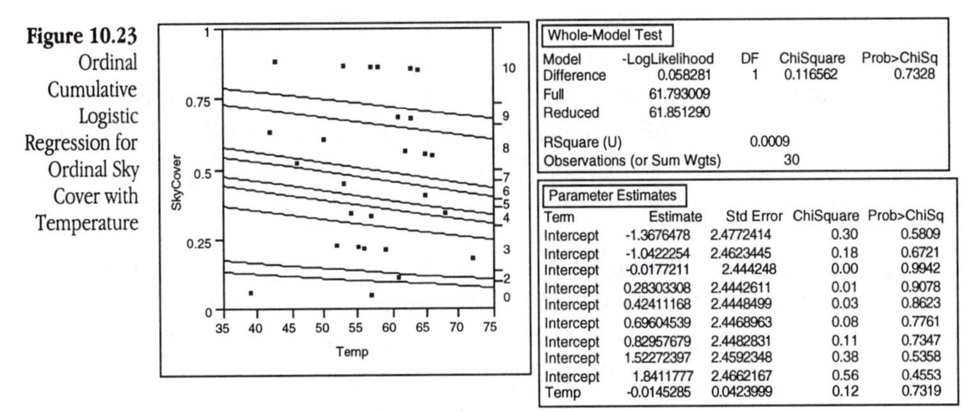

Whole-Model Test

Model	-LogLikelihood	DF	ChiSquare	Prob>ChiSq
Difference	0.058281	1	0.116562	0.7328
Full	61.793009			
Reduced	61.851290			

RSquare (U)	0.0009
Observations (or Sum Wgts)	30

Parameter Estimates

Term	Estimate	Std Error	ChiSquare	Prob>ChiSq
Intercept	-1.3676478	2.4772414	0.30	0.5809
Intercept	-1.0422254	2.4623445	0.18	0.6721
Intercept	-0.0177211	2.444248	0.00	0.9942
Intercept	0.28303308	2.4442611	0.01	0.9078
Intercept	0.42411168	2.4448499	0.03	0.8623
Intercept	0.69604539	2.4468963	0.08	0.7761
Intercept	0.82957679	2.4482831	0.11	0.7347
Intercept	1.52272397	2.4592348	0.38	0.5358
Intercept	1.8411777	2.4662167	0.56	0.4553
Temp	-0.0145285	0.0423999	0.12	0.7319

Figure 10.24 indicates that the relationship with humidity is very strong. As the humidity rises to 70%, it predicts a 50% probability of a sky cover of 10. At 100% humidity, the sky cover will be almost certainly 10. The RSquare is 29% and the likelihood ratio chi-square is highly significant.

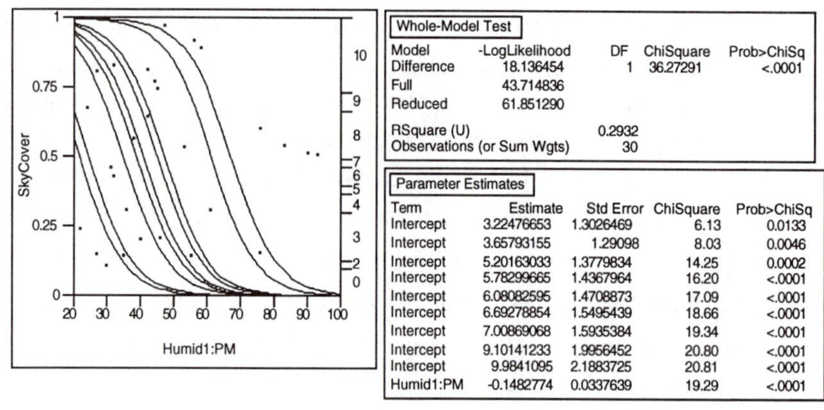

Whole-Model Test

Model	-LogLikelihood	DF	ChiSquare	Prob>ChiSq
Difference	18.136454	1	36.27291	<.0001
Full	43.714836			
Reduced	61.851290			

RSquare (U)	0.2932
Observations (or Sum Wgts)	30

Parameter Estimates

Term	Estimate	Std Error	ChiSquare	Prob>ChiSq
Intercept	3.22476653	1.3026469	6.13	0.0133
Intercept	3.657793155	1.29098	8.03	0.0046
Intercept	5.20163033	1.37779834	14.25	0.0002
Intercept	5.78299665	1.4367964	16.20	<.0001
Intercept	6.08082595	1.4708873	17.09	<.0001
Intercept	6.692788854	1.5495439	18.66	<.0001
Intercept	7.00869068	1.5935384	19.34	<.0001
Intercept	9.10141233	1.9956452	20.80	<.0001
Intercept	9.9841095	2.1883725	20.81	<.0001
Humid1:PM	-0.1482774	0.0337639	19.29	<.0001

Figure 10.24 Logistic Regression for Ordinal Sky Cover with Humidity

There is a useful alternative interpretation to this ordinal model. Suppose you assume there is some continuous response with a random error component that the linear model is really fitting. But for some reason you can't observe the response directly. You are given a number that indicates which of r ordered intervals contains the actual response, but you don't know how the intervals are defined. You assume that the error term follows a logistic distribution, which has a similar shape to a normal distribution. This case is identical to the ordinal cumulative logistic model, and the intercept terms are estimating the threshold points that define the intervals corresponding to the response categories.

Unlike the nominal logistic model, the ordinal cumulative logistic model is efficient to fit for even hundreds of response levels. It can even be used effectively for continuous responses when there are n unique response levels for n observations. In this case n–1 intercept parameters constrained to be in order, and there is one parameter for each regressor.

Surprise: Simpson's Paradox: Aggregate Data versus Grouped Data

Several statisticians have studied the *hot hand* phenomenon in basketball. The idea is that basketball players seem to have hot streaks, when they make most of their shots, alternating with cold streaks when they shoot poorly. The HOTHAND.JMP table contains the free throw shooting records for two Boston Celtics players over the 1980-81 and 1981-82 seasons (Tversky and Gilovich, 1989).

The null hypothesis is that two sequential free throw shots are independent. There are two directions in which they could be nonindependent, the positive relationship (hot hand) and a negative relationship (cold hand).

The HOTHAND.JMP sample data have the columns **First** and **Second** (first shot and second shot) for two players and a count variable. There are 4 possible shooting combinations, hit-hit, hit-miss, miss-hit, and miss-miss.

§ Open HOTHAND.JMP. Choose **Analyze→Fit Y by X** and select **Second** as **Y**, **First** as **X**, and **Count** as the **Freq** variable, then click **OK**. When the report appears, select **Col%** from the popup menu next to the Crosstabs table name.

The results (see **Figure 10.25**) show that if the first shot is made, then the probability of making the second is .811; if the first shot is missed the probability of making the second is .729. This tends to support the hot hand hypothesis. The two chi-square statistics are on the border of .05 significance.

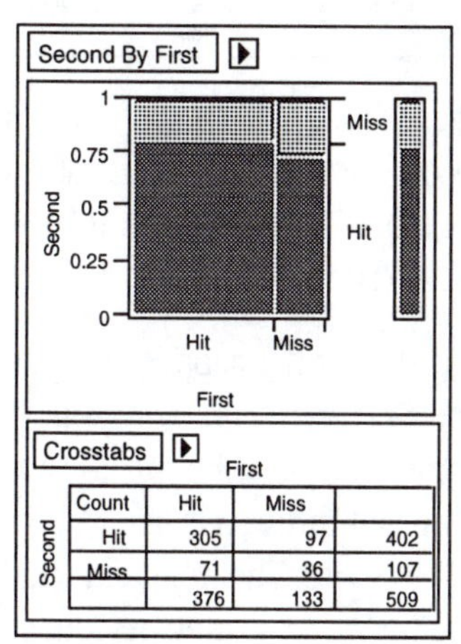

Count	Hit	Miss	
Hit	305	97	402
Miss	71	36	107
	376	133	509

Figure 10.25 Crosstabs and Tests for HOTHAND Basketball Data

Tests

Source	DF	-LogLikelihood	RSquare (U)
Model	1	1.90667	0.0073
Error	507	259.84294	
C Total	508	261.74962	
Total Count	509		

Test	ChiSquare	Prob>ChiSq
Likelihood Ratio	3.813	0.0508
Pearson	3.964	0.0465

Fisher's Exact Test	Prob
Left	0.9813
Right	0.0326
2-Tail	0.0627

Kappa	Std Err
0.087366	0.04626

Kappa measures the degree of agreement.

But does this really work? A researcher (Wardrop 1995), looked at contingency tables for each player. You can do this using the **Group/Summary** command.

§ Choose **Tables→Group/Summary**, select **Player** as the grouping variable, and click **OK**.

§ When the summary table appears, click the **By-Mode** button in the upper left corner of the table so that it says **By-Mode On**.

§ Again Choose **Analyze→Fit Y by X** with **Second** as **Y**, **First** as **X**, and **Count** as the **Freq** variable. The results for the two players are shown in **Figure 10.26**.

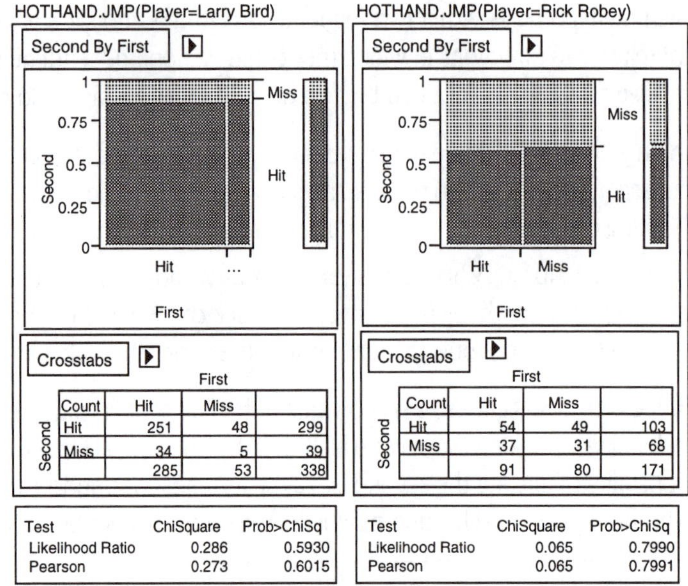

Figure 10.26 Crosstabs and Tests for grouped HOTHAND Basketball Data

Contrary to the first result, both players shot better the second time after a miss than after a hit. So how can this be when the aggregate table gives the opposite results from both individual tables? This is an example of a phenomenon called Simpson's paradox (Simpson, 1951, Yule, 1903).

In this example, it is not hard to understand what happens if you think how the aggregated table works. If you see a hit on the first throw, it is probably Larry Bird, and because he is usually more accurate he will likely hit the second basket. If you see a miss on the first throw, it is likely by Rick Robey, so the second throw will be less likely to score. The hot hand relationship is an artifact that the players are much different in scoring percentages generally and populate the aggregate unequally.

A better way to summarize the aggregate data, taking into account these background relationships, is to use a blocking technique called the Cochran-Mantel-Haenszel test.

§ Click the aggregate Fit Y by X platform to make it active and choose **Cochran-Mantel-Haenszel** from the popup menu next to the mosaic plot name. Select Player as the grouping dialog for this test.

When you click **OK**, the table in **Figure 10.27** appears. These results are more accurate because they are based on the grouping variable instead of the ungrouped data.

Figure 10.27 Crosstabs and Tests for Grouped HOTHAND Basketball Data

Cochran-Mantel-Haenszel Tests

Stratified by Player

CMH Test	Chi-Square	DF	Prob>Chisq
Correlation of Scores	0.2507	1	0.6166
Row Score by Col Categories	0.2507	1	0.6166
Col Score by Row Categories	0.2507	1	0.6166
General Assoc. of Categories	0.2507	1	0.6166

Chapter 11
Multiple Regression

Overview

Multiple regression is the technique of fitting or predicting a response variable by a linear combination of several other variables. The fitting principle is least squares, the same as with simple linear regression.

Most of the regression concepts were introduced in previous chapters, so in this chapter we concentrate on showing some new concepts not encountered in simple regression: the point-by-point picture of a hypothesis test with the leverage plot; the effect of collinearity, (the situation in which one regressor variable is closely related to another), and the case of exact linear dependencies.

Chapter 11 Contents

You can mouse along with the examples whenever you see an action symbol (§)

Parts of a Regression Model

Linear regression models are the sum of the products of coefficient parameters and factor columns. Linear models for continuous responses are usually specified with a normal error term. The parameters are given the values that minimize the sum of squared residuals; this technique is called estimation by least squares.

Figure 11.1 Parts of a Linear Model

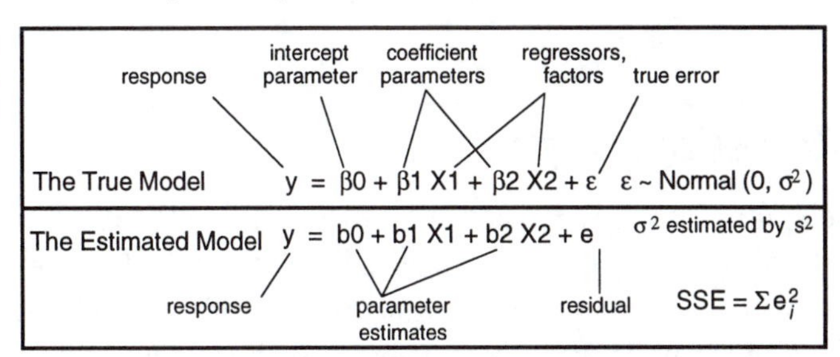

Note in **Figure 11.1** the differences in notation between the assumed true model with unknown parameters and the model you actually estimate.

response, y

The response or dependent variable is the one you want to predict using the observed response, y, in the regression model.

regressors, X's

The regressors (X) in the regression model are also called independent variables, factors, explanatory variables, and other discipline-specific effects. The regression model uses a linear combination of these effects to fit the response values.

coefficients, parameters

The fitting technique produces estimates of the parameters, which are the unknown coefficients for the linear combination that defines the regression model.

intercept term

Most models have intercept terms to fit the constant in a linear equation. This is equivalent to having a regressor variable that is always 1. The intercept is meaningful by itself only if it is meaningful to know the predicted value where all the regressors are zero. However, the intercept parameter has a strong role to play for testing the rest of the model because it carries the mean if all the other parameters (coefficients) are set to zero.

error, residual

If you can't fit perfectly, then there is error left over. Error is the difference between the actual and predicted value. When you are speaking of true parameters, it is called error. When you are using estimated parameters, then it is called residuals.

A Multiple Regression Example

Aerobic fitness can be evaluated using a special test that measures the oxygen uptake of a person running on a treadmill for a prescribed distance. However, it would be more economical to evaluate fitness with a formula that predicts oxygen uptake using simple measurements such as running time and pulse measurements.

To identify such an equation, run time and pulse measurements were taken for 31 participants who ran 1.5 miles and the oxygen uptake measurement was taken. (Rawlings 1988, data courtesy of A.C. Linnerud). **Figure 11.2** shows a partial listing of the data, with variables **Age**, **Weight**, **O2 Uptake** (the response measure), **Run Time**, **Rest Pulse**, **Run Pulse**, and **Max Pulse**.

Figure 11.2 The Oxygen Uptake Data Table

LINNERUD.JMP

7 Cols / 31 Rows	Age	Weight	O2 Uptake	Run Time	Rest Pulse	Run Pulse	Max Pulse
1	38	81.87	60.055	8.63	48	170	186
2	38	89.02	49.874	9.22	55	178	180
3	40	75.07	45.313	10.07	62	185	185
4	40	75.98	45.681	11.95	70	176	180

Suppose you want to investigate **Run Time** and **Run Pulse** as predictors of oxygen uptake (**O2 Uptake**):

§ Open the LINNERUD.JMP sample data table. Choose **Analyze→Fit Model** to see the Fit Model dialog.

§ Select the **O2 Uptake** column and click on **Y** to make it the response (Y) variable. Select **Run Time** and **Run Pulse**, and click **Add** to make them the **Effects in Model.** Your dialog should look like the one in **Figure 11.3.**

§ Click **Run Model** to launch the platform.

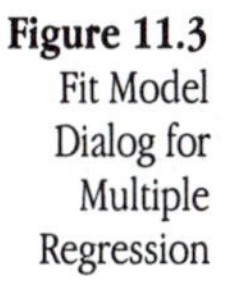

Figure 11.3 Fit Model Dialog for Multiple Regression

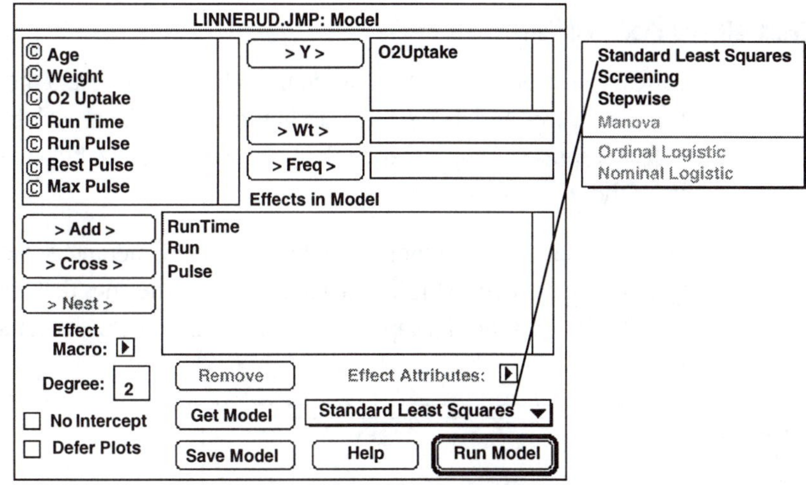

Now you have tables, as in **Figure 11.4**, that report on the regression fit:

- The Summary of Fit table shows that the model accounted for 76% of the variation around the mean (RSquare). The remaining residual error is estimated to have a standard deviation of 2.69 (Root Mean Square Error).
- The Parameter Estimates table shows Run Time to be highly significant with a negative sign (p<.0001), but Run Pulse is not significant (p=.15). Using these parameter estimates, the prediction equation is

  O2 Uptake = 93.089 – 3.14 Run Time – .0735 Run Pulse

- The Effect Test table shows details of how each regressor contributes to the fit.

Figure 11.4 Statistical Tables for Multiple Regression Example

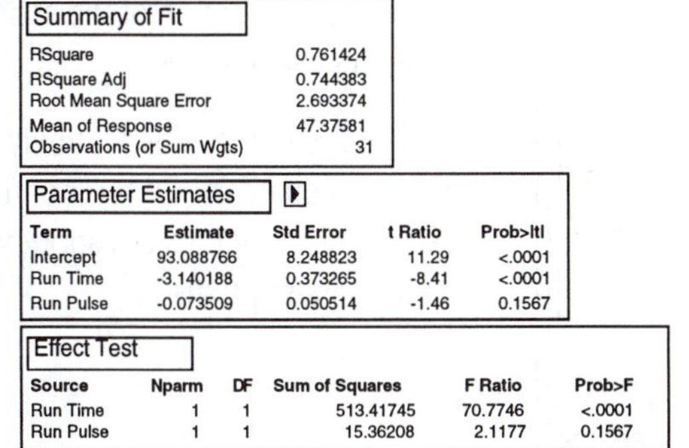

Response: O2 Uptake

Summary of Fit

RSquare	0.761424
RSquare Adj	0.744383
Root Mean Square Error	2.693374
Mean of Response	47.37581
Observations (or Sum Wgts)	31

Parameter Estimates

Term	Estimate	Std Error	t Ratio	Prob>|t|
Intercept	93.088766	8.248823	11.29	<.0001
Run Time	-3.140188	0.373265	-8.41	<.0001
Run Pulse	-0.073509	0.050514	-1.46	0.1567

Effect Test

Source	Nparm	DF	Sum of Squares	F Ratio	Prob>F
Run Time	1	1	513.41745	70.7746	<.0001
Run Pulse	1	1	15.36208	2.1177	0.1567

Residuals and Predicted Values

The residual is the difference between the actual response and the response predicted by the model. The residuals fit the error in the model. Large residuals represent points that don't fit very well. It can be helpful to look at a plot of the residuals and the predicted values:

§ Select the **Plot Residual** command in the check-mark menu to see the plot in **Figure 11.5** appended to the Whole Model report. The lines that connect the points to the reference line appear if you also select the **Connect Points to Line** option.

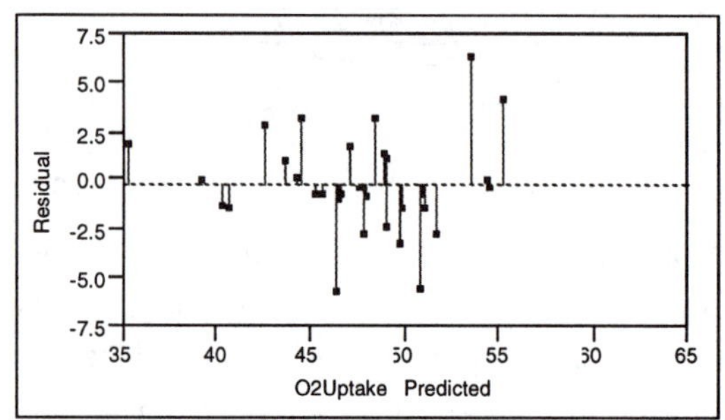

Figure 11.5 Residual Plot for Multiple Regression Example

You can save these residuals as a column in the data table:

§ Select **Save Residuals** from the Save ($) popup menu, found on the lower-left window border. The result is a new column in the data table called **Residual O2 Uptake**, which contains the residuals.

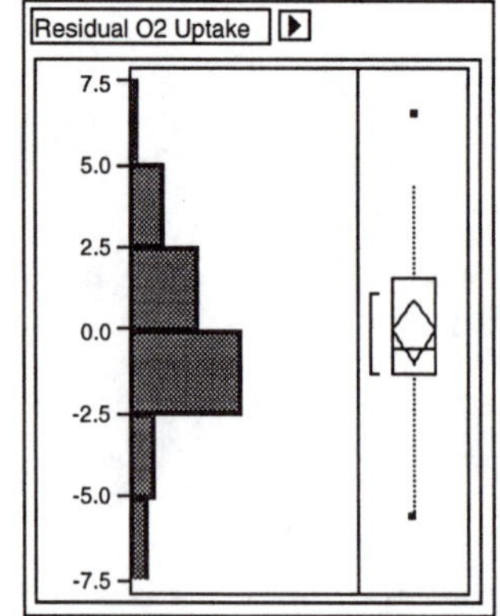

Figure 11.6 Histogram of Residuals in Multiple Regression Example

§ Choose **Analyze→Distribution of Y** to view the distribution of the residuals, as shown in **Figure 11.6**.

Many researchers do this routinely to verify that the residuals are not too nonnormal looking to warrant concern about violating normality assumptions.

You might also want to store the prediction formula from the multiple regression:

§ Select **Save Prediction Formula** from the Save ($) popup menu to create a new column in the LINNERUD data table called Pred Formula O2 Uptake. Its values are the calculated predicted values for the model.

§ To see the column's formula, click at the top of the Pred Formula O2 Uptake column to highlight it. Choose **Cols→Column Info** and click on the formula picture box in the lower left of the Column Info dialog. The calculator window opens and displays the formula,

93.0087761+–3.1410876•*Run Time*+–0.0735095•*Run Pulse*

This formula defines a plane of fit for O2 Uptake as a function of Run Time and Run Pulse. The formula stays in the column and is evaluated whenever new rows are added, or when variables used in the expression change their values. You can paste this formula into other tables, or paste it as a picture into a report.

The Analysis of Variance Table

The Whole-Model report consists of several tables that compare the full model fit to the simple mean model fit.

The Analysis of Variance table (**Figure 11.7**) lists the sums of squares and degrees of freedom used to form the whole model test:

- The Error Sum of Squares (SSE) is 203. It is the sum of squared residuals after fitting the full model.
- The C Total Sum of Squares is 851. It is the sum of squared residuals if you removed all the regression effects except for the intercept and fit only the mean.
- The Model Sum of Squares is 648. It is the sum of squares caused by the regression effects, which measures how much variation is accounted for by the regressors. It is the difference between the Total Sum of Squares and the Error Sum of Squares.

Figure 11.7 Analysis of Variance Table

Analysis of Variance

Source	DF	Sum of Squares	Mean Square	F Ratio
Model	2	648.26218	324.131	44.6815
Error	28	203.11936	7.254	Prob>F
C Total	30	851.38154		<.0001

The Error, C Total, and Model sums of squares are the ingredients needed to test the whole-model hypothesis that all the parameters in the model are zero except for the intercept (the simple mean model).

The Whole Model F Test

To form the F test,

1) Divide the Model Sum of Squares (324.13 in this example) by the number of terms (effects) in the model (except for the intercept). That divisor, 2, is found in the column labeled DF (Degrees of Freedom). The result is the Mean Square for the Model.
2) Divide the Error Sum of Squares (208.119 in this example) by the its associated degrees of freedom, 28, giving the Mean Square for Error.
3) Compute the F Ratio as the Model Mean Square divided by the Mean Square for Error.

The significance level, or p value, for this ratio is then calculated for the proper degrees of freedom (2 used in the numerator and 28 used in the denominator). The F ratio, 44.6815, in **Figure 11.7** is highly significant ($p<.0001$), which indicates that the model does fits better than a simple mean.

Whole Model Leverage Plot

There is a good way to view this whole-model hypothesis graphically using a scatterplot of actual response values against the predicted values. **Figure 11.8** shows the actual versus predicted values for the aerobic exercise example.

You can draw a 45-degree line from the origin showing where the actual response and predicted response are equal. The vertical distance from a point to the 45-degree line of fit is the difference of the actual and the predicted values–the residual error. What would the residual be if you removed all the regression effects and only fit a mean? The mean is shown by the horizontal dashed line. The distance from a point to the horizontal line at the mean is what the residual would be if you removed all the effects from the model.

The portrayal of a plot that compares residuals from the two models in this way is called a *leverage plot*. The idea is to get a feel for how much better the sloped line fits than the horizontal line.

Superimposed on the plot are the confidence curves representing the .05-level whole-model hypothesis. If the confidence curves cross the horizontal line, the whole model F test is significant.

Figure 11.8 Leverage Plot for the Whole Model (Run Time and Run Pulse)

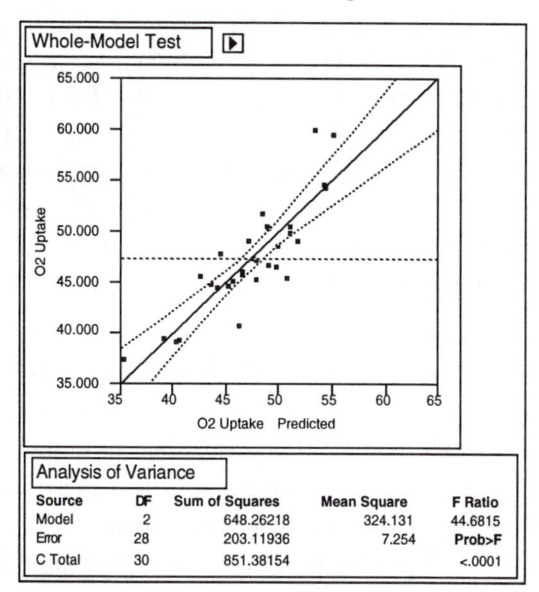

Analysis of Variance

Source	DF	Sum of Squares	Mean Square	F Ratio
Model	2	648.26218	324.131	44.6815
Error	28	203.11936	7.254	**Prob>F**
C Total	30	851.38154		<.0001

Details on Effect Tests

You can think about the significance of an effect in a model by looking at the distribution of the estimate, or by looking at the contribution of the effect to the model:

- To look at the *distribution of the estimate* you compute its standard error. The standard error can be used either to construct confidence intervals for the parameter or to make a t test that the parameter is some value, usually zero. The t tests are given in the Parameter Estimates table. Confidence intervals can also be requested.
- If you take an effect away from the model, then the sum of squared errors will higher. That difference in sums of squares can be used to construct an F test on whether the *contribution of the effect to the model* is significant. The F tests are given in the Effect Tests table.

The F tests and the t tests are equivalent. If you take the square root of the F value in the Effect Tests table, you get the same value as the t statistic in the

Parameter Estimates table. For example, the square of the t ratio (8.41) for Run Time is 70.77, which is the F ratio for Run Time.

Effect Leverage Plots

When you scroll to the right on the regression platform, you see details of how each regressor contributes to the model fit. The plots for the effect tests are also called leverage plots. The effect leverage plots (see **Figure 11.9**) show how each effect contributes to the fit after all the other effects have been included in the model. A leverage plot for an hypothesis test (an effect leverage plot) is any plot with horizontal and sloped reference lines and points laid out having the following two distance properties:

- The distance from each point to the sloped line measures the residual for the full model. The sums of squares of these residuals form the error sum of squares (SSE).
- The distance from each point to the horizontal line measures the residual for a model without the effect. The sums of squares of these residuals form the SSE for the constrained model (the model without the effect). In this way, you can see point by point how the sum of squares for the effect is formed. The difference in sums of squares of the two residual distances form the numerator for the F test for the effect.

Figure 11.9 Leverage Plots for Significant Effect (Left) and Nonsignificant Effect (Right)

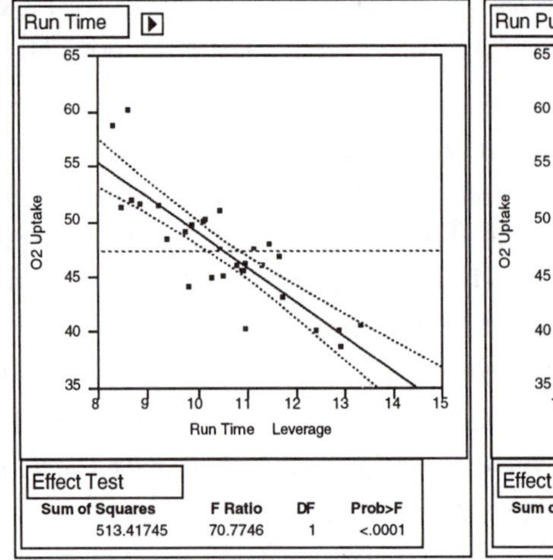

Effect Test

Sum of Squares	F Ratio	DF	Prob>F
513.41745	70.7746	1	<.0001

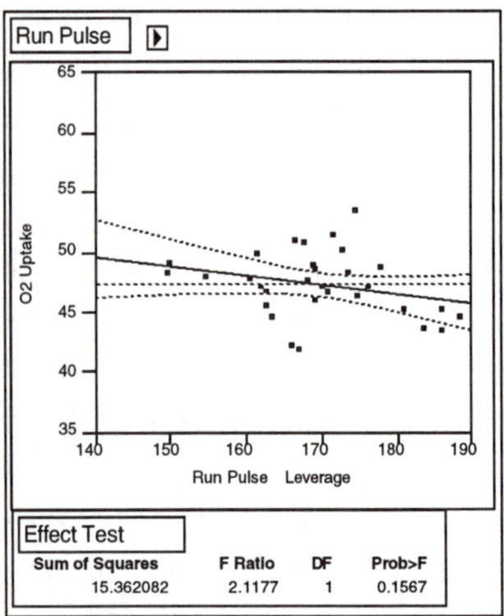

Effect Test

Sum of Squares	F Ratio	DF	Prob>F
15.362082	2.1177	1	0.1567

The leverage plot can be interpreted for an effect the same way you interpret a simple regression plot. In fact JMP superimposes a kind of 95% confidence curve on the sloped line. If the line is sloped significantly away from horizontal, then the confidence curves intersect the horizontal line that represents the constrained model. Alternatively, when the confidence curves enclose the horizontal line, the effect is not significant at the .05 level.

The leverage plots in **Figure 11.9** show that Run Time is significant and Run Pulse is not. You can see the significance by how the points support (or don't support) the line of fit in the plot and by whether the confidence curves for the line cross the horizontal line.

You can form a leverage plot for any kind of effect or set of effects in a model, or for any linear hypothesis. Leverage plots in the special case of single regressors are also known by the terms: *partial plot*, *partial regression leverage plot*, and *added variable plot*.

Special Topic: Collinearity

When you do a regression analysis and there is a close linear relationship between two or more regressors, they are said to have a collinearity problem. The problem is that the regression points do not occupy all the directions of the regression space very well; when you fit a regression plane, there are certain directions for which the fitting plane is not well supported. The fit is weak in those directions and the estimates become unstable, which means they are sensitive to small changes in the data.

In the statistical results this phenomenon translates into high standard errors for the parameter estimates and potentially high values for the parameter estimates themselves. This occurs because a small random error in the narrow direction can have a huge effect on the slope of the corresponding fitting plane. An indication of collinearity in leverage plots is when the points tend to collapse toward the center of the plot in the X direction.

To see an example of collinearity, consider the aerobic exercise example with the correlated regressors Max Pulse and Run Pulse:

§ With the LINNERUD exercise table active, choose **Analyze→Fit Model** (or click on the Fit Model dialog if it is still open).

§ Complete the Fit Model dialog by adding **Max Pulse** as an effect in the model after **Run Time** and **Run Pulse**, and click **Run Model**.

When the new analysis appears, scroll to the **Run Pulse** and **Max Pulse** leverage plots. Note in **Figure 11.10** that **Run Pulse** is very near the boundary of .05 significance, and thus the confidence curves almost line up along the horizontal axis, without actually crossing it.

Figure 11.10 Leverage Plots for Effects in Model

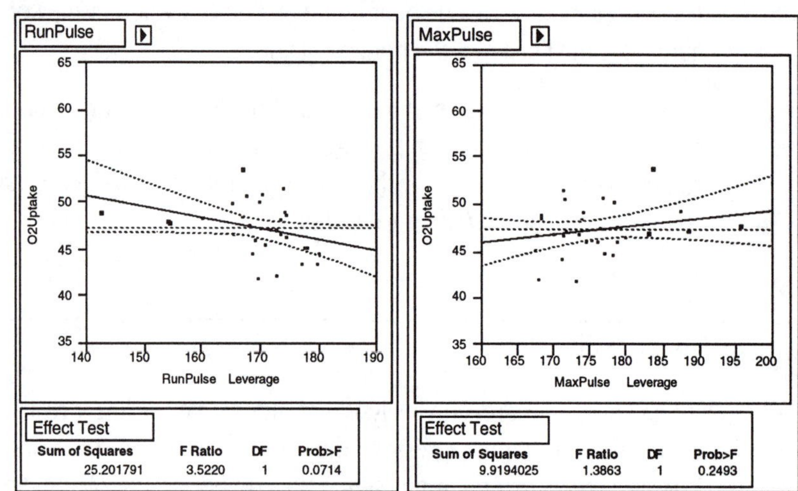

Now, as an example, let's change the relationship between the two regressors by changing a few values to cause collinearity.

§ Choose **Analyze→Fit Y by X**, selecting **Max Pulse** as the **Y** and **Run Pulse** as the **X** variable. This produces a scatterplot showing the bivariate relationship.

§ Select **Density Ellipse** from the **Fitting** popup menu and use .90 in the density ellipse submenu.

You should now see the scatterplot with ellipse shown to the left in **Figure 11.11**. The **Density Ellipse** command also generates the Bivariate table beneath the plot, which shows the current correlation to be .55.

The variables don't appear very collinear, with points scattered in all directions. However, it appears that if you exclude the four labeled points, the correlation would increases dramatically. To do this, exclude these points and rerun the analysis:

§ Highlight the points as shown in **Figure 11.11**. You can do this by highlighting rows in the spreadsheet, or you can SHIFT-click the points in the scatterplot. With these points highlighted, choose **Rows→Label** to identify them.

§ Choose **Rows→Exclude** while the rows are highlighted, and notice in the spreadsheet that they become marked with the international *not* symbol.

§ Again select **Density Ellipse** from the Fitting popup menu with the submenu value .90. Now the ellipse and the Bivariate table shows the relationship without the excluded points. The new correlation is .95, and the ellipse is much narrower, as shown in the plot on the right in **Figure 11.11**.

Figure 11.11 Effect of Excluding Points on Collinearity

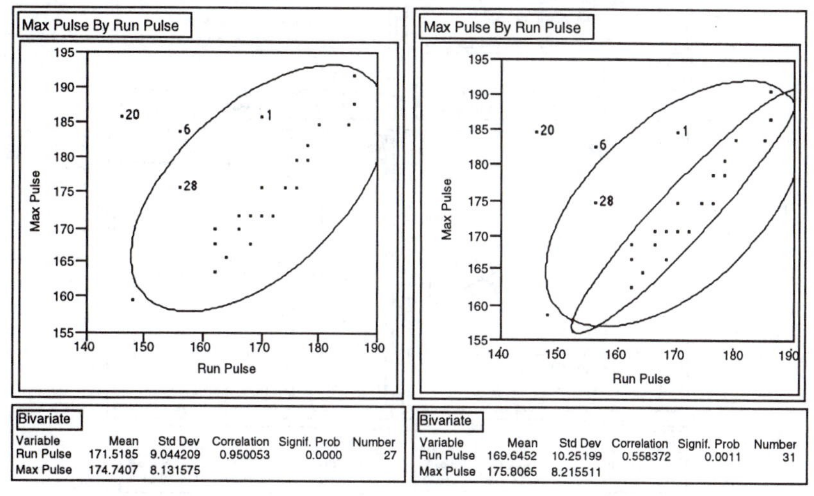

Bivariate

Variable	Mean	Std Dev	Correlation	Signif. Prob	Number
Run Pulse	171.5185	9.044209	0.950053	0.0000	27
Max Pulse	174.7407	8.131575			

Bivariate

Variable	Mean	Std Dev	Correlation	Signif. Prob	Number
Run Pulse	169.6452	10.25199	0.558372	0.0011	31
Max Pulse	175.8065	8.215511			

Also, run the regression model again to see the effect of excluding points:

§ Click **Run Model** again in the Fit Model dialog (with the same model as before). Examine both the Parameter Estimates table and the leverage plots for Run Pulse and Max Pulse, comparing them with the previous report (see **Figure 11.12**).

The parameter estimates and standard errors for the last two regressors have more than doubled in size.

The leverage plots now have confidence curves that flare out because the points themselves have shrunk towards the middle. When a regressor suffers collinearity, then the other variables have already absorbed much of that variable's variation, and there is less left to help predict the response. Another way of thinking about this is that there is less leverage of the points on the hypothesis. Points that are far out horizontally are said to have high leverage for the hypothesis test; points in the center have little leverage.

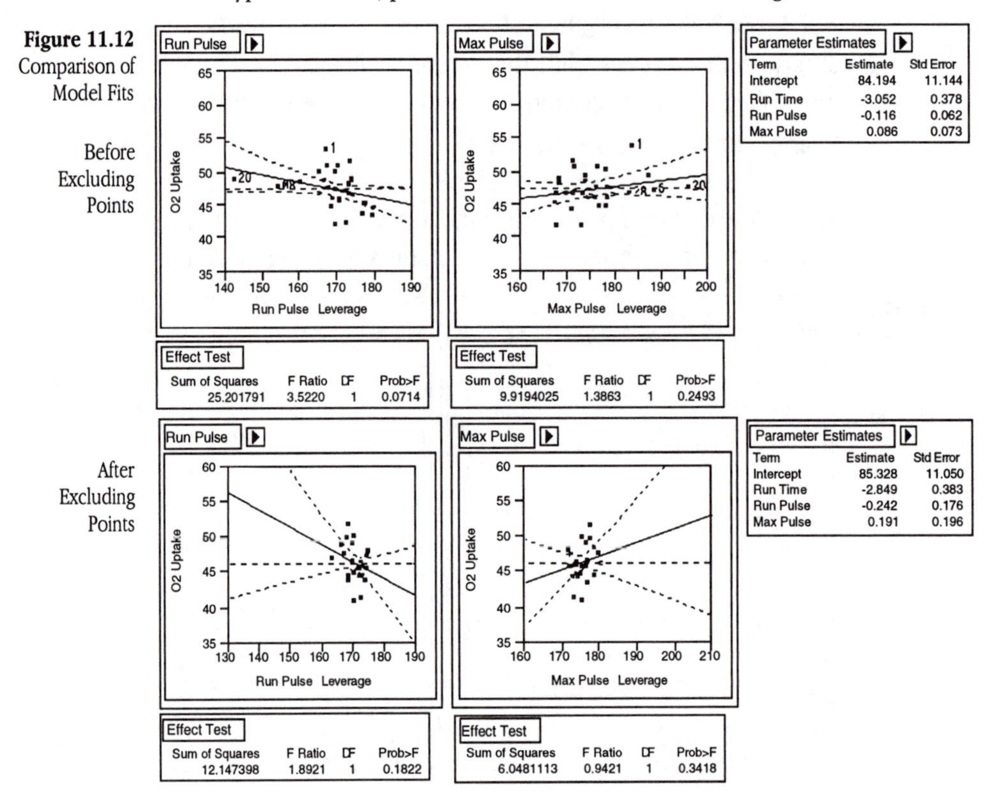

Figure 11.12 Comparison of Model Fits

Exact Collinearity, Singularity, Linear Dependency

You can construct a variable to show what happens when there is an exact linear relationship, the extreme of collinearity, among the regressors:

§ Choose **Columns→New Column**, to add a new variable (call it Run-Rest) to the data table. Use the calculator to create a formula that computes the difference between Run Pulse and Rest Pulse.

§ Now run a model of O2 Uptake against all the response variables, including the new variable Run-Rest.

The report in **Figure 11.13** shows signs of trouble. In the parameter estimates table, there are notations on Rest Pulse and Run Pulse that these estimates are *biased,* and on Run-Rest that it is *zeroed.* With exact linear dependency, the least squares solution is no longer unique, so JMP chooses the solution that zeroes out every parameter estimate corresponding to variables that are linearly dependent on previous variables. The Singularity table shows what the exact relationship is, in this case expressed in terms of **Rest Pulse**. The t tests for the parameter estimates must now be interpreted in a conditional sense. JMP refuses to make tests for the non-estimable hypotheses for **Rest Pulse**, **Run Pulse**, and **Run-Rest** and shows them with no degrees of freedom.

Figure 11.13 Report When There is a Linear Dependency

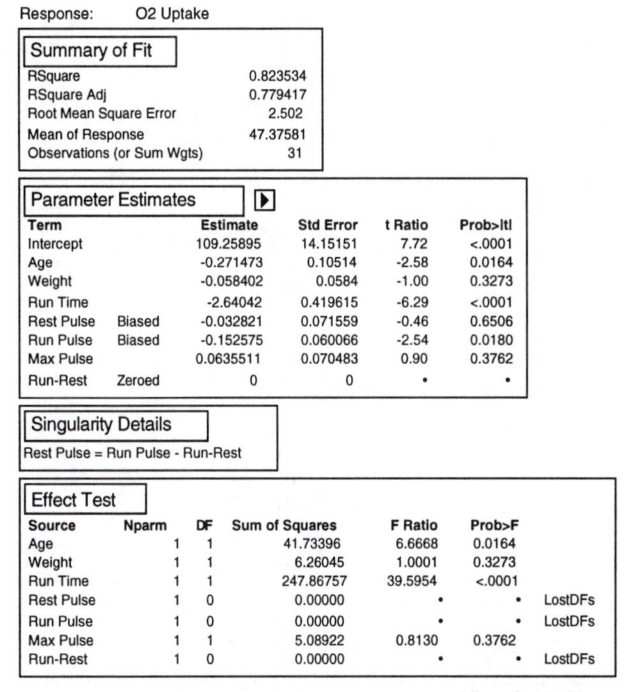
Response: O2 Uptake

Summary of Fit

RSquare	0.823534
RSquare Adj	0.779417
Root Mean Square Error	2.502
Mean of Response	47.37581
Observations (or Sum Wgts)	31

Parameter Estimates

Term		Estimate	Std Error	t Ratio	Prob>|t|
Intercept		109.25895	14.15151	7.72	<.0001
Age		-0.271473	0.10514	-2.58	0.0164
Weight		-0.058402	0.0584	-1.00	0.3273
Run Time		-2.64042	0.419615	-6.29	<.0001
Rest Pulse	Biased	-0.032821	0.071559	-0.46	0.6506
Run Pulse	Biased	-0.152575	0.060066	-2.54	0.0180
Max Pulse		0.0635511	0.070483	0.90	0.3762
Run-Rest	Zeroed	0	0	•	•

Singularity Details

Rest Pulse = Run Pulse - Run-Rest

Effect Test

Source	Nparm	DF	Sum of Squares	F Ratio	Prob>F	
Age	1	1	41.73396	6.6668	0.0164	
Weight	1	1	6.26045	1.0001	0.3273	
Run Time	1	1	247.86757	39.5954	<.0001	
Rest Pulse	1	0	0.00000	•	•	LostDFs
Run Pulse	1	0	0.00000	•	•	LostDFs
Max Pulse	1	1	5.08922	0.8130	0.3762	
Run-Rest	1	0	0.00000	•	•	LostDFs

You can see in the leverage plots for the three variables involved in the exact dependency, **Max Pulse**, **Run Pulse**, and **Run-Rest**, that the points have completely shrunk horizontally—nothing has any leverage for these effects. However, you can still test the unaffected regressors, like **Max Pulse**, and make good predicted values (see **Figure 11.14**).

Figure 11.14 Leverage Plots When There is a Linear Dependency

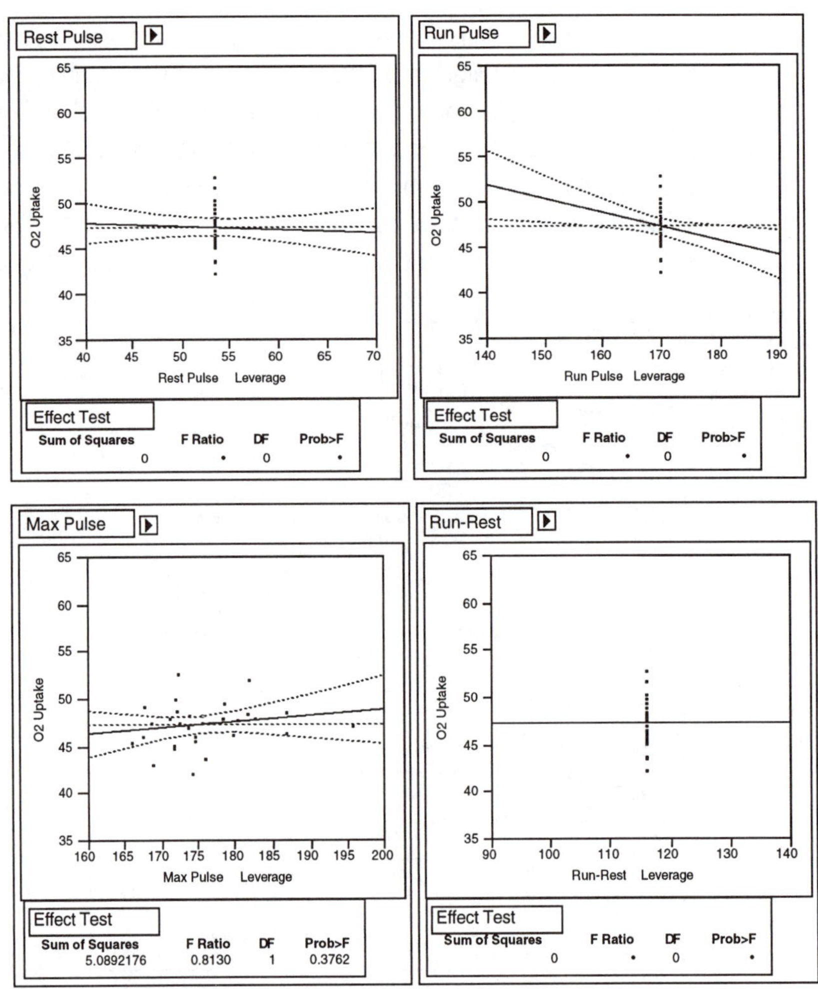

The Longley Data: An Example of Collinearity

The Longley data is famous because it is run routinely on most statistical packages to test accuracy of calculations. Why is it a challenge? Open the LONGLEY.JMP data table and take a look:

§ Choose **Analyze→Fit Model** Enter Y as Y, and all the x columns as the **Effects in Model**. Then click **Run Model** to see results shown in **Figure 11.1**.

Figure 11.15 Multiple Regression Report for Model with Collinearity

Response: y

Summary of Fit

RSquare	0.995479
RSquare Adj	0.992465
Root Mean Square Error	304.8541
Mean of Response	65317
Observations (or Sum Wgts)	16

Parameter Estimates

Term	Estimate	Std Error	t Ratio	Prob>|t|
Intercept	-3.4823e6	890420.4	-3.91	0.0036
x1	15.061872	84.915	0.18	0.8631
x2	-0.035819	0.033	-1.07	0.3127
x3	-2.02023	0.488	-4.14	0.0025
x4	-1.033227	0.214	-4.82	0.0009
x5	-0.051104	0.226	-0.23	0.8262
x6	1829.1515	455.478	4.02	0.0030

Effect Test

Source	Nparm	DF	Sum of Squares	F Ratio	Prob>F
x1	1	1	2924.0	0.0315	0.8631
x2	1	1	106306.3	1.1439	0.3127
x3	1	1	1590138.0	17.1100	0.0025
x4	1	1	2160905.5	23.2515	0.0009
x5	1	1	4748.9	0.0511	0.8262
x6	1	1	1498813.4	16.1274	0.0030

Whole-Model Test

Analysis of Variance

Source	DF	Sum of Squares	Mean Square	F Ratio
Model	6	184172402	30695400	330.2853
Error	9	836424	92936.01	Prob>F
C Total	15	185008826		<.0001

x1

Effect Test

Sum of Squares	F Ratio	DF	Prob>F
2923.9763	0.0315	1	0.8631

x2

Effect Test

Sum of Squares	F Ratio	DF	Prob>F
106306.26	1.1439	1	0.3127

x3

Effect Test

Sum of Squares	F Ratio	DF	Prob>F
1590138.0	17.1100	1	0.0025

x4

Effect Test

Sum of Squares	F Ratio	DF	Prob>F
2160905.5	23.2515	1	0.0009

x5

Effect Test

Sum of Squares	F Ratio	DF	Prob>F
4748.9482	0.0511	1	0.8262

x6

Effect Test

Sum of Squares	F Ratio	DF	Prob>F
1498813.4	16.1274	1	0.0030

Figure 11.15 shows the whole-model regression analysis. Looking at this overall picture doesn't tell you which (if any) of the six regressors are affected by collinearity. It's not obvious what the problems are in this regression until you look at leverage plots for the effects, which show that **x1**, **x2**, **x5**, and **x6** have collinearity problems. Their leverage plots appear very unstable with the points clustered at the center of the plot and the confidence lines showing no confidence at all.

Special Topic: The Case of the Hidden Leverage Point

Data was collected in a production setting where the yield of a process was related to three variables called **Aperture, Ranging**, and **Cadence**. Suppose you want to find out which of these effects are important, and in what direction:

§ Open the RO.JMP data table and choose **Analyze→Fit Model**. Enter **Yield** as **Y**, and **Aperture**, **Ranging**, and **Cadence** as the **Effects in Model**. Then click **Run Model.**

§ Select the **Plot Residuals** command from the check-mark popup menu. This creates the plot on the right in **Figure 11.16**, (**part 2**).

Everything looks fine. The Summary of Fit table in **Figure 11.17** shows an RSquare of 99.7%, which makes the regression model look like a great fit. All t statistics are highly significant—but don't stop there.

Figure 11.16 Multiple Regression Report for Model with Collinearity (continued next page)

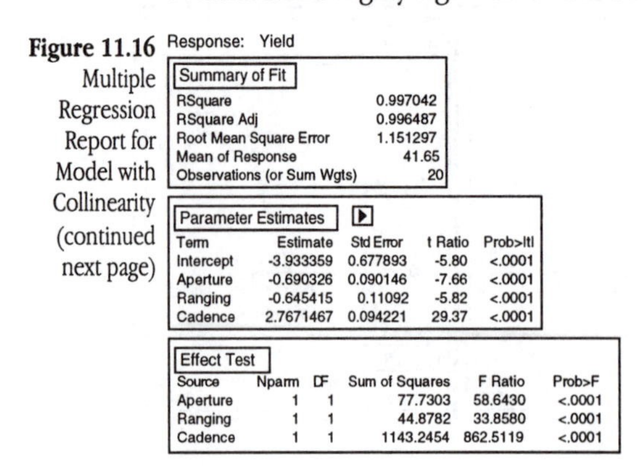

Response: Yield

Summary of Fit

RSquare	0.997042
RSquare Adj	0.996487
Root Mean Square Error	1.151297
Mean of Response	41.65
Observations (or Sum Wgts)	20

Parameter Estimates

Term	Estimate	Std Error	t Ratio	Prob>|t|
Intercept	-3.933359	0.677893	-5.80	<.0001
Aperture	-0.690326	0.090146	-7.66	<.0001
Ranging	-0.645415	0.11092	-5.82	<.0001
Cadence	2.7671467	0.094221	29.37	<.0001

Effect Test

Source	Nparm	DF	Sum of Squares	F Ratio	Prob>F
Aperture	1	1	77.7303	58.6430	<.0001
Ranging	1	1	44.8782	33.8580	<.0001
Cadence	1	1	1143.2454	862.5119	<.0001

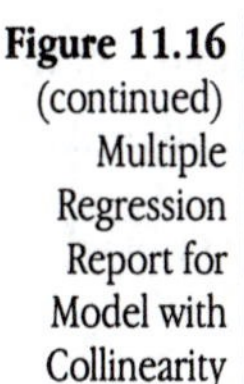

Figure 11.16 (continued) Multiple Regression Report for Model with Collinearity

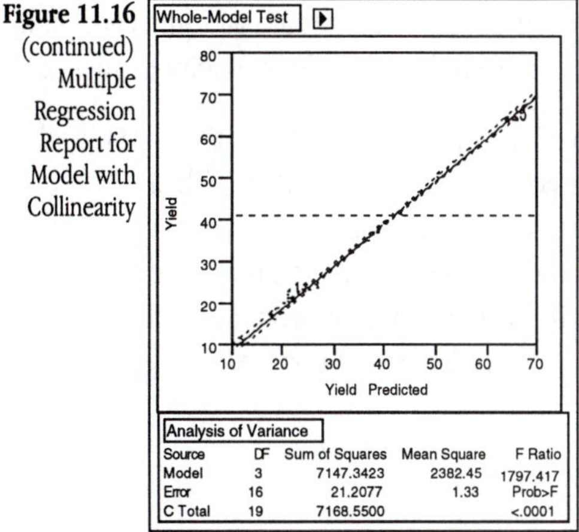

Source	DF	Sum of Squares	Mean Square	F Ratio
Model	3	7147.3423	2382.45	1797.417
Error	16	21.2077	1.33	Prob>F
C Total	19	7168.5500		<.0001

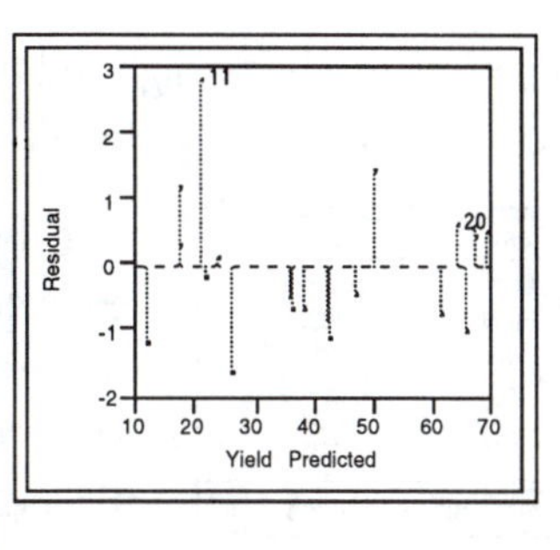

JMP produces all the standard regression results, and many more graphics. For each regression effect, there is a leverage plot showing what the residuals would be without that effect in the model. Note in **Figure 11.17** that row 20, which appeared unremarkable in the whole-model leverage and residual plots is far out into the extremes of the effect leverage plots.

It turns out that row 20 has monopolistic control of the estimates on all the parameters. All the other points appear wimpy because they track the same part of the shrunken regression space.

Figure 11.17 Leverage Plots That Detect Unusual Points

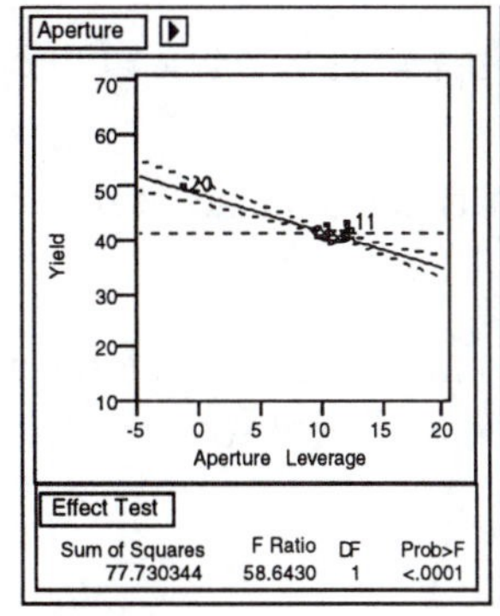

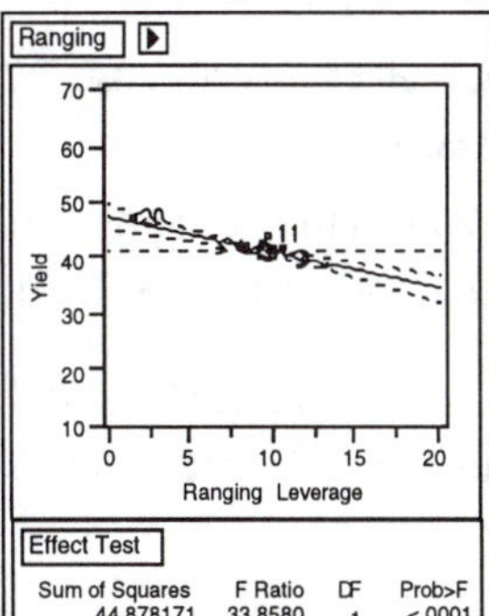

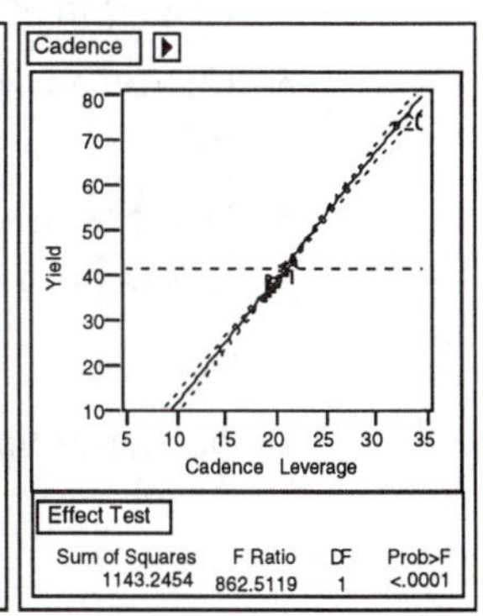

In a real analysis, you would want to give some special attention to row 20. Suppose you found that row 20 had an error and was really 32 instead of 65:

§ Change the value of Y in row 20 from 65 to 32 and run the model again.

The RSquare is again high and the parameter estimates are all significant—but every estimate is completely different even though only one point changed!

Figure 11.18 Parameter Estimates for Data with Incorrect Point (Top) and Corrected Point (Bottom)

Parameter Estimates

Term	Estimate
Intercept	-3.933359
Aperture	-0.690326
Ranging	-0.645415
Cadence	2.7671467

Parameter Estimates

Term	Estimate
Intercept	-0.035267
Aperture	1.7097579
Ranging	1.7318952
Cadence	0.2853072

The top table in **Figure 11.18** shows the parameter estimates computed from the data with an incorrect point. The bottom table has the corrected estimates. At the high response range, the first prediction equation would give very different results than the second equation.

Special Topic: Mining Data with Stepwise Regression

Let's try a regression analysis on the O2 Uptake variable with a set of 30 random columns generated by a random number generator as regressors. It seems like you should get all nonsignificant results with random regressors but that's not always the case:

§ Open the LINNRAND.JMP data table. It has 30 columns named X1 to X30 that were generated by a uniform random number generator and stored as a formula in each column.

§ Choose **Analyze→Fit Model** and use O2 Uptake as Y. Select X1 through X30 as the **Effects in Model.**

§ Select **Stepwise** from the fitting personality popup menu, as shown in **Figure 11.19**, and then click **Run Model**.

This stepwise approach launches a different regression platform, geared to playing around with different combinations of regressors.

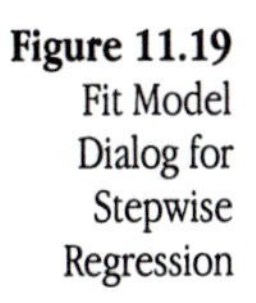

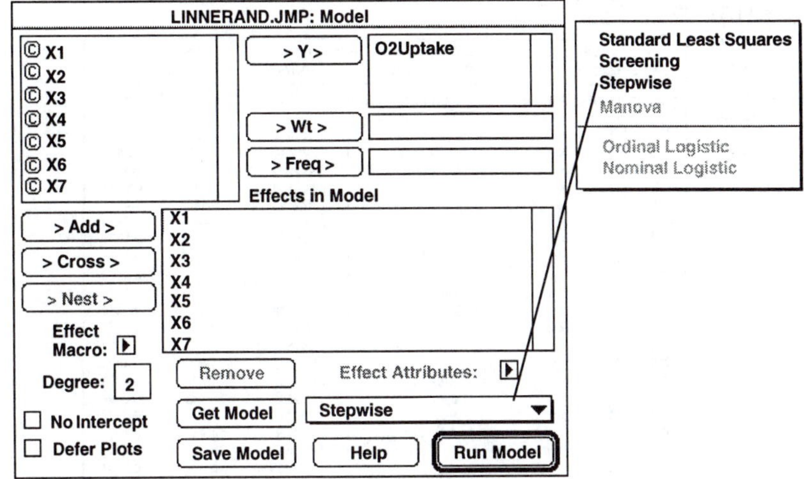

Figure 11.19 Fit Model Dialog for Stepwise Regression

To run a stepwise regression, use the control panel that appears after you run the model from the Fit Model dialog (see **Figure 11.20**).

§ Click **Go** in the Stepwise Regression Control panel to begin the stepwise variable selection process.

By default, stepwise runs a forward selection process. At each step, it adds the variable to the regression that is most significant. You can also select **Backward** or **Mixed** as the stepwise direction from the control panel popup menu.

The process selects variables that are significant at the level specified in the **Prob to Enter** and **Prob to Leave** fields on the control panel. The process stops when no more variables are significant, and displays the Current Estimates table seen in **Figure 11.21.** You will also see an Iteration table (not shown here) that lists the order in which the variables entered the model.

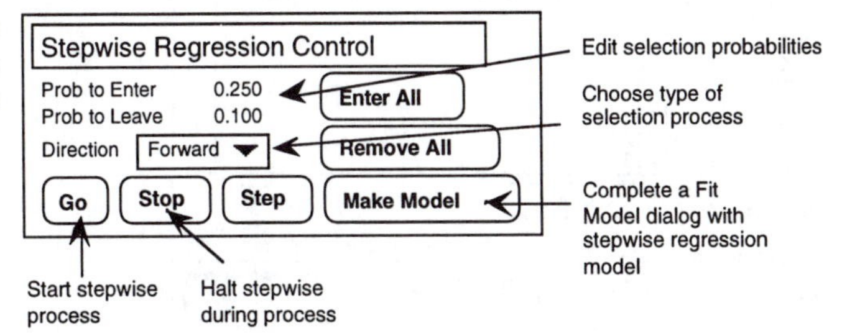

Figure 11.20 Stepwise Regression Control Panel

Figure 11.21 Current Estimates Table Showing Selected Variables

Current Estimates

SSE	DFE	MSE	RSquare	RSquare Adj	Cp	AIC
372.70308	21	17.74777	0.5622	0.3746	•	97.09064

Lock	Entered	Parameter	Estimate	nDF	SS	"F Ratio"	"Prob>F"
☒	☒	Intercept	49.6623498	1	0	0.000	1.0000
☐	☒	X1	5.71365596	1	57.30852	3.229	0.0867
☐	☒	X2	3.39144164	1	25.73404	1.450	0.2419
☐	☐	X3	•	1	0.423586	0.023	0.8816
☐	☐	X4	•	1	0.835656	0.045	0.8343
☐	☐	X5	•	1	7.673099	0.420	0.5241
☐	☐	X6	•	1	0.865353	0.047	0.8314
☐	☒	X7	4.90744798	1	35.25525	1.986	0.1733
☐	☒	X8	3.27917998	1	25.16178	1.418	0.2471
☐	☐	X9	•	1	3.19341	0.173	0.6820
☐	☒	X10	-4.0418716	1	47.17665	2.658	0.1179
☐	☐	X11	•	1	0.647898	0.035	0.8538
☐	☐	X12	•	1	12.4726	0.692	0.4151
☐	☐	X13	•	1	1.441517	0.078	0.7834
☐	☐	X14	•	1	11.26442	0.623	0.4391
☐	☒	X15	-7.2561191	1	84.99197	4.789	0.0401
☐	☒	X16	-9.4823876	1	145.2289	8.183	0.0094
☐	☐	X17	•	1	4.351117	0.236	0.6322
☐	☒	X18	-7.2277477	1	88.04206	4.961	0.0370
☐	☐	X19	•	1	1.825848	0.098	0.7569
☐	☐	X20	•	1	1.55203	0.084	0.7754
☐	☐	X21	•	1	0.183057	0.010	0.9220
☐	☐	X22	•	1	7.505798	0.411	0.5287
☐	☐	X23	•	1	0.002476	0.000	0.9909
☐	☐	X24	•	1	0.367011	0.020	0.8897
☐	☐	X25	•	1	0.421149	0.023	0.8819
☐	☐	X26	•	1	3.925378	0.213	0.6495
☐	☐	X27	•	1	0.228462	0.012	0.9129
☐	☐	X28	•	1	0.734565	0.039	0.8445
☐	☒	X29	6.56199307	1	81.90596	4.615	0.0435
☐	☐	X30	•	1	5.470934	0.298	0.5912

§ After the stepwise selection finishes selecting variables, click **Make Model** on the control panel. The Fit Model dialog shown in **Figure 11.22** then appears, and you can run a standard least squares regression with the effects that were selected by the stepwise process as most active.

Figure 11.22 Fit Model Dialog Generated by the Make Model Option

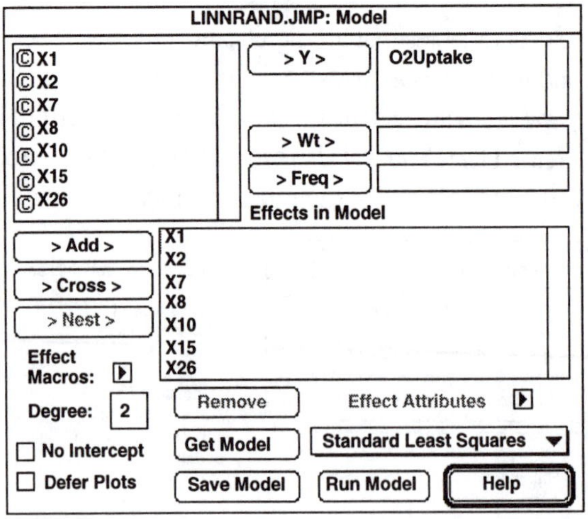

LINNRAND.JMP: Model

Ⓒ X1
Ⓒ X2
Ⓒ X7
Ⓒ X8
Ⓒ X10
Ⓒ X15
Ⓒ X26

\> Y > O2Uptake

\> Wt >

\> Freq >

Effects in Model

\> Add >
\> Cross >
\> Nest >

X1
X2
X7
X8
X10
X15
X26

Effect Macros:

Degree: 2

Remove

Effect Attributes

☐ No Intercept

☐ Defer Plots

Get Model

Standard Least Squares

Save Model

Run Model

Help

When you run the model, you get the standard regression reports, shown in **Figure 11.23**. The Parameter Estimates table shows that X16 and a number of other variables are significant regressors.

But what has just happened is that we created enough data to generate a number of coincidences, and then gathered those coincidences into one analysis and ignored the rest of the variables. This is like gambling all night in a casino, but exchanging money only for those hands where you win. When you mine data to the extreme you will get results that are too good to be true.

Figure 11.23 Results of Standard Least Squares Regression for Selected Variables

Summary of Fit

RSquare	0.562237
RSquare Adj	0.374625
Root Mean Square Error	4.21281
Mean of Response	47.37581
Observations (or Sum Wgts)	31

Parameter Estimates

Term	Estimate	Std Error	t Ratio	Prob>|t|
Intercept	49.66235	4.58537	10.83	<.0001
X1	5.713656	3.179628	1.80	0.0867
X2	3.3914416	2.816451	1.20	0.2419
X7	4.907448	3.481895	1.41	0.1733
X8	3.27918	2.754016	1.19	0.2471
X10	-4.041872	2.479082	-1.63	0.1179
X15	-7.256119	3.315792	-2.19	0.0401
X16	-9.482388	3.314843	-2.86	0.0094
X18	-7.227748	3.245112	-2.23	0.0370
X29	6.5619931	3.054569	2.15	0.0435

Chapter 12
Fitting Linear Models

Overview

First you learned to fit means. Then you learned to fit separate means to separate groups. Then you fit models to situations where the mean was a continuous function of a regressor variable. This chapter introduces a new approach involving general linear models, which will encompass all the models covered so far and extend to many more situations. They are all unified under the technique of least squares, fitting parameters to minimize the sum of squared residuals.

The techniques can be generalized even further to cover categorical response models and other more specialized applications of the linear model.

Chapter 12 Contents

You can mouse along with the examples whenever you see an action symbol (§)

The General Linear Model

Linear models are the sum of the products of coefficient parameters and factor columns. But this linear model is rich enough to encompass most statistical work. By using a coding system, you can map categorical factors to regressor columns. You can also form interactions and nested effects from products of coded terms. **Table 12.1** lists many of the situations handled by the general linear model approach. To read the model notation in **Table 12.1**, suppose that factors A, B, and C are categorical factors, and that X1, X2, and so forth, are continuous factors.

Table 12.1 Different Linear Models

Situation	Model Notation	Comments
one-way anova	Y = A	add a different value for each level
two-way anova no interaction	Y = A, B	additive model with terms for *A* and *B*.
two-way anova with interaction	Y = A, B, A*B	each combination of *A* and *B* has a unique add-factor.
three-way factorial	Y = A, B, A*B, C, A*C, B*C, A*B*C	for *k*-way factorial, 2^k–1 terms. The higher order terms are often dropped.
nested model	Y = A, B[A]	*The levels of B* are only meaningful within the context of *A* levels, e.g. *City*[*State*], read city within state
simple regression	Y = X1	an intercept plus a slope coefficient times the regressor
multiple regression	Y = X1, X2, X3, X4...	there can be dozens of regressors
polynomial regression	Y = X1, $X1^2$, $X1^3$, $X1^4$	linear, quadratic, cubic, quartic,...
quadratic response surface model	Y = X1, X2, X3 $X1^2$, $X2^2$, $X3^2$, X1*X2, X1*X3, X2*X3	all the squares and cross products of effects define a quadratic surface with a unique critical value where the slope is zero, which can be a minimum or a maximum, or a saddle point.
analysis of covariance	Y = A, X1	main effect (A), adjusting for the covariate (X1)
analysis of covariance with different slopes	Y = A, X1, A*X1	tests that the covariate slopes are different in different A groups
nested slopes	Y = A, X1[A]	separate slopes for separate groups
multivariate regression	Y1, Y2 = X1, X2...	test if a regressor affects several responses
manova	Y1, Y2, Y3 = A	test if a categorical variable affects several responses
multivariate repeated measures	sum and contrasts of (Y1 Y2 Y3)= A and so on.	the responses are repeated measurements over time on each subject.

Kinds of Effects in Linear Models

The richness of the general linear model is the result of the kind of effects you can include as columns in a coded model. You construct special sets of columns to support various effects, as in the following menagerie:

Intercept term

Most models have an intercept term to fit the constant in the linear equation. This is equivalent to having a regressor variable that is always 1. If this is the only term in the model, it serves to estimate the mean. This part of the model is so automatic that it becomes part of the background, the submodel to which the rest of the model is compared.

The only case in which intercepts are not used is the one in which the surface of fit must go through the origin. This happens in mixture models, for example. If you suppress the intercept term, then certain statistics, such as the whole-model F test and the RSquare do not apply because the question is no longer of just fitting a grand mean submodel against a full model.

Continuous effects

These values are direct regressor terms, taken into the model without modification. If all your variables are continuous effects, then the linear model is called multiple regression.

Categorical effects

The model must fit a separate constant for each level of a categorical effect. These effects lead to columns through an internal coding scheme, which is described in the next section. These are also called main effects, when contrasted with compound effects such interactions.

Interactions

These are crossings of categorical effects, in which you fit a different constant for each combination of levels of the interaction terms. Interactions are often written with an asterisk between terms, like Age*Sex.

Nested effects

Nested effects occur when a term is only meaningful in the context of another term, and thus is kind of a combined main effect and interaction with the term within which it is nested. For example, city is nested within state because if you know you're in Chicago, you also know you're in Illinois, and if you specify city alone, you will confuse Trenton, New Jersey, with Trenton, Michigan. Nested effects are written with the upper level term in parentheses or brackets, like "City[State]."

It is also possible to have combinations of continuous and categorical effects, and to combine interactions and nested effects.

Coding Scheme to Fit a One-Way Anova as a Linear Model

When you include categorical variables in a model, JMP needs to convert the categorical values (levels) into internal columns of numbers to analyze the data as a linear model. The rules to make these columns are the *coding scheme.* These columns are sometimes called *dummy* or *indicator* variables. They make up an internal *design matrix* used to fit the linear model.

Coding determines how you interpret parameter estimates, but the interpretation of parameters is different than the construction of the coded columns:

In JMP, the categorical variables in a model are such that:

- There is an indicator column for each level of the categorical variable except the last level. An indicator variable is 1 for a row that has the value represented by that indicator, is –1 for rows that have the last categorical level (for which there is no indicator variable), and zero otherwise.
- A parameter is interpreted as the comparison of that level with the average effect across all levels. The effect of the last level is the negative of the sum of all the parameters for that effect. That is why this coding scheme is often called sum-to-zero coding.

Different software packages use different codings. The coding scheme doesn't matter in many simple cases, but it does matter in more complex cases because it affects the hypotheses that are tested. This explains why different packages sometimes get different answers.

In Chapter 7, "Comparing Many Means: One-Way Analysis of Variance," you saw the DRUG.JMP data, where three drugs were compared (Snedecor and Cochran). Let's return to this data table and see how the general linear model handles a one-way Anova. It's best to start learning the new approach covered in this chapter by looking at this familiar model.

The sample table called DRUG.JMP (**Figure 12.1**) contains the results of a study that measured the response of 30 subjects after treatment by one of three drugs. First, look at the one-way analysis of variance given previously by the Fit Y by X continuous-by-nominal platform.

Figure 12.1 Partial Listing of the Drug Data Table with One-Way Analysis of Variance Report

DRUG.JMP

3 Cols, 30 Rows

	Drug	LBI	LBS
8	a	6	1
9	a	11	8
10	a	3	0
11	d	6	0
12	d	6	2
13	d	7	3

Oneway Anova

Summary of Fit

RSquare	0.227826
RSquare Adj	0.170628
Root Mean Square Error	6.070878
Mean of Response	7.9
Observations (or Sum Wgts)	30

Analysis of Variance

Source	DF	Sum of Squares	Mean Square	F Ratio
Model	2	293.6000	146.800	3.9831
Error	27	995.1000	36.856	Prob>F
C Total	29	1288.7000	44.438	0.0305

Means for Oneway Anova

Level	Number	Mean	Std Error
a	10	5.3000	1.9198
d	10	6.1000	1.9198
placebo	10	12.3000	1.9198

Std Error uses a pooled estimate of error variance

Now let's do the same analysis using the **Fit Model** command.

§ Open DRUG.JMP, which has variables called **Drug**, **LBI**, and, **LBS**; **Drug** has values "a", "d", and "placebo". **LBS** is bacteria count after treatment, and **LBI** is a baseline count.

§ Choose **Analyze→Fit Model**. When the Fit Model dialog appears, select **LBS** as the **Y** (response) variable. Select **Drug** and click **Add** to see it in the **Effects In Model** list. The completed dialog should look like the one in **Figure 12.2**.

§ Click **Run Model** to see the analysis result in **Figure 12.3**.

Figure 12.2 Fit Model Dialog for Simple One-Way Anova

DRUG.JMP: Model

© Drug
© LBI
© LBS

> Y > LBS

> Wt >

> Freq >

Effects in Model

> Add > Drug

> Cross >

> Nest >

Effect Macros:

Degree: 2

Remove

Effect Attributes:

No Intercept

Get Model

Standard Least Squares

Defer Plots

Save Model

Help

Run Model

Now compare the reports from this new analysis (**Figure 12.3**) with the One-Way Anova reports in **Figure 12.1**. Note that the statistical results are the same: the same RSquare, the same Anova F test, the same means, the same standard errors on the means. (The Anova F test is in both the Whole-Model Analysis of Variance table and in the Effect Test table because there is only one effect in the model.)

Figure 12.3 Anova Results Given by the Fit Model Platform

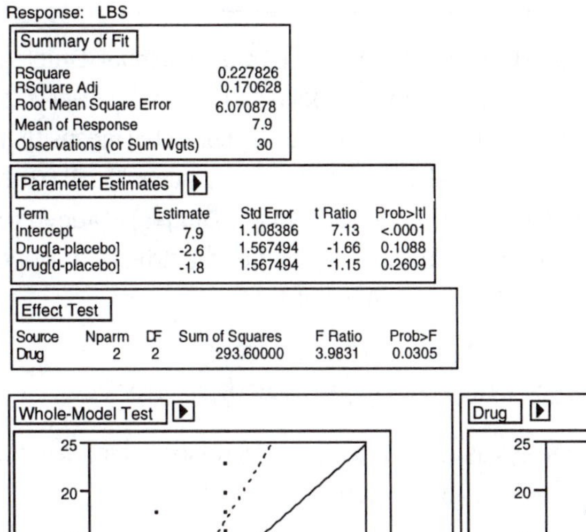

Response: LBS

Summary of Fit

RSquare	0.227826
RSquare Adj	0.170628
Root Mean Square Error	6.070878
Mean of Response	7.9
Observations (or Sum Wgts)	30

Parameter Estimates

Term	Estimate	Std Error	t Ratio	Prob>|t|
Intercept	7.9	1.108386	7.13	<.0001
Drug[a-placebo]	-2.6	1.567494	-1.66	0.1088
Drug[d-placebo]	-1.8	1.567494	-1.15	0.2609

Effect Test

Source	Nparm	DF	Sum of Squares	F Ratio	Prob>F
Drug	2	2	293.60000	3.9831	0.0305

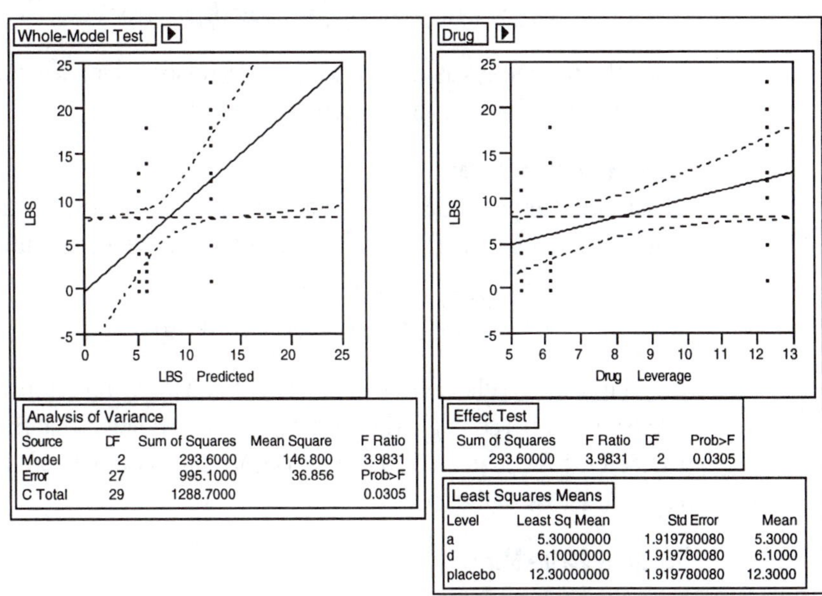

Analysis of Variance

Source	DF	Sum of Squares	Mean Square	F Ratio
Model	2	293.6000	146.800	3.9831
Error	27	995.1000	36.856	Prob>F
C Total	29	1288.7000		0.0305

Effect Test

Sum of Squares	F Ratio	DF	Prob>F
293.60000	3.9831	2	0.0305

Least Squares Means

Level	Least Sq Mean	Std Error	Mean
a	5.30000000	1.919780080	5.3000
d	6.10000000	1.919780080	6.1000
placebo	12.30000000	1.919780080	12.3000

Although the two platforms produce the same results, the way the analyses were run internally was not the same. The Fit Model analysis ran as a regression on an intercept and two regressor variables constructed from the levels of the model main effect. The next section describes how this is done.

Regressor Construction

The terms in the Parameter Estimates table (see **Figure 12.4**) are named according to how the regressor variables were constructed.

Figure 12.4 Parameter Estimates Notation

Parameter Estimates

Term	Estimate	Std Error	t Ratio	Prob>|t|
Intercept	7.9	1.108386	7.13	<.0001
Drug[a-placebo]	-2.6	1.567494	-1.66	0.1088
Drug[d-placebo]	-1.8	1.567494	-1.15	0.2609

The terms are called Drug[a-placebo] and Drug[d-placebo]. Drug[a-placebo] means that the regressor variable is coded as 1 when the level is "a", –1 when the level is "placebo", and 0 otherwise. Drug[d-placebo] means that the variable is 1 when the level is "d", –1 when the level is "placebo", and 0 otherwise. You can write the notation for Drug[a-placebo] as ([Drug=a]–[Drug=placebo]), where [Drug=a] is a one-or-zero indicator of whether the drug is "a" or not. The regression equation then looks like this:

$$y = b_0 + b_1*((\text{Drug=a})-(\text{Drug=placebo})) + \\ + b_2*((\text{Drug=d}-(\text{Drug=placebo})) + \text{error}$$

So far, the parameters associated with the regressor columns in the equation are represented by the names b_0, b_1 and so forth.

Interpretation of Parameters

What is the interpretation of the parameters for the two regressors, now named in the equation as b_1, and b_2? The equation can be rewritten like this:

$$y = b_0 + b_1*[\text{Drug=a}] + b_2*[\text{Drug=d}] + \\ + (-b_1-b_2)*[\text{Drug=placebo}] + \text{error}$$

The sum of the coefficients (1, –1, and 0) on the three indicators is always zero (sum-to-zero coding). The advantage of this coding is that the regression parameter tells you immediately how its level differs from the average response across all the levels.

Predictions are the Means

To verify that the coding system works, let's calculate the means, which are the predicted values for the levels "a", "d", and "placebo", by substituting the parameter estimates shown in **Figure 12.4** into the regression equation:

$$\text{Pred } y = b_0 + b_1*([\text{Drug=a}]-[\text{Drug=placebo}]) + \\ + b_2*([\text{Drug=d}]-[\text{Drug=placebo})$$

For the "a" level you have,

Pred y = 7.9 + –2.6*(1–0) + –1.8*(0–0)
= 5.3, which is the mean y for "a".

For the "d" level you have,

Pred y = 7.9 + –2.6*(0–0) + –1.8*(1–0)
= 6.1, which is the mean y for "d".

For the "placebo" level you have,

Pred y = 7.9 + –2.6*(0–1) + –1.8*(0–1)
= 12.3, which is the mean y for "placebo".

Parameters and Means

Now, you can substitute the means symbolically and solve for the parameters as functions of these means. First write the equations for the predicted values for the three levels, called A for "a", D for "d" and P for "placebo":

$$\text{MeanA} = b_0 + b_1*1 + b_2*0$$
$$\text{MeanD} = b_0 + b_1*0 + b_2*1$$
$$\text{MeanP} = b_0 + b_1*(-1) + b_2*(-1)$$

If you solve for the b's, you get the following coefficients:

$$b_1 = \text{MeanA} - (\text{MeanA}+\text{MeanD}+\text{MeanP})/3$$

$$b_2 = \text{MeanD} - (\text{MeanA}+\text{MeanD}+\text{MeanP})/3$$

$$(-b_1-b_2) = \text{MeanP} - (\text{MeanA}+\text{MeanD}+\text{MeanP})/3$$

Each level's parameter is interpreted as how different the mean for that group is from the mean of the means for each level.

In the next sections you will meet the generalization of this and other coding schemes, with each coding scheme having a different interpretation of the parameters.

Keep in mind that the coding of the regressors does not necessarily follow the same rule as the interpretation of the parameters. (If you study matrix algebra, you learn that the inverse of a matrix is its transpose only if the matrix is orthogonal).

Overall, analysis using this coding technique is a way to convert estimating group means into an equivalent regression model. It's all the same least-squares results using a different approach.

Analysis of Covariance: Putting Continuous and Classification Terms into the Same Model

Let's take the previous drug example that had one main effect (**Drug**), and now add the other term (**LBI**) to the model. **LBI** is a regular regressor, meaning it a continuous effect, and is called a covariate.

§ In the Fit Model dialog window used earlier, click **LBI** and then the **Add** button. Now both **Drug** and **LBI** are effects, as shown in **Figure 12.5**.

§ Click **Run Model** to see the results in **Figure 12.6**.

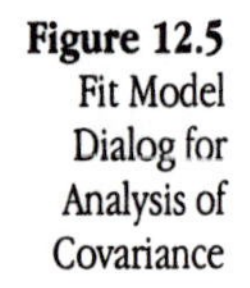

Figure 12.5 Fit Model Dialog for Analysis of Covariance

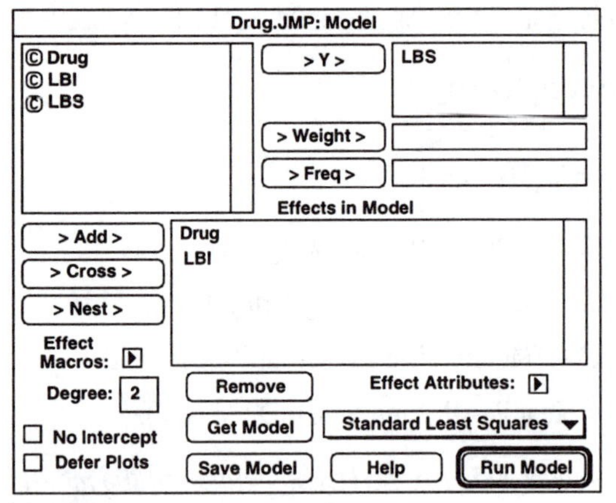

Figure 12.6 Analysis of Covariance Results Given by the Fit Model Platform

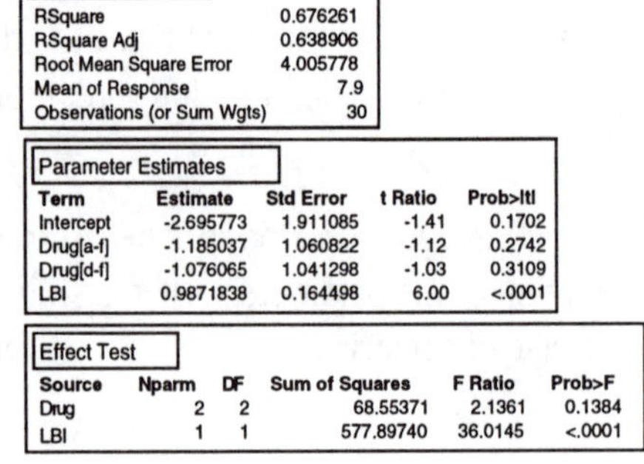

Summary of Fit

RSquare	0.676261
RSquare Adj	0.638906
Root Mean Square Error	4.005778
Mean of Response	7.9
Observations (or Sum Wgts)	30

Parameter Estimates

Term	Estimate	Std Error	t Ratio	Prob>|t|
Intercept	-2.695773	1.911085	-1.41	0.1702
Drug[a-f]	-1.185037	1.060822	-1.12	0.2742
Drug[d-f]	-1.076065	1.041298	-1.03	0.3109
LBI	0.9871838	0.164498	6.00	<.0001

Effect Test

Source	Nparm	DF	Sum of Squares	F Ratio	Prob>F
Drug	2	2	68.55371	2.1361	0.1384
LBI	1	1	577.89740	36.0145	<.0001

This new model is a hybrid between the Anova models with nominal effects and the regression models with continuous effects. Because the analysis method uses a coding scheme, you can put the categorical term into the model with the regressor.

The new results show that adding the covariate LBI to the model raises the RSquare from 22.78% to 67.62%. The parameter estimate for LBI is .987, which is not unexpected because the response is the bacteria count, and LBI is the baseline count before treatment. With a coefficient of nearly 1 for LBI, the model is really fitting the difference in bacteria counts.

The t test for LBI is highly significant. The difference in counts has a smaller variation than the absolute counts. Because the Drug effect uses two parameters, you need to refer to the F tests to see if Drug is significant. The F test is testing that both parameters are zero. The p value for Drug is now .1384.

Drug, which was significant in the previous model, is no longer significant! How could this be? The error in the model has been reduced, so it should be easier for differences to be detected. Could there be a relationship between LBI and Drug?

§ Choose **Analyze→Fit Y by X**, selecting LBI as Y and Drug as X.

§ When the one-way platform appears, select the **Means, Anova/t-Test** command from the popup menu beneath the scatterplot. Look at **Figure 12.7** to examine the relationship of the covariate LBI to Drug.

Figure 12.7 Fit Y by X Anova to Look at Relationship of Effects in Model

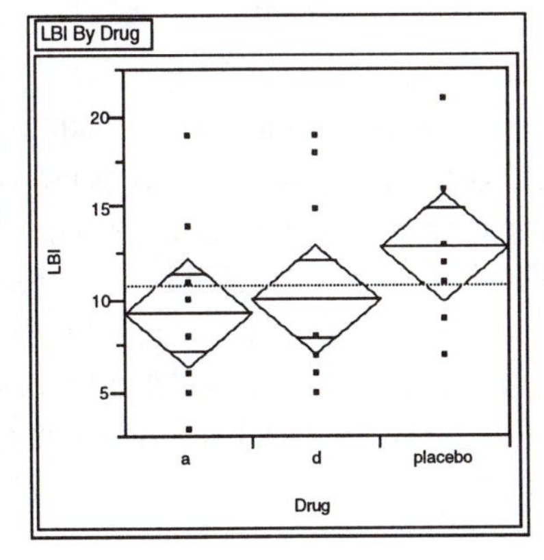

Aha! The Drugs have not been randomly assigned or if they were, they drew an unlikely unbalanced distribution. The toughest cases with the most bacteria tended to be given the inert drug “placebo.” This gave the “a” and “d” drugs a head start at reducing the bacteria count until LBI was brought into the model.

Whenever you fit models where you don't control all the factors, you may find that the factors are interrelated, and the significance of one depends on what else is in the model.

The Prediction Equation

You can make a column in the data table that contains the prediction equation generated by JMP stored as a calculator formula.

§ Close the Fit Y by X window and return to the Fit Model results. Select the **Save Prediction Formula** command in the Save ($) menu on the lower-left window border.

This command creates a new column in the data table called **Pred Formula Y**. To see its formula (the prediction formula), OPTION-click in the column heading area. This opens a calculator window with the following formula for the prediction equation:

$$-2.6957729 + \left(\begin{cases} \text{match } \textit{Drug} : & \\ -1.1850365, & \text{when "a"} \\ -1.0760652, & \text{when "d"} \\ 2.26110174, & \text{when "placebo"} \\ \bullet, & \text{otherwise} \end{cases} \right) + 0.98718381 \bullet \textit{LBI}$$

The Whole-Model Test and Leverage Plot

The whole-model test shows how the model fits as a whole compared with fitting a mean only. This is equivalent to testing that all the parameters in the linear model are zero except for the intercept. This fit has three degrees of freedom, 2 from **Drug**, and 1 from the covariate **LBI**. The F of 18.1 is highly significant (see **Figure 12.8**).

The whole-model leverage plot, which shows automatically with the whole-model test (**Figure 12.8**), is a plot of the actual value versus its predicted value. The residual is the distance from each point to the 45-degree line of fit where the actual is equal to the predicted. If you remove all the terms from the model except the intercept, then the residual is the distance from each point to the horizontal line at the mean. Thus the whole-model test is made from the model sum of squares, which is the difference in residual sum of squares between these two sets of residuals.

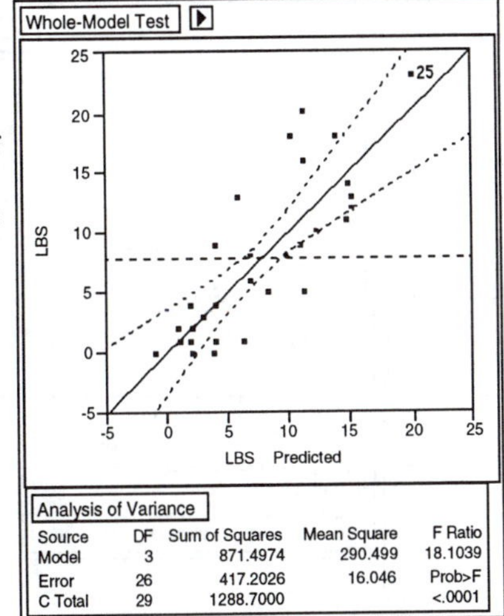

Analysis of Variance

Source	DF	Sum of Squares	Mean Square	F Ratio
Model	3	871.4974	290.499	18.1039
Error	26	417.2026	16.046	Prob>F
C Total	29	1288.7000		<.0001

Figure 12.8 Whole Model Test and Leverage Plot for Analysis of Covariance Example

The leverage plot (**Figure 12.8**) shows the hypothesis test point by point. The points that are far out horizontally (like point 25) tend to contribute more to the test because the predicted values from the two modelsdiffer more there. Points like this are called high leverage points.

Effect Tests and Leverage Plots

Now look at in **Figure 12.9** to examine the details for testing each effect in the model. Each effect test is computed from a difference in the residual sums of squares that compare the fitted model to the model without that effect.

For example, the sum of squares (SS) for LBI can be calculated by noting that the SS(Error) for the full model is 417.2, but the SS(Error) was 995.1 for the model that had only the Drug main effect (see **Figure 12.3**). So the reduction in sum of squares is the difference, 995.1–417.2 = 577.9, as you can see in the Effect Test table (**Figure 12.6**). Similarly, if you remove Drug from the model, the SS(Error) grows from 417.2 to 485.8, a difference of 68.5 from the full model.

The leverage plot shows the composition of these sums of squares point by point. The Drug leverage plot in **Figure 12.9** shows the effect on the residuals that would result from removing Drug from the model. The distance from each point to the sloped line is the residual. The distance from each point to the horizontal line is what the residual would be if Drug were removed from the model. The difference in the sum of squares for these two sets of residuals is the sum of squares for the effect, which is the numerator of the F test for the Drug effect.

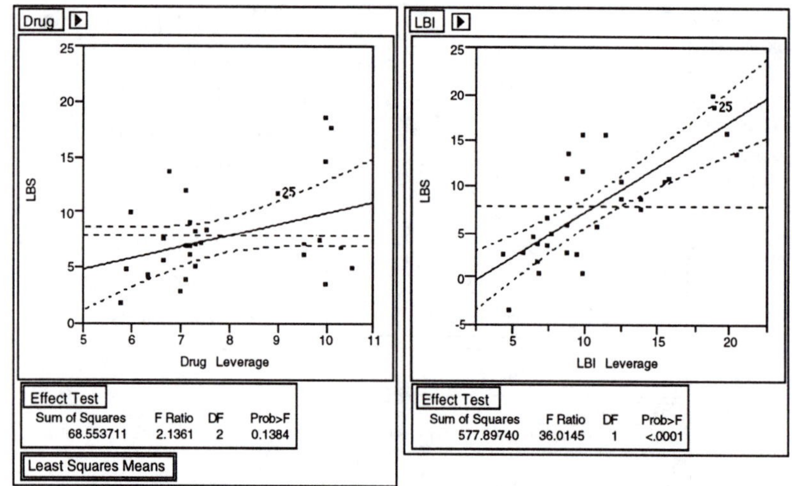

Figure 12.9 Effect Tests and Effect Leverage Plots for Analysis of Covariance Example

You can evaluate leverage plots in a way that is similar to evaluating the plot for a simple regression. The effect is significant if the points are able to support the sloped line significantly away from the horizontal. The confidence curves are placed around the sloped line to show the .05-level test. The curves cross the horizontal line if the effect is significant (as on the right in **Figure 12.9**), but they encompass the horizontal line if the effect is not significant (as on the left).

Click on points to see which ones are high leverage—away from the middle on the horizontal axis. Note whether they seem to support the test or not (whether they are on the side trying to pull the line to have a higher slope).

Figure 12.10 summarizes the elements of a leverage plot. A leverage plot for a specific hypothesis is any plot with the following properties:

- There is a sloped line representing the full model and a horizontal line representing a model constrained by that hypothesis.
- The distance from each point to the sloped line is the residual from the full model.
- The distance from each point to the horizontal line is the residual from the constrained model.

Figure 12.10 Schematic Defining a Leverage Plot

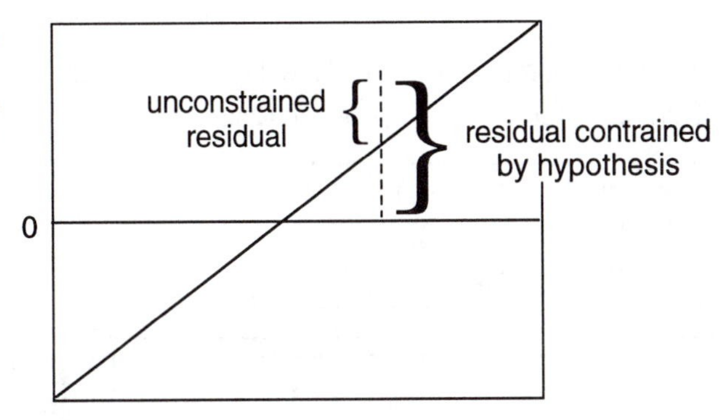

Least Squares Means

It might not be fair to make comparisons between raw cell means in data that you fit to a linear model. Raw cell means do not compensate for different covariate values and other factors in the model. Instead you construct predicted values, which are the expected value of a typical observation from some level of a categorical factor when all the other factors have been set to neutral values. These predicted values are called *least squares means*. There are other terms used for this idea: *marginal means*, *adjusted means*, and *marginal predicted values*.

The role of adjusted or least-squares means is that they allow comparisons of levels with the control of other factors being held fixed (*ceteris paribus*).

In the drug example, the least squares means are the predicted values that you expect for each of the three values of Drug given that the covariate LBI is held at some constant value. The constant value is chosen for convenience to be the mean of the covariate, which is 10.7333. The prediction equation gives the least squares means as follows:

fit equation:

–2.695 – 1.185 drug[a-placebo]–1.0760 drug[d-placebo] + .98718 LBI

for a:

$$-2.695 - 1.185\,(1) - 1.0760\,(0) + .98718\,(10.7333) = 6.71$$

for d:

$$-2.695 - 1.185\,(0) - 1.0760\,(1) + .98718\,(10.7333) = 6.82$$

for placebo:

$$-2.695 - 1.185\,(-1) - 1.0760\,(-1) + .98718\,(10.7333) = 10.16$$

To verify these results, return to the Fit Model platform and open the Least Squares Means table for the **Drug** effect (see **Figure 12.11**).

Figure 12.11 Least Squares Means for Drug

Least Squares Means

Level	Least Sq Mean	Std Error	Mean
a	6.71496346	1.288494280	5.3000
d	6.82393479	1.272468995	6.1000
placebo	10.16110174	1.315923424	12.3000

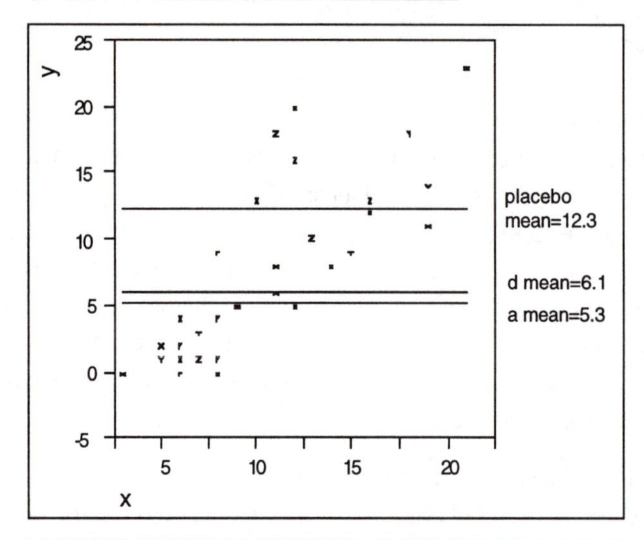

The ordinary means are taken with different values of the covariate, so it is not fair to compare them.

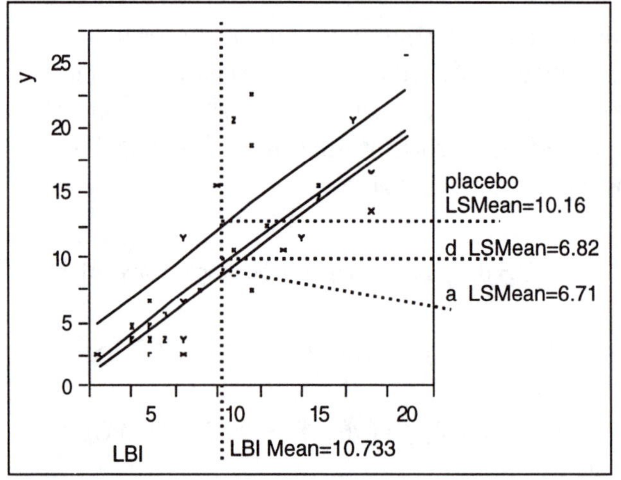

The least squares means for this model are the intersections of the lines of fit for each level with the **LBI** value of 10.733. With this data the least squares means are less separated than the raw means.

Lack of Fit

The lack-of-fit test is the opposite of the whole-model test. Where the whole-model tests whether anything you have in your model is significant, the lack-of-fit tests whether anything you left out of your model is significant. Unlike all other tests, you usually want the lack-of-fit test to be nonsignificant. If a lack-of-fit test is significant, then you are advised to add more effects to the model using higher orders of terms already in the model.

But how can you test effects that you haven't put in your model? All tests in linear models are comparisons that one model is better than a constrained or reduced version of that model. To test all the terms that are not in your model, but could be—that would be amazing!

Lack-of-fit compares the fitted model with a saturated model using the same terms. A saturated model is one that has a parameter for each combination of factor values that exist in the data. For example, a one-way analysis of variance is already saturated because it has a parameter for each level of the single factor. A complete factorial with all the higher order interactions is completely saturated. For simple regression, it is like having a separate coefficient to estimate for each value of the regressor.

Usually you want the result to be a nonsignificant lack-of-fit test. If the lack-of-fit test is significant, it means that there is some significant effect that you have left out of your model, and that effect is a function of the factors already in the model. It could be a higher-order power of a regressor variable, or some form of interaction among classification variables. If a model is already saturated, there is no lack-of-fit test possible.

The other requirement for a lack-of-fit test in continuous responses is that there be some exact replications of factor combinations in the data table. These exact duplicate rows (except for responses) allow the test to get a handle on estimating the variation to use as a denominator in the lack-of-fit F test. The error variance estimate from exact replicates is called *pure error* because it is independent of whether the model is right or wrong (assuming that it includes all the right factors).

In the drug model with covariate, the observations shown in **Table 12.2** form exact replications of data for Drug and LBI. If you take the sum of squares around the mean in each replicate group, you get the contributions to pure error shown in **Figure 12.12**.

This pure error represents the best you can ever do in fitting these terms to the model for this data because whatever you do to the model involving Drug and LBI, there will always be these replicates and this error. It is called pure error because it exists in the model regardless of the exact form of the model. **Figure 12.12** shows the Lack of Fit table for the drug example.

Table 12.2 Lack of Fit Analysis

Replicate Rows	Drug	LBI	LBS	Pure Error DF's	Contribution to Pure Error
6	a	6	4	1	$4.5 = (4-2.5)^2+(1-2.5)^2$
8	a	6	1		
1	a	11	6	1	$2.0 = (6-7)^2+(6-8)^2$
9	a	11	8		
11	d	6	0	1	2.0
12	d	6	2		
14	d	8	1	2	32.667
16	d	8	4		
18	d	8	9		
27	placebo	12	5	2	120.667
28	placebo	12	16		
30	placebo	12	20		
21	placebo	16	13	1	.5
26	placebo	16	12		
Total				8	162.333

Pure error can tell you how complete the model is. If the error variance estimate from the model is much greater than the pure error, then adding higher order effects of terms already in the model will not contribute much to the fit.

Figure 12.12 Lack of Fit Table for Drug Example

Lack of Fit

Source	DF	Sum of Squares	Mean Square	F Ratio
Lack of Fit	18	254.86926	14.1594	0.6978
Pure Error	8	162.33333	20.2917	Prob>F
Total Error	26	417.20260		0.7507
				Max RSq
				0.8740

The difference between the total error from the fitted model and pure error is called *Lack-of-Fit* error. It represents all the terms that might have been added to the model, but were not. The ratio of the lack-of-fit mean square to the pure error mean square is the F test for lack-of-fit. For the drug with covariate model, the lack-of-fit error is not significant, which is good because it is an indication that the model is adequate with respect to the terms included in the model.

Separate Slopes: When the Covariate Interacts with the Classification Effect

When a covariate model includes a main effect and a covariate regressor, the analysis uses a separate intercept for the covariate regressor for each level of the main effect.

If the intercepts are different, might not the slopes of the lines also be different? To find out, you need a way to capture the interaction of the regression slope with the main effect. This is done by introducing a crossed term, the interaction of **Drug** and **LBI**, into the model:

§ Return to the Fit Model dialog, which already has **Drug** and **LBI** as effects in the model. SHIFT-click on **Drug** and **LBI** in the column selector list so that both columns are highlighted as shown in **Figure 12.13**. Now click the **Cross** button. The result is an effect in the model called **Drug*LBI**.

§ Click **Run Model** to see the results (**Figure 12.14**).

Figure 12.13 Fit Model Dialog for Analysis of Covariance with Separate Slopes

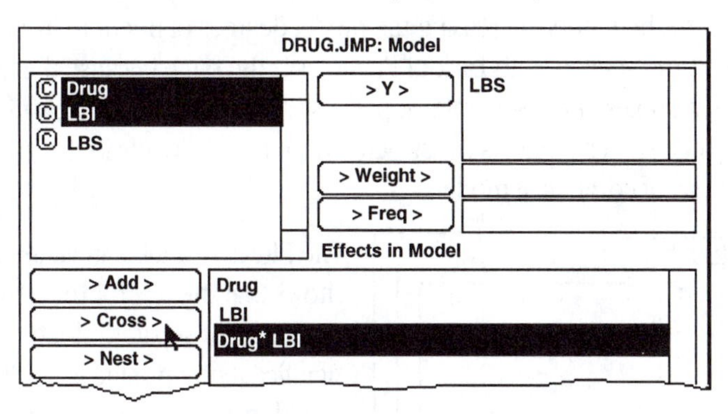

This specification adds two parameters to the linear model that allow the slopes for the covariate to be different for each **Drug** level. The new variables the product of the dummy variables for **Drug** by the covariate values.

The Summary of Fit tables in **Figure 12.14** compare this separate slopes fit to the same slopes fit, showing an increase in **RSquare** from 67.62% to 69.15%.

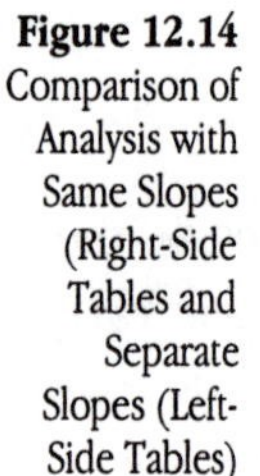

Figure 12.14 Comparison of Analysis with Same Slopes (Right-Side Tables and Separate Slopes (Left-Side Tables)

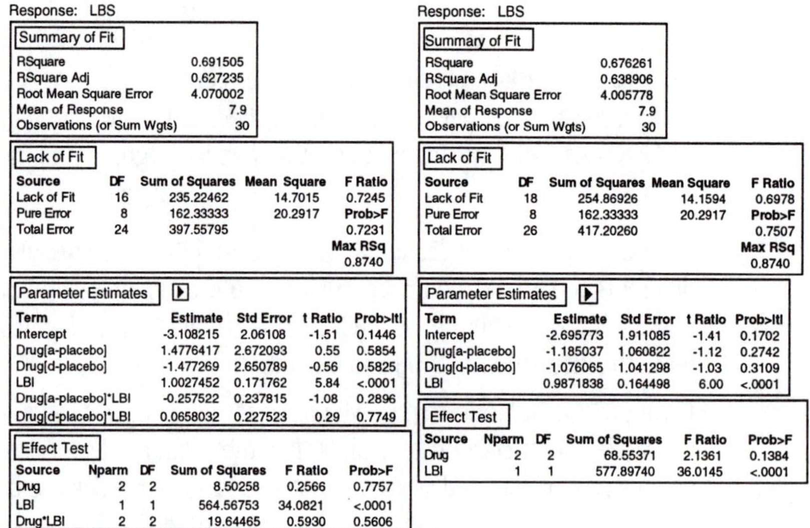

Response: LBS

Summary of Fit

RSquare	0.691505
RSquare Adj	0.627235
Root Mean Square Error	4.070002
Mean of Response	7.9
Observations (or Sum Wgts)	30

Lack of Fit

Source	DF	Sum of Squares	Mean Square	F Ratio
Lack of Fit	16	235.22462	14.7015	0.7245
Pure Error	8	162.33333	20.2917	Prob>F
Total Error	24	397.55795		0.7231
				Max RSq
				0.8740

Parameter Estimates

Term	Estimate	Std Error	t Ratio	Prob>|t|
Intercept	-3.108215	2.06108	-1.51	0.1446
Drug[a-placebo]	1.4776417	2.672093	0.55	0.5854
Drug[d-placebo]	-1.477269	2.650789	-0.56	0.5825
LBI	1.0027452	0.171762	5.84	<.0001
Drug[a-placebo]*LBI	-0.257522	0.237815	-1.08	0.2896
Drug[d-placebo]*LBI	0.0658032	0.227523	0.29	0.7749

Effect Test

Source	Nparm	DF	Sum of Squares	F Ratio	Prob>F
Drug	2	2	8.50258	0.2566	0.7757
LBI	1	1	564.56753	34.0821	<.0001
Drug*LBI	2	2	19.64465	0.5930	0.5606

Response: LBS

Summary of Fit

RSquare	0.676261
RSquare Adj	0.638906
Root Mean Square Error	4.005778
Mean of Response	7.9
Observations (or Sum Wgts)	30

Lack of Fit

Source	DF	Sum of Squares	Mean Square	F Ratio
Lack of Fit	18	254.86926	14.1594	0.6978
Pure Error	8	162.33333	20.2917	Prob>F
Total Error	26	417.20260		0.7507
				Max RSq
				0.8740

Parameter Estimates

Term	Estimate	Std Error	t Ratio	Prob>|t|
Intercept	-2.695773	1.911085	-1.41	0.1702
Drug[a-placebo]	-1.185037	1.060822	-1.12	0.2742
Drug[d-placebo]	-1.076065	1.041298	-1.03	0.3109
LBI	0.9871838	0.164498	6.00	<.0001

Effect Test

Source	Nparm	DF	Sum of Squares	F Ratio	Prob>F
Drug	2	2	68.55371	2.1361	0.1384
LBI	1	1	577.89740	36.0145	<.0001

The separate slopes model shifts two degrees of freedom from the lack-of-fit error to the model, increasing the model degrees of freedom from 3 to 5. The Pure Error seen in both Lack-of-Fit tables is the same because there are no new variables in the separate slopes covariance model. The new effect in the separate slopes model is constructed from terms already in the original analysis of covariance model.

Figure 12.15 Leverage Plot for Interaction of Effect and Covariate

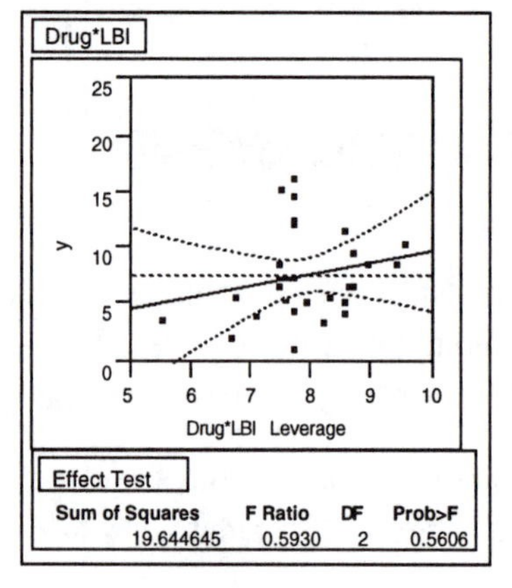

Effect Test

Sum of Squares	F Ratio	DF	Prob>F
19.644645	0.5930	2	0.5606

The Effect Test table in **Figure 12.15** shows that the test for the new term Drug*LBI for separate slopes is not significant; the p value is .56. The confidence curves on the leverage plot for the Effect Test enclose the horizontal mean line, showing that the interaction term doesn't significantly contribute to the model.

The least squares means for the separate slopes model have a more dubious value now. Previously, with the same slopes on LBI as shown in **Figure 12.11**, the least squares means changed with whatever value of LBI was used, but the separation between them did not. Now, with separate slopes as shown in **Figure 12.16**, the separation of the least squares means is also a function of LBI. The least squares means are more or less significantly different depending on whatever value of LBI is used. JMP uses the overall mean, but this does not represent any magic standard base.

Figure 12.16 Illustration of Covariance with Separate Slopes

Least Squares Means

Level	Least Sq Mean	Std Error	Mean
a	6.368152866	1.350271000	5.3000
d	6.883602151	1.300927285	6.1000
placebo	9.711994322	1.486066631	12.3000

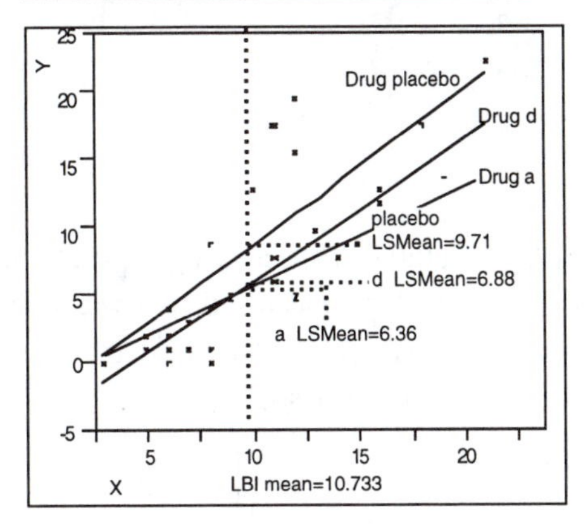

Interaction effects always have the potential to cloud the main effect, as you will see again with the two-way model in the next section.

Two-way Analysis of Variance and Interactions

How can you analyze a model in which there are two nominal or ordinal classification variables—a two-way model instead of a one-way model?

For example, a popcorn experiment was run, varying three factors and measuring the popped volume yield per volume of kernels. The goal was to see what factors gave the greatest volume of popped corn. **Figure 12.17** shows a listing of the popcorn data.

§ To see the data, open the sample table POPCORN.JMP. It has variables **type** with values "plain" and "gourmet"; **batch**, which is whether the popcorn was popped in a large or small batch; and **oil amt** with values "lots" or "little." Let's start with two of the three factors, **type** and **batch**.

Figure 12.17 Listing of the Popcorn Data Table

POPCORN.JMP

5 Cols, 16 Rows

	type (N)	oil amt (N)	batch (N)	yield (C)	trial (C)
1	plain	little	large	8.2	1
2	gourmet	little	large	8.6	1
3	plain	lots	large	10.4	1
4	gourmet	lots	large	9.2	1
5	plain	little	small	9.9	1
6	gourmet	little	small	12.1	1
7	plain	lots	small	10.6	1
8	gourmet	lots	small	18.0	1
9	plain	little	large	8.8	2
10	gourmet	little	large	8.2	2
11	plain	lots	large	8.8	2
12	gourmet	lots	large	9.8	2
13	plain	little	small	10.1	2
14	gourmet	little	small	15.9	2
15	plain	lots	small	7.4	2
16	gourmet	lots	small	16.0	2

§ Choose **Analyze→Fit Model**. When the Fit Model dialog appears select **yield** as the Y (response) variable. Select **type** and **batch** and click **Add** to use them as the **Effects in Model.** The Fit Model dialog should look like the one in **Figure 12.18**. Click **Run Model** to see the analysis.

Figure 12.18 Fit Model Dialog for Two-Factor Analysis

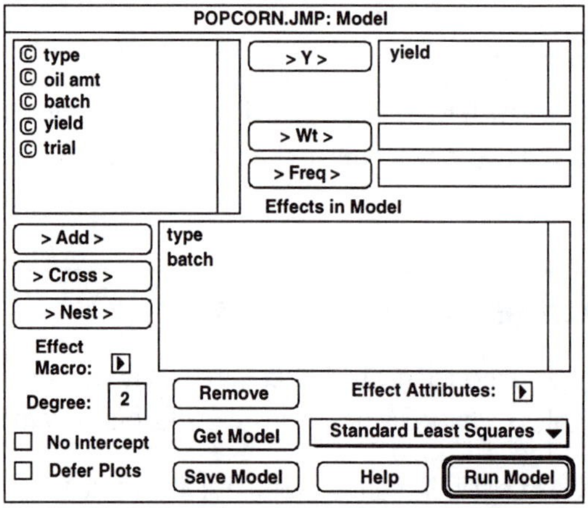

Figure 12.19 shows the analysis tables for the two-factor analysis of variance:

- The model explains 58% of the variation in yield (the RSquare).
- The remaining variation has a standard error of 2.2 48 (Root Mean Square Error).
- The significant Lack of Fit test (p value of .0019) says that there is something in the two factors that is not being captured by the model.

The factors are affecting the response in a more complex way than is shown by main effects alone. The model needs an interaction term.

- Each of the two effects has two levels, so they each have a single parameter. Thus the t test results are identical to the F test results. Both factors are significant.

Figure 12.19 Two-Factor Analysis of Variance for Popcorn Experiment

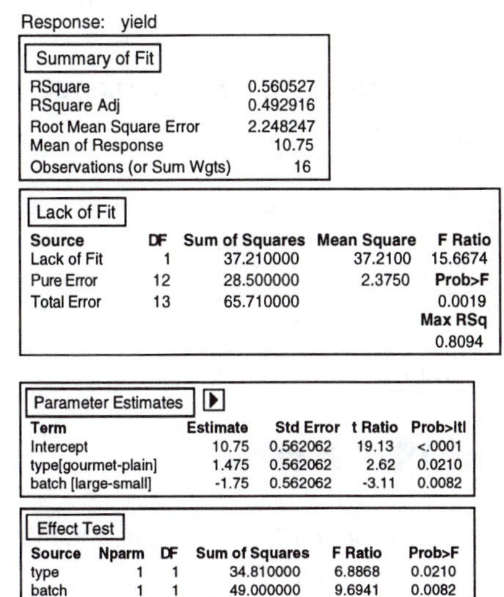

Response: yield

Summary of Fit

RSquare	0.560527
RSquare Adj	0.492916
Root Mean Square Error	2.248247
Mean of Response	10.75
Observations (or Sum Wgts)	16

Lack of Fit

Source	DF	Sum of Squares	Mean Square	F Ratio
Lack of Fit	1	37.210000	37.2100	15.6674
Pure Error	12	28.500000	2.3750	Prob>F
Total Error	13	65.710000		0.0019
				Max RSq
				0.8094

Parameter Estimates

Term	Estimate	Std Error	t Ratio	Prob>Itl
Intercept	10.75	0.562062	19.13	<.0001
type[gourmet-plain]	1.475	0.562062	2.62	0.0210
batch [large-small]	-1.75	0.562062	-3.11	0.0082

Effect Test

Source	Nparm	DF	Sum of Squares	F Ratio	Prob>F
type	1	1	34.810000	6.8868	0.0210
batch	1	1	49.000000	9.6941	0.0082

The leverage plots in **Figure 12.20** show the point-by-point detail for the fit as a whole and the fit as it is carried by each factor partially. Because this is a balanced design, all the points have the same leverage. This means they are spaced out horizontally the same in the leverage plot for each effect.

Figure 12.20 Leverage Plots for Two-factor Popcorn Experiment

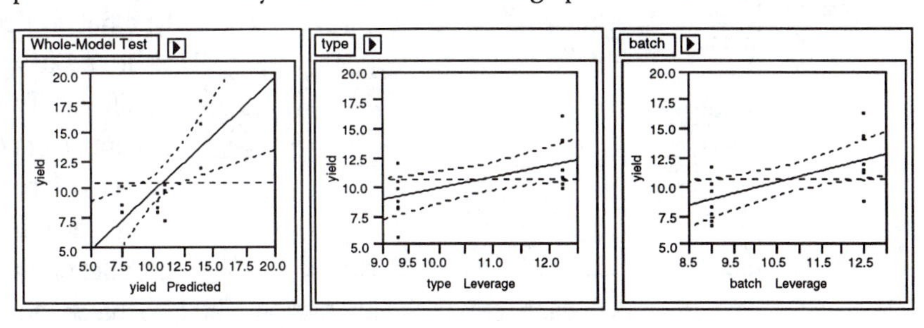

With Interaction

The Lack of Fit test shown in **Figure 12.19** suggests adding a higher order effect, such as an interaction, also called a crossed effect. An interaction means that the response is not only the sum of a separate function for each term. In addition, each term affects the response differently depending on the level of the other term in the model.

You can add the type by batch interaction to the model as follows:

§ Return to the Fit Model dialog, which already has the type and batch terms in the model. Select both type and batch in the variable selection list (click on type, and OPTION-click on batch to extend the selection).

§ Then click the **Cross** button to see the type*batch interaction effect in the Fit Model dialog as shown in **Figure 12.21**.

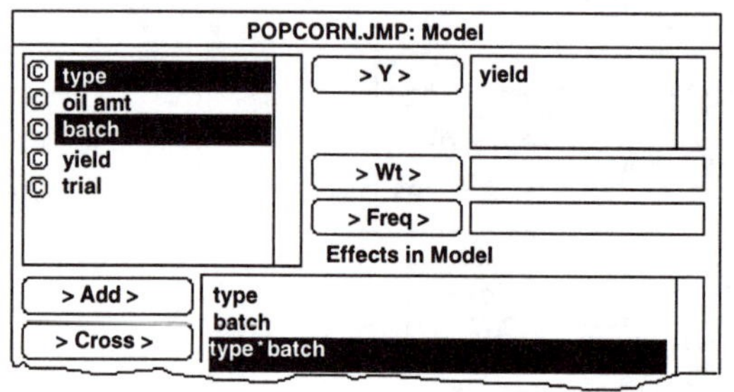

Figure 12.21 Model for Two-Factor Popcorn Experiment with Interaction

Click **Run Model** to see the tables in **Figure 12.22**. Including the interaction term increased the RSquare from 58% to 81%. The standard error of the residual (Root Mean Square Error) has gone down from 2.2 to 1.54.

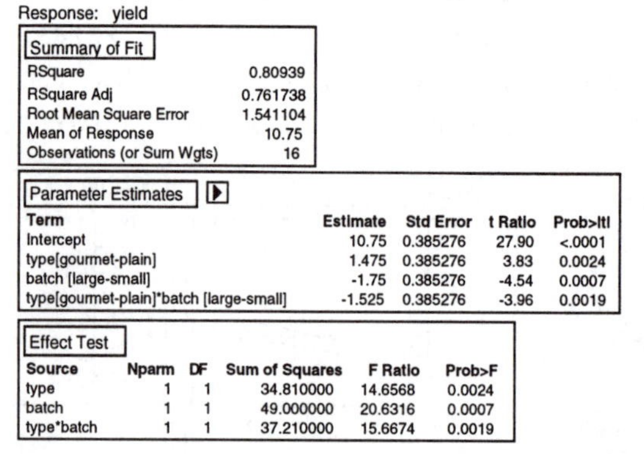

Response: yield

Summary of Fit

RSquare	0.80939
RSquare Adj	0.761738
Root Mean Square Error	1.541104
Mean of Response	10.75
Observations (or Sum Wgts)	16

Parameter Estimates

Term	Estimate	Std Error	t Ratio	Prob>|t|
Intercept	10.75	0.385276	27.90	<.0001
type[gourmet-plain]	1.475	0.385276	3.83	0.0024
batch [large-small]	-1.75	0.385276	-4.54	0.0007
type[gourmet-plain]*batch [large-small]	-1.525	0.385276	-3.96	0.0019

Effect Test

Source	Nparm	DF	Sum of Squares	F Ratio	Prob>F
type	1	1	34.810000	14.6568	0.0024
batch	1	1	49.000000	20.6316	0.0007
type*batch	1	1	37.210000	15.6674	0.0019

Figure 12.22 Statistical Analysis of Two-Factor Experiment with Interaction

The Effect Test table shows that all effects are significant. The type*batch effect has a p value of .0019, highly significant. The number of parameters (and degrees of freedom) of an interaction are the product of the number of parameters of each term in the interaction. The type*batch interaction has one parameter (and one degree of freedom) because the type and batch terms each have only one parameter.

An interesting phenomenon, which is true only in balanced designs, is that the parameter estimates and sums of squares for the main effects is the same as in the previous fit without interaction. The F tests are different only because the error variance (Mean Square Error) is smaller in the interaction model. The interaction effect test is identical to the lack-of-fit test in the previous model.

Again, the leverage plots (**Figure 12.23**) show the tests in point-by-point detail. The confidence curves cross the horizontal strongly. The effects tests (not shown here) confirm that the model and all effects are highly significant.

Figure 12.23 Leverage Plots for Two-Factor Experiment with Interaction

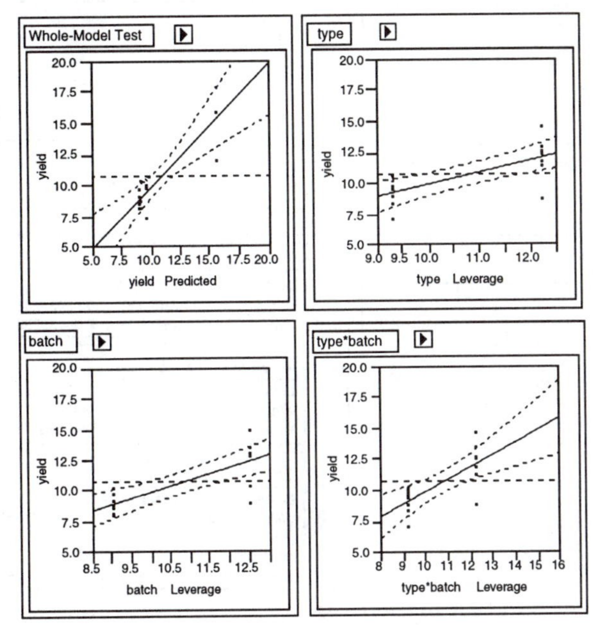

Let's see some detail on the means. They are labeled least squares means, but in a balanced design (equal numbers in each level and no covariate regressors), they are equal to the raw cell means.

§ To see profile plots for each effect, select the **Plot Effect** command in the popup menu at the top of each leverage plot. The result is a series of profile plots at the top of each effect's report. Profile plots are a graphical form of the values in the Least Squares Means table.

- The upper-left plot in **Figure 12.24** is the profile plot for the **type** main effect. The "gourmet" type popcorn seems to have a higher yield.
- The upper-right plot is the profile plot for the **batch** main effect. It looks like small batches have higher yields.

- The lower-left plot is the profile plot for the **type** by **batch** interaction effect. Looking at the effects together in an interaction plot shows that the popcorn type matters for small batches but not for big ones. Said another way, the batch size matters for gourmet popcorn, but not for plain popcorn. In an interaction profile plot, one interaction term is on the X axis and the other term forms the different lines.

Figure 12.24 Interaction Plots and Least Squares Means for Two-Factor Experiment with Interaction

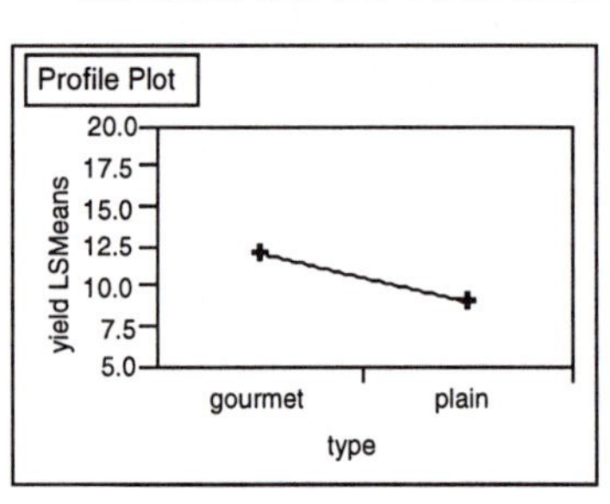

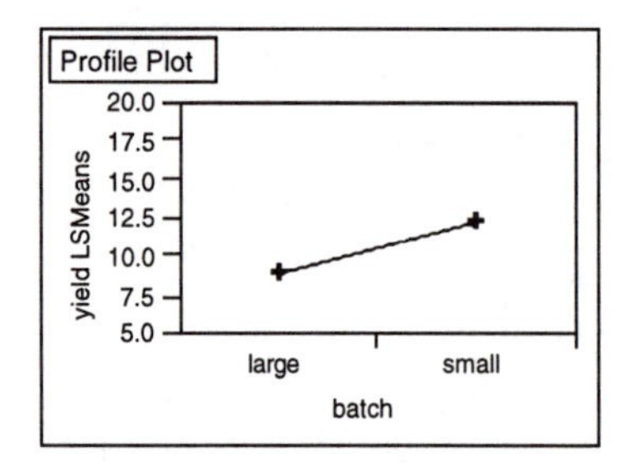

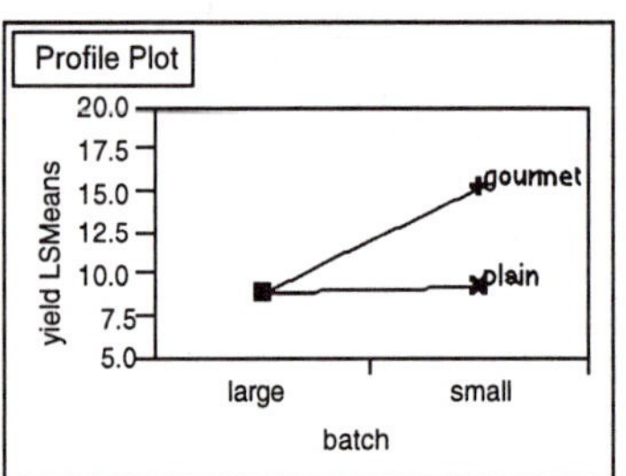

Least Squares Means

Level	Least Sq Mean	Std Error	Mean
gourmet	12.22500000	0.5448623679	12.2250
plain	9.27500000	0.5448623679	9.2750

Least Squares Means

Level	Least Sq Mean	Std Error	Mean
large	9.00000000	0.5448623679	9.0000
small	12.50000000	0.5448623679	12.5000

Least Squares Means

Level	Least Sq Mean	Std Error
gourmet,large	8.95000000	0.7705517504
gourmet,small	15.50000000	0.7705517504
plain,large	9.05000000	0.7705517504
plain,small	9.50000000	0.7705517504

Optional Topic: Random Effects and Nested Effects

This section talks about nested effects, repeated measures, and random effects mixed models. That is a large collection of topics to cover in a few pages, so hopefully this overview will inspire you enough to look to other textbooks and study these topics more completely.

As an example, consider the following situation. Six animals from two species were tracked, and the diameter of the area that each animal wandered was recorded. Each animal was measured four times, once per season. **Figure 12.25** shows a partial listing of the ANIMALS data.

Figure 12.25
Listing of the Animals Data Table

ANIMALS.JMP

4 Cols / 24 Cols	species (N)	subject (N)	miles (C)	season (N)
8	FOX	2	4	summer
9	FOX	3	4	fall
10	FOX	3	3	winter
11	FOX	3	6	spring
12	FOX	3	2	summer
13	COYOTE	1	4	fall
14	COYOTE	1	2	winter
15	COYOTE	1	7	spring

Nesting

One feature of the data is that the labeling for each subject animal is nested within species. The observations for subject 1 for species Fox are not for the same animal as subject 1 for Coyote. The way to express this in a model is to always write the subject effect as "subject(species)" which is read as "subject nested within species" or "subject within species." The rule about nesting is that whenever you refer to a subject with a given level of factor, if that implies what another factor's level is, then the factor should only appear in nested form.

When the linear model machinery in JMP sees a nested effect such as "B within A", denoted B(A), it computes a new set of A parameters for each level of B. You can specify a nested effect in the Fit Model dialog:

§ Open the ANIMALS.JMP table and choose **Analyze→Fit Model**. Use miles as the Y variable and construct the following model:

1) Add species to the **Effects in Model** list.
2) Add subject to the **Effects in Model** list.
3) Select species in the variable selection list; also select subject in the **Effects in Model** list.
4) Now click the **Nest** button. This adds the nested effect subject[species] shown in **Figure 12.26.**
5) Add season to the **Effects in Model** and click **Run Model** to see the results in **Figure 12.27**.

Figure 12.26 Nested Model for the Animals Data

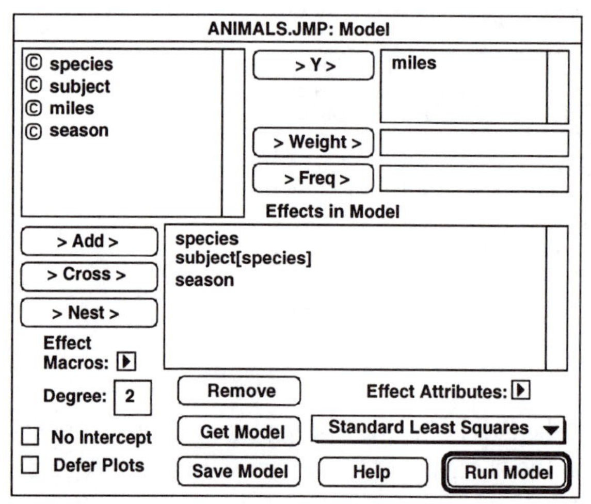

This model runs fine, but it has something wrong with it. The F tests for all the effects in the model used the residual error in the denominator. You will see what's wrong with doing that and how to fix it in a minute.

Figure 12.27 Results for Animal Data Analysis

Response: miles

Summary of Fit

RSquare	0.838417
RSquare Adj	0.75224
Root Mean Square Error	1.219062
Mean of Response	4.458333
Observations (or Sum Wgts)	24

Parameter Estimates

Term	Estimate	Std Error	t Ratio	Prob>|t|
Intercept	4.4583333	0.24884	17.92	<.0001
species[COYOTE-FOX]	1.4583333	0.24884	5.86	<.0001
species[COYOTE]:subject[1-3]	-0.666667	0.49768	-1.34	0.2003
species[COYOTE]:subject[2-3]	-0.666667	0.49768	-1.34	0.2003
species[FOX]:subject[1-3]	-1	0.49768	-2.01	0.0628
species[FOX]:subject[2-3]	0.25	0.49768	0.50	0.6227
season[fall-winter]	-0.625	0.431003	-1.45	0.1676
season[spring-winter]	1.7083333	0.431003	3.96	0.0012
season[summer-winter]	0.875	0.431003	2.03	0.0605

Effect Test

Source	Nparm	DF	Sum of Squares	F Ratio	Prob>F
species	1	1	51.041667	34.3458	<.0001
subject[species]	4	4	17.166667	2.8879	0.0588
season	3	3	47.458333	10.6449	0.0005

Note the treatment of nested effects in the model. There is one parameter for the two levels of species (Fox and Coyote). Subject is nested in species, so there is a separate set of two parameters (for three levels of subject) for subject within each level of species, giving a total of four parameters for subject. Season with four levels has three parameters. The total parameters for the model (not including the intercept) is 1 for species + 4 for subject + 3 for season = 8. If you use the **Save Prediction Formula** command and look at the calculator window for the saved formula, you see the following prediction equation using the parameter estimates:

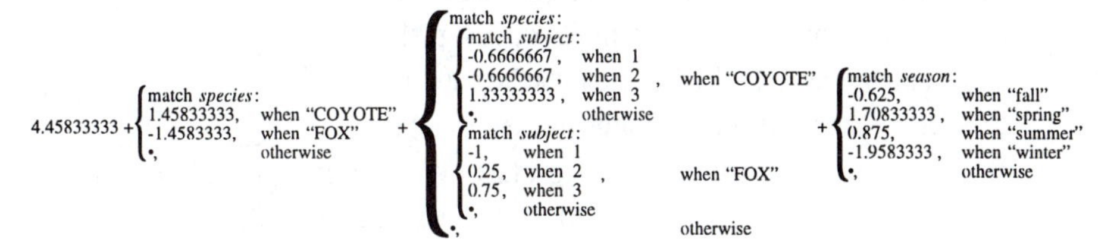

Repeated Measures

As mentioned, the previous analysis has a problem—the F test used to test the species effect is constructed using the model residual in the denominator, which isn't appropriate for a nested effect. The following sections explain this problem and outline solutions.

This problem has three different ways of understanding it, which correspond to three different (but equivalent) resolutions:

- You can declaring effects as *random*, and JMP will synthesize special F tests.
- The observations can be made to correspond to the experimental unit.
- The analysis can be viewed as a multivariate problem.

The key in each method is to focus on only one of the effects in the model. In the animals example, the effect is species—how does the wandering radius differ between "Fox" and "Coyote?" The species effect is incorrectly tested in the previous example so species is the effect that needs attention.

Method 1: Random Effects—Mixed Model

The subject effect is what is called a random effect. The animals were selected randomly from a large population, and you expect that the variability from animal to animal will be from some distribution. To generalize to the whole population, you need to study the species effect with respect to the variability.

It turns out that if your design is balanced, you can use an appropriate random term in your model as an error term instead of using the residual error to get an appropriate test. In this case subject(species), the nested effect, acts as an error term for the species main effect.

To construct the appropriate F test, you can do a hand calculation using the results from Effect Test table shown in **Figure 12.27**. Divide the mean square for species by the mean square for subject(species) as shown by the formula here. This F test has 1 numerator degree of freedom and 4 denominator degrees of freedom and evaluates to 11.89.

$$F = \frac{\left(\frac{51.041667}{1}\right)}{\left(\frac{17.16667}{4}\right)}$$

If you study random effects in your general statistics text, often described as split plots or repeated measures designs, the text will describe which mean squares need to be used to test each model effect.

Now, let's have JMP do this calculation instead of doing it by hand. JMP will give the correct tests even if the design is not balanced.

First, specify subject[species] as a random effect.

§ Click the Model window for the ANIMALS data to make it active. Click to highlight subject[species] showing in the **Effects in Model** list.

§ Now select the **Random Effect** attribute found in the **Effect Attributes** popup menu. The subject[species] effect then appears with **{Random}** appended to it as shown in **Figure 12.28**. Click **Run Model** to see the results.

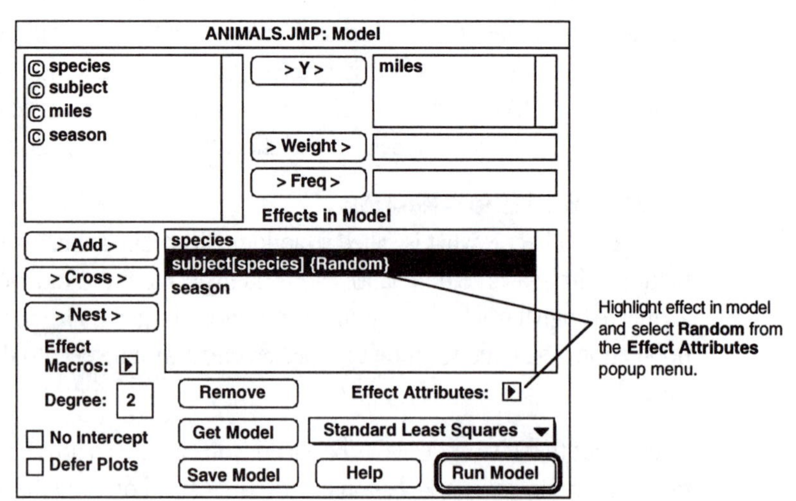

Figure 12.28 Fit Model Dialog Using a Random Effect

JMP constructs tests for random effects by the following steps:

1) First, the expected mean squares are found. These are coefficients that relate the mean square to the variances for the random effects.

Expected Mean Squares

The Mean Square per row by the Variance Component per column

EMS	Intercept	species	subject[species]	season
Intercept	0	0	0	0
species	0	12	4	0
subject[species]	0	0	4	0
season	0	0	0	6

plus 1.0 times Residual Error Variance

2) Next, the variance component estimates are found using the mean square values for the random effects and their coefficients. It is possible (but rare) for a variance component to be negative.

Variance Component Estimates

Component	Var Comp Est
subject[species]	0.701389
Residual	1.486111

These estimates based on equating Mean Squares to Expected Value.

3) For each effect in the model, JMP then determines what linear combination of other mean squares would make an appropriate denominator for an F test. This denominator is the linear combination of mean squares that has the same expectation as the mean square of the effect (numerator) under the null hypothesis.

Test Denominator Synthesis

Source	MS Den	DF Den	Denom MS Synthesis
species	4.29167	4	subject[species]
subject[species]	1.48611	15	Residual
season	1.48611	15	Residual

4) F tests are now constructed using the denominators synthesized from other mean squares. If an effect is prominent, then it will have a much larger mean square than expected under the null hypothesis for that effect.

Tests wrt Random Effects

Source	SS	MS Num	DF Num	F Ratio	Prob>F
species	51.0417	51.0417	1	11.8932	0.0261
subject[species]	17.1667	4.29167	4	2.8879	0.0588
season	47.4583	15.8194	3	10.6449	0.0005

Again the F statistic for species is 11.89 with a p value of .026. The tests for the other factors use the residual error mean square, which is the same as the tests done in the first model.

What about the test for season? Because the experimental unit for season corresponds to each row of the original table, the residual error in the first model is appropriate. The F of 10.64 with a p value of .0005 means that the miles (the response) does vary across season. If you include an interaction between species and season in the model, it would also be correctly tested using the residual mean square.

Note 1: There is an alternate way to define the random effects that produces slightly different expected mean squares and variance component estimates, but gives the same F tests. The argument over which method of parameterization is proper has been raging for 40 years.

Note 2: When you have random effects, then it is not only the F tests that you need to refigure. Standard deviations and contrasts on least squares means may need to be adjusted, depending on details of the situation. Consult an expert if you need to delve into the details of an analysis.

Method 2: Reduction to the Experimental Unit

There are only 6 animals, but are there 24 rows in the data table because each animal is measured 4 times. But taking 4 measurements on each animal doesn't mean that you can count each measurement as an observation. That would mean you could measure each animal millions of times and throw all that data into the computer to get an extremely powerful test.

The experimental unit that is relevant to **species** is the individual animal, not each measurement of that animal. When testing effects that only vary subject to subject, the experimental unit should be a single response per subject, instead of the repeated measurements.

One way to handle this situation is to group the data to find the means of the repeated measures, and analyze these means instead of the individual values:

§ With the ANIMALS table active, choose **Tables→Group/Summary,** which displays the dialog shown at the top of **Figure 12.29**.

§ Pick both **species** and **subject** as grouping variables.

§ Highlight the **miles** variable as shown and select **Mean** from the **Stats** popup menu on the Group/Summary dialog to see **Mean(Miles)** in the dialog. This notation indicates you want to see the mean (average) miles for each subject within each species. Click **OK** to see the summary table shown at the bottom of **Figure 12.29**.

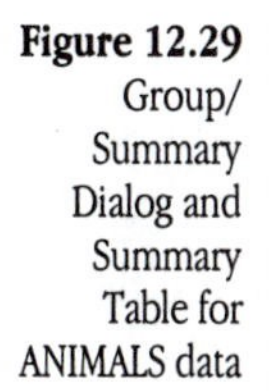

Figure 12.29
Group/Summary Dialog and Summary Table for ANIMALS data

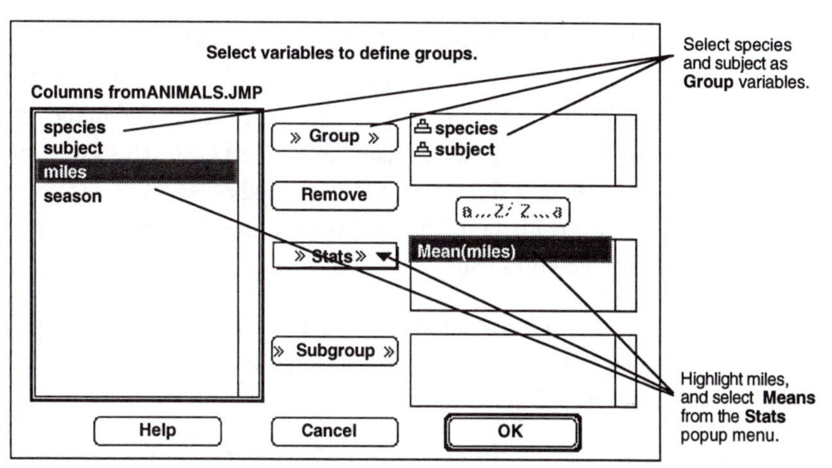

ANIMALS.JMP by species subject

4 Cols, 6 Rows, By-Mode Off

	species	subject	N Rows	Mean(miles)
1	COYOTE	1	4	5.25
2	COYOTE	2	4	5.25
3	COYOTE	3	4	7.25
4	FOX	1	4	2
5	FOX	2	4	3.25
6	FOX	3	4	3.75

Now fit a model to the summarized data

§ With the ANIMALS.JMP **by species subject** table (the Summary table) active, choose **Analyze→Fit Model**. You want to use **Mean(miles)** as **Y**, and **species** as the **X (Effects in Model)** variable.

§ Note that the variables in the summary table have analysis roles automatically assigned. **Mean(miles)** automatically appears as the **Y** variable in the Fit Model dialog, and both **species** and **subject** appear as **Effects in Model**. Remove the **subject** effect from the **Effects in Model** list.

§ Click **Run Model** to see the proper F test of 11.89 for **species**, with a p value of .0261 (**Figure 12.30**). Note that this is the same result as the hand calculation shown previously.

Figure 12.30
Effect Test for Species

Effect Test

Source	Nparm	DF	Sum of Squares	F Ratio	Prob>F
species	1	1	12.760417	11.8932	0.0261

Method 3: Correlated Measurements—Multivariate Model

In the animal example there were multiple (four) measurements for the same animal. These measurements are likely to be correlated in the same sense as two measurements are correlated in a paired t test. This situation of multiple measurements of the same subject is called *repeated measures*, or a *longitudinal* situation. This kind of experimental situation can be looked at as a multivariate problem.

To use a multivariate approach, the data table must be rearranged so that there is only one row for each individual animal, with the four measurements on each animal in four columns:

§ To rearrange the ANIMALS table, choose **Tables→Split Columns**. Complete the Split Columns dialog by assigning the variables as follows:

1) miles is the Split variable
2) season is the Col ID variable—its values become the new column names
3) species and subject arc Group variables

§ Click **OK** to see a new untitled table like the one shown in **Figure 12.31**.

Figure 12.31 Rearrangement of the ANIMALS data

Untitled 1

6 Cols / 6 Rows	species	subject	fall	spring	summer	winter
1	COYOTE	1	4	7	8	2
2	COYOTE	2	5	6	6	4
3	COYOTE	3	7	8	9	5
4	FOX	1	0	5	3	0
5	FOX	2	3	5	4	1
6	FOX	3	4	6	2	3

You then fit a multivariate model with four Y variables and a single response:

§ Choose **Analyze→Fit Model** and select fall, spring, summer, and winter as Y variables. Select species as the model effect.

§ Select **Manova** from the fitting personality popup menu as shown in **Figure 12.32**. Then click **Run Model** to see the analysis results.

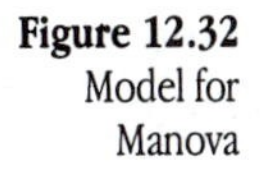

Figure 12.32 Model for Manova

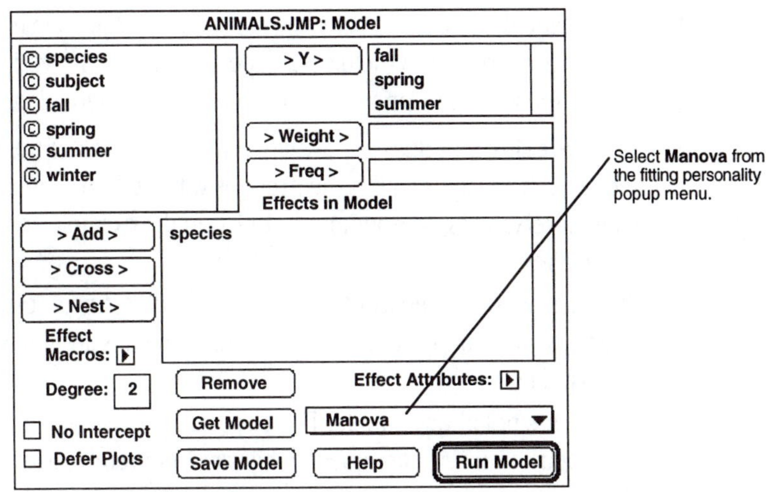

The resulting fit includes the report for species shown in **Figure 12.33**. The report for the species effect shows four different multivariate tests, but each reduces to an exact F test. Note that you get the same F test of 11.89 with a p value of .0261 as you did with the other methods.

Figure 12.33 Multivariate Analysis of Repeated Measures Data

species

Test	Value	Exact F	DF Num	DF Den	Prob>F
Wilk's Lambda	0.2516799	11.8932	1	4	0.0261
Pillai's Trace	0.7483201	11.8932	1	4	0.0261
Hotelling-Lawley	2.973301	11.8932	1	4	0.0261
Roy's Max Root	2.973301	11.8932	1	4	0.0261

Varieties of Analysis

In the previous cases, all the tests resulted in the same F test for species. However, it is not generally true that different methods produce the same answer. For example, if species had more than two levels, the four multivariate tests (Wilk's lambda, Pillai's trace, Hotelling-Lawley trace, and Roy's maximum root) each produce a different test results, and none of them would agree with the mixed model approach discussed previously.

Two more tests involving adjustments to the univariate method can be obtained from the multivariate fitting platform. If you had an unequal number of measurements per subject, then you couldn't use the multivariate approach. With unbalanced data, the mixed model approach also plunges into a diversity of methods that offer different answers from the *Method of Moments* that JMP uses.

If you are using the residual error to form F statistics, ask yourself if the row in the table corresponds to the unit of experimentation. Are you measuring variation in the way appropriate for the effect you are examining?

- When the situation does not live up to the framework of the statistical method, the analysis will be incorrect, as was method 1 (treating data table observations as experimental units) in the example above for the species test.
- Statistics offers a diversity of methods (and a diversity of results) for the same question. In this example the different results are not wrong, they are just different.
- Statistics is not always simple. There are many ways to go astray. Educated common sense goes a long way, but there is no substitute for expert advice when the situation warrants it.

Chapter 13
Bivariate and Multivariate Relationships

Overview

This chapter explores the relationship between two variables, the correlation between them, and the relationships between more than two variables. You look for patterns and you look for points that don't fit the patterns. You see where the data points are located, where the distribution is dense, and which way it is oriented.

Detective skills are built with the experience of looking at a variety of data, and learning to look at them in a variety of different ways. As you become a better detective, you also develop better intuition for understanding more advanced techniques.

It is not easy to look at lots of variables, but the increased range of the exploration will help you make more interesting and valuable discoveries.

Chapter 13 Contents

You can mouse along with the examples whenever you see an action symbol (§)

Bivariate Distributions

Previous chapters covered how the distribution of a response can vary depending on factors and groupings. This chapter returns to distributions as a simple unstructured batch of data. However, instead a single variable, the focus is on the joint distribution of two or more responses.

Density Estimation

As with univariate distributions, the question is: where are the data? What regions of the space are dense with data, and what areas are relatively vacant? The histogram forms a simple estimate of the density of a univariate distribution. If you want a smoother estimate of the density, JMP has an option that takes a weighted count of a sliding neighborhood of points to produce the smooth curve. This idea can be extended to several variables.

One of the most classic multivariate data sets in statistics contains the measurements of iris flowers that R. A. Fisher analyzed. Fisher's iris data are in the data table called IRISMKB.JMP, with variables **Sepal length**, **Sepal width**, **Petal length**, and **Petal width**. First, look at the variables one at a time:

§ Open IRISMKB.JMP and choose **Analyze→Distribution of Y** on **Sepal Length** and **Petal Length**.

§ When the report appears, select the **Smooth Curve** option from the check-mark popup menu at the lower left of the platform window. When the smooth curve appears, drag the density slider beneath the histogram to see the effect of using a wider or narrower smoothing distribution (**Figure 13.1**).

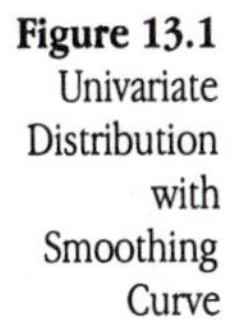

Figure 13.1 Univariate Distribution with Smoothing Curve

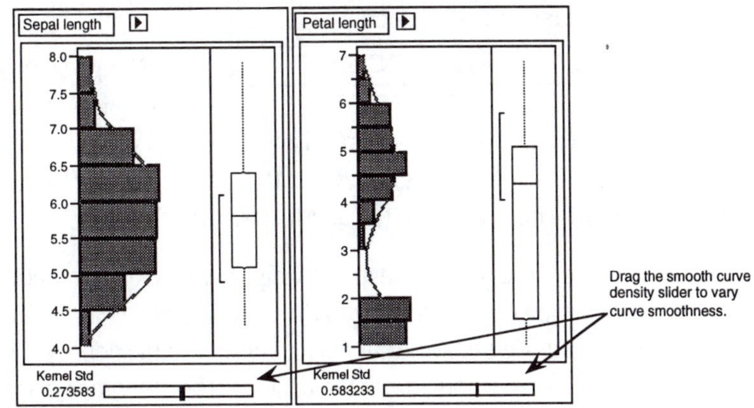

Notice in **Figure 13.1** that Petal length has an unusual distribution with two modes and a vacant area in the middle of the range. There are no petals with a length in the range from 2 to 3.

Bivariate Density Estimation

JMP has an implementation of a smoother that works for two variables to show bivariate densities. The goal is to draw lines around areas that are dense with points. Continue with the iris data and look at Petal length and Sepal length together:

§ Choose **Analyze→Fit Y by X** with Petal length as Y and Sepal Length as X. When the scatterplot appears, select **Nonpar Density** from the Fitting popup menu beneath the plot. (The density estimation is time intensive and can take a minute.)

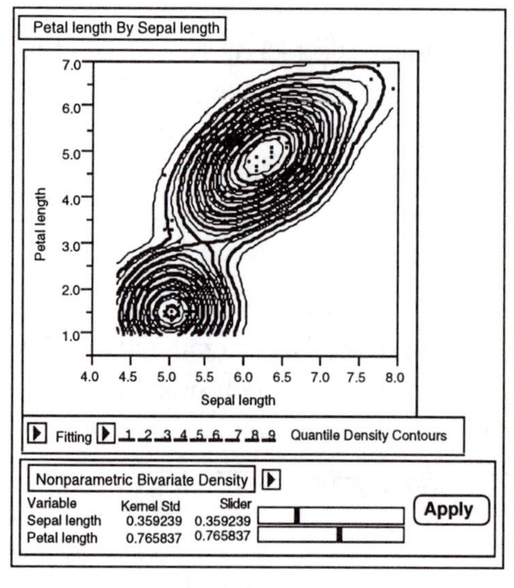

Figure 13.2 Bivariate Density Estimation Curves

The result (**Figure 13.2**) is a contour graph, where the various contour lines show paths of equal density. The density is estimated for each point on a grid by taking a weighted average of the points in the neighborhood, where the weights decline with distance. Estimates done in this way are called kernel smoothers.

The Nonparametric Bivariate Density table beneath the plot has slider controls for the vertical and horizontal width of the smoothing distribution. Because it can take a while to calculate densities, you have to click the **Apply** button before the densities are reestimated.

The density contours form a map showing where the data are most dense. The contours are done according to the quantiles, where a certain percent of the data lie outside each contour curve. These quantile density contours show where each 5% and 10% of the data are. The inner-most narrow contour line encloses the densest 5% of the data. The heavy line just outside shows the densest 10% of the data. It is labeled the .9 contour because 90% of the data lie outside it. Half the data distribution is inside the solid green lines, the 50% contours. Only about 5% of the data is outside the outermost 5% contour.

One of the features of the iris data is that there seem to be several local peaks in the density. There are two islands of data, one in the lower-left and one in the upper-right of the scatterplot.

These groups of locally dense data are called *clusters*, and the peaks of the density are called *modes*.

§ Now, select **Mesh Plot** from the popup menu at the top of the Nonparametric Bivariate Density table. This produces a 3-D surface of the density. Click and drag on the mesh plot to rotate it.

Figure 13.3 Mesh Plot of the Density for Iris Petal Length and Sepal Length

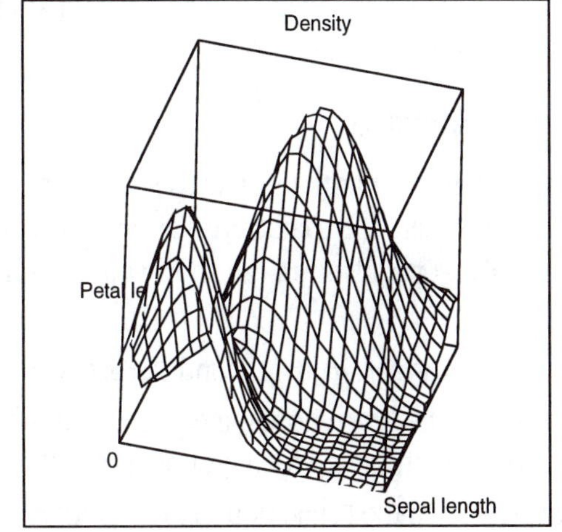

Mixtures, Modes, and Clusters

Multimodal data often comes from a mixture of several groups. If you look at the iris data table, you see that it is actually a collection of three species of iris: virginica, versicolor, and setosa.

If you do a bivariate density for each group you get the bivariate density plots in **Figure 13.4**. These plots have their axes adjusted to show the same scales.

Figure 13.4 Spinning Plot of Bivariate Density Curves

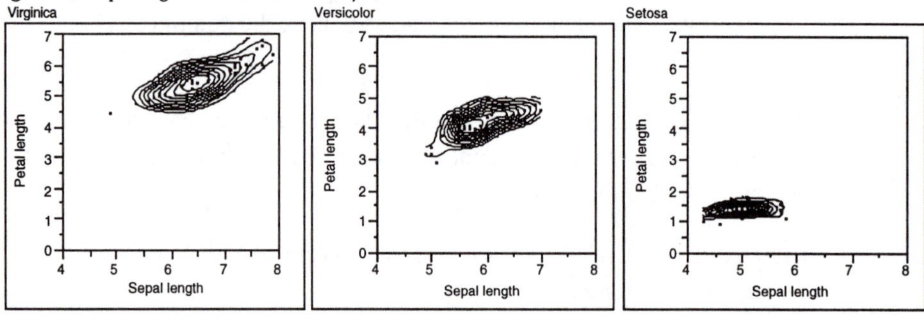

If you wanted to classify an observation (an iris) into one of these three groups, a natural procedure would be to compare the density estimates corresponding to the petal and sepal length of your specimen over the three groups, and assign it to the one with the highest density. That kind of statistical method is called *discriminant analysis*.

The Elliptical Contours of the Normal Distribution

Notice that the contours of the distributions on each species are elliptical in shape. It turns out that ellipses are the characteristic shape for a bivariate normal distribution. The Fit Y by X platform can show you a graph of these normal contours:

§ Choose **Analyze→Fit Y by X** for Petal length as Y and Sepal length as X.

§ Select **Grouping Variable** from the Fitting popup menu beneath the scatterplot and use Species when you are prompted for a grouping variable.

§ Then select **Density Ellipses** from the Fitting popup menu with .50 as the level for the ellipse.

The result of theses steps is shown in **Figure 13.5**. When there is a grouping variable in effect, there is a separate estimate of the bivariate normal density (or any fit you select) for each group. The normal density ellipse for each group enclose the densest 50% of the estimated distribution.

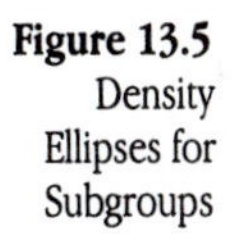

Figure 13.5 Density Ellipses for Subgroups

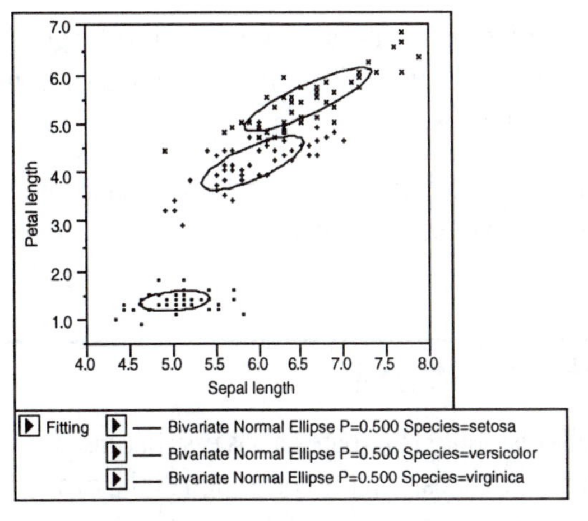

Bivariate Species=setosa

Variable	Mean	Std Dev	Correlation	Signif. Prob	Number
Sepal length	5.006	0.35249	0.267176	0.0607	50
Petal length	1.462	0.173664			

Bivariate Species=versicolor

Variable	Mean	Std Dev	Correlation	Signif. Prob	Number
Sepal length	5.936	0.516171	0.754049	0.0000	50
Petal length	4.26	0.469911			

Bivariate Species=virginica

Variable	Mean	Std Dev	Correlation	Signif. Prob	Number
Sepal length	6.588	0.63588	0.864225	0.0000	50
Petal length	5.552	0.551895			

Notice that the two ellipses toward the top of the plot are fairly diagonally oriented, while the one at the bottom is not. The reports beneath the plot show the means, standard deviations, and correlation of **Sepal length** and **Petal length** for the distribution of each species. Note that the correlation is low for setosa, and high for versicolor and virginica. The diagonal flattening of the elliptical contours is a sign of strong correlation. If variables are uncorrelated, then their normal density contours appear to have a nondiagonal shape.

One of the main uses of a correlation is to see if variables are related. You want to know if the distribution of one variable is a function of the other. When the variables are normally distributed and uncorrelated, then the univariate distribution of one variable is the same no matter what the value of the other variable is. When the density contours have no diagonal aspect, then the density across any slice is the same no matter where you take that slice (after you normalize the slice to have an area of one so it becomes a univariate density).

The **Density Ellipses** command in the Fit Y by X platform also gives a significance test on the correlation, which shows the p value for the hypothesis that the correlation is 0.

The bivariate normal is so common and is very basic for analyzing data, so let's cover it in more detail with simulations.

Correlations and the Bivariate Normal

Describing normally distributed bivariate data is easy because you need only the means, standard deviations, and the correlation of the two variables to completely characterize the distribution. If the distribution is not normal, you might need a good deal more to summarize it.

Correlation is a measure, on a scale of –1 to 1, of how close two variables are to being linearly related. If you can draw a straight line through all the points of a scatterplot, then the correlation is one, or minus one if the slope of the line is negative.

Simulation Exercise

As in earlier chapters, it is useful to examine simulated data that you create with formulas. This simulated data provides a reference point when you move on to analyze real data.

§ Open CORRSIM.JMP. This table has no rows, but contains formulas to generate correlated data. Choose **Rows→Add Rows** and enter 1000 when prompted for the number of rows you want.

The formulas evaluate to give simulated correlations (**Figure 13.6**). There are two independent standard normal random columns, labeled X and Y.00. The remaining columns (y.50, y.90, y.99, and y.1.00) have formulas constructed to produce the level of correlation indicated in the column names (.5, .9, .99, and 1). The formula for generating a correlation r with variable X is to make the linear mix with coefficient r for X and the square root of $(1–r^2)$ for Y.

Figure 13.6 Partial Listing of Simulated Values

CORRSIM.JMP

6 Cols / 1000 Rows	x	y.00	y.50	y.90	y.99	y1.00
1	-1.66357	0.519119	-0.38222	-1.27094	-1.5737	-1.66357
2	0.266931	-1.4974	-1.16332	-0.41246	0.053027	0.266931
3	0.859231	0.154887	0.563751	0.840822	0.872488	0.859231
4	-0.1067	0.124094	0.054118	-0.04194	-0.08813	-0.1067
5	1.102657	0.840377	1.279116	1.358703	1.21018	1.102657

You can use the Fit Y by X platform to examine the correlations:

§ Choose **Analyze→Fit Y by X** with X as the X variable, and all the Y columns as the Ys.

§ Hold down the COMMAND (or CONTROL key) and select **Density Ellipse** from the Fitting popup menu beneath the plot. Also, choose .9 as the density level. Holding down the COMMAND or CONTROL key causes the command to apply to all the open plots in the Fit Y by X window simultaneously.

§ Do the previous step twice more with .95, and .99 as density parameters.

These steps make normal density ellipses (**Figure 13.7**) containing 90%, 95%, and 99% of the bivariate normal density, using the means, standard deviations, and correlation from the data.

Figure 13.7 Examples of Density Ellipses for Various Correlation Coefficients

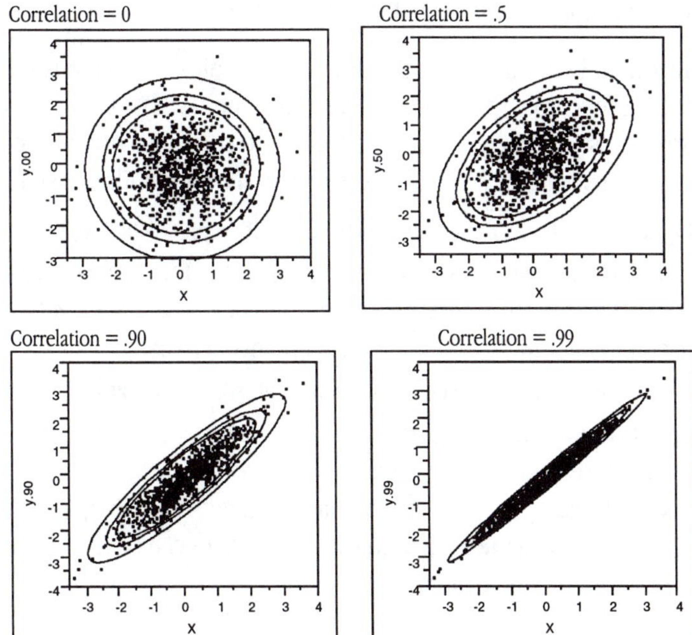

(continued next page)

Figure 13.7 (continued) Ellipses for Various Correlation Coefficients

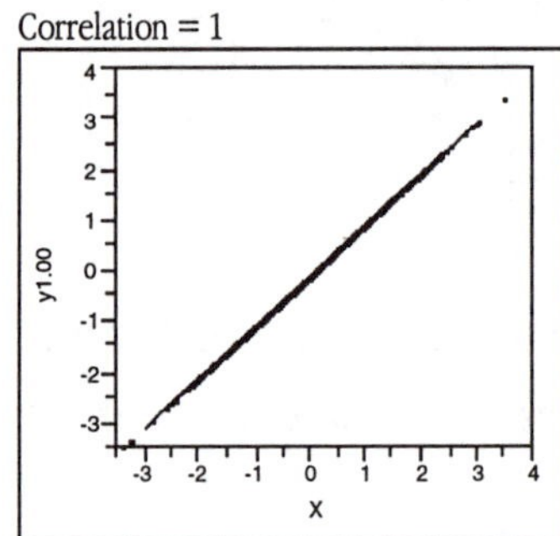

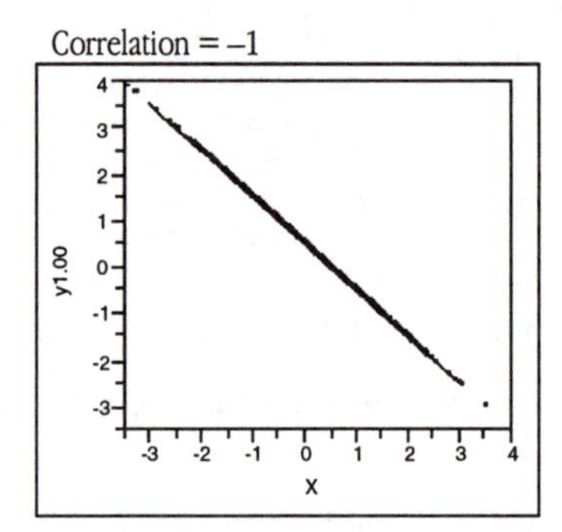

As the correlation grows from 0 to 1, the relationship between the variables gets closer and closer. The normal density contours are circular at correlation 0 (if the axes are scaled by the standard deviations) and collapse to a line at correlation 1.

Correlations Across Many Variables

Let's look at six variables. All you need to characterize the distribution of a six-variate normal distribution is the means, the standard deviations, and the bivariate correlations of all the pairs of variables.

In a chemistry study, the solubility of 72 chemical compounds was measured with respect to 6 solvents (Koehler and Dunn, 1988). One purpose of the study was to see if any of the solvents were correlated—that is, to identify any pairs of solvents that acted on the chemical compounds in a similar way.

§ Open SOLUBIL.JMP. to see variables **Label**, **eth**, **oct**, **ccl4**, **c6c6**, **hex**, and **chc13**.

§ Choose the **Label** option from the popup at the top of the **Labels** column name. An "L" shows in the popup box, which tells JMP to use the values in this column to label points in plots.

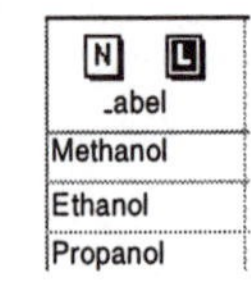

§ Choose **Analyze→Correlation of Y's** with all solvent variables as Ys.

§ Select **Scatterplot Matrix** from the check-mark menu found on the lower left of the platform window and expand the window to see as much as you can of the results. Initially, your scatterplot matrix will look like a larger version of the one shown in **Figure 13.8**. Each small scatterplot can be identified by the name cells of its row and column.

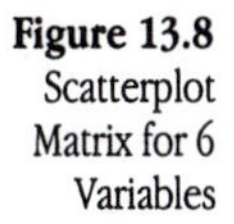

Figure 13.8 Scatterplot Matrix for 6 Variables

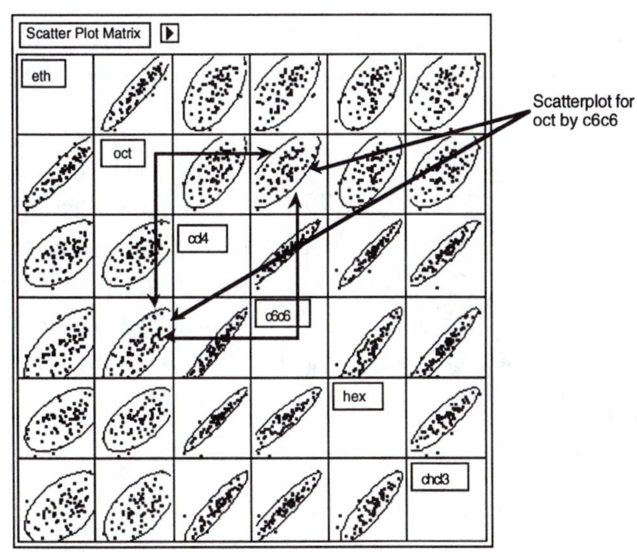

You can resize the whole matrix by resizing one of its small scatterplots; click in any cell and notice a small resize box in the lower-right corner of the cell. Drag in the resize box to change the matrix cell sizes. Also, you can change the row and column location of a variable in the matrix by dragging its name on the diagonal with the hand tool.

Close the scatterplot matrix (it can take time to redraw), but keep the Correlation of Y's platform for six variables open to use again later in this chapter.

Bivariate Outliers

Let's switch platforms to get a closer look at the relationship between ccl4 and hex using a set of density contours:

§ Choose **Analyze→Fit Y by X** with cc14 as Y and hex as X.

§ Select **Density Ellipses** from the Fitting popup menu four times for arguments .5, .9, .95, and .99 to add four density contours to the plot, as in **Figure 13.9**.

On the assumption that the data are distributed bivariate normal, the inside ellipse contains half the points, the next ellipse 90%, then 95%, and the outside ellipse contains 99% of the points.

Note that there are two points that are outside even the 99% ellipse:

§ Click and then SHIFT-click to highlight the two outside points. With the points highlighted, choose **Rows→Label** to label them, as shown on the plot in plot in **Figure 13.9**.

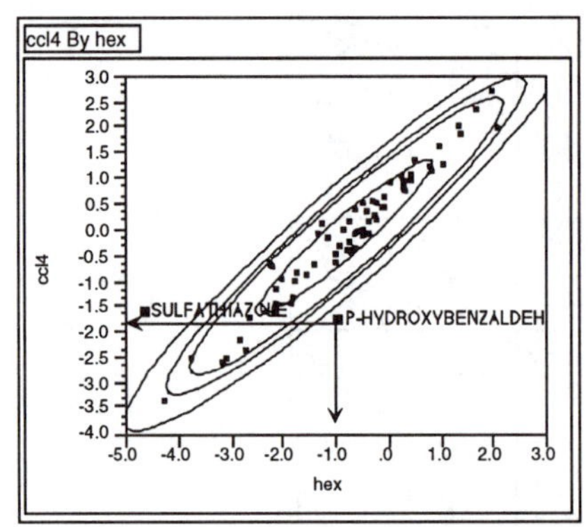

Figure 13.9 Density Ellipses Showing 2-Dimensional Outliers

The labeled points are *outliers*. Consider the following definition of *outlier* for the bivariate case: A point is an *outlier* if its bivariate normal density contour is associated with a very low probability.

Note that "P-hydroxybenzaldehyde" is not an outlier for either variable individually. In the scatterplot it is near the middle of the **hex** distribution, and is barely outside the 50% limit for the **ccl4** distribution. However it is a bivariate outlier because it falls outside the correlation pattern, which shows most of the points in a narrow diagonal elliptical area.

The technical name for this outlier distance is *Mahalanobis distance.* The Mahalanobis distance is computed with respect to the correlations as well as the means and standard deviations of both variables.

§ Click the Correlation of Y's platform to make it the active window and select **Outlier Analysis** from the check-mark popup menu. This command gives the Mahalanobis Distance outlier plot shown in **Figure 13.10**.

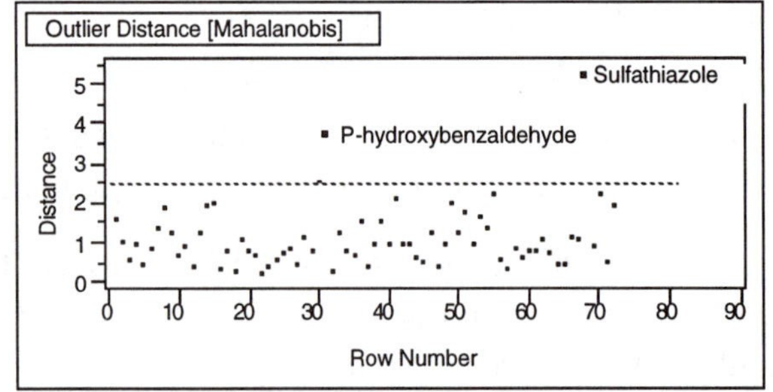

Figure 13.10 Outlier Analysis with Mahalanobis Outlier Distance Plot

The reference line is drawn using an F quantile and shows the estimated distance that contains 95% of the points. In agreement with the ellipses, "Sulfathiazole" and "P-hydroxybenzaldehyde" show as prominent outliers.

§ Click the scatterplot matrix to activate it and select the brush tool from the **Tools** menu. Try dragging the brush tool over these two plots to confirm that the points near the central ellipse have low outlier distances and the points outside are greater distances.

§ Close this Correlation of Y's window.

Three and More Dimensions

To consider three variables at a time, consider the first three variables, **eth**, **oct**, and **ccl4**. You can see the distribution of points with a spinning plot:

§ Choose **Graph→Spinning Plot** and select **eth**, **oct**, and **ccl4** as variables to spin. When the 3-D spinning plot appears, use the hand tool found in the **Tools** menu, or use the buttons on the spin control panel to spin the plot and look for three-variate outliers.

The spin orientation in **Figure 13.11** shows three points, including "Caffeine," that appear to be outlying from the rest of the points with respect to the ellipsoid-shaped distribution.

Figure 13.11 Outliers as Seen in Spinning Plot and in Outlier Distance Plot

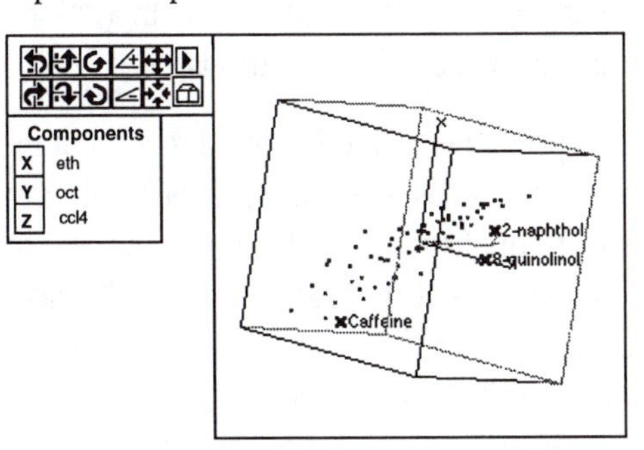

As before, you can also see these outliers with a Mahalanobis distance plot:

§ Choose **Analyze→Correlation of Y's** with **eth**, **oct**, and **ccl4** as **Y** variables. Then select **Outlier Analysis** from the check-mark popup menu to see the Outlier Distance plot in **Figure 13.12.**

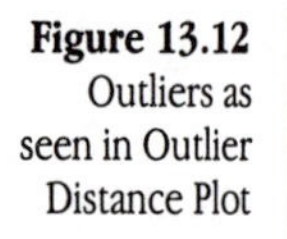

Figure 13.12 Outliers as seen in Outlier Distance Plot

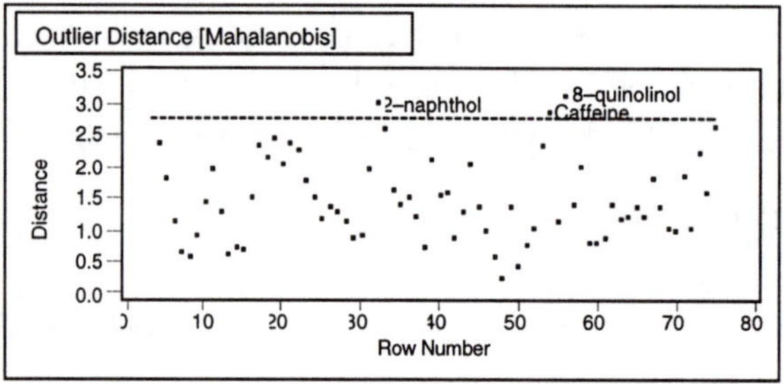

Principal Components

As you spin the points, notice that some directions in the space show a lot more variation in the points than other directions. This was true in two dimensions when variables were highly correlated. The long axis of the normal ellipse had the most variance; the short axis had the least. Now in three dimensions, there are three axes on a three-dimensional ellipsoid for trivariate normal. The solubility data does seem to have the ellipsoidal contours characteristic of normal densities, except for a few outliers.

The directions of the axes of the normal ellipsoids are called the *principal components*. They were mentioned in the Galton example in Chapter 8, "Fitting Curves Through Points: Regression."

The first principal component is defined as the direction of the linear combination of the variables that has maximum variance, subject to being scaled so the sum of squares of the coefficients is one. In a spinning plot, it is easy to rotate the plot and see which direction this is.

The second principal component is defined as the direction of the linear combination of the variables that has maximum variance, subject to it being at right angles (*orthogonal*) to the first principal component, and so on. There are as many principal components as there are variables. The last principal component has little or no variance if there is substantial correlation among the variables. This means that there is a direction for which the normal density hyper-ellipsoid is very thin.

The spinning plot platform is good at showing principal components. Click the Spinning Plot platform shown in **Figure 13.11** to make it active:

§ Select the **Principal Components** option found in the check-mark menu of the Spinning Plot window. This adds three principal components to the variables list and creates three new rays in the spinning plot, as shown in **Figure 13.13**.

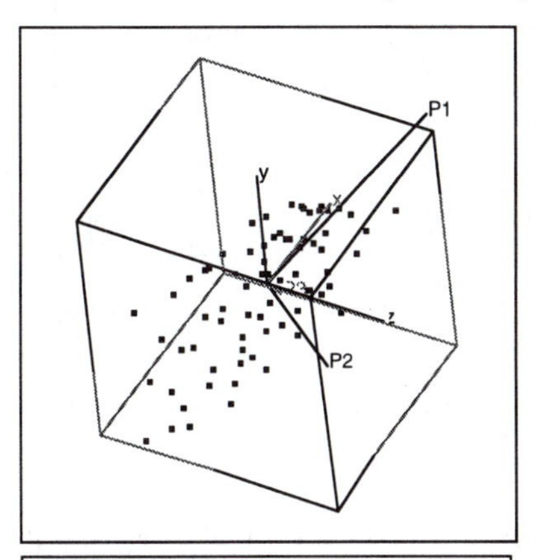

Figure 13.13 Biplot Showing First Principal Component and Principal Components Report

Principal Components			
EigenValue:	2.4079	0.5257	0.0664
Percent:	80.2631	17.5233	2.2136
CumPercent:	80.2631	97.7864	100.0000
Eigenvectors:			
eth	0.60321	-0.42153	0.67709
oct	0.61833	-0.28908	-0.73083
ccl4	0.50379	0.85951	0.08627

The directions of the principal components are shown as rays from the origin, labeled P1 for the first principal component, P2 for the second, and P3 for the third. As you rotate the plot, you see that the principal component rays correspond to the directions in the data for more or less variance. You can also see that the principal components are at right angles in three-dimensional space.

The Principal Components report in **Figure 13.13** shows what portion of the variance among the variables is carried by each principal component. In this example, 80% of the variance is carried by the first principal component, 17% by the second, and 2% by the third. It is the correlations in the data make the principal components interesting and useful. If the variables are not correlated, then the principal components will all carry the same variance.

Principal Components for Six Variables

Now let's move up to more dimensions than most humans can see. Click the SOLUBIL.JMP data table to make it the active window and look at principal components for all six variables in the data table:

§ Proceed as before. Choose **Graph→Spinning Plot** and select the six solvent variables as spin components. Then select **Principal Components** in the check-mark menu, which adds six principal component axes to the spinning plot and produces a principal component analysis.

§ Open the Principal Components table (**Figure 13.14**). There is a column in the table for each of the six principal components. Note in the **Cumpercent** row that the first three principal components work together to account for 97.8% of the variation in six dimensions.

Figure 13.14 Principal Components Shown in Biplot

Principal Components						
EigenValue:	4.7850	0.9408	0.1427	0.0630	0.0464	0.0221
Percent:	79.7503	15.6804	2.3779	1.0495	0.7738	0.3680
CumPercent:	79.7503	95.4307	97.8087	98.8582	99.6320	100.0000
Eigenvectors:						
eth	0.34887	0.64095	0.17653	0.63780	-0.12841	0.11419
oct	0.37314	0.56277	-0.17684	-0.67519	0.23816	0.01361
ccl4	0.43187	-0.29274	0.17821	-0.20373	-0.46785	0.66003
c6c6	0.44552	-0.14827	-0.21629	-0.01927	-0.52424	-0.67642
hex	0.42231	-0.27033	0.68072	-0.00820	0.47344	-0.24696
chcl3	0.41919	-0.30285	-0.62900	0.30884	0.45709	0.18056

§ Drag the Spinning Plot axis labels to be like those in the components panel shown in **Figure 13.15**: click on **X** in the column of boxes beside the variable names and drag it down to the box by **Prin Comp 1**, drag **Y** to **Prin Comp 2**, and **Z** to **Prin Comp 3.**

The first three principal components now become the spinning plots X, Y, and Z axes (**Figure 13.15**). As you spin the points, remember that you are seeing 97.8% of the variation in six dimensions as summarized by the first three principal components. This is the best 3-dimensional view of six dimensions.

Figure 13.15 Principal Components Report for 6 Variables

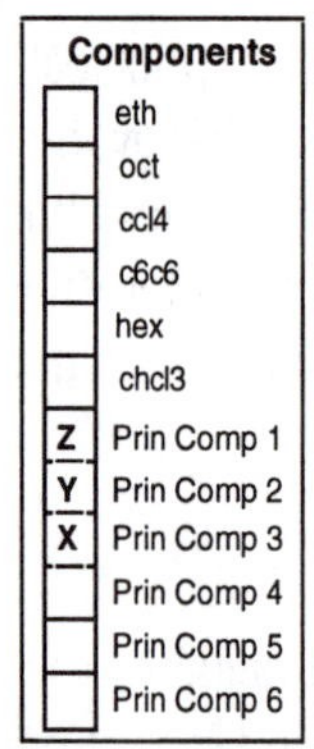

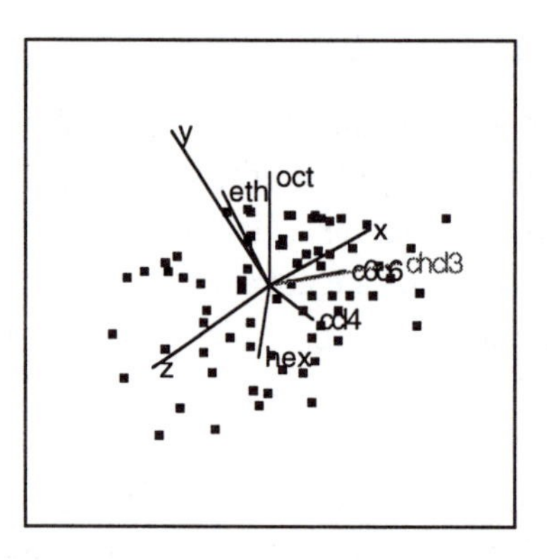

Notice that there are now rays labeled with the names of the variables. You can actually measure the coordinates of each point along these axes. The measurement will be accurate to the degree that the first principal components capture the variation in the variables. This technique of showing the points on the same plot that shows directions of the variables was pioneered by Ruben Gabriel; the plot is called a Gabriel biplot. The rays are actually formed from the eigenvectors in the report, which are the coefficients of the principal components on the standardized variables (and vice versa since they form an orthogonal matrix where the inverse is the transpose).

Correlation Patterns in Biplots

Note that the rays for **eth** and **oct** have a very narrow angle between them. However, these two rays are at near right angles (in 3-D space) to the other four rays. Narrow angles between principal component rays are a sign of correlation. Principal components try to squish the data from six dimensions down to three dimensions. To represent the points most accurately in this squish, the directions for correlated variables are close together because they represent most of the same information. Thus the Gabriel biplot shows the correlation structure of high-dimensional data.

Refer back to **Figure 13. 8** to see the scatterplot matrix for all six variables. The scatterplot matrix confirms the fact that **eth** and **oct** are highly correlated. The other four variables are also highly correlated, but there is much less correlation between these two sets of variables.

You might also consider a simple form of factor analysis, in which the components are rotated to positions so that they point in directions that correspond to clusters of variables. In JMP, this can be done in the Spinning Plot platform with the **Rotated Components** command.

Outliers in Six Dimensions

The Ft Y by X and the Spinning plot platforms revealed extreme values in one, two, and three dimensions. These outliers will also show in six dimensions. But there could be additional outliers that violate the higher dimensional correlation pattern. In six dimensions some of the directions of the data are very flat, and there could be an outlier in that direction that wouldn't be revealed in an ordinary scatterplot.

§ In the Spinning Plot platform with six variables, drag the X, Y, and Z axis labels to the last three principal components as shown in **Figure 13.16.**

Now you are seeing the directions in six-dimensional space that are least prominent for the data. The data points are crowded near the center because you are looking at a small percentage of the variability.

§ Use the zoom-out button (the one with the four arrows pointing outward) on the spin control panel to expand the point cloud away from the center.

§ Then, highlight and label the points that seem to stand out. These are the points that are in the directions that are most unpopular. Sometimes they are called *class B outliers*.

Figure 13.16 Biplot Showing Six Principal Components

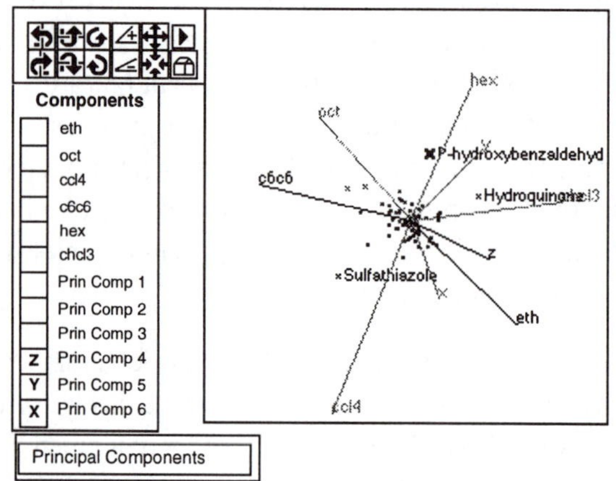

All the outliers from one dimension to six dimensions should show up on an outlier distance plot that measures distance with respect to all six variables.

§ Click (to activate) the Correlation of Y's platform previously generated for all six variables. Select **Outlier Analysis** from the check-mark menu to see the six-dimensional outlier distance plot in **Figure 13.17**. (If you closed the correlation window, choose **Analyze→Correlation of Y's** with all six responses as Y variables, and then select **Outlier Analysis** from the check-mark menu.

Figure 13.17 Outlier Distance Plot for 6 Variables

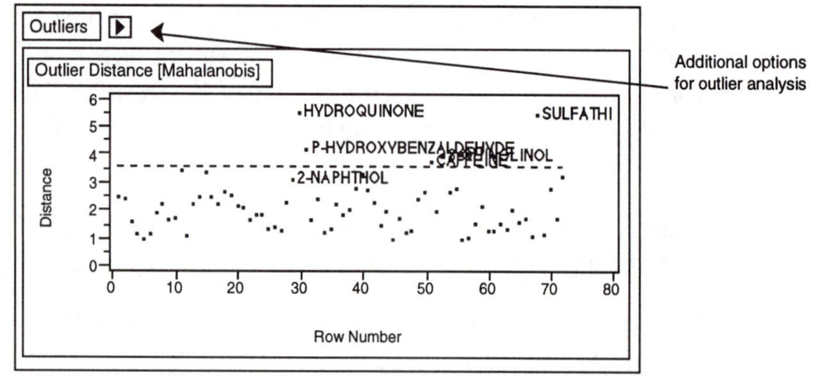

There is a refinement to the outlier distance that can help to further isolate outliers. When you estimate the means, standard deviations, and correlations, all points, including outliers, are included in the calculations and affect these estimates, causing an outlier to disguise itself. Suppose that as you measure the outlier distance for a point, you exclude that point from all mean, standard deviation, and correlation estimates. This technique is called *jack-knifing*. The jackknifed distances often make outliers stand out better.

§ To see the Jackknifed Distance plot in **Figure 13.18**, select the **Jackknife Distances** option from the popup menu at the top of the Outliers analysis.

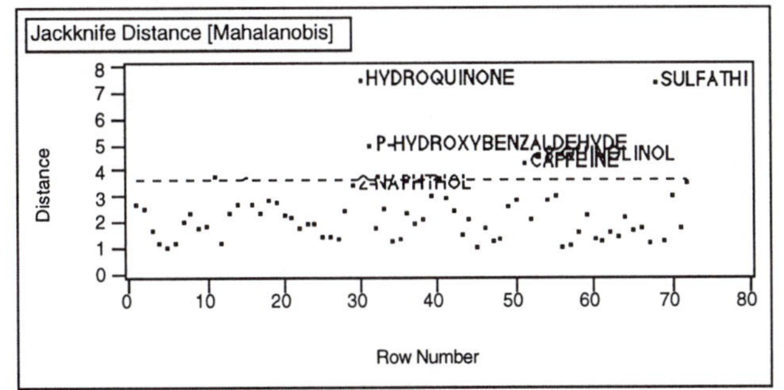

Figure 13.18 Jack-knifed Outlier Distance Plot for 6 Variables

Strategies for High-Dimensional Exploration

When you have more than three variables, the relationships among them can get very complicated. Many things can be easily found in one, two, or three dimensions; but it is hard to visualize a space of more than three dimensions.

The histogram provides a good one-dimensional look at the distribution, with a smooth curve option for estimating the density. The scatterplot provides a good two-dimensional look at the distribution, with normal ellipses or bivariate smoothers to study it. In three dimensions the spinning plot provides the third dimension. To look at more than three dimensions, you must be creative and imaginative.

One good basic strategy for high-dimensional exploration is to take advantage of correlations to reduce the number of dimensions. The technique for this is principal components, and the graph is the Gabriel biplot, which shows all the original variables as well as the points in principal component space.

You can also use highlighting tools to brush across one distribution and see how the points highlight in another view.

The hope is that you either find patterns that help you understand the data, or points that don't fit patterns. In both cases you can make valuable discoveries.

Chapter 14
Design of Experiments

Overview

When you have a problem in industry, you have to do a little science to fix it. When you need something to perform better, you have to do a little science to improve it. When your product is competing in a world market, you need to do a little science to make it the best. When you need evidence to prove your product is effective, you need to do a little science. Sometimes you can use deductions and calculations from known science directly, but more often you will need to experiment.

When you experiment, you control the environment so that the only variables are the factors that you want to vary and the response that you want to measure under the different settings. Then you can find out which conditions lead to what response. Sometimes you don't care why certain factor levels make the response better—you are happy enough just to find it out what factor level is best. In an age where quality and performance are increasingly important, experimentation has become indispensable throughout industry.

The search for factors and their optimal settings is expensive. The field of *experimental design* provides ways of making the search process efficient and effective.

This chapter covers how JMP can be used to construct experimental designs and then analyze the results. The tools to do this are the **Design Experiment** command in the **Tables** menu and the Screening personality of the Fit Model platform.

Chapter 14 CONTENTS

You can mouse along with the examples whenever you see an action symbol (§)

Introduction

Experimentation is the fundamental tool of the scientific method. In an experiment a response of interest is measured as some factors are changed systematically, while other factors are held as constant as possible. The goal is to determine if and how the factors affect the response. Each combination of factor settings is a *design point*. An individual response value with its factor settings is sometimes referred to as a *run*.

The challenge is to determine the fewest number of runs needed to gain adequate information for optimizing quality.

JMP IN has tools that allow you to produce several kinds of experimental designs:

Screening Designs for Scouting Many Factors
: Screening designs examine many factors to see which have the greatest effect on the results of a process. To economize on the number of runs needed, each factor is usually set at only two levels, and response measurements are not taken for all possible combinations of levels. Screening designs are a prelude to further experiments.

Response Surface Designs for Optimization
: Response surface experiments try to focus on the optimal values for a set of continuous factors. They are modeled with a curved surface so that the maximum point of the surface (optimal response) can be found mathematically.

Complete Factorial Designs
: A complete factorial includes a run for all possible combinations of factor levels.

Note: The designs in this chapter are commonly used in industrial settings, but there are many other designs used in other settings. The professional version of JMP can produce additional designs such as mixture, mixed-level and D-optimal designs. JMP can analyze other designs such as split plots, repeated measures, crossovers, balance incomplete blocks, and Latin squares, but has no built-in features for generating runs for these designs.

Generating an Experimental Design in JMP

Experimental designs are generated by JMP in the following steps:

1) Select the **Design Experiment** command from the **Tables** menu. This displays the Choose Design Type dialog shown in **Figure 14.1**.
2) Optionally, click **Define Factors and Levels** to give each factor a name, data type, and upper and lower values.
3) Choose one of the design types in the Choose Design Type dialog. Each choice automatically displays its own specific design choice dialog.
4) When the specific design choice dialog appears, specify the number of factors and give information specific to the design type.
5) Select from the generated list of designs.
6) Optionally, request randomized runs.
7) Optionally, request a Fit Model window that will contain the appropriate model for the design you specified.
8) Generate the specified design.

Figure 14.1 The Design Experiment Dialog

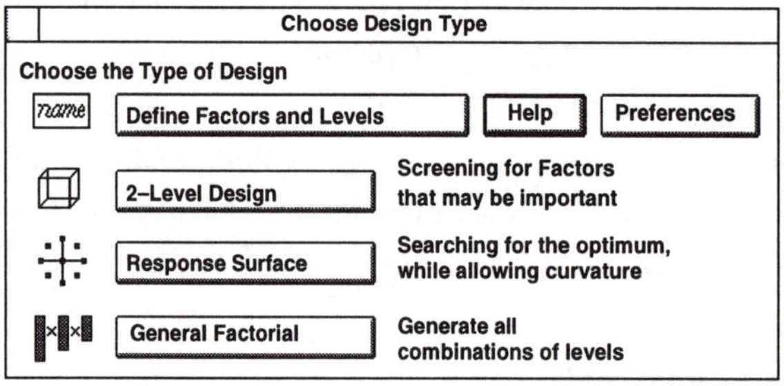

Two-Level Screening Designs

Two-level designs are commonly used to examine many factors with a minimum number of runs. Screening experiments can help identify which of many factors most affect a response. Often screening designs are a prelude to further experiments.

Complete two-level factorials are composed of all combinations of m factors at two levels. For m factors, there are 2^m runs in a full factorial. Because 2^m

becomes very large as m increases, a fraction of the full factorial is often employed as the experimental design.

Fractional Factorial Designs

For a fractional factorial with 2^{m-p} runs, you take the first $m-p$ factors and make a full factorial design, with a run for each combination of high and low levels for these $m-p$ factors. The other p factors are set to the values of some higher-order interaction of the first $m-p$ factors. In screening designs, the higher-order factor interactions are assumed to have minimal effect on the response in the real world, so they can be *confounded* with main effects and lower-order interactions.

Two effects are said to be *confounded* when the design cannot distinguish which effect affects the response. Confounding also means that an effect is a linear combination of other effects, so that if all the effects are put into the model, there would not be a unique least squares solution for the estimates.

Resolution Number: The Degree of Confounding

The focus of attention is usually on how the second-order interactions are confounded, because interactions above second-order are rare. The *resolution number* is a way to describe the degree of confounding.

Resolution 3
: means that two-factor interactions are confounded with main effects. Only main effects are included in the model, and the two-factor interactions are assumed to be negligible.

Resolution 4
: means that main effects are not confounded with two-factor interactions, but some two-factor interactions can be confounded with each other. Some two-factor interactions can be modeled without being confounded. Other two-factor interactions can be modeled with the understanding that they are confounded with two-factor interactions included in the model. Three-factor interactions are assumed to be negligible.

Resolution 5 or more
: means that all two-factor interactions are estimable (are not confounded). The model can contain any or all of the two-factor interactions.

Plackett-Burman Designs

Plackett-Burman designs are an alternative to fractional factorial designs. Unlike the two-level fractional factorials, where the number of runs is always a power of 2, the number of runs in Plackett-Burman designs is usually 12, 20, 24, 28, or 36 .

Screening for Main Effects: The Flour Paste Experiment

Acme Pinata Corporation discovered that its pinatas were too easily broken. The company wants to perform experiments to discover what factors might be important for the peeling strength of flour paste. "Strength" refers to how well two pieces of paper that are glued together resist being peeled apart.

The Factors

Batches of flour paste were prepared to determine the effect of the following nine factors on peeling strength:

Flour 1/8 cup of white unbleached flour or 1/8 cup of whole wheat flour

Sifted flour was sifted or not sifted

Type water-based paste or milk-based paste

Temp mixed when liquid was cool or when liquid was warm

Salt formula had a dash of salt or had no salt.

Liquid 4 teaspoons of liquid or 5 teaspoons of liquid

Clamp pasted pieces were tightly clamped together or not clamped during drying

Sugar formula contained 1/4 teaspoon or no sugar

Coat whether the amount of paste applied was thin or thick

The following sections describe how to use JMP IN to generate a design for the flour paste experiment and then analyze the results.

The Experimental Procedure

A preliminary study was undertaken first to establish a workable experimental procedure. For a run, two 3.5 by 5 inch note cards were overlapped by half to make the paste joint. After 12 hours of drying they were peeled back to within 1.5 inches of the paste joint and then attached to a spring mechanism. The paste joint was pulled, stretching the spring as a pencil marked the motion on paper. When the joint gave way the paper was marked and the length of the pencil track measured. Assuming the spring was well behaved, this distance is linearly related to the force acting to peel the paste joint.

Name the Factors

The first step in JMP is to specify the factors and levels. The names you enter at this step will label the experimental plan and the results of the analyses.

§ Choose **Tables→Design Experiment** to see the Choose Design Type dialog shown previously in **Figure 14.1**.

§ Click Define Factors and Levels. This displays the panel in **Figure 14.2**, which initially shows default names X1, X2,... and so forth. Enter the factor names and levels as shown in **Figure 14.2**. To ensure proper entry of the factors and levels, it is best to select the data Type (Character or Numeric) before you enter the other information for an effect.

§ Click **OK** to return to the Choose Design Type dialog (**Figure 14.1**).

Figure 14.2
Define Factors and Levels

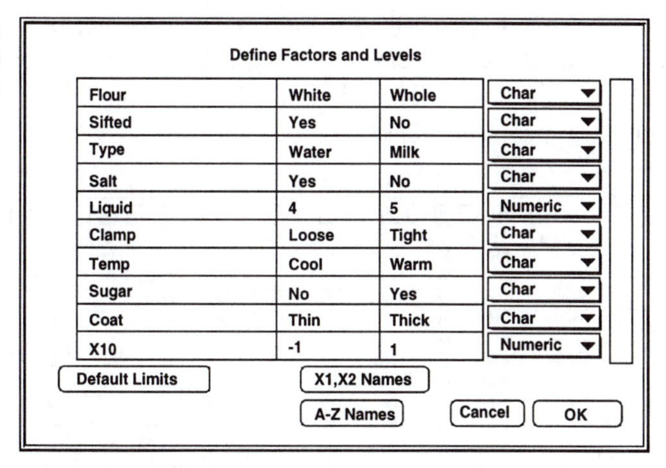

Select the Design and Generate the Table

§ In the Choose Design Type dialog (**Figure 14.1**), click **2-level Design**.

§ When the Two-Level Design Selection dialog appears, type the number of flour paste experiment factors (9), and then click **Search for Designs**. JMP lists the designs for the number of factors you specified. Your dialog should look like the one in **Figure 14.3**.

§ For the flour paste experiment, select the 16-run **Fractional Factorial** design and click **Generate Selected Design.**

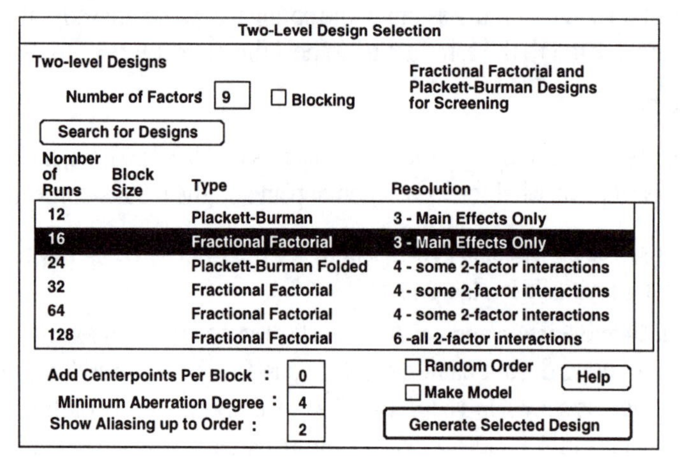

Figure 14.3 The Two-Level Design Dialog

The generated design now appears in a JMP data table (**Figure 14.4**). It lists 16 runs (observations) out of the $2^9 = 512$ runs possible from combinations of nine factors. Save this data table (use **File→Save As**), naming the file MYPASTE.JMP. If you were actually conducting an experiment, you would now print a copy of the design table and use Y column for entering the results you got from the experimental runs.

Figure 14.4 The Flour Paste Experimental Data

MYPASTE.JMP (11 Cols, 16 Cols)

	Pattern	Flour	Sifted	Type	Salt	Liquid	Clamp	Temp	Sugar	Coat	Y
1	−−−−+−−+−	White	Yes	Water	Yes	5	Loose	Cool	Yes	Thin	.
2	−−−+−++−+	White	Yes	Water	No	4	Tight	Warm	No	Thick	.
3	−−+−−++−−	White	Yes	Milk	Yes	4	Tight	Warm	No	Thin	.
4	−−+++−−++	White	Yes	Milk	No	5	Loose	Cool	Yes	Thick	.
5	−+−−−+−++	White	No	Water	Yes	4	Tight	Cool	Yes	Thick	.
6	−+−++−+−−	White	No	Water	No	5	Loose	Warm	No	Thin	.
7	−++−+−+−+	White	No	Milk	Yes	5	Loose	Warm	No	Thick	.
8	−+++−+−+−	White	No	Milk	No	4	Tight	Cool	Yes	Thin	.
9	+−−−−−+++	Whole	Yes	Water	Yes	4	Loose	Warm	Yes	Thick	.
10	+−−+++−−−	Whole	Yes	Water	No	5	Tight	Cool	No	Thin	.
11	+−+−++−−+	Whole	Yes	Milk	Yes	5	Tight	Cool	No	Thick	.
12	+−++−−++−	Whole	Yes	Milk	No	4	Loose	Warm	Yes	Thin	.
13	++−−++++−	Whole	No	Water	Yes	5	Tight	Warm	Yes	Thin	.
14	++−+−−−−+	Whole	No	Water	No	4	Loose	Cool	No	Thick	.
15	+++−−−−−−	Whole	No	Milk	Yes	4	Loose	Cool	No	Thin	.
16	+++++++++	Whole	No	Milk	No	5	Tight	Warm	Yes	Thick	.

Confounding Structure

When you generate a fractional factorial design, JMP displays a text-edit window like the one in **Figure 14.5**. This report lists how the design you requested was generated and what factors are confounded. The confounding rules are discussed later in this chapter

Figure 14.5
The Two-Level Design Dialog

```
Fractional Factorial Structure
Factor Confounding Rules
Liquid = Flour*Sifted*Type*Salt
Clamp = Sifted*Type*Salt
Temp = Flour*Type*Salt
Sugar = Type*Salt
Coat = Flour*Sifted*Salt

Aliasing Structure
        Flour          = Liquid*Clamp = Temp*Sugar
        Sifted         = Liquid*Temp = Clamp*Sugar
        Type           = Salt*Sugar = Liquid*Coat
        Salt           = Type*Sugar
        Liquid         = Flour*Clamp = Sifted*Temp = Type*Coat
        Clamp          = Flour*Liquid = Sifted*Sugar
        Temp           = Flour*Sugar = Sifted*Liquid
        Sugar          = Flour*Temp = Sifted*Clamp = Type*Salt
        Coat           = Type*Liquid
        Flour*Sifted   = Salt*Coat = Liquid*Sugar = Clamp*Temp
        Flour*Type     = Salt*Temp = Clamp*Coat
        Flour*Salt     = Sifted*Coat = Type*Temp
        Flour*Coat     = Sifted*Salt = Type*Clamp
        Sifted*Type    = Salt*Clamp = Temp*Coat
        Salt*Liquid    = Sugar*Coat
```

In this example, you want to look at many factors in a very few runs, so you select a resolution 3 design. This means that main effects are confounded with two-way interactions.

The design first included all possible combinations of the first four factors, which resulted in the 16 runs and the 2^4 (16) combinations (Note the values in the first four columns in the data table shown in **Figure 14.4**). The first four factors are listed in standard order (actually a randomized order is usually recommended and a check box on the dialog provides for this). There are eight runs for each Flour type. Within each Flour type there are four rows for each sifted value. Within sifted there are two runs per Type. Then Salt is alternated every run. So for the first four factors, there is a complete factorial design.

The last five factors are generated by multiplying internal codes for combinations of the first four factors together, the same as coding for interactions. The codes for factor levels are –1 and 1, so any product of the codes is also –1 or 1. For example, the Liquid factor is the product of Flour*Sifted*Type*Salt; Clamp is the product Sifted*Type*Salt, and so on for the remaining factors, as shown by the **Factor Confounding Rules** in **Figure 14.5**. If you do not enter factor names and levels into the experimental design, these codes (–1, 1) would show in the design data table (**Figure 14.4**) instead of the level names.

By generating the design with these confoundings, there are other relationships between factors and interactions, as shown in the **Aliasing Structure**. Basically, all main effects are confounded with two-factor interactions. For example, if an analysis indicates a significant main effect for Clamp, it could be a real main effect, or an interaction between Flour and Liquid, an interaction between Sifted and Sugar, any combination of these effects, or higher order interactions not shown.

The resolution 3 design assumes there are no interactions of consequence, which makes it possible to use a small number of runs to look at a large number of factors.

§ Verify that the design is balanced. Choose **Analyze→Distribution of Y** for all nine factor variables. Now click in each histogram bar to see that the distribution is flat for all the other variables. The highlighted area representing the distribution for a factor level is equal in each of the other histograms, as shown by the example for Flour in **Figure 14.6**.

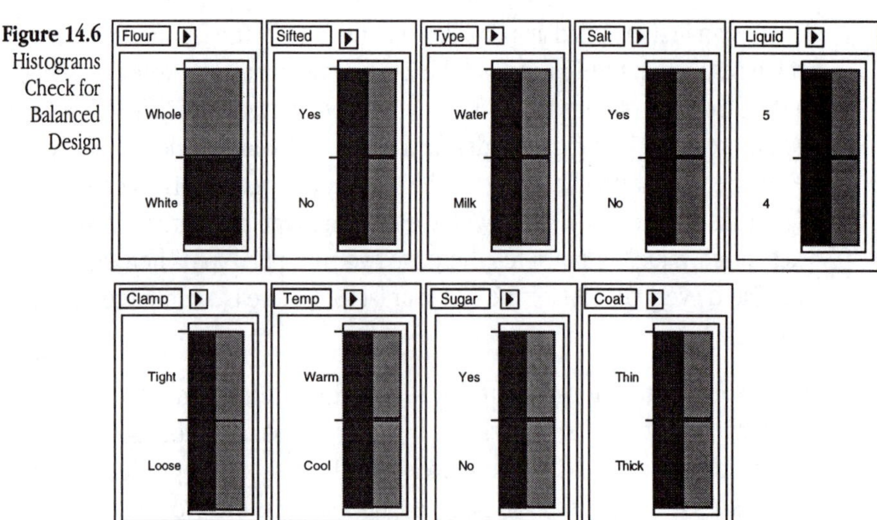

Figure 14.6 Histograms Check for Balanced Design

Perform Experiment and Enter Data

§ Suppose you complete the experiment and collect and enter the response (Y) data shown in **Figure 14.7**. Rather than reenter the data into the generated design data table, open the sample data table FLRPASTE.JMP.

Figure 14.7
JMP Table for the Flour Paste

FLRPASTE.JMP

11 Cols / 16 Rows	Pattern	Flour	Sifted	Type	Salt	Liquid	Clamp	Temp	Sugar	Coat	Y
1	—+—+-	White	Yes	Water	Yes	5	Loose	Cool	Yes	Thin	7.8
2	—++++	White	Yes	Water	No	4	Tight	Warm	No	Thick	17.4
3	—+—++–	White	Yes	Milk	Yes	4	Tight	Warm	No	Thin	8.2
4	—+++—++	White	Yes	Milk	No	5	Loose	Cool	Yes	Thick	3.7
5	+—+++	White	No	Water	Yes	4	Tight	Cool	Yes	Thick	6.5
6	+++++–	White	No	Water	No	5	Loose	Warm	No	Thin	7.1
7	++++++	White	No	Milk	Yes	5	Loose	Warm	No	Thick	4.3
8	++++++-	White	No	Milk	No	4	Tight	Cool	Yes	Thin	6.1
9	+——+++	Whole	Yes	Water	Yes	4	Loose	Warm	Yes	Thick	4.8
10	+—+++—	Whole	Yes	Water	No	5	Tight	Cool	No	Thin	3.5
11	+++++—+	Whole	Yes	Milk	Yes	5	Tight	Cool	No	Thick	4.3
12	+++—++-	Whole	Yes	Milk	No	4	Loose	Warm	Yes	Thin	2.9
13	++—++++-	Whole	No	Water	Yes	5	Tight	Warm	Yes	Thin	3.7
14	+++—+	Whole	No	Water	No	4	Loose	Cool	No	Thick	9
15	+++——	Whole	No	Milk	Yes	4	Loose	Cool	No	Thin	1.8
16	+++++++++	Whole	No	Milk	No	5	Tight	Warm	Yes	Thick	4.1

You are now ready to analyze the data.

Examine the Response Data

As usual, a good place to start is by examining the distribution of the response, Y, which is the peel strength in this example.

§ Choose **Analyze→Distribution of Y** for the variable Y.

§ When the histogram appears, select **Normal Quantile** from the check-mark popup menu at the lower-left window border.

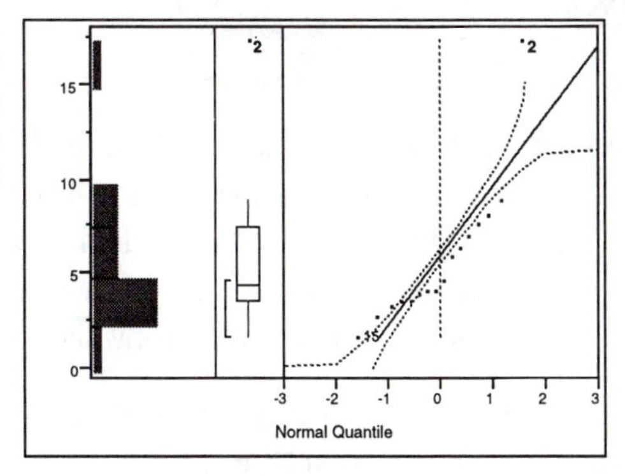

Figure 14.8
Distribution of Y Platform for the Response Variable

You should now see the plots in **Figure 14.8**. Click on the highest point to identify it. The box plot and the normal quantile plot are useful for identifying runs that have extreme values. In this case run 2 has an unusually high peeling strength.

Run a Screening Model

Of the nine factors in the flour paste experiment, there may be only a few that stand out in comparison with the others. The goal of the experiment is to find the factor combinations that optimize the predicted response (peeling strength), rather than show statistical significance of model effects. This kind of experimental situation lends itself to the JMP screening facility.

§ To run a screening model, choose **Analyze→Fit Model**. When the Fit Model dialog appears, select Y as the response (Y) variable. SHIFT-click all the columns from Flour to Coat in the variable selection list and click **Add** to make nine effects in the model. **Figure 14.9** shows the completed dialog.

§ Select the **Screening** option from the fitting personality popup menu on the dialog and then click **Run Model.**

Figure 14.9 Fit Model Dialog for Flour Paste Screening Model

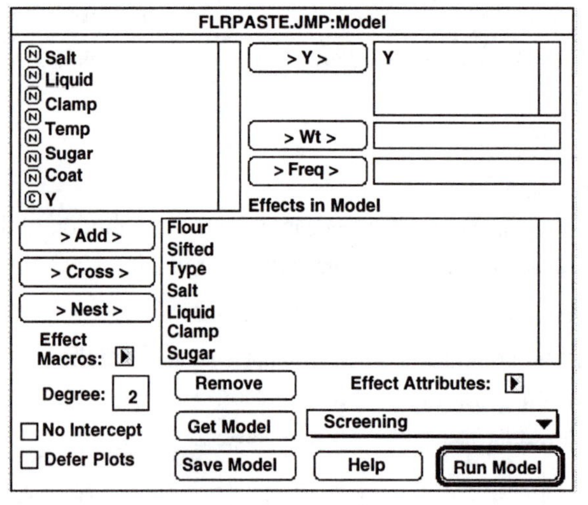

The Analysis of Variance table in **Figure 14.10** shows that the model as a whole is not significant (p=.1314). The most significant factors are Flour and Type (of liquid). Note in the Summary of Fit table that the standard deviation of the error (Root Mean Square Error) is estimated as 2.6, a high value relative to the scale of the response. The RSquare of .79 is not particularly high.

Double-click on columns in the report to specify the number of decimals to display if you want to duplicate the report in **Figure 14.10**.

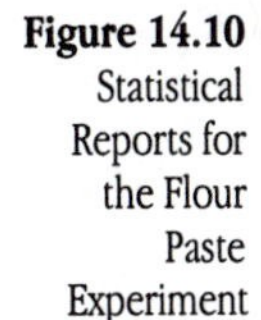

Figure 14.10 Statistical Reports for the Flour Paste Experiment

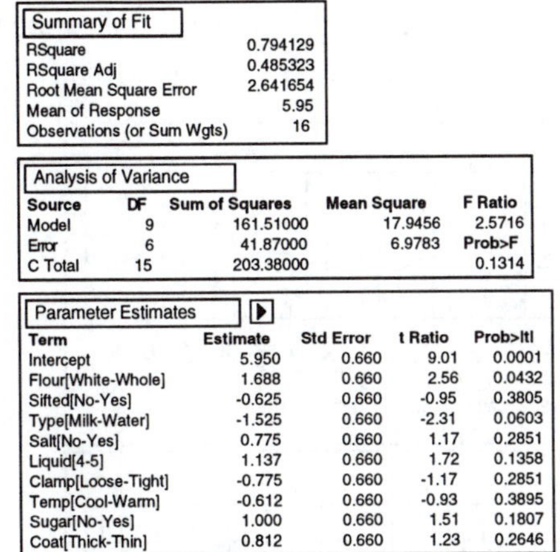

Summary of Fit

RSquare	0.794129
RSquare Adj	0.485323
Root Mean Square Error	2.641654
Mean of Response	5.95
Observations (or Sum Wgts)	16

Analysis of Variance

Source	DF	Sum of Squares	Mean Square	F Ratio
Model	9	161.51000	17.9456	2.5716
Error	6	41.87000	6.9783	Prob>F
C Total	15	203.38000		0.1314

Parameter Estimates

Term	Estimate	Std Error	t Ratio	Prob>|t|
Intercept	5.950	0.660	9.01	0.0001
Flour[White-Whole]	1.688	0.660	2.56	0.0432
Sifted[No-Yes]	-0.625	0.660	-0.95	0.3805
Type[Milk-Water]	-1.525	0.660	-2.31	0.0603
Salt[No-Yes]	0.775	0.660	1.17	0.2851
Liquid[4-5]	1.137	0.660	1.72	0.1358
Clamp[Loose-Tight]	-0.775	0.660	-1.17	0.2851
Temp[Cool-Warm]	-0.612	0.660	-0.93	0.3895
Sugar[No-Yes]	1.000	0.660	1.51	0.1807
Coat[Thick-Thin]	0.812	0.660	1.23	0.2646

Following the statistical report you will see a *Prediction Profiler*, which shows the predicted Y response for each combination of factor settings. **Figure 14.11** shows three manipulations of the Prediction Profiler for the flour experiment. The settings of each factor are connected by a line, called the *prediction trace* or *effect trace*. You can grab and move each vertical dotted line in the Prediction Profile plots to change the factor settings. The predicted response automatically recomputes and shows on the vertical axis, and the prediction traces are redrawn.

The Prediction Profiler lets you look at the effect on the predicted response of changing one factor setting while holding the other factor settings constant. It can be useful for judging the importance of the factors.

The effect traces in the plot at the top of **Figure 14.11** show a larger response for "White" than for "Whole wheat" and for "Water" than for "Milk", which indicates that changing them to their *up* positions increases peel strength.

The second plot in **Figure 14.11** shows what happens if you click and move the effect trace to the high response of each factor; the predicted response changes from 7.825 to 14.9. This occurs with sifted white flour, 4 teaspoons warm water, no salt, no sugar, pasted applied thickly, and clamped tightly while drying.

If you had the opposite settings for these factors (lowest plot in **Figure 14.11**), then the linear model would predict a surface strength of –3 (a clearly impossible response). The prediction is off the original scale, though you can OPTION-click in the Y axis area to change the Y scale of the plot.

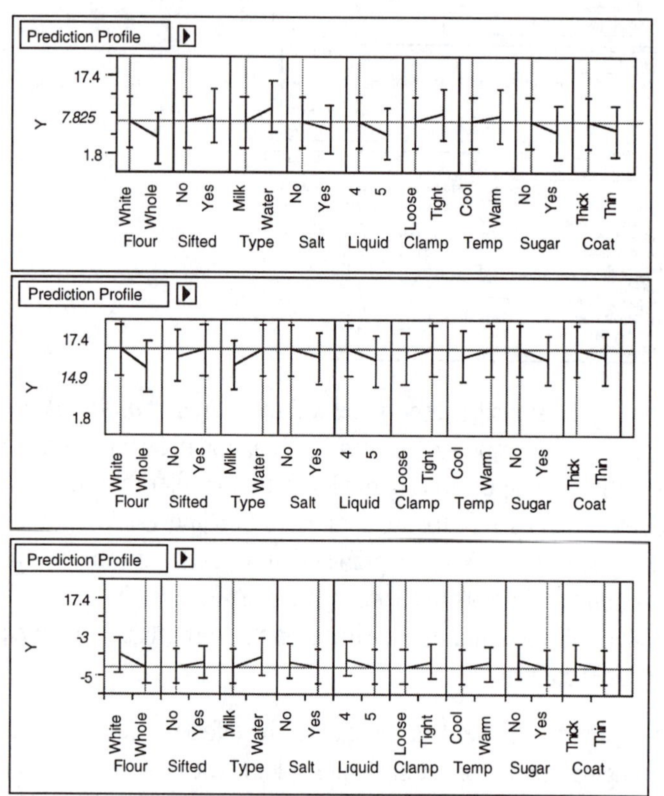

Figure 14.11 Screening Model Prediction Profiler for the Flour Paste Experiment

Commands in the check-mark popup menu on the lower-left window border give you other analyses:

§ Select the **Effect Screening** option from the check-mark popup menu to see the Normal Plot of the parameter estimates as shown in **Figure 14.12**.

The Normal Plot is a normal quantile plot (Daniel 1959), which shows the parameter estimates on the vertical axis and the normal quantiles on the horizontal axis. In a screening experiment you expect most of the effects to be inactive; you expect them to have little or no effect on the response. If that is true then the estimates for those effects will be a realization of random noise centered at zero. What you want is a sense of the magnitude of an effect you should expect when it is truly active instead of just noise. On a Normal

Plot, the active effects appear as outliers that lie away from the line that represents normal noise.

Looking at responses in this way is a valid thing to do for two-level balanced designs because the estimates are uncorrelated and all have the same variance. The Normal Plot is a useful way (and about the only way) to evaluate the results of a saturated design with no degrees of freedom for estimating the error.

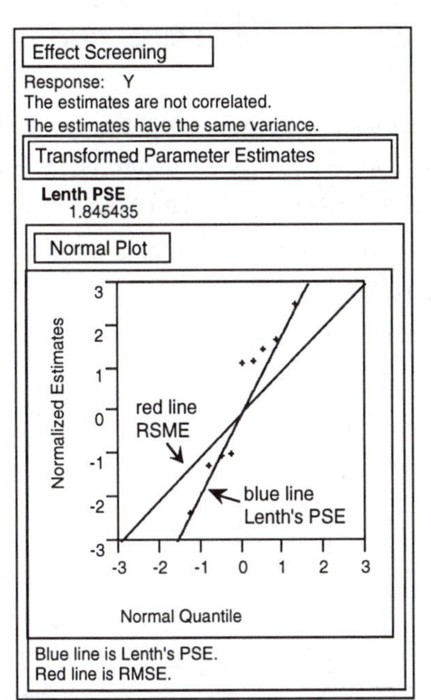

Figure 14.12 Effect Screening to See Most Influential Effect

The Normal Plot (**Figure 14.12**) also shows the straight line with slope equal to the Lenth PSE (*pseudo standard error*) estimate (Lenth 1989). This estimate is formed by taking 1.5 times the median absolute value of the estimates after removing all the estimates greater than 2.75 times the median absolute estimate in the complete set of estimates. Lenth's PSE is computed using the normalized estimates and disregards the intercept. Effects that deviate substantially from this normal line are automatically labeled on the plot.

The distribution of the effects for the flour experiment is disappointing. Usually most effects have small values and a few have larger values. In this case, there appears to be only noise because none of the effects separate from the normal lines more than would be expected from a normal distribution of the estimates. This experiment actually has the opposite of the usual distribution. There is a vacant space near the middle of the distribution—none of the estimates is very near zero. (Note in your analysis tables that the Whole Model F test shows that the model as a whole is not significant.)

So far, the screening experiment has not been conclusive for most of the factors, although white flour does appear to give better results than whole wheat flour. The experiment might bear repeating with better control of variability and a better experimental procedure.

There are 16 runs and 9 variables in this example, which gives (16 –9) –1=6 degrees of freedom for error (see the Analysis of Variance table in **Figure 14.10**). As long as there are degrees of freedom for error you can look at the residuals to see if anything interesting is happening.

§ Return to the Fit Model dialog used previously (if you have closed it, choose **Analyze→Fit Model** to open a new dialog). Use the same model as before, with all factors as effects in the model.

§ Change from **Screening Fit** to **Standard Least Squares** and click **Run Model**. The standard least squares analysis gives the left plot in **Figure 14.13**, which shows the Whole-Model leverage plot and Analysis of Variance table.

§ Select the **Plot Residuals** command from the check-mark popup menu to see the plot on the right in **Figure 14.13**.

Figure 14.13 Distribution of Residuals from Paste Experiment

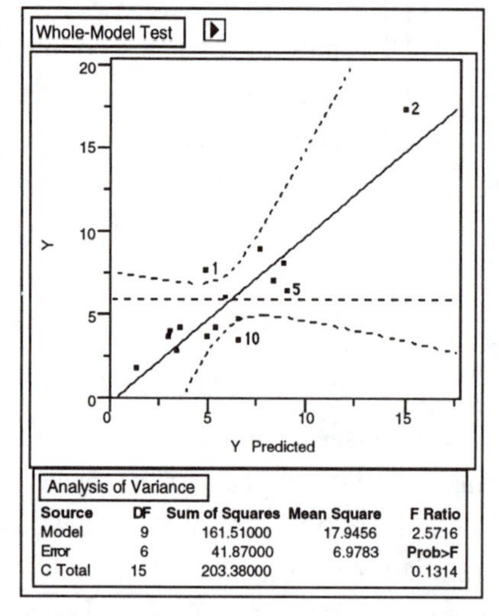

Source	DF	Sum of Squares	Mean Square	F Ratio
Model	9	161.51000	17.9456	2.5716
Error	6	41.87000	6.9783	Prob>F
C Total	15	203.38000		0.1314

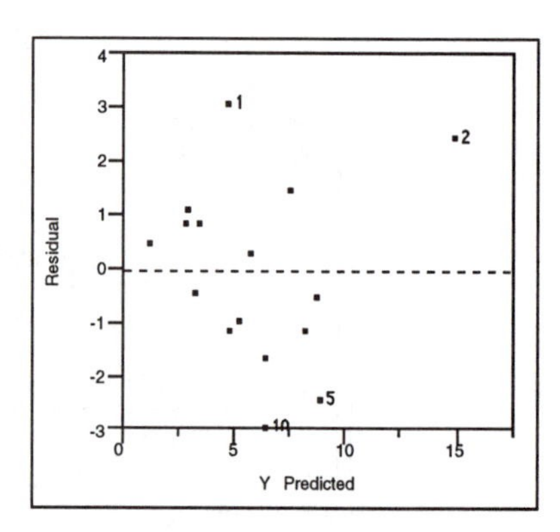

Residual values for runs 1, 2, 5, and 10 lie the farthest from the dotted reference line at zero. However, if you save the residuals and look at their distribution, they don't appear that unusual.

§ Select the **Save Residuals** command from the Save ($) popup menu on the lower left of the window to save the residuals as a column in the data table.

§ Choose **Analyze→Distribution of Y** to look at the distribution of the Residual Y column (**Figure 14.14**).

§ Select **Normal Quantile Plot** from the check-mark menu to see the results shown in **Figure 14.14**.

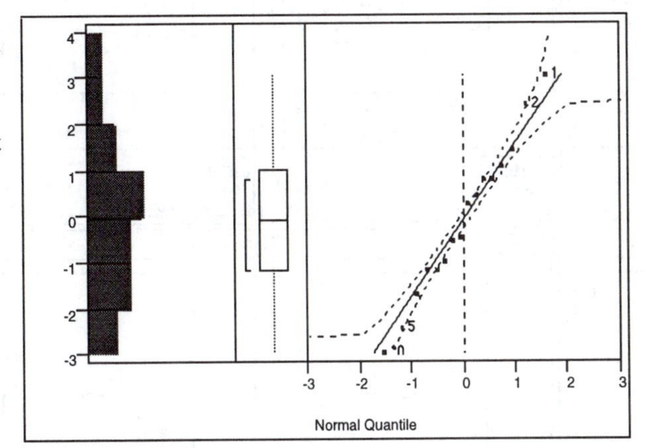

Figure 14.14 Distribution of Residuals from Paste Experiment

Figure 14.15 shows the leverage plots from the standard least squares analysis for the nine flour paste effects. The leverage plots show the characteristic pattern of leverage for a balanced design. All the points are at the ends of the scale. The actual means are the same as the least squares means. Only the Flour effect is significant.

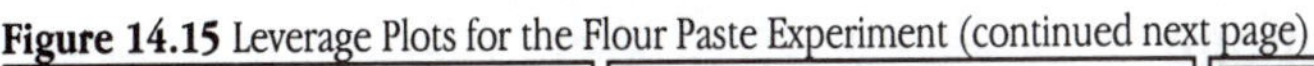

Figure 14.15 Leverage Plots for the Flour Paste Experiment (continued next page)

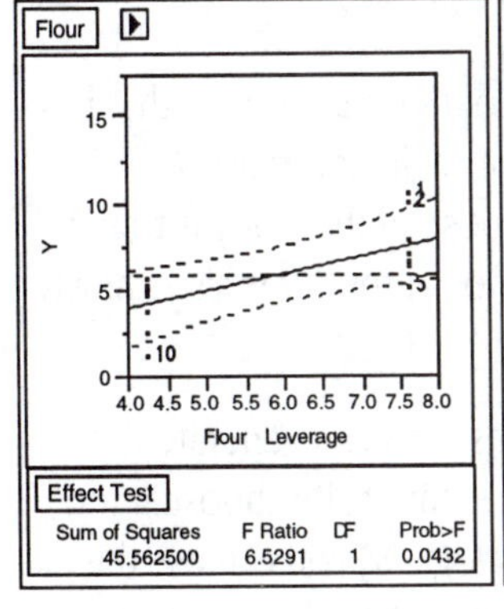

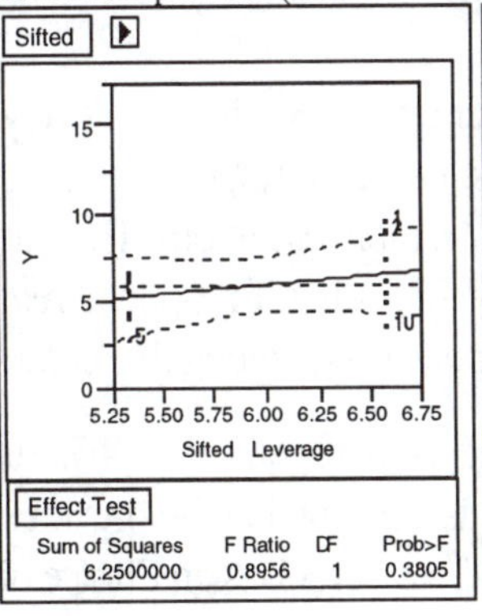

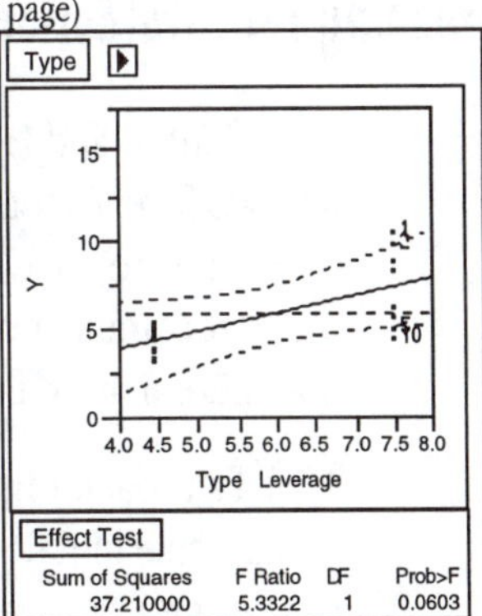

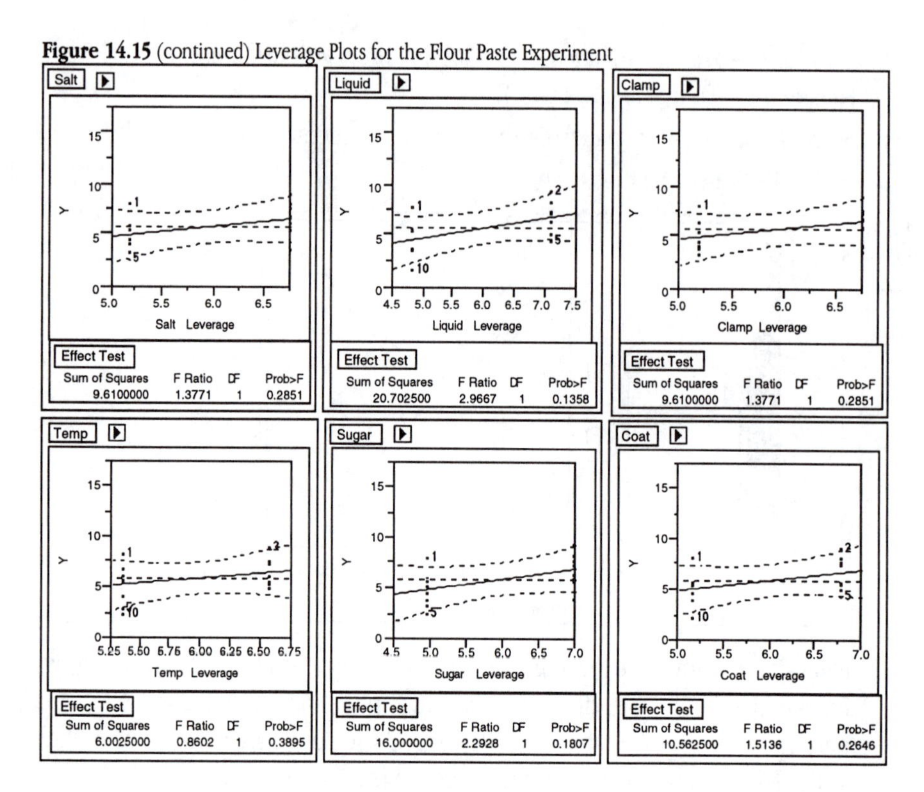

Figure 14.15 (continued) Leverage Plots for the Flour Paste Experiment

Screening for Interactions

Box, Hunter, and Hunter (1978) discuss a study of nuclear reactors that has five two-level factors, F (feed rate), Ct (catalyst), A (agitation rate), T (temperature), and Cn (concentration). The purpose of the study is to find the best combination of settings for optimal reactor output. It is also known that there may be interactions among the factors.

A full factorial for five factors requires $2^5 = 32$ runs. You can generate the design table in JMP, as described in the previous example, by choosing the 32-run design from the Two-Level Design Choice dialog. Or, you can use the **Full Factorial** selection as your design choice and simply define the five factors.

§ To analyze the reactor data, open the table called REACTOR.JMP. **Figure 14.16** shows a partial listing of the data.

Figure 14.16
Design Table and Data for the Reactor Example

REACTOR.JMP

8 Cols, 32 Rows

	Run	F	Ct	A	T	Cn	Pattern	Y
1	1	-1	-1	-1	-1	-1	-----	61
2	2	1	-1	-1	-1	-1	+----	53
3	3	-1	1	-1	-1	-1	-+---	63
4	4	1	1	-1	-1	-1	++---	61
5	5	-1	-1	1	-1	-1	--+--	53
6	6	1	-1	1	-1	-1	+-+--	56

It is useful to begin the analysis with a quick look at the data.

§ Choose **Analyze→Distribution of Y** to look at the distribution of the response variable Y. Select the **Normal Quantile Plot** option from the check-mark popup menu to see the Normal Plot shown in **Figure 14.17.**

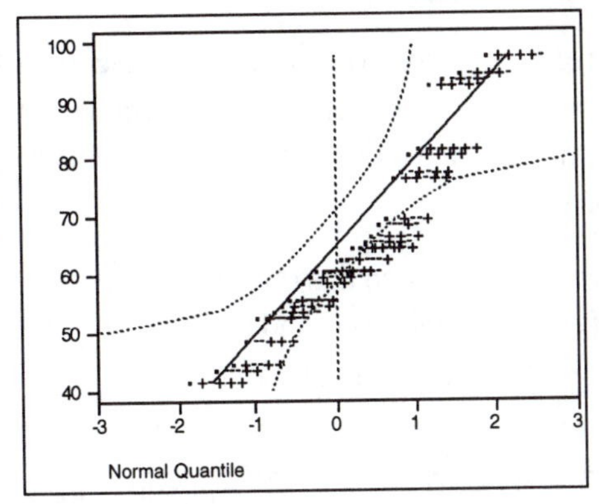

Figure 14.17
Normal Quantile Plot of the Response Variable in the Reactor Study

For each run, the **Pattern** variable in the design table shows the factors' values as a string of minus's and plus's. When this column is used as a **Label** variable, you can see the pattern of factor values by clicking on the point, or by highlighting rows and selecting **Label/Unlabel** in the **Rows** menu.

Next, you can use the Fit Model platform for a screening fit of the full factorial model:

§ Choose **Analyze→Fit Model.** When the Fit Model dialog appears, use Y as the response (Y) and specify a second degree factorial analysis as follows:

1) Highlight the five factor columns (F, Ct, A, T, and Cn) in the variable selection list.

2) Be sure the value in the **Degree** box is 2, and choose **Factorial to Degree** in the **Effects Macro** popup menu. This automatically builds the five-factor factorial model with main effects and all two-factor interactions, as illustrated in **Figure 14.18.**

§ Select **Screening** from the popup menu on the Fit Model dialog and click **Run Model.**

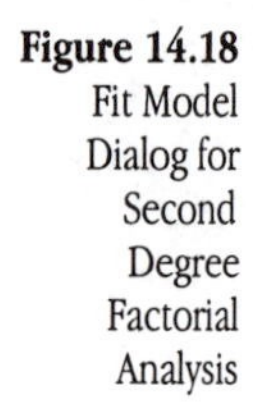

Figure 14.18 Fit Model Dialog for Second Degree Factorial Analysis

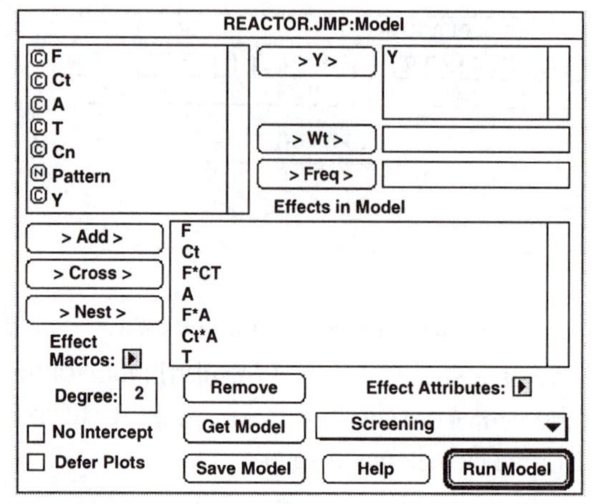

When you look at the resulting Prediction Profile plot in **Figure 14.19** you can see that feed rate (F) and agitation (A) appear flat. Concentration (Cn) and temperature (T) are somewhat weak. Catalyst (Ct) seems to have the strongest effect over its range.

However, if there are interactions, then this one-at-a-time profile plot may be misleading. The traces can shift their slope and curvature as you change current values. That is what interaction is all about.

Figure 14.19 Fit Model Dialog for Second Degree Factorial Analysis

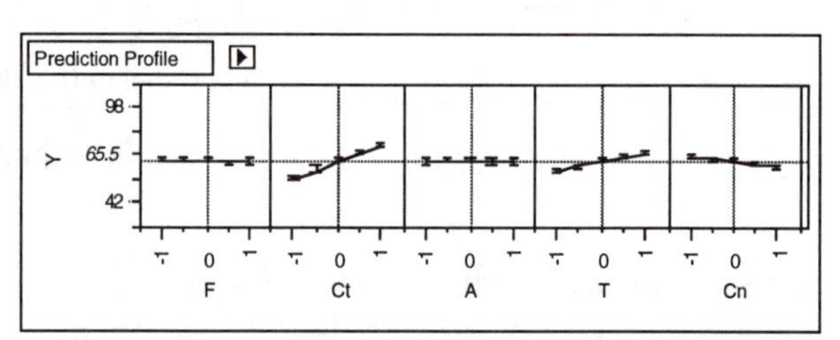

You can find out more about possible interactions with the **Effect Screening** option in the check-mark popup menu:

§ Select **Effect Screening** from the check-mark popup menu to see what a normal plot of the parameter estimates has to say (see **Figure 14.20**).

If all effects are due to random noise, they follow a straight line with slope σ, the standard error. The line with slope equal to the Lenth's PSE estimate shows in blue. If there are degrees of freedom for error, a line with slope equal to the root mean squared error (RMSE) shows in red.

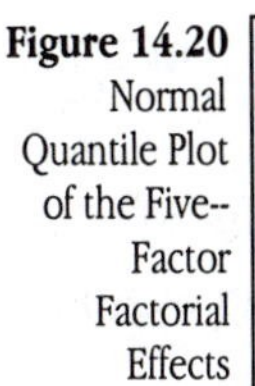

Figure 14.20 Normal Quantile Plot of the Five-- Factor Factorial Effects

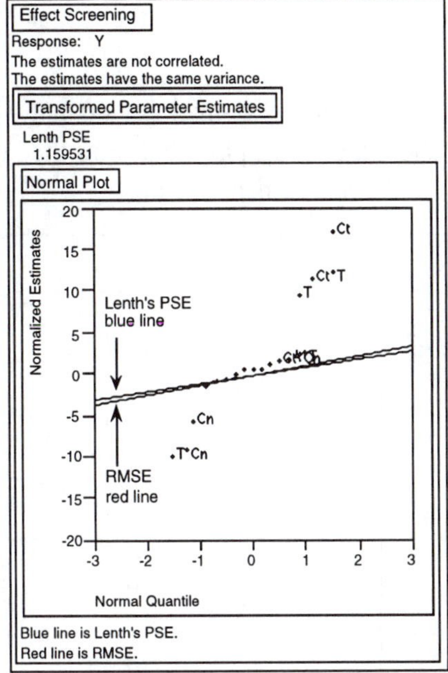

The Normal plot for the Reactor experiment helps you pick out effects that deviate from the normal lines. In this example, the A factor does not appear important. However, not only are the three factors (T, Ct, and Cn) active, but T (temperature) also appears to interact with both Ct and Cn. Let's focus on the interaction of concentration and temperature. In the profile plot, click the levels of T repeatedly to alternate its setting from –1 to +1. Now watch the slope on the profile for Cn. The slopes change dramatically as temperature is changed, which indicates an interaction. When there is no interaction, only the heights of the profile would change, not the slopes. Watch the slope for catalyst (Ct) too, because it also interacts with temperature.

The Prediction Profile plots at the top of **Figure 14.21** show responses when temperature is at its low setting. The lower set of plots show what happens when temperature is at its higher setting.

Figure 14.21 Change in T Changes Slope of Ct and Cn Due to Interaction

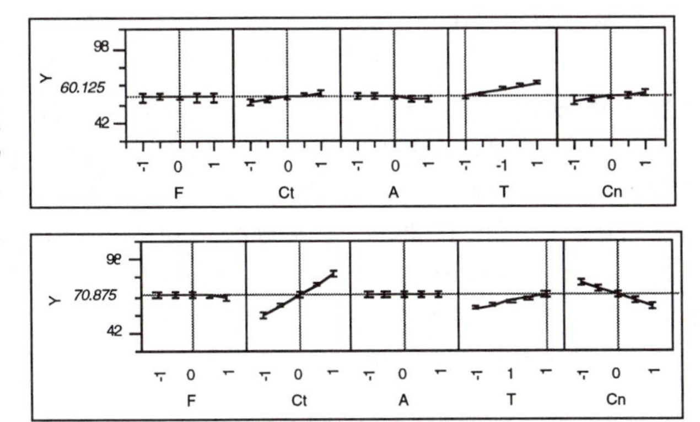

This slope change that is caused by interaction can be seen for all interactions in one picture as follows:

§ Use the **Interaction Plots** command in the check-mark popup menu to see the matrix of plots **Figure 14.22.** These are profile plots that show the interactions between all pairs of variables in the reactor experiment.

Figure 14.22 Interaction Plots for Five-factor Reactor Experiment

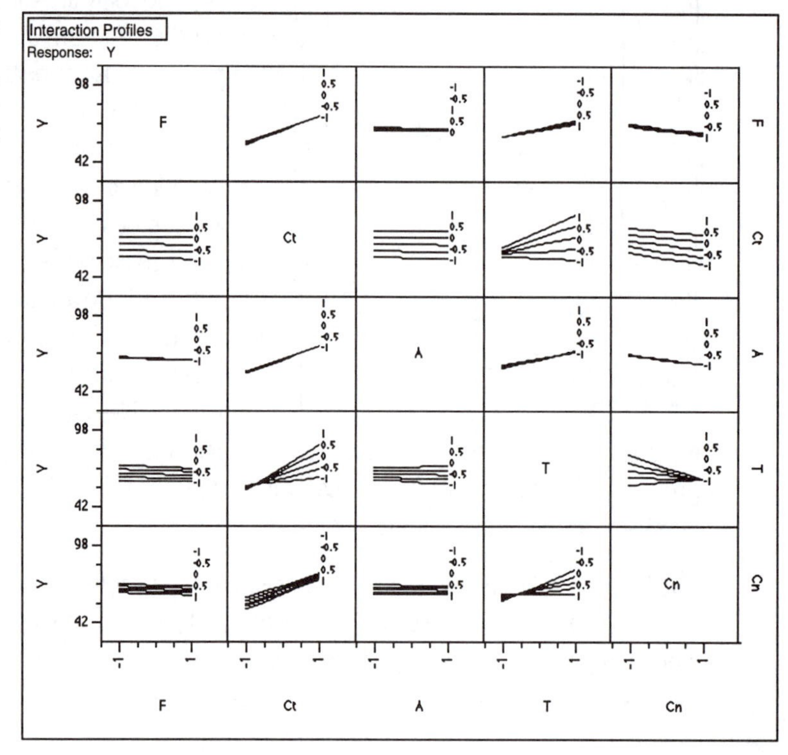

In an interaction plot, the Y axes are the response. Each small plot shows the effect of two factors on the response. One factor (associated with the column of the matrix of plots) is on the X axis. This factor's effect shows as the slope of the lines in the plot. The other factor becomes multiple prediction profiles (lines) as it varies from low to high. This factor shows its effect on the response as the vertical separation of the profile lines. If there is an interaction, then the slopes are different for different profile lines, like those in the T by Ct plot.

The Prediction Profile plots in **Figure 14.21** indicated that T is active and A is inactive. In the matrix of Interaction Profile plots, look at column A for row T;

T causes the lines to separate, but doesn't determine the slope of the lines. Look at the plot when the factors are reversed (row A and column T); the lines are sloped, but they don't separate.

Recall that temperature interacted with catalyst and concentration. This is evident by the differing slopes showing in the T by Ct and the T by Cn Interaction Profile plots.

Response Surface Designs

Response surface designs are built for fitting a curved surface (quadratic) to continuous factors so that the factor values for the minimum or maximum response can be sought. Standard two-level designs cannot fit curved surfaces because they have only two levels.

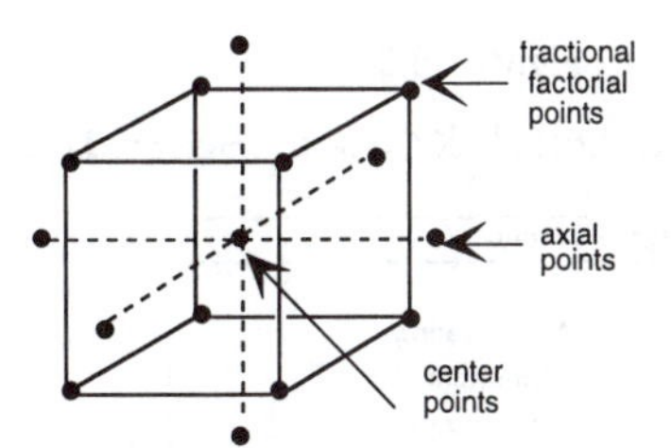

The most popular response surface design is the central composite design, illustrated by the diagram to the left. It combines a two-level fractional factorial and two other kinds of points defined as follows:

- *Center points*, for which all the factor values are at the midrange value.
- *Axial* (or *star*) *points*, for which all but one factor are set at midrange and one factor is set at outer (axial) values.

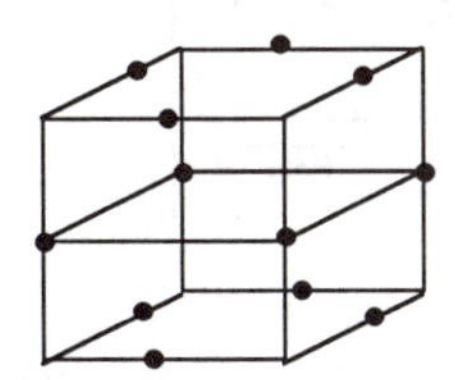

The Box-Behnken design is an alternative to central composite designs. The illustration to the left is of a Box-Behnken design with three factors. It combines a fractional factorial with incomplete block designs in such a way as to avoid the extreme vertices and to present an approximately rotatable design with only three levels per factor.

The Response Surface Design Dialog

§ To generate a response surface design, begin as before by choosing **Tables→Design Experiment**.

§ Click the **Response Surface Design** selection in the Choose Design Type window to see the dialog shown in **Figure 14.23**.

When you generate a response surface design list, you can choose from a Box-Behnken design and two types of central composite design, called *uniform precision* and *orthogonal*. These properties relate to the number of center points in the design and to the axial values:

- Uniform precision means that the number of center points is chosen so that the prediction variance at the center is approximately the same as at the design vertices.
- For orthogonal designs, the number of center points is chosen so that the second order parameter estimates are minimally correlated with the other parameter estimates.

§ To complete the dialog, enter the number of factors (as many as eight) and click **Search for Designs**. For this example, enter 3 as the **Number of Factors.**

§ Select the 20-run design, as shown in **Figure 14.23**.

§ Also, click the **Make Model** box, and then click **Generate Selected Design**.

Figure 14.23 The Search for Design Dialog for Response Surface Designs

Response Surface Design Selection

Response Surface Designs

Number of Factors: 3

Search for Designs

Axial Scaling: ◉ Rotatable ○ Orthogonal ○ On Face ○ Specified 1.6818 ☐ Inscribe

Number of Runs	Block Size	Number of Center Points	Type
15		3	Box-Behnken
20		6	Central Composite-Uniform
20		6	Central Composite-Orthog
23		9	Central Composite-Orthog

☐ Random Order
☒ Make Model
Generate Selected Design
Help

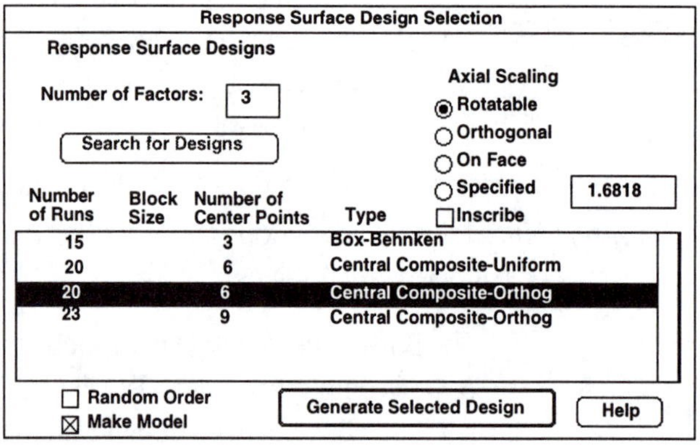

The Design Table

The generated design shows in a JMP data table (**Figure 14.24**) that lists the runs for the design specified in **Figure 14.23**. Again note that the design table automatically includes a column called Y for recording experimental results.

The column called Comment identifies the type of point with these codes:

- FF if it is generated from the fractional factorial.

- **Center-FF** if it is a center point in the fractional factorial blocks.
- **Axial** if it is an axial point.
- **Center-Ax** if it is a center point in the axial block.

Figure 14.24 Design Table for a Central Composite Response Surface Design

CCDesign1

7 Cols / 20 Rows	Pattern	Block	X1	X2	X3	Comment	Y
1	---	1	-1	-1	-1	FF	•
2	--+	2	-1	-1	1	FF	•
3	-+-	2	-1	1	-1	FF	•
4	-++	1	-1	1	1	FF	•
5	+--	2	1	-1	-1	FF	•
6	+-+	1	1	-1	1	FF	•
7	++-	1	1	1	-1	FF	•
8	+++	2	1	1	1	FF	•
9	000	1	0	0	0	Center-FF	•
10	000	1	0	0	0	Center-FF	•
11	000	2	0	0	0	Center-FF	•
12	000	2	0	0	0	Center-FF	•
13	-00	3	-1.68179	0	0	Axial	•
14	+00	3	1.681793	0	0	Axial	•
15	0-0	3	0	-1.68179	0	Axial	•
16	0+0	3	0	1.681793	0	Axial	•
17	00-	3	0	0	-1.68179	Axial	•
18	00+	3	0	0	1.681793	Axial	•
19	000	3	0	0	0	Center-Ax	•
20	000	3	0	0	0	Center-Ax	•

The Table Info (choose **Tables→Table Info**) for the design table describes the type of design. In this example the description is "[CCDesign1] has 8 factorial runs in 2 blocks of size 4, resolution=6."

The Make Model Option

When you check the **Make Model** box at the bottom of the Design Selection dialog (see **Figure 14.23**), JMP opens a Fit Model window and lists the appropriate effects for the model you selected. You can click the **Save Model** button on the Fit Model dialog (**Figure 14.25**) to save the model specification in a file for use later after you have performed the experiment and entered the data.

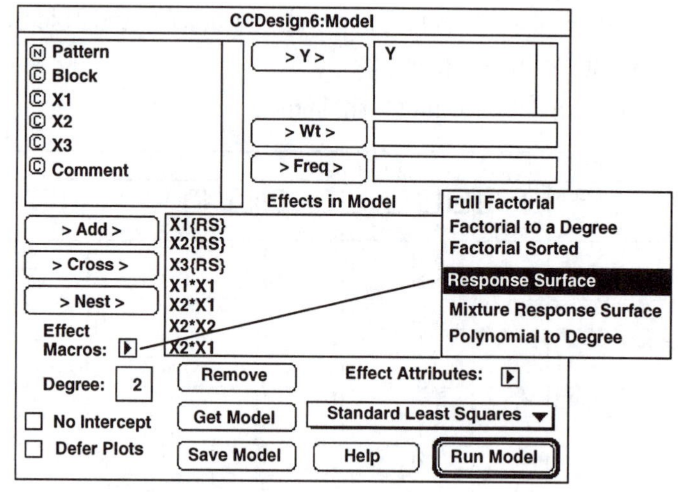

Figure 14.25 Fit Model Dialog for Response Surface Design

A Box-Behnken Design: The Odor Control Example

The objective of an industrial experiment is to minimize the unpleasant odor of a chemical. It is known that the odor varies with temperature (**temp**), gas-liquid ratio (**gl ratio**), and packing height (**ht**). The experimenter wants to collect data over a wide range of values for these variables to see if a response surface can identify values that give a minimum odor (John, 1971).

First, generate a response surface experimental design:

§ Choose **Tables→Design Experiment**, and click the **Response Surface** button in the Choose Design Type dialog . You will then see the Response Surface Design Selection dialog (**Figure 14.26**).

§ Enter "3" as the **Number of Factors** and then click **Search for Designs**.

§ When the list of response surface designs appears, click on the 15-run **Box-Behnken** design, then click **Generate Selected Design**.

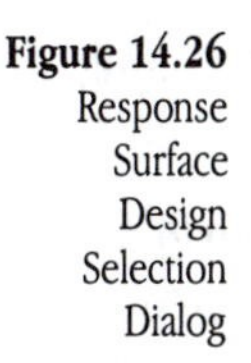

Figure 14.26 Response Surface Design Selection Dialog

The Box-Behnken design selected for three effects generates the design table of 15 runs. When the experiment is conducted, the responses are entered into the JMP table.

§ Suppose the factor names have been edited, the experiment has been conducted, and the results entered into the Odor column, as shown in **Figure 14.27**. Open ODOR.JMP to see the finished data table ready for analysis.

Figure 14.27 JMP Table for a Three-Factor Box-Behnken Design and Finished Table with Data for Analysis

ODOR.JMP

5 Cols / 15 Rows	Run	odor	temp	gl ratio	ht
1	1	66	-1	-1	0
2	2	58	-1	1	0
3	3	65	1	-1	0
4	4	-31	1	1	0
5	5	39	0	-1	-1
6	6	17	0	-1	1
7	7	7	0	1	-1
8	8	-35	0	1	1
9	9	43	-1	0	-1
10	10	-5	1	0	-1
11	11	43	-1	0	1
12	12	-26	1	0	1
13	13	49	0	0	0
14	14	-40	0	0	0
15	15	-22	0	0	0

Before you analyze the data, look at the design structure with the Spinning Plot platform:

§ With the ODOR.JMP as the current data table, choose **Graph→Spinning Plot** for the three factors **temp**, **gl ratio**, and **ht**. To see the Spinning Plot shown in **Figure 14.28**, rotate the plot and use the **Box** and the **Rays** options found in the popup menu on the spin control panel.

As you can see from the spinning plot, this design has points that are extreme in two factors, but midrange in the third. For a three-factor design, this has the effect of locating points midway along the edges connecting the extreme vertices of the factor space.

Figure 14.28 Spinning Plot of a Box-Behnken Design for Three Effects

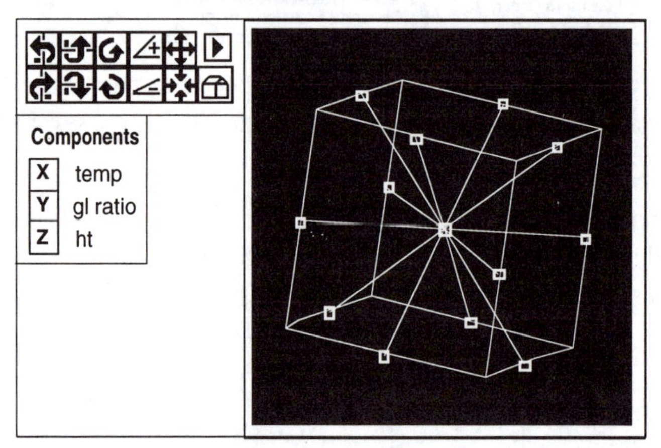

§ To analyze the data, choose **Analyze→Fit Model.** Complete the Fit Model dialog as shown in **Figure 14.29**.

1) Select **odor** as the response variable (Y) in the Fit Model dialog.
2) SHIFT-click to select the variables **temp**, **gl ratio**, and **ht** together in the variable selection list.
3) Then click the **Effect Macros** popup menu, and select **Response Surface.**

The effects appear in the **Effects in Model** list with the **{RS}** notations on the main effects (**temp**, **gl ratio**, and **ht**), which indicates to the modeling system that these terms are to be subjected to a curvature analysis.

§ Use the popup menu at the bottom of the dialog to change the fitting personality from **Standard Least Squares** to **Screening**.

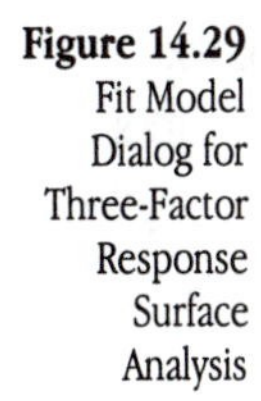

Figure 14.29 Fit Model Dialog for Three-Factor Response Surface Analysis

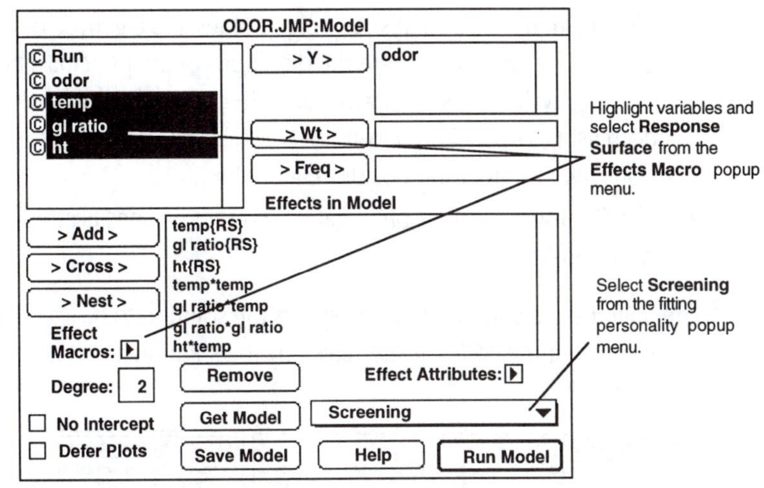

§ When you click **Run Model** the analysis tables in **Figure 14.30** appear, which will be discussed later.

Figure 14.30 Analysis of Three-Factor Response Surface Analysis

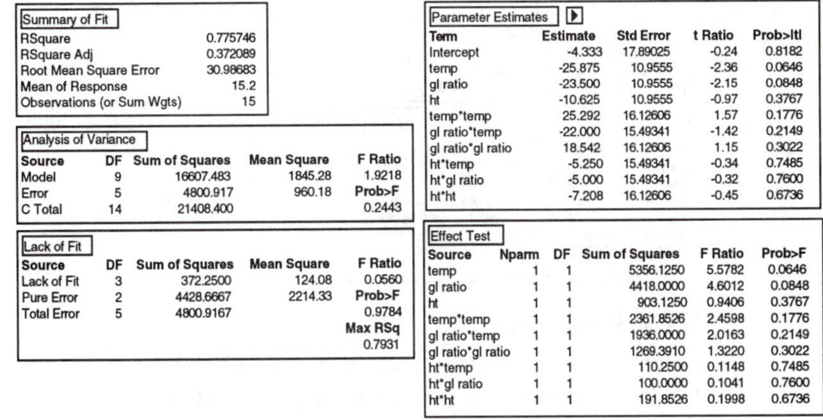

Summary of Fit

RSquare	0.775746
RSquare Adj	0.372089
Root Mean Square Error	30.98683
Mean of Response	15.2
Observations (or Sum Wgts)	15

Analysis of Variance

Source	DF	Sum of Squares	Mean Square	F Ratio
Model	9	16607.483	1845.28	1.9218
Error	5	4800.917	960.18	Prob>F
C Total	14	21408.400		0.2443

Lack of Fit

Source	DF	Sum of Squares	Mean Square	F Ratio
Lack of Fit	3	372.2500	124.08	0.0560
Pure Error	2	4428.6667	2214.33	Prob>F
Total Error	5	4800.9167		0.9784
				Max RSq
				0.7931

Parameter Estimates

Term	Estimate	Std Error	t Ratio	Prob>|t|
Intercept	-4.333	17.89025	-0.24	0.8182
temp	-25.875	10.9555	-2.36	0.0646
gl ratio	-23.500	10.9555	-2.15	0.0848
ht	-10.625	10.9555	-0.97	0.3767
temp*temp	25.292	16.12606	1.57	0.1776
gl ratio*temp	-22.000	15.49341	-1.42	0.2149
gl ratio*gl ratio	18.542	16.12606	1.15	0.3022
ht*temp	-5.250	15.49341	-0.34	0.7485
ht*gl ratio	-5.000	15.49341	-0.32	0.7600
ht*ht	-7.208	16.12606	-0.45	0.6736

Effect Test

Source	Nparm	DF	Sum of Squares	F Ratio	Prob>F
temp	1	1	5356.1250	5.5782	0.0646
gl ratio	1	1	4418.0000	4.6012	0.0848
ht	1	1	903.1250	0.9406	0.3767
temp*temp	1	1	2361.8526	2.4598	0.1776
gl ratio*temp	1	1	1936.0000	2.0163	0.2149
gl ratio*gl ratio	1	1	1269.3910	1.3220	0.3022
ht*temp	1	1	110.2500	0.1148	0.7485
ht*gl ratio	1	1	100.0000	0.1041	0.7600
ht*ht	1	1	191.8526	0.1998	0.6736

The **Response Surface** button opens the additional group of tables shown in **Figure 14.31**:

- The first table is a summary of the parameter estimates.
- The Solution table lists the critical values of the surface and tells the kind of solution (maximum, minimum, or saddlepoint). The critical values are where the surface has a slope of zero, which could be an optimum depending on the curvature.

- The EigenStructure table shows eigenvalues and eigenvectors of the effects. The eigenvectors are the directions of the principal curvatures. The eigenvalue associated with each direction tells whether it is decreasing slope, like a maximum (negative eigenvalue), or increasing slope, like a minimum (positive eigenvalue).
- The Contour Specification panel lets you set up specifications for a contour plot.

The Solution table in this example shows the solution to be a saddlepoint and also warns that the critical values given by the solution are outside the range of data values.

Figure 14.31 Response Surface Solution Tables and Response Surface Grid Specification

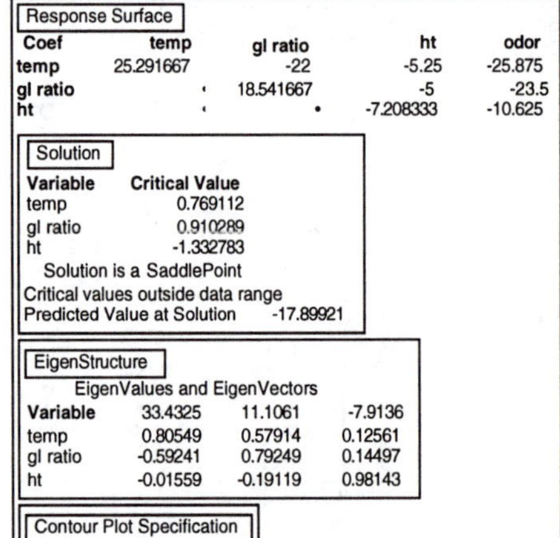

Response Surface

Coef	temp	gl ratio	ht	odor
temp	25.291667	-22	-5.25	-25.875
gl ratio	.	18.541667	-5	-23.5
ht	.	.	-7.208333	-10.625

Solution

Variable	Critical Value
temp	0.769112
gl ratio	0.910289
ht	-1.332783

Solution is a SaddlePoint
Critical values outside data range
Predicted Value at Solution -17.89921

EigenStructure

EigenValues and EigenVectors

Variable	33.4325	11.1061	-7.9136
temp	0.80549	0.57914	0.12561
gl ratio	-0.59241	0.79249	0.14497
ht	-0.01559	-0.19119	0.98143

Contour Plot Specification

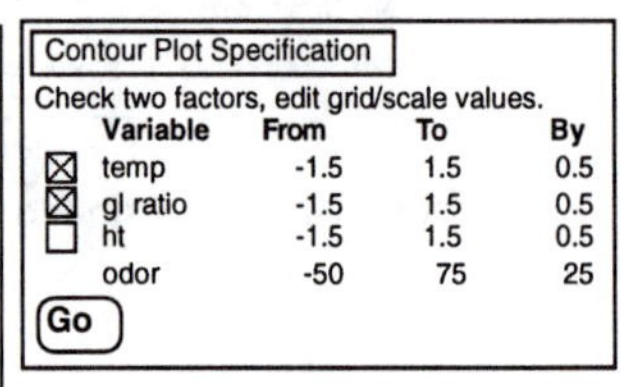

Contour Plot Specification

Check two factors, edit grid/scale values.

	Variable	From	To	By
☒	temp	-1.5	1.5	0.5
☒	gl ratio	-1.5	1.5	0.5
☐	ht	-1.5	1.5	0.5
	odor	-50	75	25

Go

Plotting Surface Effects

If there are more than two factors, you can see a contour plot of any two factors at intervals of a third factor by using the Contour Plot Specification panel shown in **Figure 14.31**. It has editable fields in the columns From, To, and By. The grid values you enter let you set the number of *slices* you want to see. To continue with this example, use the default grid specifications:

§ When you click **Go** in the Contour Plot Specification panel, contour plots like those shown in **Figure 14.32** appear.

You can move (*flip*) through the frames forward or backward, one at a time, or continuously by using the first three control buttons at the top of the plot.

The fourth button is a popup menu that lets you see all the contour frames at the same time arranged horizontally or vertically.

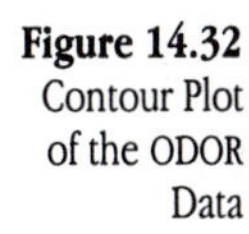

Figure 14.32 Contour Plot of the ODOR Data

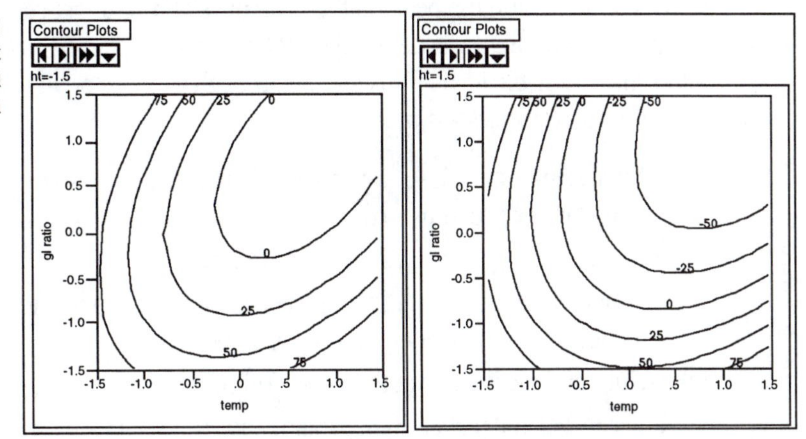

You can also get contour plots from a screening fit through the *Contour Profiler*:

§ Select the **Contour Profiler** command from the check-mark popup menu.

The **Contour Profiler** command gives a panel that lets you use interactive sliders to vary one factor's values and observe the effect on the other two factors. You can also vary one factor and see the effect on a mesh plot of the other two factors. The mesh plots in **Figure 14.33** show the response surface in the three combinations of the three factors taken two at a time.

Figure 14.33 Mesh Plots of All Three Combinations of the Factors

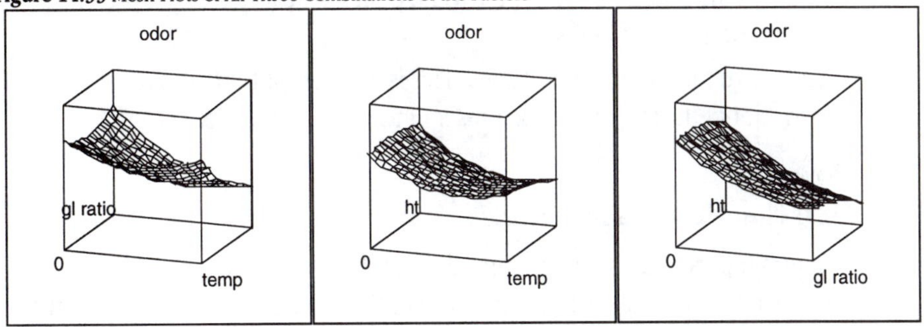

Design Issues

So far, the discussion has been on the particulars of how to get certain designs. But why are these designs good? Here is some general advice that will seem reasonable if you think about it, though all these points have limitations or drawbacks as well.

1) Keep the design balanced. Allocate an equal number of runs to each factor value. There are two reasons to do this:
   - A balanced design achieves the most power.
   - A balanced design keeps the estimates uncorrelated so that tests on one parameter are independent of tests on another.
2) Take values at the extremes of the range. This is done for two reasons:
   - It maximizes the power. By separating the values, you are increasing the parameter for the difference, making it easy to distinguish it from zero.
   - It keeps the applicable range of the experiment wide. When you use the prediction formula outside the range of the data, its variance is high and it is unreliable for other reasons.
3) Put runs at all combinations of extreme levels. If you can't afford this, try to come close to this by designing so that many subsets of variables cover all combinations of levels.
4) Put a few runs in the center too, if this makes sense.
   - Now you can estimate curvature.
   - If you have several runs at the same point you can estimate pure error and do a lack-of-fit test.
5) Randomize the assignment of runs. This is an attempt to neutralize any inadvertent effects that may be present in a given sequence.

Balancing

Let's illustrate the balancing issue:

In the simplest case of two groups, the standard error of the difference between the two means is proportional to the square root of the sum of the reciprocals of the sample size, as shown in the formula below.

$$\sqrt{\frac{1}{n1}+\frac{1}{n2}}$$

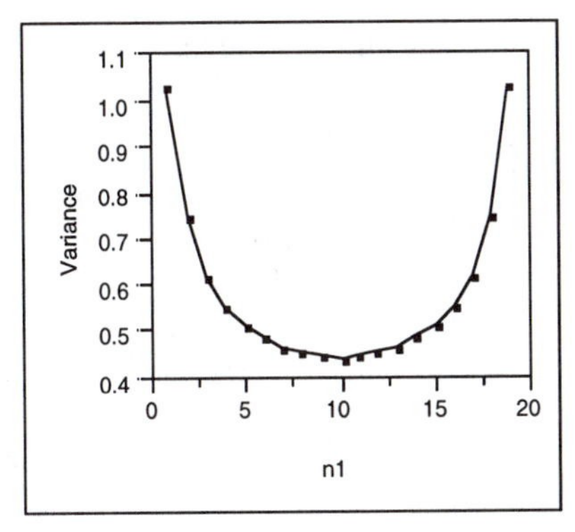

Figure 14.34 Plot of Variance Versus Sample Size

The plot in **Figure 14.34** is a graph of this formula as a function of *n1*, where *n1* + *n2* is kept constant at 20. You can see that the smallest variance occurs when the design is balanced; where *n1* = *n2* = 10. So if the design is balanced, you minimize the standard errors of differences, and thus maximize the power of the tests.

Wide Range

Let's illustrate why taking a wide range is good.

Figure 14.35 Data Table

Untitled

12 Rows / 2 Cols		X	Y
+	1	-4	-1
+	2	-4	-3
+	3	-2	-2
+	4	-2	0
+	5	0	-1
+	6	0	-1
+	7	0	1
+	8	0	1
+	9	2	0
+	10	2	2
+	11	4	1
+	12	4	3

The data table shown here in **Figure 14.35** is simple arrangement of 12 points. There are 4 points at extreme values of X, 4 points at moderate range, and 4 points at the center. Let's look at following subsets of 8 out of the 12 points, to see which ones perform better:

- Exclude the extremes of ±4
- Exclude the intermediate points of ±2
- Exclude the center points

Figure 14.36 compares the confidence interval of the regression line in the three subsets. Each situation has the same number of observations, the same error variance, and the same residuals. The only difference is in the spacing of the X values, but the subsamples show the following differences:

- The fit excluding the points at the extremes of ±4 has the widely flared confidence curves. The confidence curves cross the horizontal line at the mean, indicating that it is significant at .05.
- The fit excluding the intermediate range points of ±2 has much less flared confidence curves. The confidence curves do not cross the horizontal line at the mean, indicating that it is not significant at .05.
- The fit excluding the center points were the best, though not much different than the previous case.

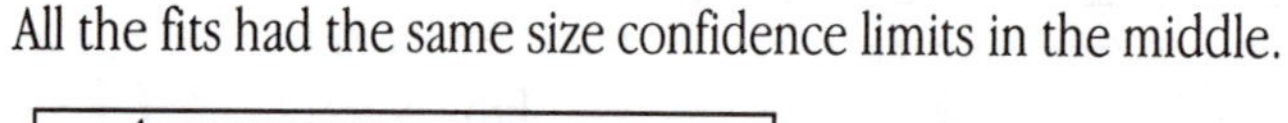

All the fits had the same size confidence limits in the middle.

Figure 14.36
Comparison of Samples with Different Ranges

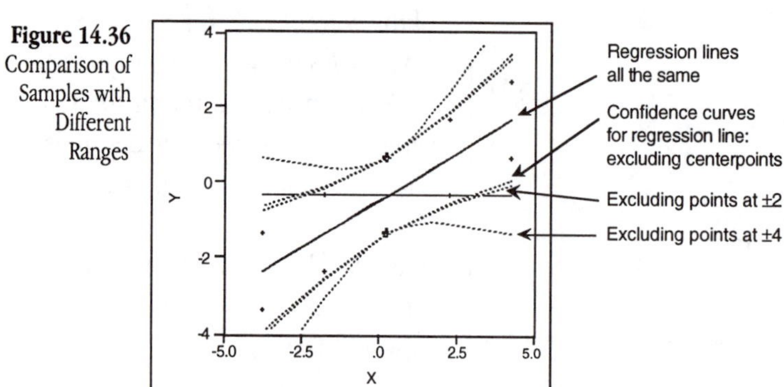

For given sample size, the better design is the one that spaces out the points farther.

Center Points

Using the same data as before, in the same three scenarios, look at a quadratic curve instead of a line (**Figure 14.37**). Note the confidence limits for the curve that excluded the center points have a much wider interval at the middle. Curvature is permitted by the model but is not well supported by the design.

Figure 14.37
Comparison of Samples with Different Ranges

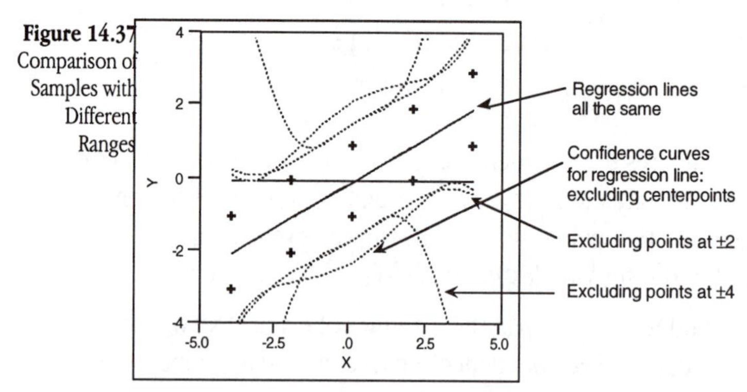

Centerpoints are also usually replicated points that allow for an independent estimate of pure error, which can be used in a lack-of-fit test. For the same number of points, the design with center points can detect lack of fit and nonlinearities.

Chapter 15 Statistical Quality Control

Overview

Some statistics are for proving things. Some statistics are for discovering things. And some statistics are to keep an eye on things, watching to make sure something stays within specified limits.

The *watching* statistics are needed mostly in industry for systems of machines in production processes that sometimes stray from proper adjustment. These statistics monitor variation, and their job is to distinguish the usual random variation (called *common causes)* from abnormal change (called *special causes*).

These statistics are usually from a time series, and the patterns they exhibit over time are clues to what is happening to the production process. If they are to be useful, the data for these statistics need to be collected and analyzed promptly so that any problems they detect can be fixed.

This whole area of statistics is called Statistical Process Control (SPC) or Statistical Quality Control (SQC). The most basic tool is a graph called a *control chart* (or Shewhart control chart, after the inventor, Walter Shewhart). In some industries, SQC techniques are taught to everyone—the engineers, the mechanics, the shop floor operators, even the managers.

The use of SQC techniques became especially popular in the 1980s as industry began to understand the issues of quality better, after the pioneering effort of Japanese industry and under the leadership of Deming and Juran.

Another graphical tool used to monitor a process is the Pareto chart. A *Pareto chart* is a bar chart that displays severity (frequency) of problems in a quality-related process or operation.

This chapter gives you a survey of the SQC tools available in JMP.

Chapter 15 Contents

Control Charts and Shewhart Charts

Control charts are a graphical and analytical tool for deciding whether a process is in a state of statistical control. Control charts in JMP are automatically updated when rows are added to the current data table. In this way control charts can be used to monitor an ongoing process.

Figure 15.1 shows a control chart example that illustrates the following characteristics of most control charts:

- Each point represents a summary statistic computed from a subgroup sample of measurements of a quality characteristic.
- The vertical axis of a control chart is scaled in the same units as the summary statistics specified by the type of control chart.
- The horizontal axis of a control chart identifies the subgroup samples.
- The center line on a Shewhart control chart indicates the average (expected) value of the summary statistic when the process is in statistical control.
- The upper and lower control limits, labeled UCL and LCL, give the range of variation to be expected in the summary statistic when the process is in statistical control.
- A point outside the control limits signals the presence of a special cause of variation.

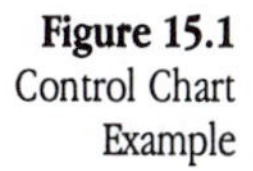

Figure 15.1
Control Chart Example

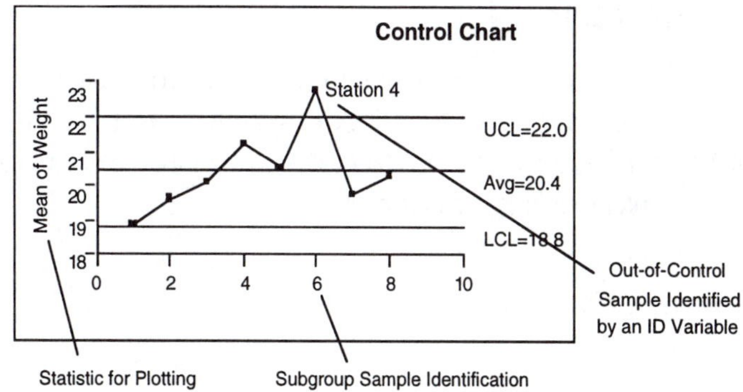

Control charts are broadly classified as *variables charts* and *attributes charts*.

Variables Charts

Control Charts for variables (variables charts) are used when the quality characteristic to be monitored is measured on a continuous scale. There are different kinds of variables control charts based on the subgroup sample summary statistic plotted on the chart, which can be the mean, the range, and the standard deviation; an individual measurement; or a moving range. For quality characteristics measured on a continuous scale, it is typical to analyze both the process mean and its variability by showing a Mean chart aligned above its corresponding R (range) or S (standard deviation) chart. If you are charting individual response measurements, the Individual Measurement chart is aligned above its corresponding Moving Range chart. JMP automatically arranges charts in this fashion.

Attributes Charts

Control Charts for attributes (attributes charts) are used when the quality characteristic of a process is measured by counting the number or proportion of nonconformities (defects) in an item or by counting the number or proportion of nonconforming (defective) items in a subgroup sample.

The Control Chart Dialog

When you select **Control Charts** from the **Graph** menu, you see the Control Chart dialog in **Figure 15.2**. You use this dialog to specify the type of control chart you want. You can think of the control chart dialog as a composite of four panels in which you enter four kinds of information:

- Process information
- Chart type information
- Requests for tests
- Limits specification

Specific information shown in the dialog varies according to the kind of chart you request. Through interaction with the dialog, you specify exactly how you want the charts created. The next sections describe the four kinds of information needed to complete the Control Chart dialog.

Figure 15.2 Control Chart Dialog

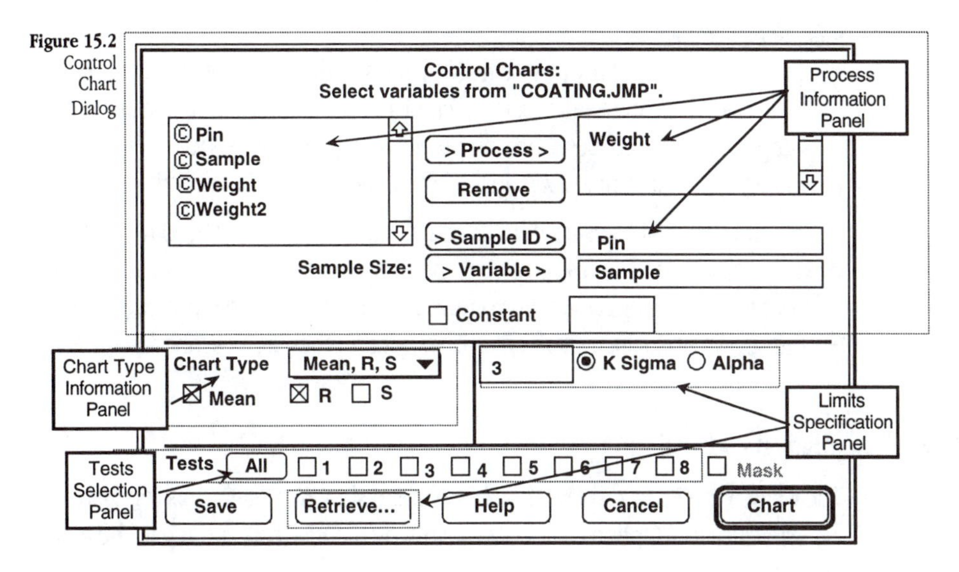

Process Information Panel

The Process Information Panel displays a list of all columns in the current data table and buttons to specify the variables to be analyzed, the subgroup sample size, and optionally, the subgroup sample ID. You use items in the process information panel as follows:

Process

identifies variables for charting.

- For variables charts, specify measurements as the process.
- For attributes charts, specify the defect count or defective proportion as the process.

Sample ID

enables you to specify a variable whose values label the horizontal axis. If no sample ID variable is specified, the subgroup samples are identified by their sequence number.

Sample Size: Variable

displays the variable you select to identify the rows that define subgroup samples. You need a sample size variable only when there is an unequal number of rows in the samples or when some samples have missing values. For attributes charts the sample size variable contains the subgroup sample size. In variables charts the sample size variable identifies the sample.

Sample Size: Constant

gives the number of rows in each subgroup sample when the samples are the same size. If the **Constant** box is checked and the sample size is specified, a sample size variable (described on previous page) is not needed. However, if you do specify a sample size variable, the **Constant** entry is ignored.

For an Individual Measurement chart type, the **Constant** box becomes the **Range Span** box. The *range span* is the number of consecutive measurements from which moving ranges are computed. For C and U chart types, the sample size shows as **# of Inspection Units**. These chart types are described in more detail later.

Chart Type Information Panel

The Chart Type Panel displays the popup menu shown in **Figure 15.3** for chart-type specification. Options below the chart-type menu vary according to the type of chart you select.

- Variables charts are generated by the **Mean**, **R**, **S**, and the **Individual** menu selections. **Mean**, **R**, and **S** stand for Means, Range, and Standard Deviation charts. These selections have check-box options to request or suppress each chart. The **Individual** selection has check-box options for the **Individual Measurement** and the **Moving Range** chart.
- Attributes chart selections are the **P**, **NP**, **C**, **and U** charts. There are no additional specifications for attributes charts.
- The uniformly weighted moving average (**UWMA**) and exponentially weighted moving average (**EWMA**) selections are special kinds of variables charts for charting subgroup sample means.

Figure 15.3 Chart Types Menu with Variables and Individual Chart Options

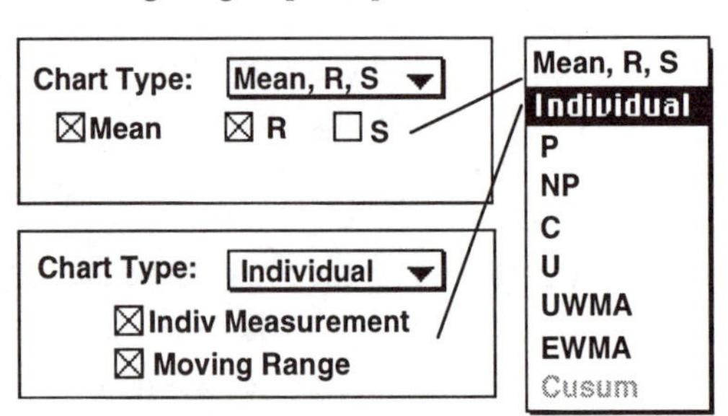

Descriptions and examples of specific kinds of charts are given later in this chapter.

Limits Specification Panel

The Limits Specification Panel allows you to specify control limits computations by entering a value for k (**K Sigma**) or by entering a probability α (**Alpha**), or by retrieving a limits value from a previously created Limits Table (discussed later).

There must be a specification of either **K Sigma** or **Alpha**.

K Sigma

allows specification of control limits in terms of a multiple of the sample standard error. **K Sigma** specifies control limits at k sample standard errors above and below the expected value, which shows as the center line. To specify k, the number of sigmas, click **K Sigma** and enter a positive k value into the box. The usual choice for k is 3, which is the dialog default.

Alpha

specifies control limits (also called probability limits) in terms of the probability α that a single subgroup statistic exceeds its control limits, assuming that the process is in control. To specify alpha, click the **Alpha** radio button and enter the probability you want. Reasonable choices for α are .01 or .001.

Retrieve...

displays a dialog for selecting a data table, referred to as a *Limits Table*, that has columns of previously computed control limits. The Limits Table is a special table that can contain values for upper and lower control limits, the center line value, and parameters for computing limits such as a mean and standard deviation.

Save

saves the dialog specifications you entered so that the next time you bring up the dialog, all the specifications become the default selections for further analyses. You can change them at any time and save new default options.

Tests Selection Panel

The Tests Selection Panel has a set of check boxes for selecting one or more *tests for special causes*, also called the Western Electric rules. When you click one or more of the eight test boxes or click **All**, the labels for the tests specified appear on the control chart. For more details on special causes tests, see the section **Tests for Special Causes** later in this chapter.

Types of Control Charts for Variables

Control charts for variables are classified according to the subgroup summary statistic plotted on the chart:

- Mean charts display subgroup means (averages).
- R charts display subgroup ranges (maximum – minimum).
- S charts display subgroup standard deviations.
- Individual Measurement charts plot single measurements.

Mean, R, and S Charts

For quality characteristics measured on a continuous scale, a typical analysis shows both the process mean, and its variability with a Mean chart aligned above its corresponding R or S chart.

The example in **Figure 15.4** uses the COATING.JMP data (taken from the *ASTM Manual on Presentation of Data and Control Charts*). The quality characteristic of interest is the Weight column. A subgroup sample of four is chosen. The Mean chart and an R chart for the process show that sample six is above the upper control limit (UCL), which indicates that the process is not in statistical control. To check the sample values, you can click the sample summary point on either control chart and the corresponding rows highlight in the data table.

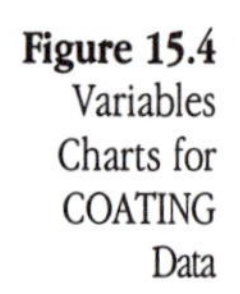

Figure 15.4 Variables Charts for COATING Data

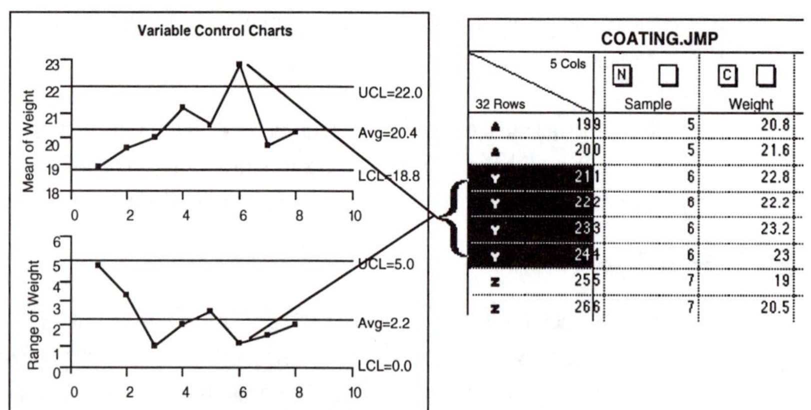

Individual Measurement and Moving Range Charts

If you are charting individual measurements, the Individual Measurement chart shows above its corresponding Moving Range chart. Follow the example dialog in **Figure 15.5** to see these kinds of control chart as follows:

§ Open the PICKLES.JMP data table. The data monitor the acid content for vats of pickles. The pickles are produced in large vats and high acidity can ruin an entire pickle vat. The acidity in four randomly selected vats was measured each day at 1, 2, and 3 PM. The data table records day (**Date**), time (**Time**), and acidity (**Acid**) measurements.

§ Choose **Graph→Control Charts** and select **Individual** from the Chart Type popup menu. Use **Acid** as the process variable, **Date** as the sample ID, and enter a range span of 2 for the Moving Range chart.

§ Click the **Chart** button on the dialog to see the individual Measurement and Moving Range charts shown at the bottom in **Figure 15.5.**

Figure 15.5 Data and Control Chart Dialog of Individual Measurement Chart

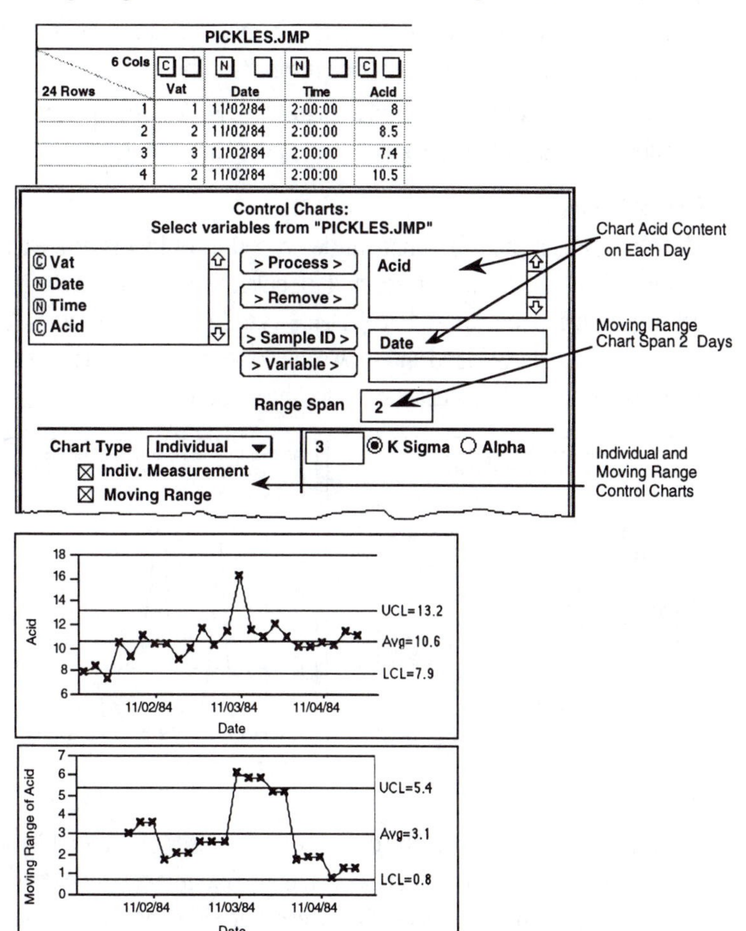

In the pickle example the **Date** variable labels the horizontal axis, which has been modified to better display the values. Tailoring axes is covered later in the chapter.

Types of Control Charts for Attributes

Attributes charts, like variables charts, are classified according to the subgroup sample statistic plotted on the chart:

- P charts display the proportion of nonconforming (defective) items in a subgroup sample.
- NP charts display the number of nonconforming (defective) items in a subgroup sample.
- C charts display the number of nonconformities (defects) in a subgroup sample that usually consists of one *inspection* unit.
- U charts display the number of nonconformities (defects) per unit in a subgroup sample with an arbitrary number of inspection units.

P and NP Charts

The WASHERS.JMP data table contains defect counts of 15 lots of 400 galvanized washers. The washers were inspected for finish defects such as rough galvanization and for exposed steel. **Figure 15.6** illustrates a P chart for the proportion of defects found and shows the corresponding NP chart to the right. These two charts are identical except for the vertical axis scale.

Figure 15.6 P and NP Chart for WASHERS Data

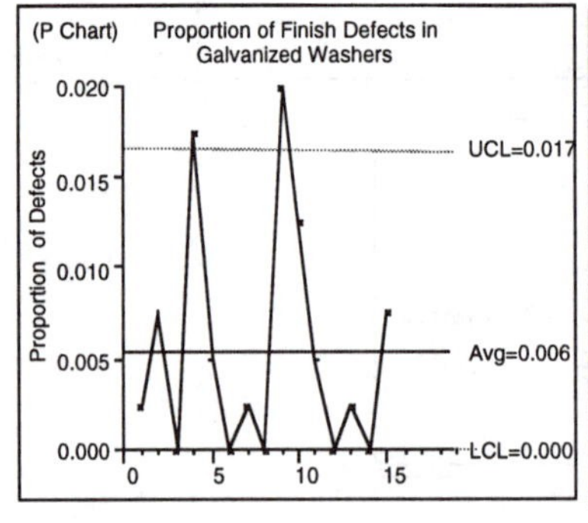

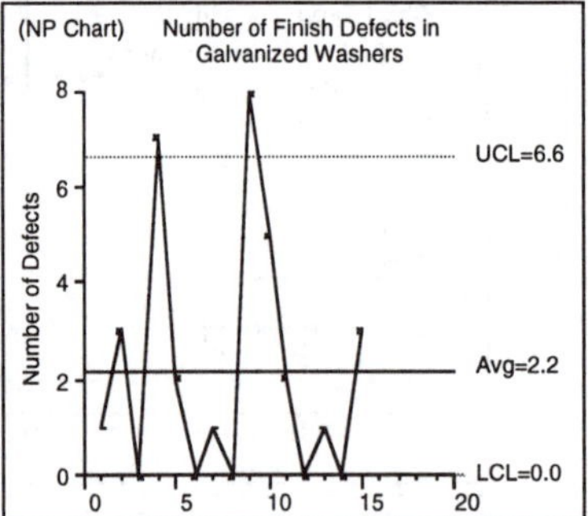

U Charts

The BRACES.JMP data records the defect count in boxes of automobile support braces. A box of braces is one inspection unit. The number of defective braces found in a day is the process variable. The subgroup sample size is the number of boxes inspected in a day, which can vary.

The U chart in **Figure 15.7** is monitoring the number of brace defects per unit. The upper and lower bounds vary according to the number of units (boxes of braces) inspected.

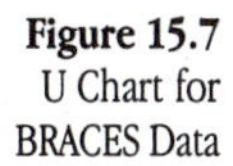

Figure 15.7
U Chart for
BRACES Data

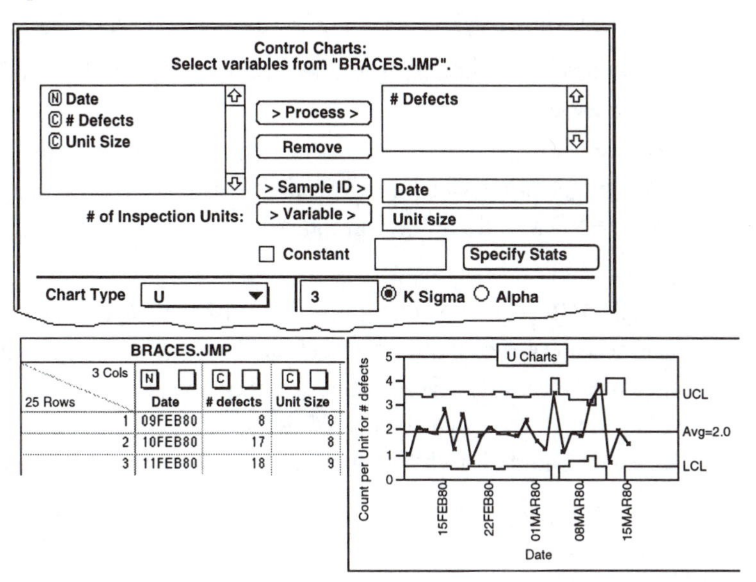

Moving Average Charts

The control charts previously discussed plot each point based on information from a single subgroup sample. The Moving Average chart is different from other types because each point combines information from the current sample and from past samples. As a result, the Moving Average chart is more sensitive to small shifts in the process average. On the other hand, it is more difficult to interpret patterns of points on a Moving Average chart because consecutive moving averages can be highly correlated (Nelson 1983).

In a Moving Average chart the quantities that are averaged can be individual observations instead of subgroup means. However, a Moving Average chart for individual measurements is not the same as a control (Shewhart) chart for individual measurements or moving ranges with individual measurements plotted.

Uniformly Weighted Moving Average (UWMA) Charts

Each point on a Uniformly Weighted Moving Average (UWMA) chart is the average of the w most recent subgroup means, including the present subgroup mean. When you obtain a new subgroup sample, the next moving average is computed by dropping the oldest of the previous w subgroup means and including the newest subgroup mean. The constant w is called the *span* of the moving average. There is an inverse relationship between w and the magnitude of the shift that can be detected. Thus, larger values of w allow the detection of smaller shifts.

§ As an example, open the data table called CLIPS1.JMP. A partial listing of the data is shown in **Figure 15.8**.

The measure of interest is the gap between the ends of manufactured metal clips. To monitor the process for a change in average gap, subgroup samples of five clips are selected daily, and a UWMA chart with a moving average span of three samples is examined.

§ To see the UWMA chart, complete the Control Chart dialog. Choose **Graph→Control Charts** and select UWMA from the Chart Type popup menu. Use Gap as the Process variable and Sample as the Sample ID. Also enter 3 as the moving average span (Ave Span) and 5 as the Constant sample size. Your completed Control Chart dialog should look like the one shown in **Figure 15.8**.

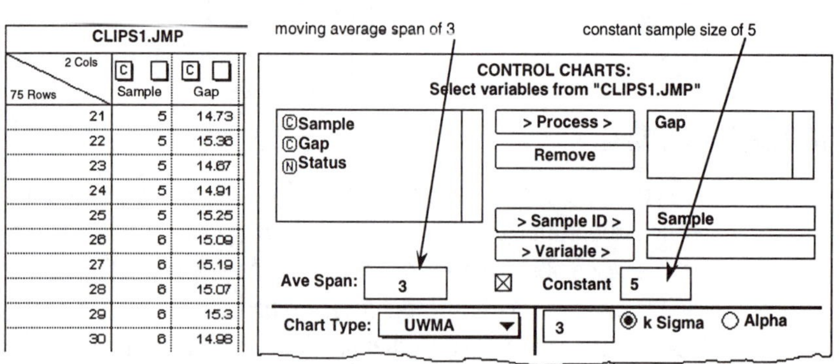

CLIPS1.JMP

75 Rows / 2 Cols	Sample	Gap
21	5	14.73
22	5	15.38
23	5	14.67
24	5	14.91
25	5	15.25
26	6	15.09
27	6	15.19
28	6	15.07
29	6	15.3
30	6	14.98

Figure 15.8 Specification for UWMA Charts of CLIPS Data

§ Click the **Chart** button to see the top chart in **Figure 15.9**. The point for the first day is the mean of the first subsample only, which consists of the five sample values taken on the first day. The plotted point for the second day is the average of subsample means for the first and second day. The points for the remaining days are the average of subsample means for each day and the two previous days.

Like all control charts, the UWMA chart updates dynamically when you add rows to the current data table.

§ Add rows to the CLIPS 1 data table as follows:

1) Open the CLIPSADD.JMP data table.

2) Click at the top of both columns, **Sample** and **Gap**, to highlight them.
3) Choose **Edit→Copy** to copy the two columns to the clipboard.
4) Click on the CLIPS 1 data table to make it the active table.
5) Click at the top of the **Sample** and **Gap** columns to highlight them in the CLIPS1 table.
6) Choose **Edit→Paste at End** to append the contents of the clipboard to the CLIPS1 table.

When you paste the new data into the table, the chart immediately updates as shown in the bottom table in **Figure 15.9**.

Figure 15.9 UWMA Charts for the CLIPS Data

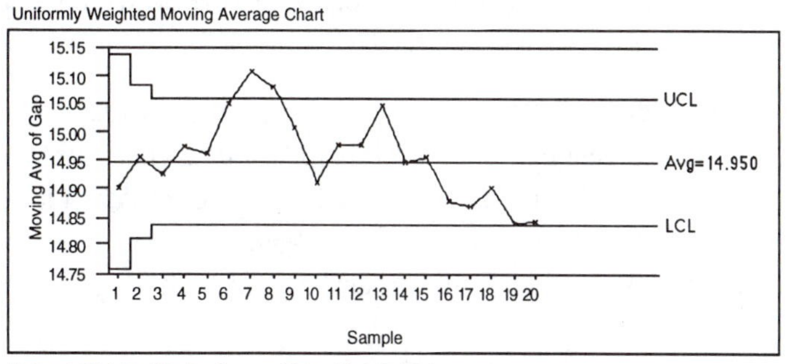

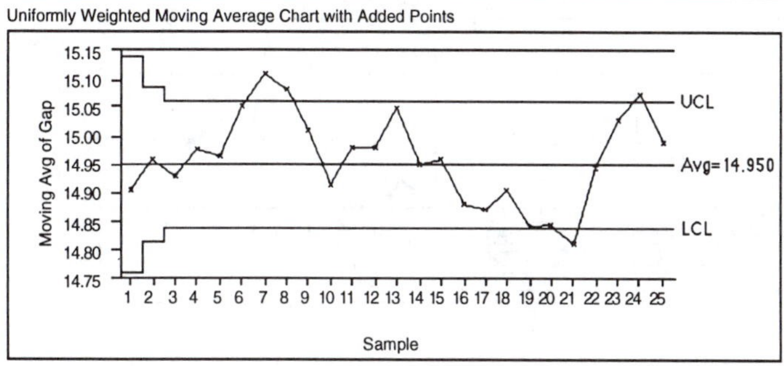

Exponentially Weighted Moving Average (EWMA) Chart

Each point on an Exponentially Weighted Moving Average (EWMA) chart, also referred to as a Geometric Moving Average (GMA) chart, is the weighted average of all the previous subgroup means, including the mean of the present subgroup sample. The weights decrease exponentially going backward in time. The weight $(0 < r < +1)$ assigned to the present subgroup sample mean is a parameter of the EWMA chart. Small values of r are used to guard against small shifts. If $r = 1$, the EWMA chart reduces to a Mean control (Shewhart) chart, previously discussed. **Figure 15.10** shows the EWMA chart for the same data used for **Figure 15.9**.

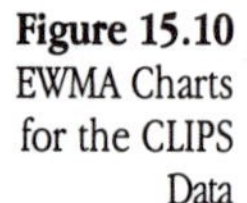

Figure 15.10 EWMA Charts for the CLIPS Data

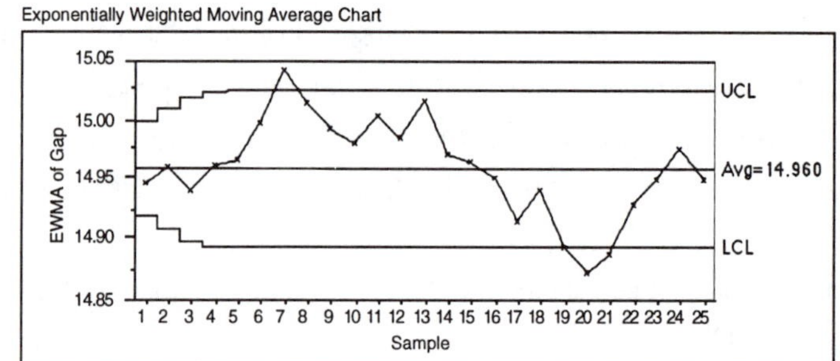

Tailoring the Horizontal Axis

When you double-click the X axis, a dialog appears that allows you to specify the number of ticks you want labeled on the X axis.

For example, the PICKLES.JMP example, done previously, lists eight measures a day for three days. **Figure 15.11** shows Individual Measurement charts for the PICKLES data. If there is no ID variable, the X axis is labeled at every tick. Sometimes this gives illegible labels, as shown to the left in **Figure 15.11**. If you specify a label for every eighth tick mark, the X axis is labeled once for each day, as shown in the plot on the right.

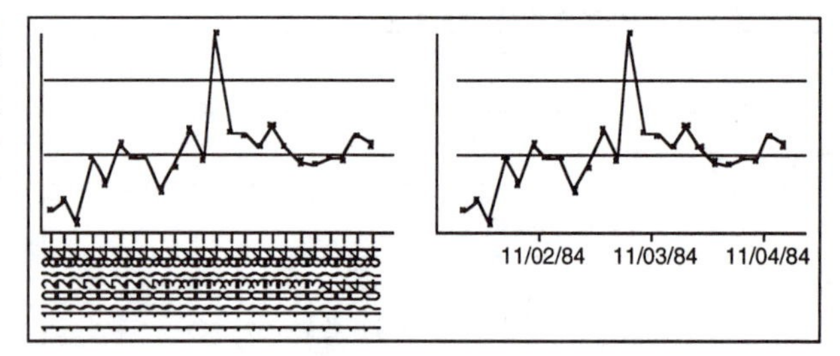

Figure 15.11 Example of Unlabeled and Labeled X-Axis Tick Marks

Tests for Special Causes

You can select one or more tests for special causes (often called the Western Electric rules) in the Control Chart dialog or with the options popup menu beneath each plot. Nelson (1984) developed the numbering notation to identify special tests on control charts.

Table 15.1 lists and interprets the eight tests, and **Figure 15.12** illustrates the tests.

The following rules apply to each test:

- The area between the upper and lower limits is divided into six zones, each with a width of one standard deviation.
- The zones are labeled A, B, C, C, B, A with Zones C nearest the center line.
- A point lies *in Zone B or beyond* if it lies beyond the line separating zones C and B.
- Any point lying on a line separating two zones lies *in or beyond* the innermost zone.

Tests 1 through 8 apply to Mean and Individual Measurement charts. Tests 1 through 4 can also apply to P, NP, C, and U charts.

Table 15.1 Description and Interpretation of Special Causes Tests (Nelson [1984, 1985])

Test 1	One point beyond Zone A	detects a shift in the mean, an increase in the standard deviation, or a single aberration in the process. For interpreting Test 1, the R chart can be used to rule out increases in variation.
Test 2	Nine points in a row in Zone C or beyond	detects a shift in the process mean.
Test 3	Six points in a row steadily increasing or decreasing	detects a trend or drift in the process mean. Small trends are signaled by this test before Test 1.
Test 4	Fourteen points in a row alternating up and down	detects systematic effects such as two alternately used machines, vendors, or operators.
Test 5	Two out of three points in a row in Zone A or beyond	detects a shift in the process average or increase in the standard deviation. Any two out of three points provide a positive test.
Test 6	Four out of five points in Zone B or beyond	detects a shift in the process mean. Any four out of five points provide a positive test.
Test 7	Fifteen points in a row in Zone C, above and below the center line	detects stratification of subgroups when the observations in a single subgroup come from various sources with different means.
Test 8	Eight points in a row on both sides of the center line with none in Zone C	detects stratification of subgroups when the observations in one subgroup come from a single source, but subgroups come from different sources with different means.

Tests 1, 2, 5, and 6 apply to the upper and lower halves of the chart separately. Tests 3, 4, 7, and 8 apply to the whole chart.

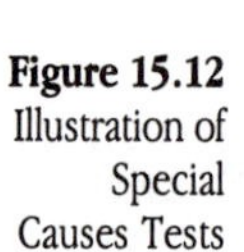

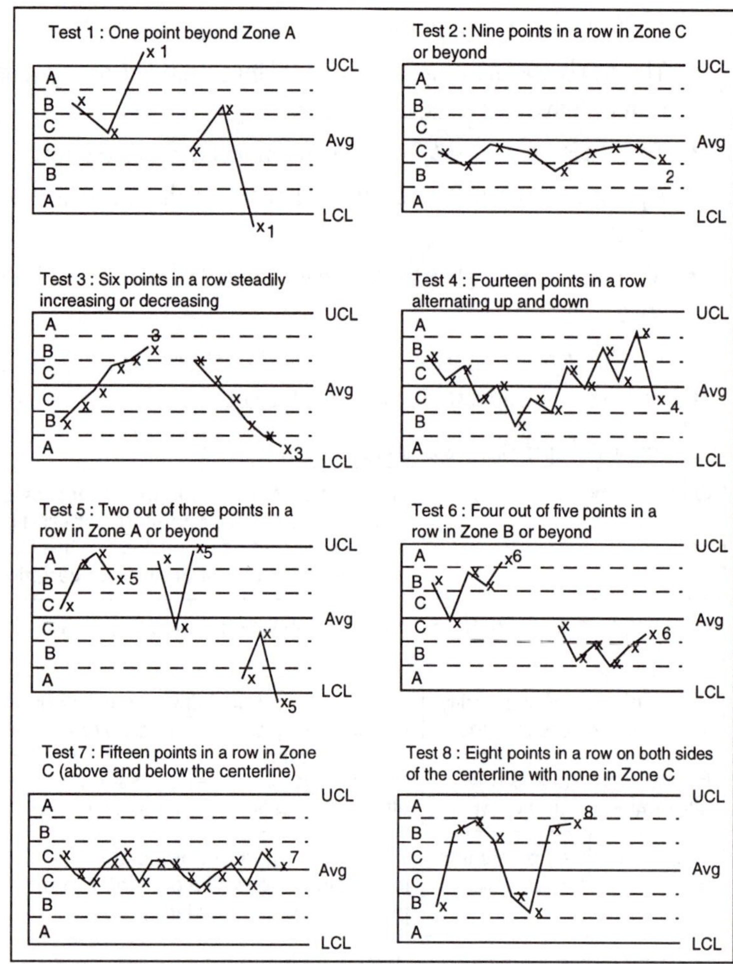

Figure 15.12 Illustration of Special Causes Tests

Nelson (1984, 1985)

Pareto Charts

The **Pareto Charts** command in the **Graph** menu creates charts that display the relative frequency or severity of problems in a quality-related process or operation. A Pareto chart is a bar chart that displays the classification of problems arranged in decreasing order. The column whose values are the frequency of a problem is assigned the Y role and is called the *process variable*.

You can also request a comparative Pareto Chart, which is a graphical display that combines two or more Pareto charts for the same process variable. The Pareto Charts platform produces a single graphical display with charts for each value in a column assigned the X role, or combination of levels from two X variables. Columns with the X role are called classification variables.

The Pareto facility does not distinguish between numeric and character variables or between modeling types. All values are treated as discrete, and bars represent either counts or percentages. The following list describes the arrangement of the Pareto graphical display:

- A Y variable with no X classification variables produces a single chart with a bar for each value of the Y variable.
- A Y variable with one X classification variable produces a row of Pareto charts. There is a chart for each level of the X variable with bars for each Y level.
- A Y variable with two X variables produces rows of Pareto charts. There is a row for each level of the second X variable listed. The rows have Pareto charts for each value of the first X variable, as described previously.

The following sections illustrate each type of Pareto chart.

A Simple Pareto Chart

The FAILURE.JMP data table shown in **Figure 15.13** lists causes of failure during the fabrication of integrated circuits. The N column contains the number of times each kind of defect occurred.

§ Open the FAILURE .JMP data table, choose **Graph→Pareto Charts**, and complete the Pareto Charts role assignment dialog. Use the **failure** column (causes of failure) as Y. It is the variable you want to inspect with Pareto charts. N column is the **Freq** variable. When you click **OK** you see the Pareto chart shown in **Figure 15.14**.

Figure 15.13 Listing of FAILURE Data

FAILURE.JMP

7 Rows / 2 Cols	failure	N
1	contamination	14
2	corrosion	2
3	doping	1
4	metallization	2
5	miscellaneous	3
6	oxide defect	8
7	silicion defect	1

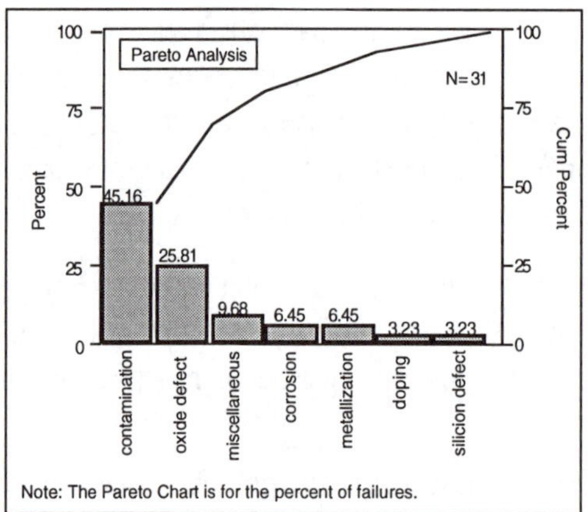

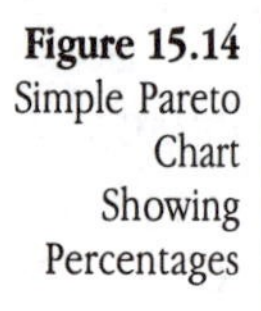

Figure 15.14 Simple Pareto Chart Showing Percentages

The height of each bar is the percent of failures in each category. For example, contamination accounts for 45% of the failures. The bars are in decreasing order with the most frequently occurring failure to the left. The curve indicates the cumulative percent of failures from left to right.

If you click on a bar to highlight it, and select **Label** from the options popup menu beneath the plot, the top of the bar is labeled with the percentage it represents, as shown in **Figure 15.14**.

The initial plot shows the percent scale on the left of the chart and the bars arranged in descending order with the cumulative curve. You can change the type of scale and arrangement of bars using display options accessed by the check mark popup menu at the lower left of the window.

Before-and-After Pareto Chart

The next example continues with the failure data, but uses an additional variable called **clean**. The failures in a sample of capacitors manufactured is recorded before cleaning a tube in the diffusion furnace and in a sample manufactured after cleaning the furnace. For each type of failure, the variable **clean** identifies the samples with the values "before" or "after."

§ To see the data, open the FAILURE2.JMP data table, shown in **Figure 15.15**.

§ Choose **Graph→Pareto Charts** and complete the dialog. Use **failure** as the **Y** variable and **clean** as the **X** grouping variable.

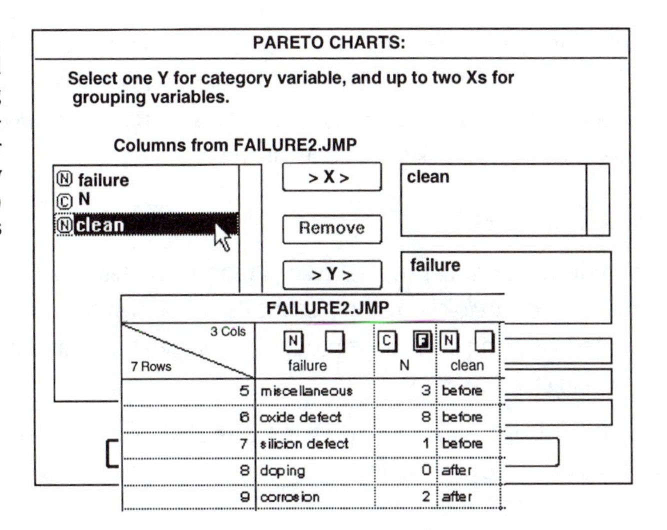

Figure 15.15 Control Charts Dialog for Before-and-After (One-Way Comparative) Pareto Charts

§ When you click **OK** you see one Pareto chart window with side-by-side charts for each value of the X variable, clean, (**Figure 15.16**).

These charts are referred to as the *cells* of a comparative Pareto chart. There is a cell for each level of the X (classification) variable. Because there is only one X variable, this is called a *one-way comparative Pareto chart.*

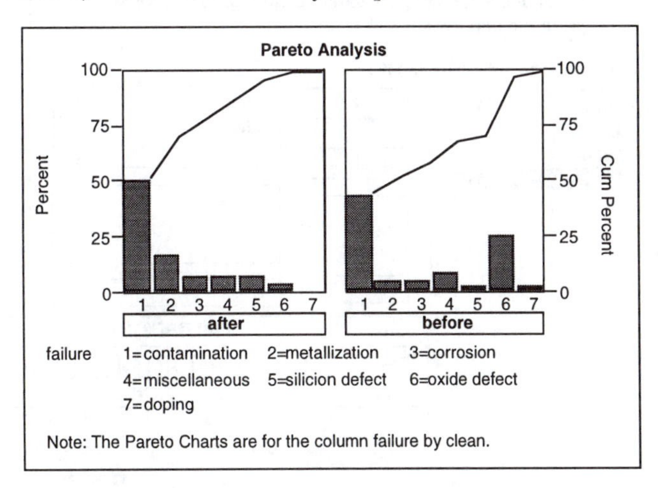

Figure 15.16 One-Way Comparative Pareto Charts

The horizontal and vertical axes are scaled identically for both charts. The bars in the first chart are in descending order of the X axis values and determine the order for all charts.

The charts themselves are in alphabetic order of the classification variable levels. The levels ("after" and "before") of the classification variable, **clean**, show beneath each chart. You can change the order of the charts with the **Reorder Horizontal** option in the check-mark popup menu.

Two-Way Comparative Pareto Chart

You can study the effects of two classification variables simultaneously with a *two-way comparative Pareto chart*. Suppose the capacitor manufacturing process from the previous example monitors production samples before and after a furnace cleaning for three days.

§ Open the FAILURE3.JMP data table, shown in **Figure 15.17**. This table has the same columns as used in the previous examples and a new column called **date** with values "OCT1," "OCT2," and "OCT3."

§ To see a two-way array of Pareto charts, choose **Graph→Pareto Charts** and complete the dialog. Again, **failure** is the Y variable. Both **clean** and **date** are the X grouping variables.

Figure 15.17 Pareto Charts with Two Grouping Variables

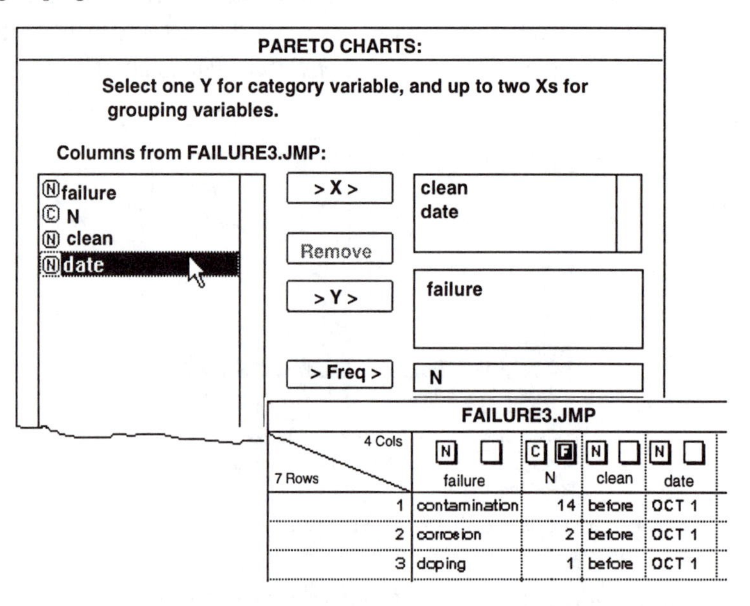

§ Click **Done** to see the Pareto charts shown in **Figure 15.18**. There is a two-way layout of charts that show each level of both X variables.

You can reorder the rows and columns of cells with the **Reorder** option in the check-mark popup menu beneath the charts. The cell that moves to the upper left corner , then becomes the key cell, and the bars in all other cells rearrange accordingly.

Also, you can click bars in the key cell and the bars for the corresponding categories highlight in all other charts. Use COMMAND-click or OPTION-click to select noncontiguous bars.

The Pareto chart shown in **Figure 15.19** illustrates highlighting the *vital few*. In each cell of the two-way comparative chart, the bars representing the two most frequently occurring problems (the vital few) are selected and given a black pattern. **Contamination** and **Oxygen Defect** are the two vital categories in all cells, but they appear to be less of a problem after furnace cleaning.

Figure 15.18 Two-Way Comparative Pareto Chart

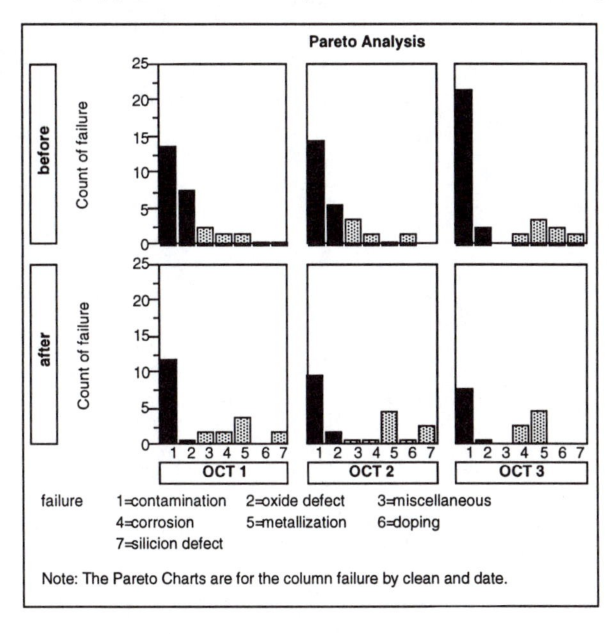

Chapter 16
Time Series

Overview

You need to use special analysis methods for data taken across time. Often the goal is to forecast future values. In JMP there are no custom tools for time series analysis, but there are more general tools that can do some basic analyses. This chapter gives a brief introduction to some common techniques for looking at time series data.

Chapter 16 Contents

You can mouse along with the examples whenever you see an action symbol (§)

Introduction

When you take data from a process in the real world, the values can be a product of everything else that happened until the time of measurement. Time series data usually is *non-experimental*: it is the work of the world rather than the product of experimentally controlled conditions. Furthermore, the data are not presented with all relevant covariates; the data are often taken alone, without any other variable to co-analyze.

So time series methods have been developed to characterize how variables behave across time and how to forecast them into the future.

There are two general approaches to fitting a time series:

- Model the data as a function of time itself.
- Model the data as a function of its past (lagged) values

The first approach emphasizes the structural part of the model, whereas the second approach emphasizes the random part of the model.

The strategy for modeling a time series is affected by the time horizon:

Long-term trend
: You want to see the long-term curve and use that for long-range forecasts. Long-term strategies usually focus on a structural part of the model, though sometimes there is a need to adjust for short-term effects.

Short-term local behavior
: You see a pattern to the local wiggles. You are interested in short-term forecasts. Short-term modelers usually focus on the random part of the model and emphasize the relationships of a value with its lagged values. Real world systems can have all sorts of short-term adjustment and feedback mechanisms that produce a variety of behaviors.

Medium-term seasonal
: You see that the variable has some kind of cyclic behavior, like buying habits that spike during Christmas season. Medium-term strategies use either time models or lag models, or a combination.

Trend strategies are usually ordinary regression models involving time, which were covered in Chapter 8, "Fitting Curves Through Points: Regression." So this chapter focuses on short-term models that emphasize the random component, and use lagged values of the time series.

Models for local short-term behavior are especially appropriate for things that are fairly unpredictable long-term, like the stock market or the weather. Suppose the question is "Will it rain a minute from now?" Faced with this question, which information would you rather know: what time it is or if it is currently raining? If rain was structurally predictable like a television schedule, you would rather know what time it is. But weather is not like that, so to predict if it will rain in a minute, you would rather know if it is currently raining.

Graphing and Fitting by Time

Creating Time Columns

To model or graph a series across time, you need a column to represent time:

§ Open SERIESA.JMP, which has one time series in a column called **Cn**.

The values in the **Cn** column are a series of chemical concentration readings, which appeared in the classic Box and Jenkins book. The Box and Jenkins book was so influential that time series modeling is often called *Box-Jenkins* analysis. The third edition of this book is now out with Reinsel as a third author.

The time series (**Cn**) values need to be identified by a time column. You can make a new time column with a formula:

§ Choose **Cols→New Column**. When the New Column dialog appears, enter **Time** as the column name and select **Formula** from the **Data Source** popup menu. When you click **OK** you will see the calculator window shown in **Figure 16.1**.

§ When you see the calculator, click on **Terms** in the function category list. Then click **i (row #)** from the list of terms. In the formula area, a single lower-case "i" appears, which is the notation for the row number.

When you close the calculator window, the new column, **Time**, fills with the row numbers as values. The data table in **Figure 16.1** is a partial listing of the SERIESA data table with the new **Time** variable.

Figure 16.1 Calculator for the Time Column and Partial Listing of the SERIESA Table

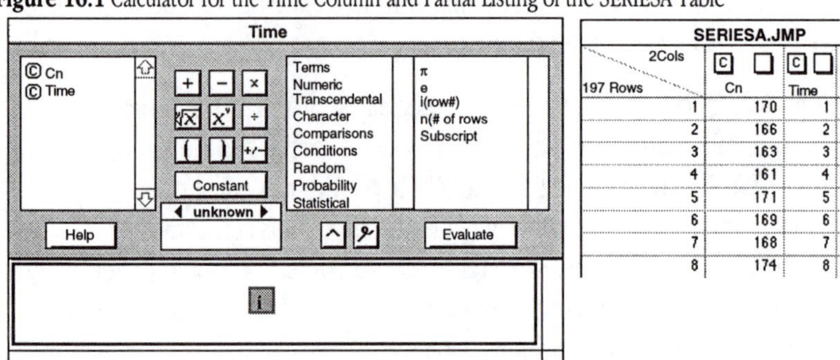

	Cn	Time
1	170	1
2	166	2
3	163	3
4	161	4
5	171	5
6	169	6
7	168	7
8	174	8

Graphing by Time

You can see how the time series behaves with an overlay plot:

§ Choose **Graph→Overlay Plot**. In the role assignment dialog, choose **Cn** as **Y** and **Time** as the **X** and click **OK**. When the plot appears, click inside the plot frame to see a small resize box in the lower right. Click and drag on the resize box to reshape the plot to be shorter and wider as in **Figure 16.2**. Time series plots generally work better with a wide aspect ratio.

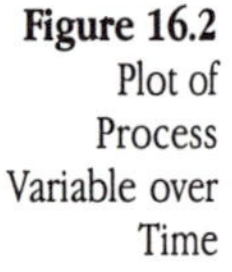

Figure 16.2 Plot of Process Variable over Time

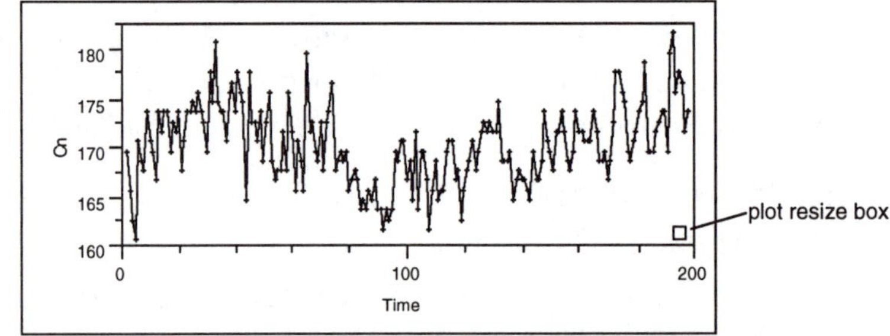

The Overlay Plots platform provides a number of features useful for time series plots, such as connecting the points with lines. But if you want to fit curves through the points, then you need to switch to the Fix Y by X platform.

Sometimes what you want to see is the longer-term trends of the model without the short-term variation. Smoothing techniques can show you trends. One of the best smoothers is the smoothing spline, which pieces together polynomials to optimize a criterion that balances the smoothness with the goodness of fit. You fit a number of models with different amounts of smoothing and choose the one that represents the amounts of smoothing that you want to see.

§ Choose **Analyze→Fit Y by X**, assign Cn as Y and Time as the X, and click **OK**. Again, stretch the plot to make it wider.

To use smoothing splines you have to specify how stiff to make the line. If it is too rigid it looks like a straight line. If you make the spline too flexible, it curves too much, trying to fit each point. Use the following spline fit options to see the plots in **Figure 16.3**:

§ Select **Fit Spline** from the Fitting popup menu with lambda=10.

§ Select **Fit Spline** again from the Fitting popup menu with lambda=100000.

Figure 16.3 Spline Fits for Time Series Data

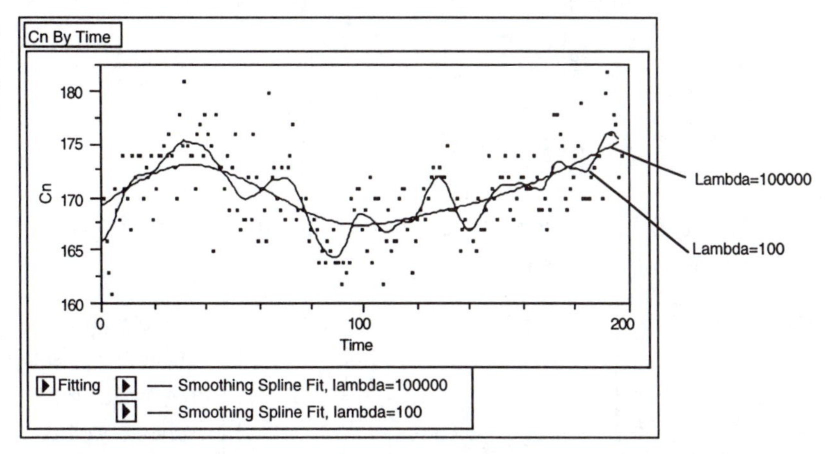

The Fit Y by X platform offers a variety of regression fits, including straight-line and polynomials. You can also save the prediction formula for some fits. Then if you extend the data table with additional rows, the Fit Y by X platform can plot the fitted values past the end of the series and produce a forecast, including the confidence limits.

Trend and Seasonal Factors

Let's fit a time model where there are two components, trend and seasonality. The data are the number of monthly airline passengers (Box and Jenkins, Series G). Airline traffic is seasonal, with some months being more popular than others.

§ Open the SERIESG.JMP sample table to see the data table shown in **Figure 16.4**.

The Time variable is the row number and serves as a time unit. The Season variable with values 1 to 12 identifies the month. (Time and Season were both created with a formula.) There are extra rows at the end with missing values for Passengers that will be used for forecasting the future.

Figure 16.4 Partial Listing of the Airline Data

SERIESG.JMP

3 Cols / 164 Rows	Passengers	Time	Season
1	112	1	1
2	118	2	2
3	132	3	3
4	129	4	4
5	121	5	5
6	135	6	6
7	148	7	7
8	148	8	8
9	136	9	9
10	119	10	10
11	104	11	11
12	118	12	12
13	115	13	1
14	126	14	2
15	141	15	3
142	461	142	10
143	390	143	11
144	432	144	12
145	•	145	1
146	•	146	2
147	•	147	3
148	•	148	4

§ To look at the time trends, Choose **Graph→Overlay Plots** and enter Passengers as X and Time as Y.

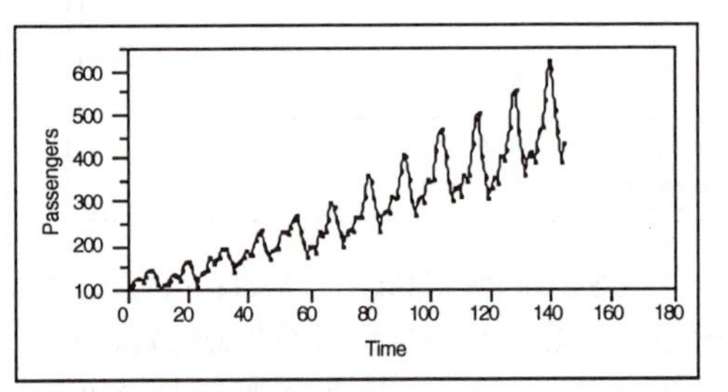

Figure 16.5 Plot of Number of Airline Passengers over Time

The overlay plot, shown in **Figure 16.5**, displays the very regular seasonal nature of the series as well as an increasing trend in the number of airline passengers over time. You can use the model fitting platform to look at a quadratic trend with Season as a nominal factor:

§ Choose **Analyze→Fit Model** and enter the terms into the model as shown in **Figure 16.6.** To build a quadratic model:

1) Select Time in the variable selection list and then click **Add.**
2) Select Time and add it to the **Effects in Model** list a second time.
3) Select Time in the variable selection list and also in the **Effects in Model** list, and click the **Cross** button.
4) Select Season and add it to **Effects in Model**, then click **Run Model.**

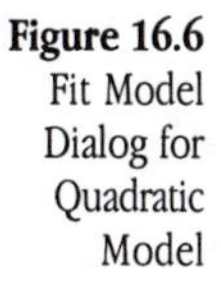

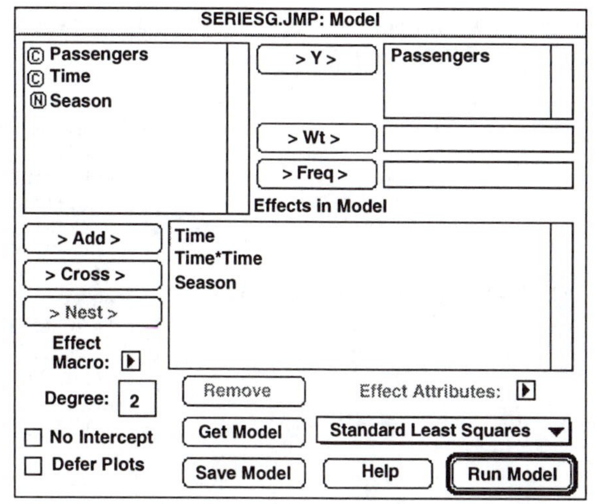

Figure 16.6 Fit Model Dialog for Quadratic Model

After the model run is complete, the Parameter Estimates table in **Figure 16.7** shows the **Time** parameter estimates and the 11 parameters for the **Season** effect (because there are 12 seasons). These parameter estimates can be saved in the prediction formula for the model. To save the prediction formula:

§ Select **Save Prediction Formula** from the Save ($) popup menu. This command creates a new column called **Predict Passengers**, with the formula using the parameter estimates shown in **Figure 16.7**.

Figure 16.7 Parameter Estimates and Prediction Formula from Quadratic Model

Parameter Estimates

Term	Estimate	Std Error	t Ratio	Prob>|t|
Intercept	112.60548	6.026922	18.68	<.0001
Time	1.6255044	0.191722	8.48	<.0001
Time*Time	0.0071367	0.001281	5.57	<.0001
Season[1-12]	-24.04764	6.568412	-3.66	0.0004
Season[2-12]	-33.3866	6.566641	-5.08	<.0001
Season[3-12]	-0.823171	6.565242	-0.13	0.9004
Season[4-12]	-6.524013	6.564204	-0.99	0.3221
Season[5-12]	-4.405796	6.563516	-0.67	0.5032
Season[6-12]	32.781482	6.563174	4.99	<.0001
Season[7-12]	69.787819	6.563174	10.63	<.0001
Season[8-12]	66.863217	6.563516	10.19	<.0001
Season[9-12]	15.507674	6.564204	2.36	0.0196
Season[10-12]	-23.02881	6.565242	-3.51	0.0006
Season[11-12]	-59.49623	6.566641	-9.06	<.0001

§ To see the formula shown in **Figure 16.8**, open the calculator for **Predict Passengers** column in the data table.

Figure 16.8 Prediction Formula From the Calculator For the Quadratic Model

112.605478+1.62550441•*Time* +*Time* •*Time* •0.00713672+ match *Season* :

-24.04764,	when 1
-33.386602,	when 2
-0.823171,	when 3
-6.5240132,	when 4
-4.4057956,	when 5
32.781482,	when 6
69.7878194,	when 7
66.8632167,	when 8
15.5076739,	when 9
-23.028809,	when 10
-59.496232,	when 11
-33.227929,	when 12
•,	otherwise

§ Choose **Graph→Overlay Plots** to plot both Passengers and Pred Formula Passengers by Time to see the overlay plot in **Figure 16.9.**

By default, the data points show and are connected. There are popup menus beneath the plot for each overlay variable. You can use options to compare the actual to predicted values and see the forecast. To produce the plot in **Figure 16.9**:

§ For the Passengers variable, select **Connect** from its popup menu to unconnect the points.

§ For the Predict Passengers variable, select **Show Points** from its popup menu to hide the points and show only the prediction line.

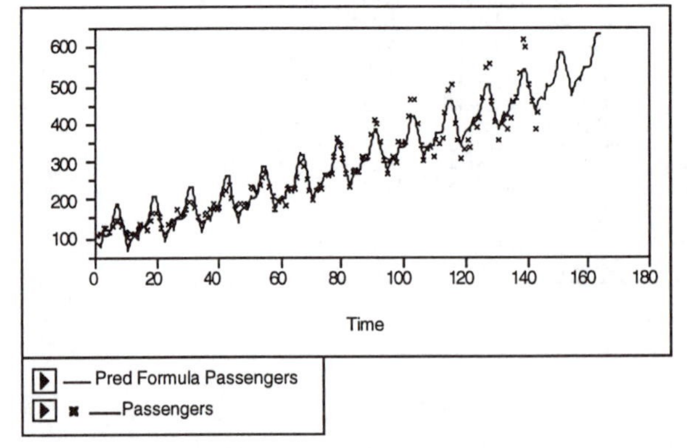

Figure 16.9 Overlay Plot of Actual and Predicted Airline Passengers

The model does capture the seasonal variation, but there is a problem. Near the beginning of the series, the seasonal factor over-estimates the variation. Near the end of the series, the seasonal factor under-estimates the variation.

There are a number of ways this problem can be addressed. You can make a nonlinear model where the seasonal factor multiplies the predicted value by a seasonal scale factor, instead of adding to it with an additive adjustment.

Another way to address this problem is to take logarithms of the passenger traffic and use that transformation as the response variable. Taking the logarithm of the Y variable is equivalent to fitting a multiplicative model because products become sums under the logarithm. Whenever the variation seems to be proportional to the Y value, taking logs is suggested.

Create a new column in the SERIESG.JMP data table that is the log of the Passengers variable :

§ Choose **Cols→New Column** and complete the new column dialog—call the new column Log Pass, select **Formula** from the **Data Source** popup menu. and click **OK** to see the calculator.

§ When the calculator appears, select Common Log from the list of Transcendental Functions, and select the Passengers variable as its argument. Then close the calculator and the New Column dialog.

§ Now do an overlay plot of Log Pass by Time. Note how the seasonal variation now seems to be scaled the same across the time series.

Figure 16.10 Overlay Plot of the Log of Passengers by Time

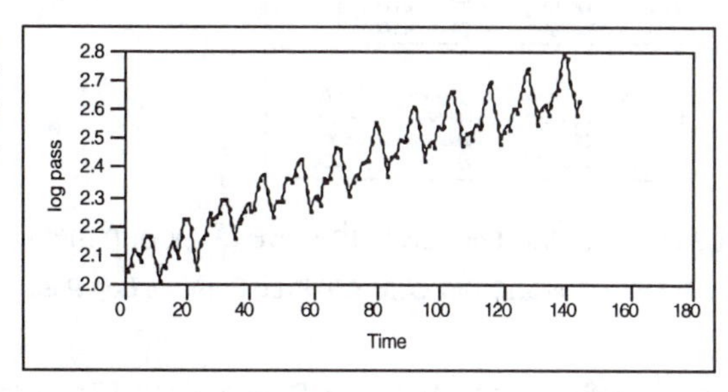

§ Continue by doing the same process as before: Fit a model of the log of Passengers by a quadratic in Time, and the nominal Season effect.

Figure 16.11 shows the Fit Model dialog for the quadratic model with Log Pass as the Y variable and the resulting report tables.

Figure 16.11 Quadratic Fit of Log of Passengers by Time and Season

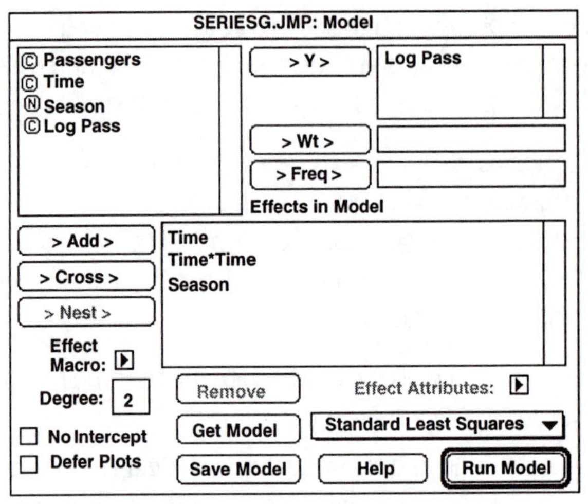

Parameter Estimates

Term	Estimate	Std Error	t Ratio	Prob>|t|
Intercept	2.0569891	0.005313	387.17	<.0001
Time	0.0057256	0.000169	33.88	<.0001
Time*Time	-0.000009	0.000001	-8.26	<.0001
Season[1-12]	-0.036921	0.00579	-6.38	<.0001
Season[2-12]	-0.046592	0.005789	-8.05	<.0001
Season[3-12]	0.0098898	0.005787	1.71	0.0899
Season[4-12]	-0.003746	0.005787	-0.65	0.5185
Season[5-12]	-0.004814	0.005786	-0.83	0.4069
Season[6-12]	0.048215	0.005786	8.33	<.0001
Season[7-12]	0.0933566	0.005786	16.14	<.0001
Season[8-12]	0.0893385	0.005786	15.44	<.0001
Season[9-12]	0.0265618	0.005787	4.59	<.0001
Season[10-12]	-0.033384	0.005787	-5.77	<.0001
Season[11-12]	-0.095725	0.005789	-16.54	<.0001

Effect Test

Source	Nparm	DF	Sum of Squares	F Ratio	Prob>F
Time	1	1	0.50290462	1147.683	<.0001
Time*Time	1	1	0.02993145	68.3068	<.0001
Season	11	11	0.42964514	89.1361	<.0001

§ As before, select **Save Prediction Formula** in the save ($) popup menu to save the prediction formula. This adds another column, **Pred Formula Log Pass** to the data table.

§ Choose **Graph→Overlay Plots** to plot both Log Pass and Pred Formula Log Pass by Time to see the overlay plot in **Figure 16.12.** The model fits well in this log scale.

§ Make another new column (call it Pred) by taking 10 to the power Pred Formula Log Pass.

$$10^{\textit{Pred Formula log pass}}$$

Now the prediction is converted to the original scale. Plot the actual (Passengers) and new converted predicted columns (Pred). Note in **Figure 16.13** that the transformed regression not only fits the data well, but also makes a sensible prediction into the future.

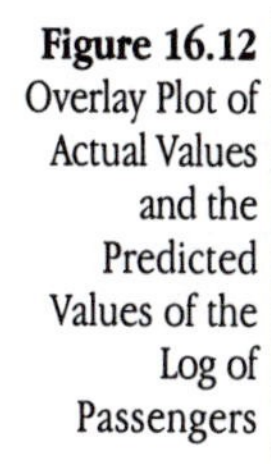

Figure 16.12
Overlay Plot of Actual Values and the Predicted Values of the Log of Passengers

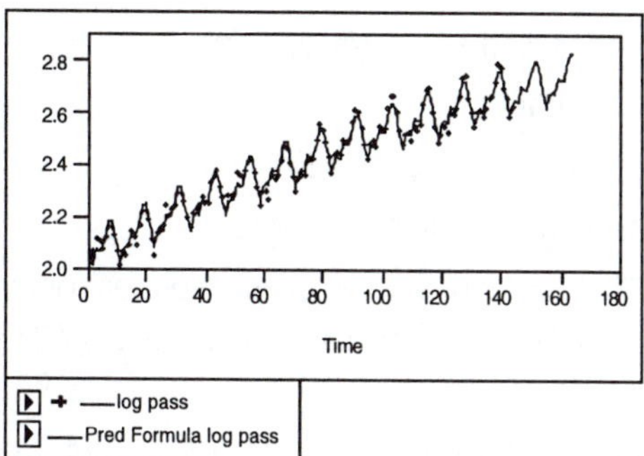

Figure 16.13
Overlay Plot of Actual Values and the Predicted Values Transformed to Original Scale

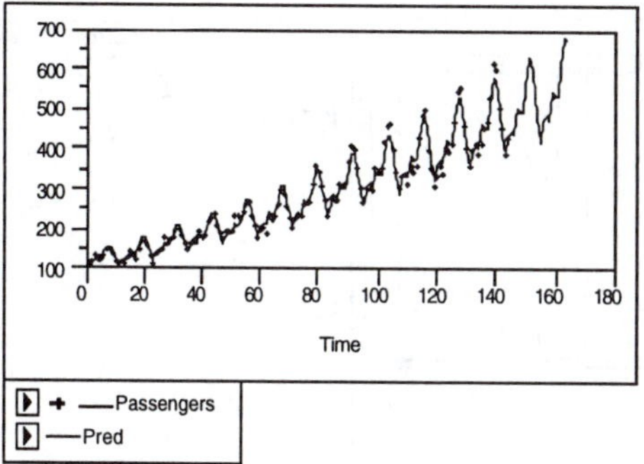

Lagging and Autocorrelation

The rest of this chapter focuses on models that use lagged values of the process variable.

Creating Columns with Lagged Values

First you need to generate variables containing lagged values. A lag variable is a column whose rows have values that are the previous row's values of another column.

§ Close all analysis and graph windows to clean up the desktop, and click the SERIESA.JMP table to make it the active window.

§ Create another new column. As before, choose **Cols→New Column**. Enter Lag Cn as the column name and select **Formula** from the **Data Source** popup menu. Then Click **OK**.

§ When its calculator appears, build the lag formula as follows:

1) Click **Cn** in the list of columns.
2) Select **Subscript** from the **Terms** function list.
3) Click the minus sign on the calculator work panel.
4) type "1" in the constant area and click the **Constant** button (or type "1" on your keyboard and press RETURN). The lag formula should look like the one in the calculator in **Figure 16.14**.
5) Click the **Evaluate** button on the calculator, or close the calculator to see the data table in **Figure 16.14.**

Figure 16.14 Formula to Compute a Lag Variable

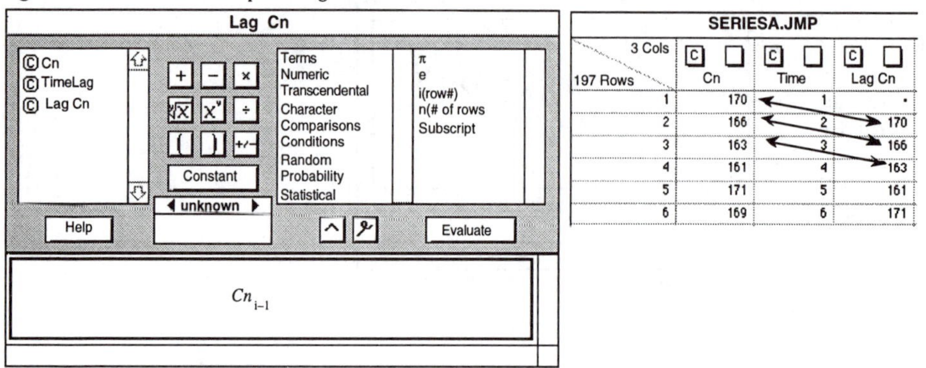

	Cn	Time	Lag Cn
1	170	1	•
2	166	2	170
3	163	3	166
4	161	4	163
5	171	5	161
6	169	6	171

(If you are in a hurry or have trouble with the calculator formula, open the SERIESA2.JMP data table, which has all the formulas.)

Autocorrelation

Autocorrelation is the correlation of a variable with its lagged values. You can look for autocorrelation by plotting a variable together with its lag as follows:

§ Choose **Analyze→Fit Y by X**, with Cn as Y, and Lag Cn as X.

§ When the scatterplot appears, select **Density Ellipses** from the Fitting popup menu and use **.95** from its submenu.

The diagonal orientation of the density ellipse in **Figure 16.15** indicates correlation; the text report shows a correlation of .57.

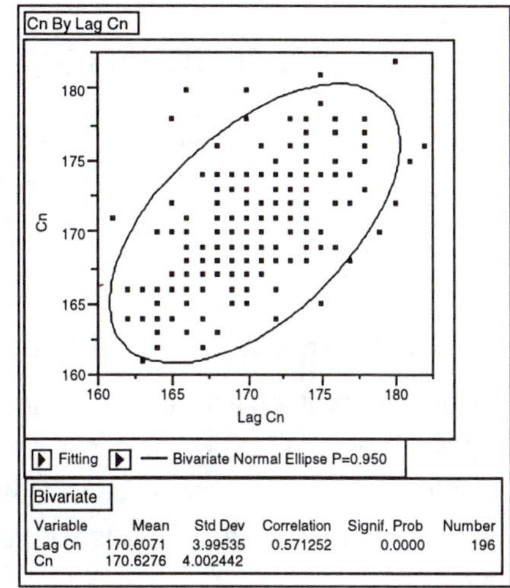

Figure 16.15 Density Ellipse Indicates Correlation

Positive correlation is very common in time series data. You can see it in the time series plots by looking at local behavior (a short span of a time series) to see if the points tend to be near neighboring points rather than randomly scattered across a trend line.

Positive autocorrelation is natural because things tend to change slowly over time if you look at short time intervals. Economic data is usually positively autocorrelated, coming from a complex system that responds at varying rates to stimuli. Negative autocorrelation is more rare, and usually occurs only if the data are sampled at the right rate so that the cycles of some feedback mechanism are being detected.

By lagging more than one row, you can measure autocorrelation across different time differences. A graph of this is called a correlogram.

Optional Topic: Correlograms

This section shows how to create a correlogram. Correlograms are not built in to JMP, but can be constructed as follows:

1) Create a data table with a series a lag columns.
2) Find the correlations between the lag columns.
3) Transfer the correlations to a JMP data table.
4) Plot the correlations.

§ Open the SERIESAL.JMP sample table, which already has eight lag variables calculated for SERIESA using formulas, as shown in **Figure 16.16.** You can verify the lag formulas by looking in the Column Info dialog for the lag columns.

Figure 16.16 Table with Eight Lag Variables

SERIESAL.JMP

10 Cols / 197 Rows	Cn	Time	Lag1	Lag2	Lag3	Lag4	Lag5	Lag6	Lag7	Lag8
1	170	1	•	•	•	•	•	•	•	•
2	166	2	170	•	•	•	•	•	•	•
3	163	3	166	170	•	•	•	•	•	•
4	161	4	163	166	170	•	•	•	•	•
5	171	5	161	163	166	170	•	•	•	•
6	169	6	171	161	163	166	170	•	•	•
7	168	7	169	171	161	163	166	170	•	•

Cn_{i-3}
Cn_{i-2}
Cn_{i-1}
and so forth

§ Choose **Analyze→Correlation of Y's**, with Cn and all eight lag variables as Y's.

When you click **OK**, the Correlation of Y's platform first presents a standard Pearson correlation matrix, as shown in **Figure 16.16.** A correlogram is a plot of the first two columns of the correlation matrix. To plot these columns in JMP, you can either key this information into a new table or use the following features to copy the information into a new table:

§ Use the scissors tool found in the **Tools** menu and drag to encompass the correlation matrix as in **Figure 16.17.**

§ Hold down the OPTION or ALT key and then choose **Edit→Copy as Text**. The text of the report is then copied to the clipboard and displayed in a Text Edit window.

Figure 16.17 Correlation Matrix of the Cn Variable and Eight Lag Variables

SERIESAL.JMP:Correlations

Correlations

Variable	Cn	Lag1	Lag2	Lag3	Lag4	Lag5	Lag6	Lag7	Lag8
Cn	1.0000	0.5724	0.5080	0.4116	0.3868	0.3549	0.3857	0.4334	0.3553
Lag1	0.5724	1.0000	0.5674	0.5023	0.4076	0.3658	0.3381	0.3746	0.4334
Lag2	0.5080	0.5674	1.0000	0.5692	0.5029	0.4094	0.3724	0.3421	0.3748
Lag3	0.4116	0.5023	0.5692	1.0000	0.5619	0.4945	0.3967	0.3610	0.3454
Lag4	0.3868	0.4076	0.5029	0.5619	1.0000	0.5461	0.4752	0.3800	0.3653
Lag5	0.3549	0.3658	0.4094	0.4945	0.5461	1.0000	0.5509	0.4729	0.3779
Lag6	0.3857	0.3381	0.3724	0.3967	0.4752	0.5509	1.0000	0.5380	0.4818
Lag7	0.4334	0.3746	0.3421	0.3610	0.3800	0.4729	0.5380	1.0000	0.5469
Lag8	0.3553	0.4334	0.3748	0.3454	0.3653	0.3779	0.4818	0.5469	1.0000

8 rows not used due to missing values.

Drag scissors tool to select report text

To copy correlations to a JMP data table, hold down the OPTION or ALT key and choose Copy as Text from the Edit menu.

§ Choose **File→New** to create a new untitled data table.

§ With this new table active, hold down the OPTION or ALT key and choose **Edit→Paste at End.** The text of the report is then copied from the clipboard to the new table, creating the rows and columns as needed.

The first line of the text is used as the column names. A portion of this new **Untitled** table is shown in **Figure 16.8.** Now, plotting the first two columns will show the correlation of the Cn variable over time.

§ Choose **Graph→Bar/Pie** with Cn as Y and Variable as X. The default display for the Bar/Pie platform is a bar chart.

§ To see the correlogram as it is shown to the right in **Figure 16.18**:

1) Use the Column Info dialog for the **Cn** column to change the number of decimals displayed to 2.
2) select the **Needle** option from the check-mark popup menu at the lower left of the Bar/Pie platform window.
3) Highlight all the rows in the **Untitled** table used to create the chart and choose **Rows→Label/Unlabel.**

Figure 16.18 Data Table of Correlations and Correlogram Plot

Untitled 1

10 Rows / 10 Cols	Variable	Cn	Lag1	Lag2	La
1	Cn	1.00	0.5724	0.508	0.4
2	Lag1	0.57	1	0.5674	0.5
3	Lag2	0.51	0.5674	1	0.5
4	Lag3	0.41	0.5023	0.5692	
5	Lag4	0.39	0.4076	0.5029	0.5
6	Lag5	0.35	0.3658	0.4094	0.4
7	Lag6	0.39	0.3381	0.3724	0.3
8	Lag7	0.43	0.3746	0.3421	0.

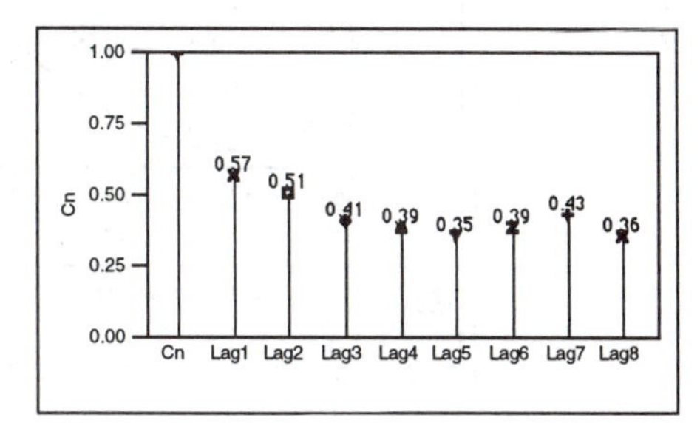

Autoregressive Models

One way to construct a time series model is to regress the series on its own lagged values. This is called an autoregressive model. You can do this in JMP using the Fit Y by X platform.

§ Return to or recreate the window for Fit Y by X of **Cn** by **Lag Cn** (**Figure 16.15**). Now select the **Fit Line** command from the **Fitting** popup menu beneath the plot to fit the straight-line regression.

The fitted regression line, shown in **Figure 16.19**, is a simple linear prediction of **Cn** by its lagged value. To use this model, you need to capture its prediction equation:

§ In the popup icon for **Linear Fit**, select the **Save Predicteds** command.

The **Save Predicteds** command stores the predicted values as a new column, **Predicted Cn**, in the SERIESA data table. It also stores the regression equation with the column. To verify the prediction equation, OPTION-click at the top of the **Pred Cn** column to see a calculator window with the formula:

$$72.9948599 + 0.57226614 \bullet \textit{Lag Cn}$$

Figure 16.19 Regression Line Gives a Prediction Equation

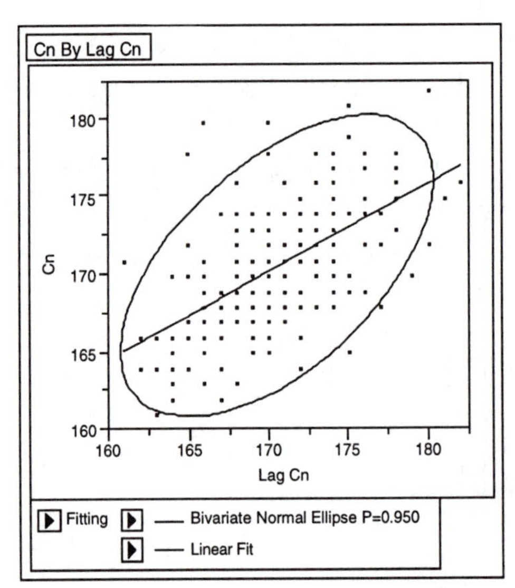

The plot in **Figure 16.20** shows how the predicted values correspond to the sample values. You can see this plot by using the **Overlay Plots** command:

§ Choose **Graph→Overlay Plots**, with Time as X and both Predicted Cn and Cn as Y's. When you click **OK**, the first plot in **Figure 16.20** appears. Using options, you can modify the plot to clearly distinguish between the observed and predicted values. To see the second plot in **Figure 16.20**:

1) Drag the resizing rectangle in the lower right corner to make the plot wide.
2) In the popup for Cn select the **Connect** option to toggle off (remove) the lines connecting the Cn points.
3) In the popup menu for Predicted Cn select the **Show Points** option to hide the points for Predicted Cn.

Now the points show the data and the connected line shows the 1-step forecast.

Figure 16.20 Actual and Predicted Cn Points Connected

(continued next page)

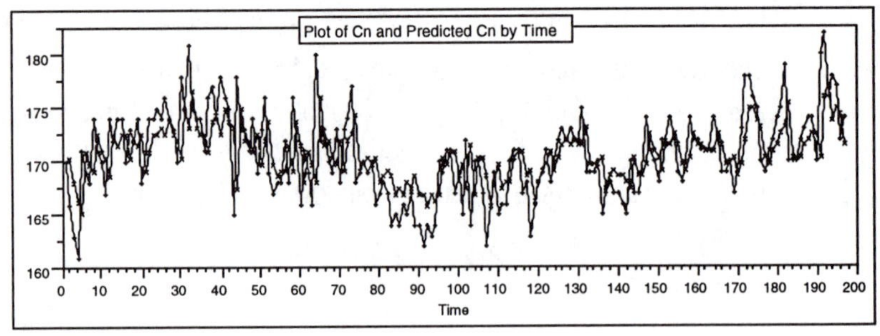

Figure 16.20 (continued) Actual **Cn** Points and Predicted Cn Points Connected

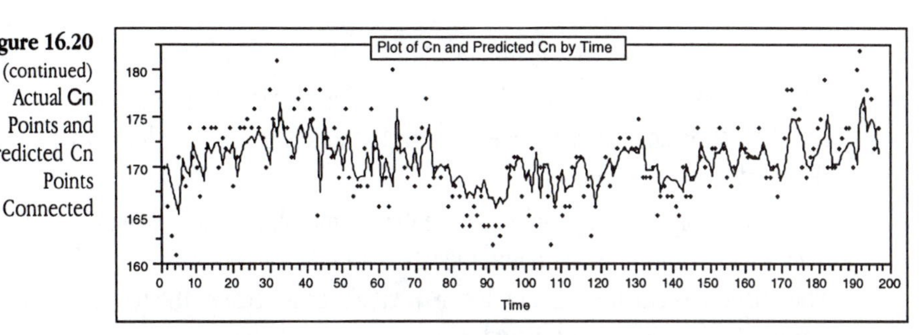

To predict future values, you need to extend the data table. But using the current formula, this will only predict one step into the future because the lagged values become missing past the end of the series. However, you can change the formula to substitute forecast values when the actual values are missing. This will be done in the next section, using a different mechanism for estimation.

Special Topic: Fitting an AR Model in the Nonlinear Platform

The Nonlinear platform can fit formulas directly. For time series models, you use formulas with lagging expressions.

Entering formulas is tedious, so it is more convenient to generate data using a table that has the formulas already built. This kind of table is called a table template. It is a table stored with its columns but no rows. The columns have formulas so when you add rows, values are automatically computed. There is a sample table template that has built-in formulas to fit a simple autoregressive model.

§ Open the FITAR1.JMP template to see the variables **Time**, **W**, and **AR1 Pred**.

§ The model is stored as a formula in the column **AR1 Pred**. OPTION-click on the heading for **AR1 Pred** column to see the model formula:

$$\mathbf{Mu}+\mathbf{Phi}\cdot\begin{cases} W_{i-1}-\mathbf{Mu}, & \text{if } W_{i-1}\neq\bullet \\ AR1\ Pred_{i-1}-\mathbf{Mu}, & \text{otherwise} \end{cases}$$

The formula uses a conditional clause that tests if the data is missing. When the lagged data is missing, the computation uses the predicted values instead of the actual (missing) values, to extend a forecast over missing data.

The terms **Mu** and **Phi** in the model are the mean and the autoregressive parameters to be estimated.

Next, copy the **Cn** values from the original table to the template table:

§ Go back to SERIESA.JMP and highlight the **Cn** column (click at the top of the column). Then choose **Edit→Copy** to copy the values of the **Cn** column to the clipboard.

§ Go to the FITAR1 table and highlight the **W** column. Choose **Edit→Paste** to paste the values from the clipboard into the **W** column. The FITAR1 table fills with data. Note that the prediction column for **AR1** has all zeros because the parameters **Mu** and **Phi** do not yet have estimated values.

§ Now choose **Analyze→Nonlinear Fit**. Pick W as the Y and AR1 Pred as the X. Then click **OK**. When the nonlinear fitting platform appears click the **Go** button to start the estimation.

When the nonlinear fitting process converges, you will see the Solution table shown in **Figure 16.21.**

Figure 16.21
Solution Table for Nonlinear Fit

Solution

SSE	DFE	MSE	RMSE
2104.4212857	194	10.847532	3.2935592

Parameter	Estimate	ApproxStdErr	Lower CL	Upper CL
Mu	170.65485515	0.55004084	•	•
Phi	0.5722661404	0.0590328	•	•

Apart from a different (better) definition of the intercept, the new results are the same as with the simple regression platform. But there is a prediction formula to forecast into the future. Furthermore, there was no need to generate the lagged variable because the formula does that work. The parameter estimates are automatically stored back into the formula for the model to give the predicted values. Now you are ready to forecast values:

§ Choose **Rows→Add Rows**, and click **OK** to add 20 rows to the series. The **Untitled** table now has 217 rows, and there are predictions for the twenty steps ahead.

§ Choose **Graph→Overlay Plots** and use Time as X and both W and AR1 Pred as Ys. The Overlay plot in **Figure 16.22** shows predictions for the twenty steps ahead. Past the end of the data, the predicted values gradually damp down to 170.65, which is the mean of the series.

Figure 16.22 Overlay Plot Showing Predicted Values

Untitled 2

3 Cols / 217 Rows	Time	W	AR1 Pred
196	196	172	174.286
197	197	174	171.4246
198	198	•	172.5692
199	199	•	171.7504
200	200	•	171.2818
201	201	•	171.0136
202	202	•	170.8602
203	203	•	170.7723
204	204	•	170.7221

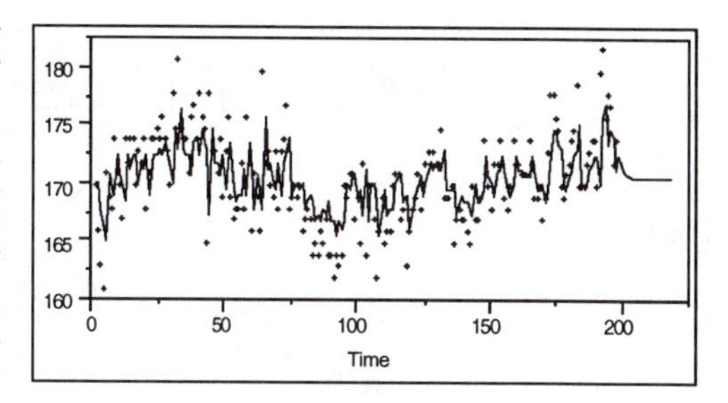

Special Topic: Moving Average and ARMA Models

The *moving average* model fits the series with a linear combination of lagged noise values (instead of a linear combination of lagged series values). It turns out that a first order moving average model is equivalent to a long autoregressive model where the coefficients are declining exponentially (the parameter to the power of the lag).

Autoregressive Model: $z_t = \phi z_{t-1} + e_t$

Moving Average Model: $z_t = e_t - \theta e_{t-1}$

There are no built-in ways of estimating moving-average models with JMP. Even the nonlinear platform does not handle them gracefully, because they need certain recursive derivatives that are not currently supported. However, you can get reasonably good estimates using the following scheme.

First you want to estimate the error term in the moving average process. Because a moving average can be approximated by a high-order autoregressive process, fit an high-order autoregressive process and then obtain its residual. For example, for Series A, you could use the eight lag columns in SERIESAL.JMP to perform the regression, regressing **Cn** against **Cn1** to **Cn8**. Then save the residuals and create a column of lagged residuals as shown in **Figure 16.23**. Substitute zeroes for missing values at the beginning of the series. Then make a column with lagged values for the residual.

Figure 16.23 Table with Saved Residual Columns from Regression on Lagged Variables

SERIESAL.JMP

11 Cols, 197 Rows

		Cn	Cn1	Cn2	Cn3	Cn4	Cn5	Cn6	Cn7	Cn8	Residual Cn	Lag Resid
+	1	170	•	•	•	•	•	•	•	•	•	•
+	2	166	170	•	•	•	•	•	•	•	•	•
+	3	163	166	170	•	•	•	•	•	•	•	•
+	4	161	163	166	170	•	•	•	•	•	•	•
+	5	171	161	163	166	170	•	•	•	•	•	•
+	6	169	171	161	163	166	170	•	•	•	•	•
+	7	168	169	171	161	163	166	170	•	•	•	•
+	8	174	168	169	171	161	163	166	170	•	•	•
+	9	171	174	168	169	171	161	163	166	170	1.07282	•
+	10	170	171	174	168	169	171	161	163	166	0.734974	1.07282

Now regress **Cn** against the lagged residual. The result is an estimate of .35 for the moving average parameter (see **Figure 16.24**).

Figure 16.24 Estimate for Moving Average Parameter

Parameter Estimates

Term	Estimate	Std Error	t Ratio	Prob>|t|
Intercept	170.62882	0.276981	616.03	0.0000
Lag Resid	0.3503742	0.09451	3.71	0.0003

A combination of autoregressive and moving average models is called an ARMA model. If it is applied to a lag-differenced series, it is called an ARIMA model, where the "I" stands for *Integrated*.

Special Topic: Simulating Time Series Processes

If you would like to try generating data from various ARMA models, you can use table templates included in the sample data. As with previous simulations, add rows to the table template to generate the data.

The table in the file TIMEAR1.JMP has a column, **AR1**, that uses the formula shown below for simulating a first order autoregressive process.

$$\begin{array}{l} p1 \Leftarrow 0.8 \\ \text{results} \begin{cases} ?\text{normal} \cdot \sqrt{\dfrac{1}{1-p1^2}}, & \text{if } i=1 \\ p1 \cdot AR1_{i-1} + ?\text{normal}, & \text{otherwise} \end{cases} \end{array}$$

The first clause in the formula gives the name *p1* to the autoregressive parameter. For the first observation, the data is normal, but needs an adjustment to have the right variance. From then on, it is a combination of lagged values and new noise.

The table in the file TIMEMA1.JMP has a column, MA1, with a formula for simulating a first order moving average process, using the formula shown here:

$$t1 \Leftarrow 0.8$$

$$\text{results}\begin{cases} Noise \cdot \sqrt{t1^2+1}, & \text{if } i=1 \\ t1 \cdot Noise_{i-1} + Noise, & \text{otherwise} \end{cases}$$

The moving average parameter is named *t1*. A second column called Noise is needed to generate the normal random error.

TIMEARMA provides a simulation of a combined autoregressive moving average (ARMA) process. The ARMA column uses the formula shown here, where *t1* is the moving average parameter and *p1* is the autoregressive parameter.

$$t1 \Leftarrow 0.6$$

$$p1 \Leftarrow 0.3$$

$$\text{results}\begin{cases} Noise \cdot \sqrt{\dfrac{(1+t1^2-2 \cdot t1 \cdot p1)}{(1-p1^2)}}, & \text{if } i=1 \\ p1 \cdot ARMA_{i-1} + t1 \cdot Noise_{i-1} + Noise, & \text{otherwise} \end{cases}$$

Autoregressive Errors in Regression Models

It is common when doing regression fitting with time series data to discover that the residuals in the model are autocorrelated.

In the Standard Least Squares version of Fit Model, a Durbin-Watson test is provided to help test for the presence of autocorrelation. Most econometrics textbooks have tables of Durbin-Watson significance values.(The professional version of JMP provides a time-intensive, but exact p value for the Durbin-Watson).

If you do find significant autocorrelation, here are several ways to regard the situation.

Diagnostic

In some situations, autocorrelation is a symptom that you have an inadequate regression model. For example, if you fit the wrong model to almost any function of time, you will get autocorrelation.

Problem Solving

There are a number of ways to adjust for autocorrelation, most of them beyond the scope of the JMP product.

Significance

In research, the problem is often to make some test significant. If you can drive down the error variance by introducing autoregressive terms in the model, this is a big advantage in making estimates more significant.

Chapter 17
Machines of Fit

Overview

This chapter is an essay on fitting for those of you who are mechanically inclined. If you have any talent for imagining how springs and tire pumps work, you can put it to work here in a fantasy in which all the statistical methods are visualized in simple mechanical terms.

The goal is to not only remember how statistics works, but also train your intuition so you will be prepared for new statistical issues.

Chapter 17 Contents

Machines of Fit

Here is an illuminating trick that will help you understand and remember how statistical fits really work. It involves pretending that statistical fitting is performed by machines. If we can figure out the right machines and visualize how they behave, we can reconstruct all of statistics by putting together these simple machines into arrangements appropriate to the situation. We need only two machines of fit, the spring for fitting continuous normal responses and the pressure cylinder for fitting categorical responses.

Springs for Continuous Responses

How does a spring behave? As you stretch the spring, the tension increases linearly with distance. The energy that you need to pull a spring a given distance is the integral of the force over the distance, which is proportional to the square of the distance. Take $1/\sigma^2$ as the measure of how stiff the spring is. Then the graph and equations for the spring are as shown in **Figure 17.1**.

Figure 17.1 Behavior of Springs

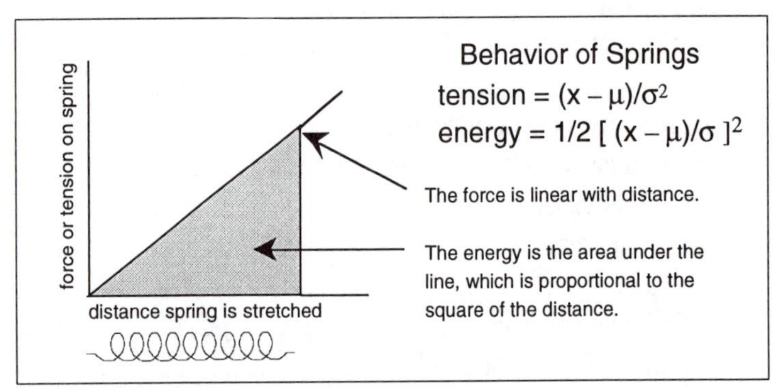
Behavior of Springs
tension = (x – μ)/σ²
energy = 1/2 [(x – μ)/σ]²
The force is linear with distance.
The energy is the area under the line, which is proportional to the square of the distance.
force or tension on spring
distance spring is stretched

In this way springs will help us visualize least squares fits. They also help us do maximum likelihood fits when the response has a normal distribution.

The formula for the log of the density of a normal distribution is identical to the formula for energy of a spring centered at the mean, with a spring constant equal to the reciprocal of the variance. A spring stores and yields energy in exactly the way that normal deviations get and give log-likelihood. So, maximum likelihood is equivalent to least squares, which is equivalent to minimizing energy in springs.

Fitting a Mean

How do you fit a mean by least squares? Imagine stretching springs between the data points and the line of fit (see **Figure 17.2**) Then you move the line of fit around until the forces acting on it from the springs balance out. That will be the point of minimum energy in the springs. For every minimization problem, there is an equivalent balancing (or orthogonality) problem, in which the forces (tensions, relative distances, residuals) add up to zero.

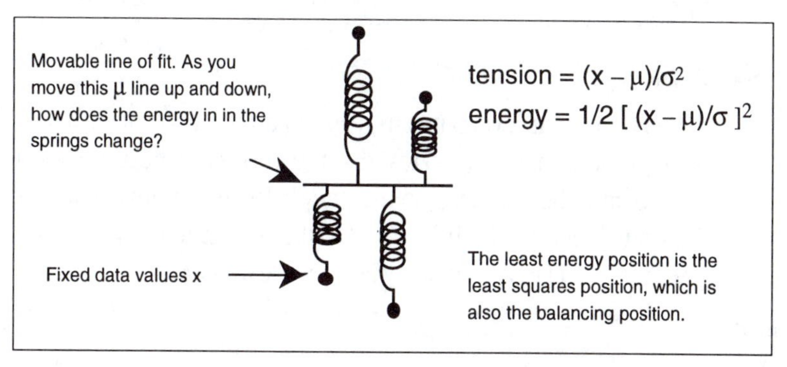

Movable line of fit. As you move this μ line up and down, how does the energy in in the springs change?
tension = (x – μ)/σ2
energy = 1/2 [(x – μ)/σ]2
Fixed data values x
The least energy position is the least squares position, which is also the balancing position.

Figure 17.2 Fitting a Mean by Springs

Testing a Hypothesis

If you want to test a hypothesis that the mean is some value, you force the line of fit to be that value and measure how much more energy you had to add to the springs (how much more the sum of squared residuals was) to constrain the line of fit. This is the sum of squares that is the main ingredient of the F test. To test that the mean is (not) the same as a given value, find out how hard it is to move it there (see **Figure 17.3**).

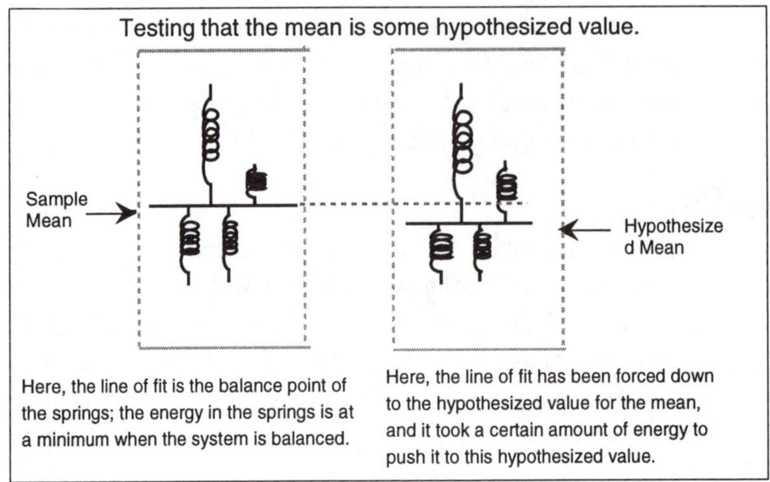

Testing that the mean is some hypothesized value.
Sample Mean
Hypothesize d Mean
Here, the line of fit is the balance point of the springs; the energy in the springs is at a minimum when the system is balanced.
Here, the line of fit has been forced down to the hypothesized value for the mean, and it took a certain amount of energy to push it to this hypothesized value.

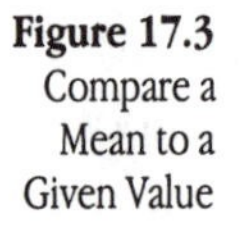
Figure 17.3 Compare a Mean to a Given Value

One-Way Layout

If you want to fit several means, you can do so by balancing a line of fit with springs for each group. To test that the means are the same, you force the lines of fit to be the same, so that they balance as a singlc line, and measure how much energy you had to add to the springs to do this (how much greater the sum of squared residuals was). See **Figure 17.4.**

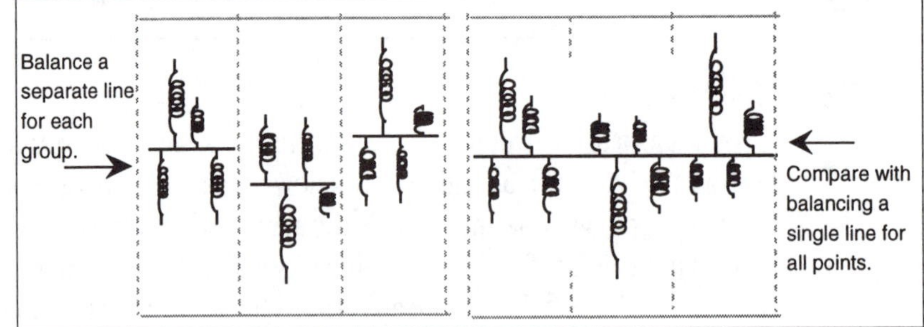

Compare with balancing a single line for all points.

Figure 17.4 Means and the One-Way Analysis of Variance

Sample Size's Effect on Significance

When you have a larger sample, there are more springs holding on to each mean estimate, and it is harder to pull them together. Larger samples lead to a greater energy expense (sum of squares) to test that the means are equal. The spring example in **Figure 17.5** show how sample size affects the sensitivity of the hypothesis test.

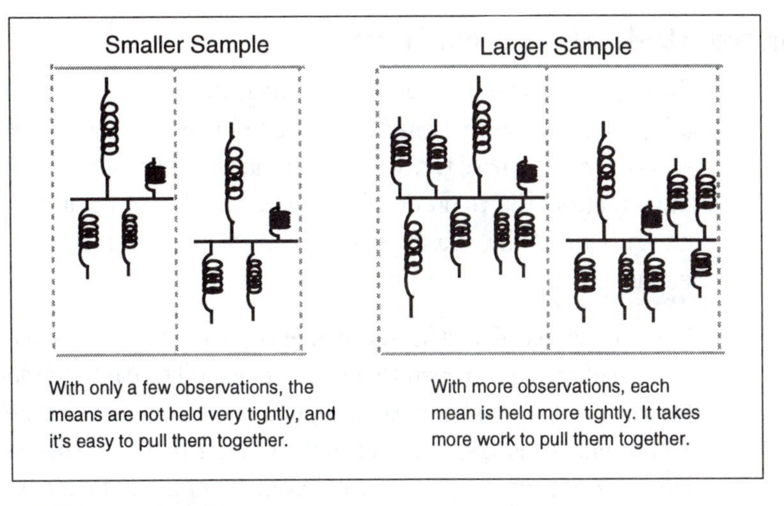

Smaller Sample
Larger Sample
With only a few observations, the means are not held very tightly, and it's easy to pull them together.
With more observations, each mean is held more tightly. It takes more work to pull them together.

Figure 17.5 Taking a Larger Sample Helps Make Hypothesis Tests More Sensitive

Error Variance's Effect on Significance

The spring constant is the reciprocal of the variance. Thus if the residual error variance is small, the spring constant is bigger, the springs are stronger, it takes more energy to bring the means together, and the test is thus more significant. The springs in **Figure 17.6** illustrate the effect of variance size.

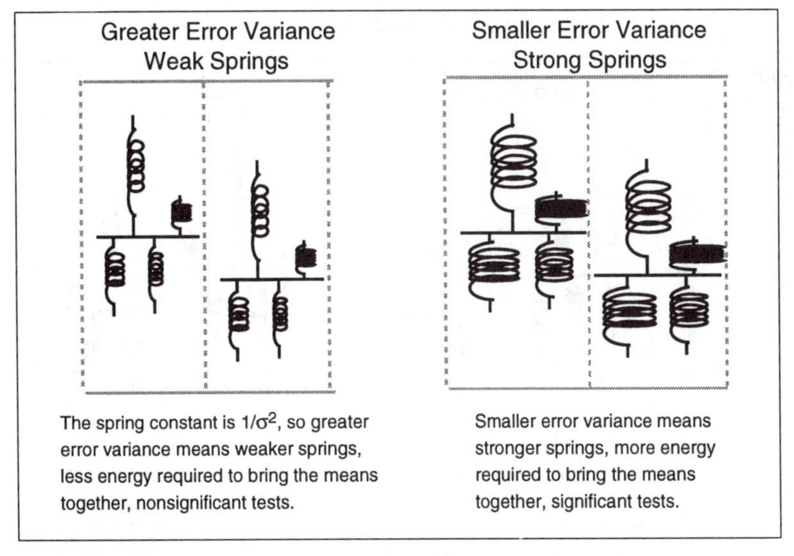

Greater Error Variance
Weak Springs
Smaller Error Variance
Strong Springs
The spring constant is $1/\sigma^2$, so greater error variance means weaker springs, less energy required to bring the means together, nonsignificant tests.
Smaller error variance means stronger springs, more energy required to bring the means together, significant tests.

Figure 17.6 Reducing the Residual Error Variance Helps Make Hypothesis Tests More Sensitive.

Experimental Design's Effect on Significance

If you have two groups, how do you arrange the points between the two groups to maximize the sensitivity of the test that the means are equal? Suppose that you have two sets of points loading two lines of fit, as in the one-way layout shown previously in **Figure 17.4**. The test that the true means are equal is done by measuring how much energy it takes to force the two lines together.

Suppose that one line of fit is suspended by a lot more points that the other. The line of fit that is suspended by few points will be easily movable and can be stretched to the other mean without much energy expenditure. The lines of fit would be more strongly separated if you had more points on this loosely sprung side, even at the expense of having fewer points on the more tightly sprung side. It turns out that to maximize the sensitivity of the test for a given number of observations, it is best to allocate points in equal numbers between the two groups. In this way both means are equally tight, and the effort to bring the two lines of fit together is maximized.

So the power of the test is maximized in a statistical sense by a balanced design, as illustrated in **Figure 17.7**.

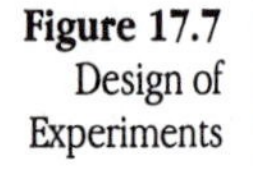

Figure 17.7 Design of Experiments

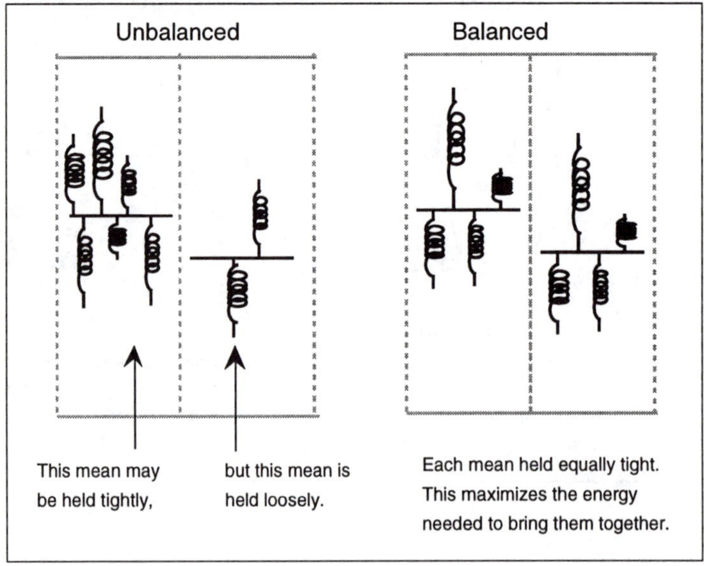
Unbalanced
Balanced
This mean may be held tightly,
but this mean is held loosely.
Each mean held equally tight. This maximizes the energy needed to bring them together.

Simple Regression

If you want to fit a regression line through a set of points, you fasten springs between the data points and the line of fit, such that the springs stay vertical. Then let the line be free so that the forces of the springs on the line balance, both vertically and rotationally (see **Figure 17.8**). That is the least-squares regression fit.

Figure 17.8 Fitting a Regression Line with Springs

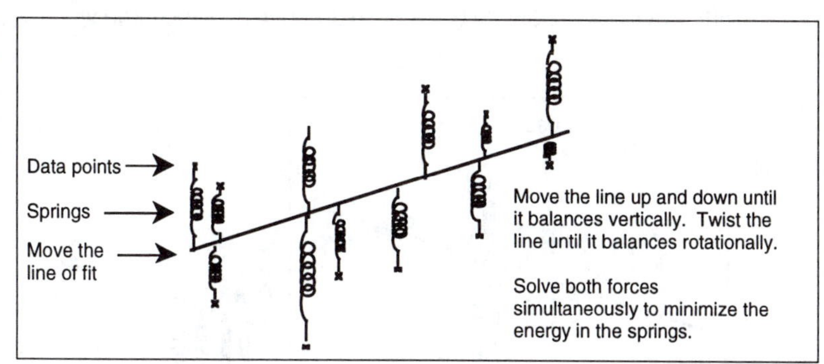

If you want to test that the slope is zero, you force the line to be horizontal so that you're just fitting a mean and measure how much energy it took to constrain the line (the sum of squares due to regression). (See **Figure 17.9**)

Figure 17.9 Testing the Slope Parameter for the Regression Line

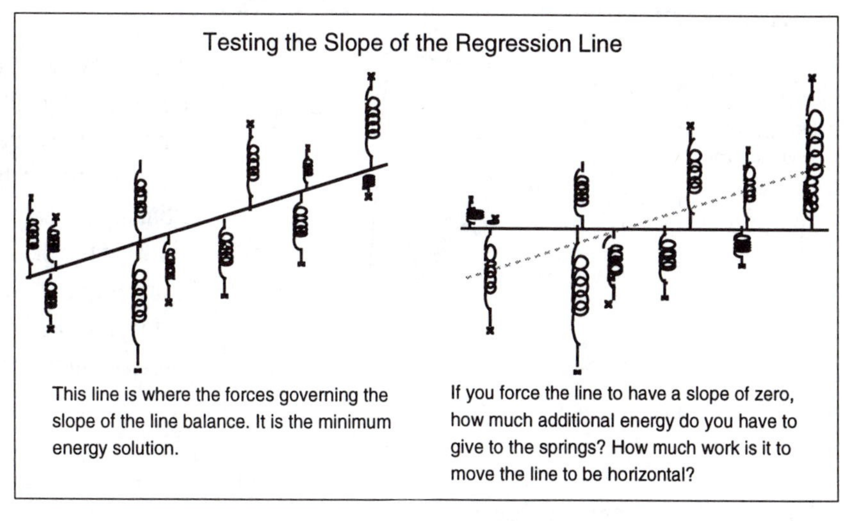

Testing the Slope of the Regression Line
This line is where the forces governing the slope of the line balance. It is the minimum energy solution.
If you force the line to have a slope of zero, how much additional energy do you have to give to the springs? How much work is it to move the line to be horizontal?

Leverage

If most of the points that are suspending the line are near the middle, then the line can be rotated without much effort to change the slope within a given energy budget. If most of the points are near the end, the slope of the line of fit is pinned down with greatest resistance to force. That is the idea of leverage in a regression model. Imagine trying to twist the line to have a different slope. Look at **Figure 17.10** and decide which line would be easier to twist?

Figure 17.10 Leverage with Springs

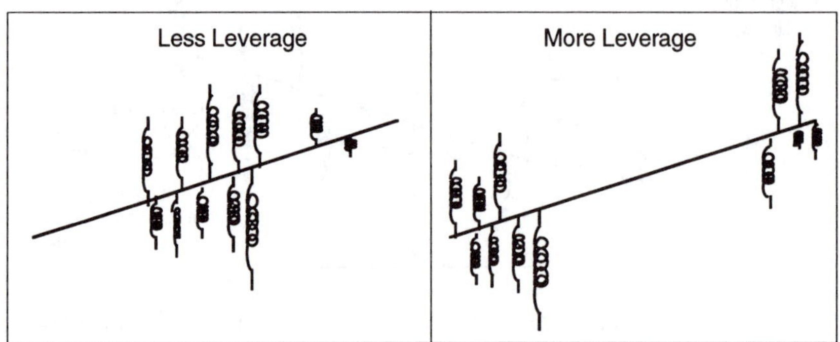

Multiple Regression

Figure 17.11 Three-Dimensional Plot of Two Regressors and Fitted Plane

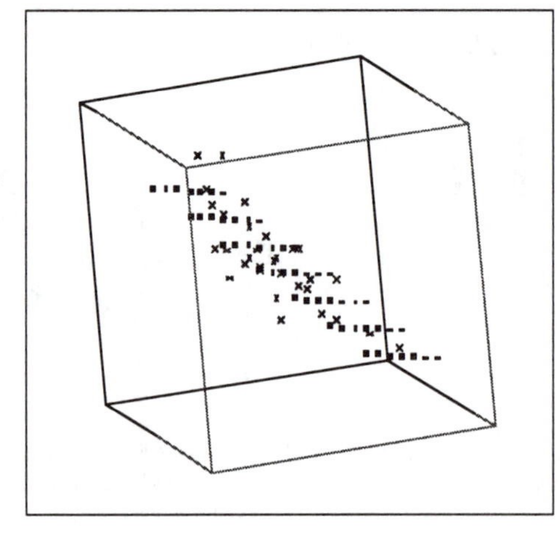

The same idea works for fitting a response to two regressors; the difference is that the springs are attached to a plane rather than a line. Estimation is done by adjusting the plane so that it balances in each way. Testing is done by constraining the plane.

Summary: Significance and Power

To get a stronger (more significant) fit, in which the line of fit is suspended more tightly, you must either have stiffer springs (have smaller variance in error), or use more data (have more points to hang springs from), or move your points farther out on both ends of the X-axis (more leverage). The power of a test is how likely it is that you will be unable to move the line of fit given a certain energy budget (sum of squares) determined by the significance level.

Machine of Fit for Categorical Responses

Just as springs are analogous to least-squares fits, gas pressure cylinders are analogous to maximum likelihood fits for categorical responses (see **Figure 17.12**.)

How Do Pressure Cylinders Behave?

Using Boyle's law of gases, that pressure times volume is a constant, the pressure in a gas cylinder is proportional to the reciprocal of the distance from the bottom of the cylinder to the piston. The energy is the force integrated over the distance (starting from a distance, p, of 1), which turns out to be –log(p).

Figure 17.12 Gas Pressure Cylinders to Equate –log(probability) to Energy

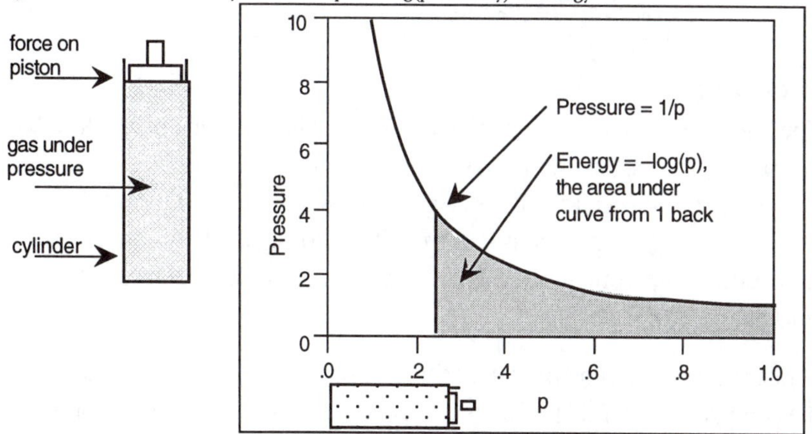

Now that you know how pressure cylinders work, start thinking of the distance from the bottom of the cylinder to the piston as the probability that some statistical model attributes to some response. The height of 1 will mean

no stored energy, no surprise, a probability of 1. The height of zero will mean infinite stored energy, an impossibility, a probability of zero.

When stretching springs, we measured energy by how much work it took to pull a spring, which turned out to the be square of the distance. Now we measure energy by how much work it takes to push a piston from distance 1 to distance p, which turns out to be –log(p), the logarithm of the probability. We used the logarithm of the probability before in categorical problems when we were doing maximum likelihood. The maximum likelihood method estimates the response probabilities so as to minimize the sum of the negative logarithms of the probability attributed to the responses that actually occurred. This is the same as minimizing the energy in gas pressure cylinders, as illustrated in **Figure 17.13**.

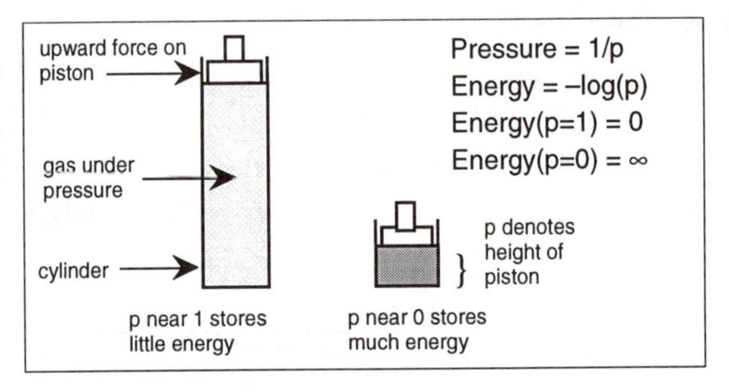

Figure 17.13 Gas Pressure Cylinders Equate –log(probability) to energy

Estimating Probabilities

Now we want to estimate the probabilities by minimizing the energy stored in pressure cylinders. First we need to build a partitioned frame with a compartment for each response category and add the constraint that the sum of the heights of the partitions is 1. We will be moving the partitions around so that the compartments for each response category can get bigger or smaller (see **Figure 17.13**).

For each observation on the categorical response, put a pressure cylinder into the compartment for that response. After you have all the pressure cylinders in the compartments, start moving the partitions around until the forces acting on the partitions balance out. This will be the solution to minimize the energy stored in the cylinders. It turns out that the solution for the minimization is to make the partition sizes proportional to the number of pressure cylinders in each compartment.

For example, suppose you did a survey in which you asked 13 people what brand of car they preferred, and 5 chose Asian, 2 chose European, and 6 chose American brands. Then you would stuff the pressure cylinders into the frame as in **Figure 17.14**, and the partition sizes that would balance the forces would work out to 5/13, 2/13, and 6/13, which sums to 1.

To test that the true probabilities are some specific values, you move the partitions to those values and measure how much energy you had to add to the cylinders.

Figure 17.14 Gas Pressure Cylinders to Estimate Probabilities for a Categorical Response

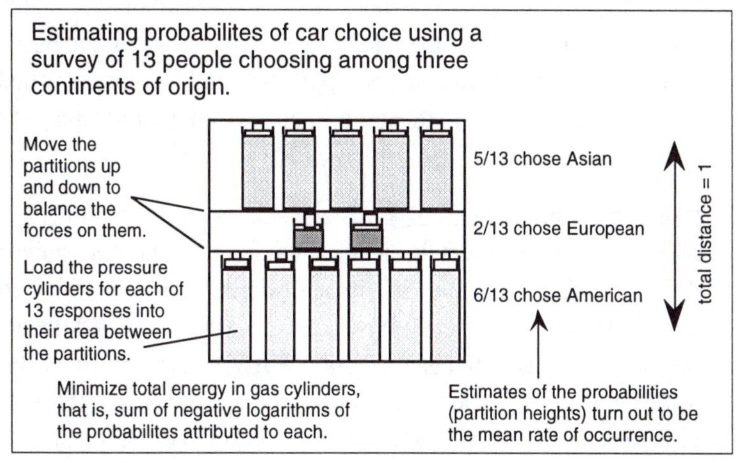

One-Way Layout for Categorical Data

If you have different groups, you can fit a different response probability to each group. The forces acting on the partitions balance independently for each group. The mosaic plot shown in **Figure 17.15** helps maintain the visualization of pressure compartments. As an alternative to pressure cylinders, you can visualize with free gas in each cell.

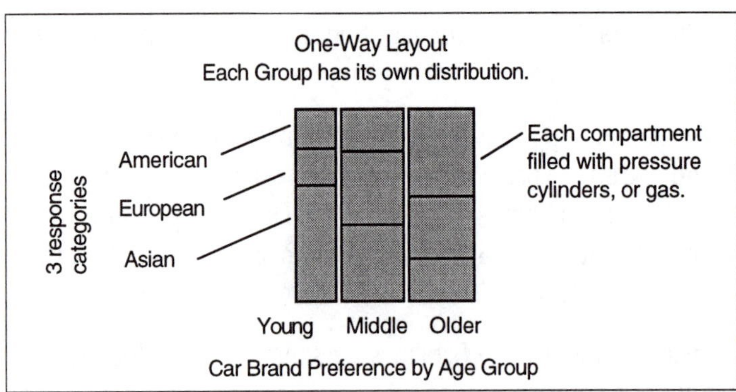

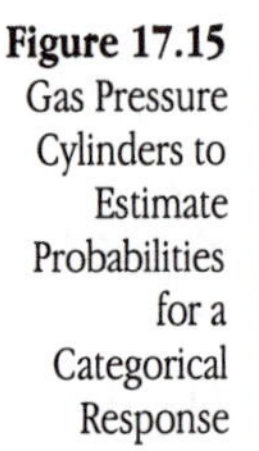

Figure 17.15 Gas Pressure Cylinders to Estimate Probabilities for a Categorical Response

How do you test the hypothesis that the true rates are the same in each group and that the observed differences can be attributed to random variation? You just move the partitions so that they are in the position corresponding to the ungrouped population and measure how much more energy you had to add to the gas-pressure system to force the partitions to be in the same positions.

Figure 17.16 shows the three kinds of results you can encounter, corresponding to perfect fit, significant difference, and nonsignificant difference. To the observer, the issue is whether knowing which group you are in will tell you which response level you will have. When the fit is near perfect, you know with near certainty. When the fit is intermediate, you have more information if you know the group you are in. When the fit is inconsequential, knowing which group you are in doesn't matter. To a statistician, though, what is interesting is how firmly the partitions are held by the gases, how much energy it would take to move the partitions, and what consequences would result from removing boundaries between samples and treating it as one big sample.

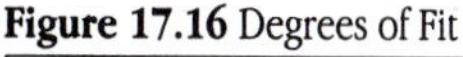

Figure 17.16 Degrees of Fit

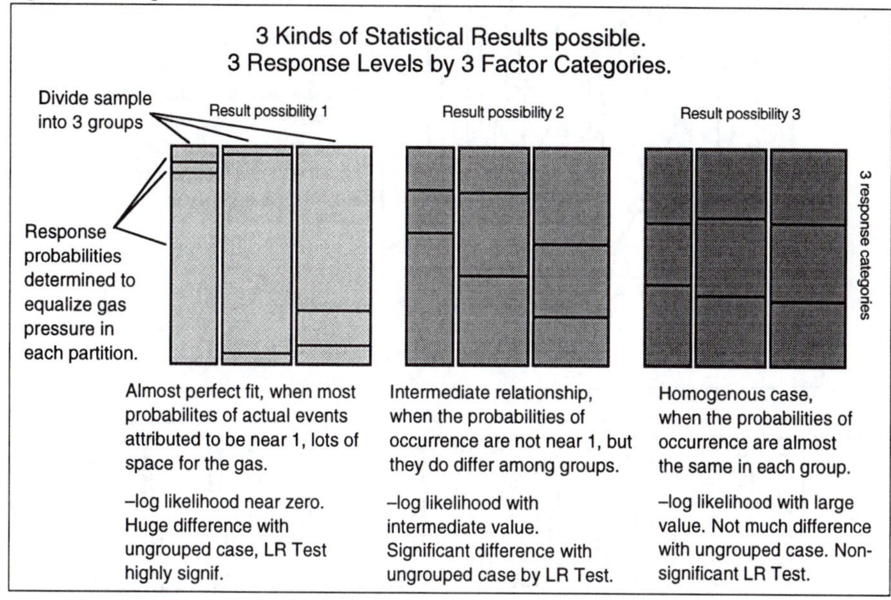

Logistic Regression

Logistic regression is the fitting of probabilities over a categorical response to a continuous regressor. Logistic regression can also be visualized with pressure cylinders (see **Figure 17.17**). The difference with contingency tables is that the partitions change the probability as a continuous function of the X axis. The distance between lines is the probability for one of the responses. The distances sum to a probability of 1. **Figure 17.18** shows what weak and strong relationships look like.

Figure 17.17 Logistic Regression as the Balance of Cylinder Forces

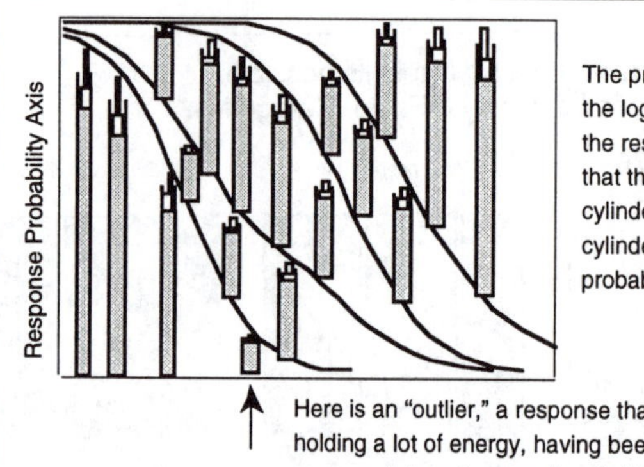

The pressure cylinders push on the logistic curves that partition the responses, moving them so that the total energy in the cylinders is minimized. The more cylinders in an area, the more probability room it pushes for.

Here is an "outlier," a response that is holding a lot of energy, having been squeezed down to a small probability.

Figure 17.18 Strengths of Logistic Relationships

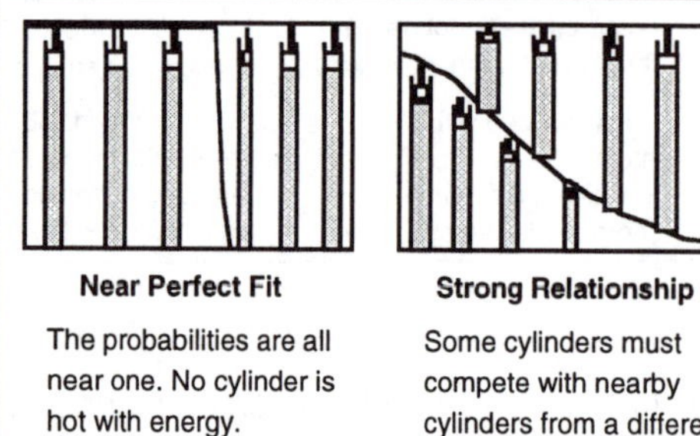

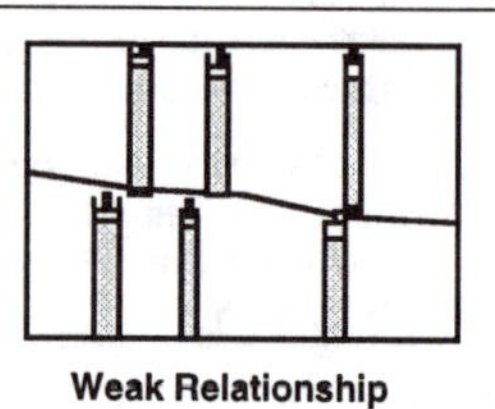

Near Perfect Fit

The probabilities are all near one. No cylinder is hot with energy.

Strong Relationship

Some cylinders must compete with nearby cylinders from a different response. More energy in some cylinders.

Weak Relationship

The probabilities are squeezed to near the horizontal case of homogeneity. All cylinders are warm with energy.

Appendix A
The JMP Main Menu

Overview

JMP Main Menu

File **Edit** **Tables** **Rows** **Cols** **Analyze** **Graph** **Tools** **Window** **Help**

When you start JMP IN you see an empty JMP data table and the main menu shown above, with menus that perform the following functions:

- The **File** menu has commands that perform file management or affect the JMP environment. **File** commands open data files and create new data tables, close windows, close and save data tables, retrieve a saved data table, set global options, print JMP windows, and end a JMP session.
- The **Edit** menu contains standard commands such as cut, copy, clear, and paste. Other commands such as **Paste at End, Save as Test, Journal,** and **Search** perform specialized JMP actions.
- The **Tables** menu commands modify or create a new JMP table from one or more existing tables. You can summarize, subset, join tables, append tables, and split and stack columns.
- The **Rows** menu commands add new rows, move existing rows, find specific rows, delete unwanted rows, and assign special characteristics to selected rows.
- The **Cols** menu commands let you assign or clear a column's analysis role, create or insert new columns, access the JMP calculator to compute column values, rearrange the order of columns, hide columns temporarily, change column characteristics, and delete unwanted columns.
- The **Analyze** and **Graph** means are the heart and soul of JMP. You assign characteristics and analysis roles to columns of data and select an analysis or graph platform. Each platform produces a window that shows statistical text reports with supporting graphical displays.
- The **Tools** menu is a palette of special tools that determine the effect of mouse actions. The default tool is the arrow. Other tools cut or copy all or part of a report, access help windows, manipulate graphical displays, and highlight points in a plot. These tools function in report windows but not on the spreadsheet.
- The **Window** menu helps you organize the windows produced during a JMP session.
- The Windows version of JMP has the standard Windows **Help** menu. Windows users will find the menu items similar to those in other applications.

This appendix contains documentation for all the commands in the JMP menu bar. It is organized for easy reference; The menus and commands are presented in their order on the menu bar. Each menu title shows beside the page header to help you locate specific commands quickly.

Appendix A Contents

The File Menu

The **File** menu has commands that perform file management or affect the JMP environment.

File→New

The **New** command creates and displays a new empty untitled data table with one column labeled Column 1. To create more columns use **Row s** and **Cols** menu commands and type or paste data into the spreadsheet. Use **Save As** from the **File** menu to name the table and save it as a JMP file. Use the **Open** command to reopen the file after it has been saved.

File→Open

The **Open** command displays the standard open file dialogs. You can open an existing JMP file using this command, or you can open a file by double clicking it. When you open a file, you have access to the spreadsheet view of the data table.

File→Import

The **Import** command lets you import raw data and create a new JMP data table by reading data from a text-format file or a SAS transport file, which are options on the Import command submenu. The **Import** command popup selections display modified open file dialogs appropriate for the selection you choose from the Import dialog.

The Text/Others submenu command displays a dialog that lets you request a new JMP data table by reading data from a text-format file, or files for which there is an XTND text translator. JMP provides the EXCEL 4 and EXCEL 5 translators.

The Text/Others submenu dialog has the following items:

Table A.1 Items in the Import Text/Others Dialog

Data Only	By default, the Text/Others option in the **Import** command reads a text file of rows and columns without header information (Data Only). The default names of the JMP data table columns are Column 1, Column 2, and so on.
Labels	tells JMP that the incoming data are from a text-format file that contain column labels as the first record of the data. These labels become column names in the new JMP data table.
Header	tells JMP that the incoming data file contains standard header information for a JMP data table.
End of Field	designates the character that identifies the end of each field in the incoming file. This delimiter can be a tab, one or more blank spaces, any character, or any combination of these. When you choose **Other**, an editable text box appears where you can enter a character or its hexadecimal representation.
End of Line	selects the characters or keys that identifies the end of each line. You can use the return key, the line feed character, a semicolon, or another character of your choice. The **Other** option works the same as described above.
Strip Enclosing Quotes	tells JMP that each data value is enclosed in single or double quotes. JMP removes the quotes before importing the data.

Note: To enter a hexadecimal value, precede the value with '0x'.

Note: If you have the ClearAccess query tool on your Macintosh, three additional **ClearAccess** commands also show in the **Import** command submenu.

File→Close

The **Close** command closes the active window. When you close a spreadsheet window while any of its analysis windows are open, a dialog asks if you want to close the data table only or all related analysis windows as well. If you close the spreadsheet, a dialog asks you whether to save or discard changes to the table. The **Close** command is the same as clicking the close box of the active window.

File→Save

The **Save** command writes the active data table to a file. If the data table has been saved before, it is rewritten to the same file name and location, replacing the old information. If the data table is new, the **Save** command has the same effect as the **Save As** command. Saving a data table does not automatically close it.

Note: JMP report windows are not saved with the data table. However, you can use JMP tools to copy reports to other applications, or you can use the journaling feature to save reports in text or word processing format.

File→Save As

The **Save As** command writes the active data table to a file after prompting you for a name and disk location. It saves the data table as a JMP file or converts it to another format if you click a different radio button in the dialog.

The Save As dialog has these options:

Table A.2 Items in the **Save As** Dialog

JMP format	saves the JMP data table in JMP format.
Pre-Version 3 compatible format	saves the JMP table in a form that previous releases of JMP can read. If this option is not checked, the saved table is accessible only by JMP Version 3.
SAS Transport	converts a JMP data table to SAS transport file format and saves it in a SAS transport library. **Ann additional option,** Append To, appends the active data table to an existing SAS transport library. If you don't use the **Append To** option, a new SAS transport library location you give. Also, if you do not specify a new file name, the SAS transport library replaces the existing JMP data table.
Text Format	converts data from a JMP file to standard text format, maintaining the rows and columns. For both the Macintosh and Windows the Text Format option displays a dialog with choices described previously for the **Import** command in Table A.1.

File→Revert

The **Revert** command restores the data table to the form it had when last saved. If it was last saved as a SAS Transport file or a text format file, the **Revert** command is not available. In this case, use the **Import** command to restore the data table.

File→Preferences

The **Preferences** command displays the dialog that lets you set and save a tailored JMP environment.

You can check any combination or all of the following preferences:

- The background for all JMP graphical displays except the Spinning Plot platform can be either white or black. You change them with the **Background** icons. The Spinning platform has an independent background setting.
- On the Macintosh the **Font** and **Size** popup menus let you specify any font, font style, and font size available in your system for the text, headings, and titles of JMP tables and report windows. Under Windows, the **Font** button lets you select font, font style, and font size, and shows sample text of these selections.
- On the Macintosh, the first three preference check boxes display memory information in the lower-left corner of your desktop.
- Under Windows, you can request a tool bar that displays beneath the main menu bar. It has icons for the most commonly used **File** menu commands, and icons for all **Analyze**, **Graph**, and **Tools** menu commands
- Under Windows, you can request a status bar that shows memory information, a brief sentence of help, and the current time across the bottom of your JMP window.
- **Thin Line for Postscript printing** reduces the lines used in JMP reports from 1 point to .5 points. The line thickness reduction does not show on your screen, but appears when you print JMP results.
- **Square box for reveal titles** changes reveal/conceal buttons and the boundary frames of graphical displays and text reports from round rectangles to rectangles.
- **Extra large markers** increases the size of all markers in all plots.
- **Prompt to journal window on close** displays a dialog that prompts you to journal your report whenever you close an analysis report window.
- **Always journal window on close** automatically journals a report window when you close it.
- **Include popup controls in copy/print** copies to the clipboard or prints popup menu icons with reports. By default, the **Copy** and **Print** commands do not show these icons.

Note: The **Analyze** button on the Preferences dialog displays the **Analyze** menu commands in a popup menu. When you select an **Analyze** command, a dialog lists that platform's text reports and graphical displays. To tailor the initial display of an analysis platform, select the options you want, and then click **Save**. Then, whenever you use that platform, the window components you saved as preferences automatically display.

File→Page Setup...

The **Page Setup** command displays the standard dialog for setting printed page characteristics. The form of the dialog depends on your current printer driver.

File→Print

The **Print** command prints the active window. It displays the standard dialog for printing. The appearance of this dialog depends on your printer driver.

File→Quit

On the Macintosh, the **Quit** command closes all JMP windows (prompting you to save changes) and quits the JMP application.

File→Exit

Under Windows, the **Exit** command closes all JMP windows (prompting you to save changes) and quits the JMP application. Also, following the **Exit** command, the **File** menu lists the four JMP tables most recently opened. When you click on a table's name in this list, JMP re-opens the table if it is closed or gives you a new spreadsheet view if it is currently open.

The Edit Menu

The **Edit** menu contains standard commands including **Cut** and **Copy.** These commands operate on selected rows and columns, on selected areas of reports, and on selected formula elements in the calculator.

In the data table spreadsheet, you can select rows, columns, or both rows and columns at the same time. **Edit** commands operate on entire rows if no columns are selected. Likewise, they operate on whole columns if no rows are selected. If you select both rows and columns, **Edit** commands affect the subset of values defined by the intersection of those rows and columns.

To select a row in a spreadsheet, click the space that contains the row number. To select a column, click the white background area. To select multiple rows or columns, drag across them or SHIFT-click the first and last rows or columns of a range. To make a discontiguous selection use COMMAND-click on the Macintosh or CONTROL-click under Windows to highlight the rows and columns you want. To select a block of cells formed by the intersection of rows and columns, drag the cross cursor diagonally across the subset of table cells.

Note: You can also select special subsets of rows using the **Select** command in the **Rows** menu, described later in this appendix.

You can also use the **Edit** menu in conjunction with JMP tools to copy all or part of active analysis report windows. See the **Tools** menu section for more information.

Edit→Undo

The **Undo** command cancels the effect of the most recent reversible **Edit**, **Rows**, or **Cols** command. If **Undo** is available, its selection in the **Edit** menu appears as **Undo** ***command*** where ***command*** is the most recent action. Most destructive spreadsheet operations (such as cutting, pasting, or deleting rows) are reversible. **Undo** dims when the most recent command is irreversible. Once you select **Undo**, its menu item changes to **Redo** ***command***, where ***command***, s the action that was undone. **Redo** cancels the effect of **Undo**.

Edit→Cut

The **Cut** command copies selected fields from the active spreadsheet to the clipboard and replaces them with missing values. It is equivalent to **Copy**, then **Clear**. You can also use the **Cut** command to copy all or part of a report window. However, **Cut** works like **Copy** in graphical displays (it does not clear the copied image).

Edit→Copy

The **Copy** command copies the values of selected data cells from the active data table to the clipboard. If you do not select columns, **Copy** copies entire rows. Likewise, you can copy values from whole columns if no rows are selected. If you select both rows and columns, **Copy** copies the subset of cells defined by their intersection.

You can also use the **Copy** command to capture graphical displays or text reports defined by the scissors tool in the **Tools** menu.

Note: Data you cut or copy to the clipboard can be pasted into JMP tables or into other applications. Pictures can be pasted into any application that accepts graphics.

Edit→Copy as Text

The **Copy as Text** command copies all text from the active report window (no graphical displays) and stores the text in the clipboard.

If you select **Copy as Text** while holding down the option key (OPTION-**Copy as Text** on the Macintosh or ALT-**Copy as Text** under Windows), JMP data table column names are written to the clipboard as the first line of information. If you use OPTION or ALT with **Copy as Text** to copy a text report, column names in the report are written to the clipboard as the first line of information.

Edit→Paste

The **Paste** command copies data from the clipboard into a JMP data table. **Paste** can be used with the **Copy** command to duplicate rows, columns, or any subset of cells defined by selected rows and columns.

To duplicate an entire row or column,

1) Select and **Copy** the row or column to be duplicated.
2) Select an existing row or column to receive the values.
3) Use the **Paste** command.

To duplicate a subset of values defined by selecting both rows and columns. Follow the previous steps but select the same arrangement of rows and columns to receive the copied values as originally contained them. If you paste data with fewer rows into a destination with more rows, the source values recycle until all receiving rows are filled.

Edit→Paste at End

The **Paste at End** command extends a JMP data table by adding rows and columns to a data table as needed to accept values from the clipboard.

If you select rows before choosing this command, the effect is similar to joining data tables by row number. If you highlight columns, the **Paste at End** command adds cells to the bottom of the data table filling them with values from the clipboard.

To transfer data from another application into a JMP data table, first copy the data to the clipboard from within the application. Then use the **Paste at End** command to copy the values to a JMP data table. Rows and columns are automatically created as needed when you **Paste at End**.

If you select **Paste at End** while holding down the OPTION or ALT key, the first line of information on the clipboard is used as the column names in the new JMP data table.

Edit→Clear

The **Clear** command clears all selected cell values from the active data table and replaces them with missing values. The values are not copied to the clipboard.

Edit→Journal

The **Journal** command copies all information from the active data table or report window to an open journal window. Each subsequent **Journal** command appends a page break followed by the information from the active JMP window to the current journal file. You can journal as much information as you need into a single open journal window. The scissors tool in the **Tools** menu and the **Cut** and **Paste** commands in the **Edit** menu let you cut and paste sections of JMP analysis reports into the journal window. You can also type your own notes at the end of the open journal file at any time.

When you select the **Save As** command for an active journal window, you give a name and disk location for that journal file. The Save As dialog has a popup menu that lists the available word-processing formats for saving journal information.

Edit→Search

The **Search** command gives you the ability to find and replace text in the way found in most word processing and editing programs. Find and replace actions deal only with character strings. Numbers are treated as text, and appear to the **Search** command as they show in the data table. All dialog actions also respond to the keystroke shortcuts shown on the **Search** popup menu. For multiple Search actions, it is sometimes much more convenient to use the keystroke shortcuts instead of the Search dialog.

If there is no value in the Search dialog **Find What** box, all **Search** popup menu items except **Don't Find, Cancel** and **Help** are inactive (dimmed).

Table A.3 describes how to use the **Search** dialog.

Table A.3 Items in the **Search** dialog

Find	displays a dialog that prompts you to enter a *find value* in the **Find What** box. Optionally, you can enter a replace value in the **Replace With** box. After you enter a *find value* and click **Find**, JMP searches the active table for the *find value*. The search begins in the focused cell. A cell is focused when it is highlighted or contains the blinking vertical bar that indicates an insertion point. By default (if there is no focused cell), the search begins with the first cell in the first column. The search covers every table cell until it locates a *find value* or reaches the end of the table. The *find value* highlights when located. You hear a beep when you reach the end of the table, or return to the cell where you began your search. The **Find** command does not find values in hidden columns or locked tables. To find these values, you must unhide the column or unlock the table.
Find Next	continues to search for the *find value* by selecting **Find Next** in the **Search** popup menu again, by clicking Find in the dialog, or by using COMMAND-G on the Macintosh or CONTROL-G under Windows
Replace	replaces the currently highlighted cell value with the contents of the **Replace With** box (the *replace value*) in the **Find** dialog. If the *replace value* is missing and you select Replace or use COMMAND-R on the Macintosh or CONTROL-R under Windows, the currently highlighted cell content becomes a missing value.
(continued next page)	

Table A.3 (continued) Items in the **Search** dialog

Replace & Find Next	functions the same as **Replace**, but continues to search for the *find* value.
Replace All	replaces all occurrences of the *find value* with the *replace value* except in locked columns.
Match Case	allows you to request a case sensitive search, useful for locating proper nouns or other capitalized words.
Match whole words only	is an additional option found on the Find and Replace dialog. Blanks count as characters, which means you can search for a series of words in a character column or locate strings with unwanted leading or trailing blanks. This option can also locate words with at least one leading and one trailing blank.
Search by row Search by column	searches the data table row by row from left to right, to the rightmost cell in the last row or until you stop the search. Likewise, **Search by column** searches the table column by column, from top to bottom until it reaches the last cell in the rightmost column, or until you stop the search.

Note: the **Undo** command works with **Replace**, **Replace & Find Next**, and **Replace All.**

The Tables Menu

The **Tables** menu commands modify or create a new JMP table from one or more existing tables.

Tables menu commands perform a variety of data management tasks on JMP data tables. These commands allow you to sort, subset, stack, or split table columns, join two tables side by side, concatenate multiple tables end to end, and transpose tables. You can also create summary tables for group processing and summary statistics and an attributes table for easy table maintenance.

This section gives you an overview of each **Tables** menu command.

Tables→Group Summary

The **Group/Summary** command creates a JMP window that contains a *summary table*. This table summarizes columns from the active data table, called its *source table*. The summary table has a single row for each level of a grouping variable you specify. When there are several grouping variables, the summary table has a row for each combination of levels of all grouping variables. The summary table is linked to its source table. When you highlight rows in a summary table, the corresponding source table rows highlight.

The Group/Summary Dialog

When you select **Group/Summary**, you see the Group/Summary dialog. To complete the Group/Summary dialog select variables and click the appropriate button. To create a summary table, use components and options in the Group/Summary dialog. Table A.4 describes the items in the Group/Summary dialog.

Table A.4 Items in the **Group/Summary** Dialog

Data Only	By default, the Text/Others option in the **Import** command reads a text file of rows and columns without header information (Data Only). The default names of the JMP data table columns are **Column 1, Column 2**, and so on.
Labels	tells JMP that the incoming data are from a text-format file that contain column labels as the first record of the data. These labels become column names in the new JMP data table.
Header	tells JMP that the incoming data file contains standard header information for a JMP data table.
End of Field	designates the character that identifies the end of each field in the incoming file. This delimiter can be a tab, one or more blank spaces, any character, or any combination of these. When you choose **Other**, an editable text box appears where you can enter a character or its hexadecimal representation.
End of Line	selects the characters or keys that identifies the end of each line. You can use the return key, the line feed character, a semicolon, or another character of your choice. The **Other** option works the same as described above.
Strip Enclosing Quotes	tells JMP that each data value is enclosed in single or double quotes. JMP removes the quotes before importing the data.

Initially, the summary table lists only the grouping values (levels) and their frequency in the source table. However, you can use the summary table for the following purposes:

Create a table of summary statistics

The **Stats** popup menu on the Group/Summary dialog lists standard univariate descriptive statistics. With this menu you can add columns of descriptive statistics to the summary table for any numeric column in the source table. You can also access the Group/Summary dialog from an existing summary table with the **Add Summary Cols** command in the popup commands found in the **$** (dollar sign) menu at the lower left of the summary table window.

Analyze subsets of data

An active summary table has two *modes*, By-Mode Off and By-Mode On, accessed by the popup menu at the upper-left corner of the table:

- When By Mode is off, the **Analyze** and **Graph** menu commands and the **Transpose** command in the **Tables** menu apply to the active summary table itself instead of its source table.
- When By Mode is on, highlighted rows in the active summary table identify subsets in the source table. Commands on the **Analyze** and **Graph** menus, and the **Transpose** command in the **Tables** menu recognize these subsets. A single **Analyze** or **Graph** menu command produces a separate report window for each subset identified by a highlighted row in this summary table.

Plot and chart data

The **Bar/Pie Charts** command in the **Graph** menu displays summarized data. The Bar/Pie Charts platform requires that an X variable or combination of X variables identify unique groups. You may need to process data tables with the **Group/Summary** command and produce summary tables to use this platform.

The Summary Table

The summary table, like analysis results, is not saved when you close it. However, the **Save As** command saves the summary table as a standard JMP table. Or, if you select the **Subset** command when the summary table is the active window, the result is a standard untitled JMP data table. This new untitled table is a duplicate of the summary table, but it is not linked to the source table.

See Chapter 2, "JMP Data Tables," for examples of how to group data and create columns of summary statistics.

Tables→Subset

The **Subset** command produces a new JMP table that is a subset of the active data table. The new table has the rows and columns defined by the highlighted rows and columns from the active spreadsheet. See the **Rows** menu and the **Columns** menu section later in this appendix for descriptions of how to select rows and columns.

Note: When you hold down the SHIFT key and select **Subset**, the subset table that results and any plot or graph of that subset table remain linked to the original table. Highlighting rows in this kind of subset table highlights the corresponding rows in the original table and in all its plots and graphs.

Tables→Sort

The **Sort** command sorts a JMP data table by one or more columns. The **Sort** command displays a dialog that allows you to specify columns as sort fields. Use components and options in the **Sort** dialog as described in Table A.5.

Table A.5 Items in the **Sort** Dialog

Columns from *table name*	Select sort fields from the **Columns** variable selection list and add them to **Sort By** list with the **Add** button. The order in which you build this list establishes the sorting order.
Sort By	The **Sort By** list shows the columns to be sorted in order of precedence. To remove sort fields, select them in the sort list and click the **Remove** button.
a...Z/Z...a	To sort a column in either ascending or descending order, by select it in the **Sort By** list and clicking **a...Z/Z...a**. By default, columns sort in ascending order.
Replace Original Table	By default, **Sort** creates a new data table. You can replace the original data table by checking **Replace Original Table**.

Note: Formulas are not preserved in the new sorted table.

Tables→Stack Columns

The **Stack Columns** command creates a new data table from the active table by stacking specified columns into a single new column. The values in other columns are preserved in the new data table. In addition, **Stack** creates an ID column to identify each row in the new table. Table A.6 describes the elements of the Stack dialog.

Table A.6 Items in the **Stack Cols** Dialog

Columns from *table name*	Select columns to be stacked from this list and add them to the **Columns to Stack** list by clicking or by dragging them to the list box (Macintosh only).
Columns to Stack	All columns you select to be stacked, stack or pile one upon another, row by row, in the order they appear in this list. You can remove a column from the list by selecting it and clicking **Remove** on the dialog. The number of rows in the new table is the total number of stacked columns multiplied by the number of rows.
Name of Stacked Cols	Enter a name in the **Stacked Column Name** box to name the new column of stacked values. By default, this column name is **_Stacked_**. If you delete the default name and do not specify a new name, the new stacked column is not written to the data table.
Name of ID Column	Enter a name in the **Type Column Name** box to name the column whose values identify the original column for each stacked cell in the new table. By default, this column name is **_ID_**. If you delete the default name and do not specify a new name, the new ID column is not written to the data table.
Output Table Name	Optionally, you can specify a table name to use instead of the default name, **Untitled**.

Tables→Split Columns

The **Split Columns** command creates a new data table from the active table by splitting one or more columns to form multiple columns. The new columns correspond to the values (levels) of an ID variable. **Split Columns** displays a dialog for you to specify the columns to be split and the ID variable. **Split Columns** also requires one or more columns whose combined values identify each row in the new table. Optionally, the values in the other columns can be preserved in the new data table.

Tables→Transpose

The **Transpose** command creates a new JMP table that is the *transpose* of the active data table. The columns of the original (active) table are the rows of the new table, and the original table rows are the new table's columns. The new table has an additional column called **Labels,** whose values are the column names of the active table. If there is a label variable in the active table, the values of that column are column names in the new transposed table. If there is no label column, the column names in the transposed table are **Row1**, **Row2**, ...**Row*n***, where *n* is the number of rows in the original table. The **Transpose** facility has the following properties:

- The columns of the original table must be either all character or all numeric, except for a **Label** column and for columns in a summary table used for grouping.
- **Transpose** can transpose any selected subset of rows.
- **Transpose** can transpose groups of rows. For example, subsets defined by a summary table created with the **Group/Summary** command in the **Tables** menu (described earlier) transpose independently and stack to form a new transposed table.

Tables→Concatenate

The **Concatenate** command appends two or more tables end to end. The arrangement of the new data table depends on the columns in the original tables. **Concatenate** creates one column in the new table for each unique column name in all appended tables. Column names that are the same in multiple tables stack into a single column.

The Concatenate dialog lists all the open JMP data tables in the table selection list. You select the tables you want to concatenate. When you finish selecting data tables, click the **Concat** button. This creates a new untitled data table that consists of all rows in the first selected table followed by all rows from the second table, and so on.

Tables→Join

The **Join** command creates a new data table by joining two tables side by side. Radio buttons in the Join dialog give you the joining options to use any of the matching specification options in Table A.7.

Table A.7 Items in the **Join** Dialog

<table>
<tr><td>By Row Number</td><td>Joining by row number joins tables side by side, matching them by row number. If the tables don't have the same number of rows, columns from the shorter table have extra cells with missing values.</td></tr>
<tr><td>Cartesian</td><td>If you choose the Cartesian join, each row in the Join table joins with every row of the With table. The number of rows in the resulting table is the product of the rows in the two original tables.</td></tr>
<tr><td>By Matching Cols</td><td>To join rows only when column values match, click By Matching Cols and complete the dialog that appears.</td></tr>
<tr><td colspan="2">If you click By Matching Cols in the Join dialog, you are prompted to select a column in each table selection list whose values must match to complete a join.

Note: Matching columns do not require the same names and do not have to be in the same relative column position in both tables.

After you choose columns whose values must match, additional matching options show for both tables. If you do not select any of these options, a Cartesian join occurs within each group of matching column values. Matching columns has the following two options:</td></tr>
<tr><td>Drop Multiples</td><td>You can specify Drop Multiples in either or both of the data tables being joined. If you drop multiples in both tables, only the first match is written to the new table. If you specify this option on only one table, the first match value joins with all matches in the other table.</td></tr>
<tr><td>Include Non-Matches</td><td>When you Include Non-Matches from a data table, the new data table includes each row from that data table even if there is no matching value. If there is no matching row in the other data table, cells for columns from the other data table have missing values for that row. You can specify this option for either or both data tables.</td></tr>
</table>

If you don't check **Select Columns** in the Join dialog all columns from both tables are included in the joined table. If you check **Select Columns**, you are prompted to select columns from the original tables to include in the new table. If a name is unique, it is written directly to the new table. Columns with the same names in both tables are not overwritten. The new table includes both columns, named by appending the original column name to its data table name.

See Chapter 2, "Managing Data Tables," for examples of the **Join** command.

Tables→Table Info

The **Table Info** command displays a dialog for you to make notes about the current data table or to examine its attributes. You can **Lock** the data table if you have not modified it since you last saved it. If you have made changes, the **Lock** box is dimmed and the **Dirty** box is checked. You can also use the Table Info dialog to create a new data table by simply entering a new table name. The new data table is a duplicate of the original table. The original data table remains as it was when last saved, and the new table is not saved until you select the **Save** or **Save As** command.

The Table Info dialog can also be accessed from an attributes table, described next.

Tables→Attributes

The **Attributes** command creates a new table called an *attributes table* from the active data table, called its *source table*. An attributes table has a row for each column in its source table and a column for each type of column characteristic.

You can modify the column characteristics of a source table by editing values in its corresponding attributes table row. Changing a column's characteristics by editing a row in an attributes table is the same as changing characteristics in the Column Info dialog for that column. The advantage of using an attributes table is that you can change the characteristics of many columns at the same time.

At any time while you are editing an attributes table, you can update its source table by selecting the **Update Source** command in the ($) popup menu at the lower left of the attributes table. Other commands in the dollar menu are as follows:

Sort Attributes

accesses the Sort dialog used by the **Sort** command. You can sort the attributes table by any of its variables. The columns in the updated source table are rearranged by the sort you requested.

Add Col Notes

adds a character column to the attributes table called **Col Notes**. By default, it has a length of 64. You can change the length of this column using its Column Info dialog. If there are any currently existing column notes, they appear in the new **Col Notes** column. Any entry you make into the **Col Notes** column appears in the Column Info dialog for the corresponding source table column.

Max Col Note Width

lets you specify a maximum width allowed for each column's notes, to a maximum width of 127.

See Chapter 2, "Managing Data Tables" for more information about the **Attributes** command.

Tables→Design Experiment

Design Experiment accesses the Design of Experiments module in JMP. Design of experiments in JMP is documented in Chapter 14 "Design of Experiments."

The Rows Menu

The first five commands in the **Rows** menu assign *row state* characteristics to selected rows. To select (highlight) a row, click the space in the row number area to the left of the row. This area is called the *row selection area*. To highlight multiple contiguous rows, drag down the row numbers or SHIFT-click the first and last rows of a desired range. To make a discontiguous selection, use COMMAND-click on the Macintosh or CONTROL-click under Windows to select the rows you want.

Rows are also highlighted (and selected) when you highlight points or bars in a corresponding graphical display. Because the JMP data table and any graphical display of that table are linked, selected points and rows always highlight simultaneously. To deselect all rows, click in the upper-left corner of the data table where the number of rows is shown. To deselect a single row, COMMAND-click that row's number (use CONTROL-click under Windows).

You can choose from a variety of **Rows** menu commands that affect highlighted rows such as

- Excluding rows from further analysis
- Hiding points in current graphical displays
- Assigning special colors or markers for graphical display of points
- Selecting all rows, or selecting rows assigned specific row state characteristics
- Hiding points in current graphical displays
- Automatic color and marker assignment
- Adding or moving rows
- Deleting rows

All of the **Rows** menu commands are described next.

Rows→Exclude/Include

Exclude/Include is a toggle command used to exclude selected rows from statistical analyses. Data remain excluded until you choose **Exclude/Include** is again for selected rows. The **Exclude/Include** status is a row state characteristic and can be saved permanently with the data table in a special row state column.

Rows→Hide/Unhide

Hide/Unhide is a toggle command that suppresses the display of points in all scatter plots. To hide data, highlight their points in any plot or in the spreadsheet and select **Hide/Unhide**. Data remain hidden until you choose **Hide/Unhide** again for selected rows. The **Hide/Unhide** status is a row state characteristic and can be saved permanently with the data table in a special row state column.

Warning. Hidden points can be included in statistical analyses even though they do not display in plots. Likewise, points can be excluded from an analysis but not hidden. These conditions could cause misleading plots and analyses.

Rows→Label/Unlabel

Label/Unlabel is a toggle command that labels points on all scatter plots. To label points, highlight them in any plot or in the spreadsheet, and select **Label/Unlabel**. By default, the row number is used as the label value on plots. However, if you designate a column in the spreadsheet as a **Label** column, its values show as labels in plots instead of the row numbers. To assign a column the label role in the spreadsheet choose **Label** from the role assignment popup menu at the top of the column.

Data remain labeled until you choose **Label/Unlabel** again for selected rows. The **Label/Unlabel** status is a row state characteristic. It can be saved permanently with the data table in a special row state column.

Rows→Colors

Colors changes highlighted points in all scatterplots to the colors you choose. To color a group of points, select the appropriate rows in the spreadsheet or select the points in a plot. Then choose a color from the **Colors** palette. On the Macintosh you can *tear off* the color palette and drag it anywhere on your desktop. To tear off the palette, click and drag it to the place you want it. Under Windows the **Show Colors Palette** command in the **Window** menu displays a floating colors palette that you can drag and place wherever you want. The floating colors palette is visible until you close it.

There are 65 colors including shades of gray, for distinguishing data points in JMP. The default color is black. The **Colors** status is a row state characteristic. It can be saved permanently with the data table in a special row state column.

Rows→Markers

Markers assigns a plot character to replace the standard points in scatterplots and spinning plots. To assign one of the markers, select the rows in the spreadsheet or select the corresponding points in a plot and choose a marker from the **Markers** submenu.

On the Macintosh you can *tear off* the markers palette and drag it anywhere on your desktop. To tear off the palette, click and drag it to the place you want it. Under Windows, the **Show Markers Palette** command in the **Window** menu displays a floating markers palette that you can drag and place wherever you want. The floating markers palette remains visible until you close it.

There are eight different markers, including the default dot, for distinguishing data points in JMP. The **Markers** status is a row state characteristic. It can be saved permanently with the data table in a special row state column.

Rows→Select

The **Select** command has a submenu with options that allow you to select all rows in a data table, or to select a subset of rows based on row states. Use the **Select** submenu as described in Table A.8.

Table A.8 Items in the **Select** Dialog

All Rows	selects the entire JMP data table, the same as SHIFT-clicking the first and last rows in the spreadsheet. Select Excluded selects all excluded rows regardless of their current selection status and deselects all included rows.
Select Hidden	selects all hidden rows regardless of their current selection status and deselects all rows that are not hidden.
Select Labeled	selects all labeled rows regardless of their current selection status and deselects all unlabeled rows.
Where	lets you search for a specific value in a column and selects all rows where that value is found. Where displays a dialog that prompts you to select a column, a comparison operation from the Where dialog popup menu, and a selection criterion value. When you click OK, the Where command highlights all rows that meet the search criterion. The Where command looks only at text strings. Before it completes the comparison, Where converts a numeric value to the character string as it appears in the table cell. The Where popup menu, shown to the right, lists the comparison operators. Multiple searches with the Search selected rows only box checked is the same as a logical *And.*

Rows→Invert Selection

The **Invert Selection** command deselects all currently selected rows and selects all unselected rows.

Rows→Clear Row States

The **Clear Row States** command clears all active row states in the data table. All rows become included, visible, unlabeled, and show in plots as black dots. It does not affect row states saved in columns.

Rows→Color/Marker by Col

The **Color/Marker by Col** command lets you specify a variable whose levels are used to color or mark points in plots.

Rows→Locate Next

Rows→Locate Previous

The **Locate Next** command locates the first selected row after the *current row* and makes it blink briefly. The current row is the first row when you open a data table. The current row changes to the most recent row you edited or identified by a **Locate Next** or **Locate Previous** command. You can set the current row by OPTION-clicking anywhere in a row (ALT-click under Windows). Each time you choose the **Locate Next** command, the next selected row is found and blinks. A beep tells you when the last selected row is located.

The **Locate Previous** command behaves the same as **Locate Next** but locates the first selected row before the *current row* and makes it blink briefly.

Rows→Add Rows

The **Add Rows** command displays the dialog that prompts you to enter the number of rows to add and to specify their location in the table. Click the appropriate radio button to add rows at the beginning of the table (At Start), the end of the table (At End), or after a row number you specify (After Row #). The new rows appear in the table when you click the Add button on the dialog, and have missing values that can be filled by typing or pasting in data.

Rows→Move Rows

The **Move Rows** command operates on highlighted rows. It moves highlighted rows to the location you specify You can move highlighted rows to the beginning of the table, to the end of the table, or after the row number you specify.

Rows→Delete Rows

The **Delete Rows** command deletes all selected rows from a JMP data table. Use the **Undo** command on the **File** menu to undo an accidental deletion.

The Cols Menu

Cols menu commands act on selected columns in the current data table. To select a column, click the background area above the column name. This area is called the *column selection area*. To highlight multiple columns, drag across their column selection areas or SHIFT-click the first and last columns of a desired range. Use COMMAND-click on the Macintosh or CONTROL-click under Windows to make a discontiguous selection.

Commands in the **Cols** menu allow you to create , insert , or rearrange columns, assign variable roles, assign or change column characteristics, access the column calculator, and hide or delete columns.

Cols→Assign Roles

The **Assign Roles** command displays the dialog for assigning the following analysis roles to columns:

X identifies columns as independent or predictor variables that are regressors, model effects, or classification variables that divide the rows into sample groups.

Y identifies columns as response or dependent variables whose distributions are to be studied.

Weight identifies a column whose values supply weights for each response variable.

Freq identifies a column whose values assign a frequency to each row for an analysis.

Label identifies a column whose values specify labeling values for plotted points.

You can also specify variable roles by using the role assignment box at the top of each column in the spreadsheet or by selecting an analysis platform and responding to an assign roles prompt.

If you use the **Assign Roles** command or the role assignment boxes in the spreadsheet, your role assignments remain in effect until you specifically change them. If you assign a sufficient number of roles, the **Analyze** menu platforms use these assignments to determine the type of analysis to do and automatically completes the analysis. If you haven't assigned roles or you choose an analysis that requires more roles than you provided, a dialog prompts you for more role assignment information. Role assignments in response to the specific analysis dialog affect only that analysis.

Cols→Clear All Roles

The **Clear All Roles** command removes role assignments from all variables in the active data table.

Cols→New Column

The **New Column** command lets you define one or more new data table columns. The New Columns dialog asks you to name the new column and provide column characteristics. If you need to compute values for a new column, the Formula selection in the Data Source popup menu gives you access to the calculator when you click OK. Table A.9 describes the items in the New Column dialog.

Table A.9 Items in the **New Column** Dialog

Table Name	displays the name of the active data table and cannot be changed.
Col Name	lets you type in the new column's name using as many as 32 keyboard characters. If you choose a long name, you can expand the width of the column in the spreadsheet by dragging the column boundary.
Lock	renders a column *uneditable*. If you use a formula to compute values for the column, it is automatically locked. Also, you can click in the Lock box to protect any column's values from modification.
Validation	lets you set up a table of acceptable values or an acceptable range of values for a column. The List Check radio button lets you enter a list of valid values for a column. If you click the Range Check radio button, you can specify a range of values and range limit conditions.
Data Type	assigns a data type to the new column according to your selection from its menu. Numeric columns must contain numbers. Character columns can contain any characters including numbers. In character columns, numbers are treated as discrete values instead of continuous values. Row State columns contain special information that affects the appearance of graphical displays. Row states include color, marker, selection status (highlighted or not highlighted), include/exclude status, hide/unhide status, and label/unlabel status.
Data Source	tells the source of the column values: • No Formula values are imported, pasted, or keyed into a column. These values are editable. • Formula specifies that values of a column are calculations. You use the column's calculator window to construct a formula that computes these values. The formula can include existing columns, constants, conditional logic, and a variety of

(continued next page)

Table A.10 (continued) Items in the **New Column** Dialog

	functions. To see a column's calculator, choose **Formula** from the popup menu and click **OK**. A column of computed values is locked and cannot be edited. However, you can disassociate a column from its formula by selecting **No Formula** as the data source in its Column Info dialog.
Modeling Type	specifies the way you want JMP to use column values. The modeling type is a column characteristic that JMP uses to determine how to analyze data. You can change a column's modeling type as long as it corresponds to one of the following data types: • **Continuous** columns must contain numeric values. • **Ordinal** columns can contain characters or numbers. The analysis platforms treat ordinal values as discrete categorical values that have an order. If the values are numbers, the order is the numeric magnitude. If the values are character, the order is the collating sequence. • **Nominal** columns can contain either numeric or character values. All values are treated by analysis platforms as though they are discrete values.
Field Width	specifies the field width needed to accommodate the largest number of digits or characters you plan to enter in the new column. The maximum field width is 40 for numeric values and 255 for character values.
Format	specifies a format to display numeric column values: • **Fixed Decimal** displays a value rounded to the number of decimals you specify. This is especially useful for showing dollars and cents amounts. • **Best** is the default format. This means JMP considers the precision of the values and chooses the best display for them. • **Date & Time** assumes the numeric value in a column represents the *number of seconds since midnight, January 1, 1904*, and displays a corresponding date format. **Date & Time** has an additional popup menu for choosing a specific date representation. The following examples show formats for the date December 31, 1995 (its unformatted value is 2903212800). The **Short** date format displays a date as dd/mm/yy, giving 12/31/95. The Long date format displays a date value as weekday, month day, year, giving Sunday, December 31, 1995. The Abbrev date format display is the same as the Long format except that weekday and month have three-character abbreviations, giving Sun, Dec 31, 1995. The Date:HH:MM and Date:HH:MM:SS formats display a date value as a Short date followed by the number of hours, minutes, and seconds after midnight of that date. This format does not display hours or minutes. The formatted values for this example are 12/31/95 12:00 AM and 12/31/95 12:00:00 AM. The :Days:HH:MM and :Days:HH:MM:SS formats show the number of days, hours, Minutes, and seconds since January 1, 1904. The results for December 31, 1995 are :33602:0:0 and :33602:0:0:0. Note: The decimal points for the fixed decimal and best formats, month and day names,

(continued next page)

Table A.10 (continued next page) Items in the **New Column** Dialog

	the date field separators, and the order of the date elements depend on the localized system software in your country. Also, if you are using the Macintosh operating System 7.1 or later, JMP uses the date forms you define in the Date and Time control panel for Long, Short, and Abbreviated.
Notes	is a text entry area you can use to document information about a new column. The text entry can contain as many as 255 characters.
Next	creates a new column with the current New Column dialog information, and makes the dialog ready to accept information for another new column.

Cols→Add Columns

The **Add Columns** command displays a dialog that lets you add more than one column at a time to a table. You specify the number of columns to add, their location, field width, and type. Check the appropriate radio button to add columns at the beginning of the table (**At Start**), after the rightmost column (**At End**) or inserted between columns (**After Selected Col**).

By default, the new column names are **Column 1**, **Column 2,** and so forth. You can enter any value in the **Col Name Prefix** box and that prefix is used in place of **Column**. Other column characteristics are the same for all the new columns. You can change the column names and characteristics by editing them in the spreadsheet, in an attributes table, or in the Column Info dialog. See the **New Column** command discussed previously for more information about specifying column characteristics.

Cols→Move to First

Cols→Move to Last

Cols→Move Columns

Move to First, **Move to Last**, and **Move Cols** operate on highlighted rows. **Move to First** moves columns to the left-most position on the spreadsheet. **Move to Last** moves columns to the right-most position. **Move Cols** moves columns to the location you specify. If you mistakenly move one or more columns, use the **Undo** command in the **Edit** menu to restore the previous order.

Cols→Hide Columns

The **Hide** command hides selected column(s) in the spreadsheet, but does not remove them from the data table. Use the **Unhide** command to reshow hidden columns.

Note: The number of columns showing in the upper-left corner of the data table does not change when you hide columns. However, when there are Hidden columns the **Unhide** command is not dim.

Cols→Unhide

The **Unhide** command displays a dialog that lists all hidden columns. You can select any subset of the hidden columns to reshow on the spreadsheet. When you unhide a column, it becomes the last column in the spreadsheet.

Cols→Column Info

The **Column Info** command displays the dialog used by the **New Column** command, except there is no **Next** button for adding new columns. You can use the Column Info dialog at any time to change a single column's attributes. For example. If you select **Formula** in the **Data Source** popup menu on the dialog for a column that has **No Formula** as its data source, the calculator window opens after you click **OK**. If you click the picture of an existing formula (at the lower-left corner in the Column Info dialog), the calculator window opens so you can modify the formula.

Cols→Delete Columns

The **Delete Columns** command removes the selected columns from the data table. If you accidentally remove columns, you can use the **Undo** command in the **Edit** menu to restore them.

The Analyze Menu

Each Analyze platform produces a window that shows statistical text reports with supporting graphical displays. Statistical platforms require variable role assignments. You assign roles with the role assignment popup menu at the top of each column in the spreadsheet or with the **Assign Roles** command in the **Rows** menu. If roles are not assigned, the Analyze platforms prompt you to assign them.

Analyze→Distribution of Y

Distribution of Y describes a distribution of values with histograms and other graphical and textual reports:

- Continuous columns display a Histogram, an Outlier Box plot, a Quantiles Box plot, and a Normal Quantile plot. You can change the width of the histogram bars using the hand tool from the **Tools** menu. Optional commands let you compare the distribution mean to constant, set specification limits for a Capability Analysis, and test the distribution for normality.
- Nominal or ordinal columns are shown with a histogram of relative frequency for each level of the ordinal or nominal variable and a mosaic (stacked) bar chart. Optionally you can compare the observed distribution with any set of probabilities you specify.

Text reports support each of the distribution plots. The reports show selected quantiles and moments of continuous values. Tables of counts and proportions support nominal and ordinal values.

Save commands let you save information such as rank, level number, standardized values, and other statistics as new columns in the data table.

Analyze→Fit Y by X

Fit Y by X studies the relationship of two variables. This platform shows plots with accompanying analyses for each pair of X and Y variables. The kind of analysis done depends on the modeling types (continuous, nominal, or ordinal) of the X and Y columns:

- If both X and Y have continuous modeling types, **Fit Y by X** displays a scatterplot. Using options, you can explore various regression fits for the data and choose the most suitable fit for further analysis. Each fit is accompanied by tables with supporting statistical analyses and parameter estimates.
- If X is nominal or ordinal and Y is continuous **Fit Y by X** plots the distribution of Y values for each discrete value of X. You can use popup menu options to see a means diamonds and a box plot for each value of X and to compare group means with comparison circles. An accompanying text report shows a one-way analysis of variance table. Optionally, you can request nonparametric analyses, see multiple comparisons, and test homogeneity of variance.
- If X has continuous values and Y has nominal or ordinal values, **Fit Y by X** performs a logistic regression and displays a family of logistic probability curves. Accompanying tables show the log likelihood analysis and parameter estimates for each curve.

Note: Logistic regressions of ordinal columns are parameterized differently from logistic regressions of nominal columns and therefore produce different results.

- If both X and Y are nominal or ordinal values, a contingency table and a mosaic chart are shown. A mosaic chart consists of side-by-side divided bars for each level of X. Each bar is divided into proportional segments representing the discrete Y values. The contingency table shows statistical tests, and frequency, proportion, and chi-square values for each cell. Optionally, you can request a correspondence analysis.

Analyze→Fit Model

Fit Model displays a dialog that lets you tailor an analysis using a model specific for your data. When you choose the **Fit Model** command, you see the dialog in Figure A.1. The variable selector list in the dialog lists all columns in the current JMP data table. You select columns, assign roles, and build the model to fit by adding effects to the **Effects in Model** list.

Fit model performs a fit of one or more Y variable by the X variables selected. You can select the kind of model appropriate to your data from the popup menu of fitting *personalities* in the Fit Model dialog. The fitting personalities available depend on the kind of responses you select.

Figure A.1 The Fit Model Dialog and Fitting Techniques

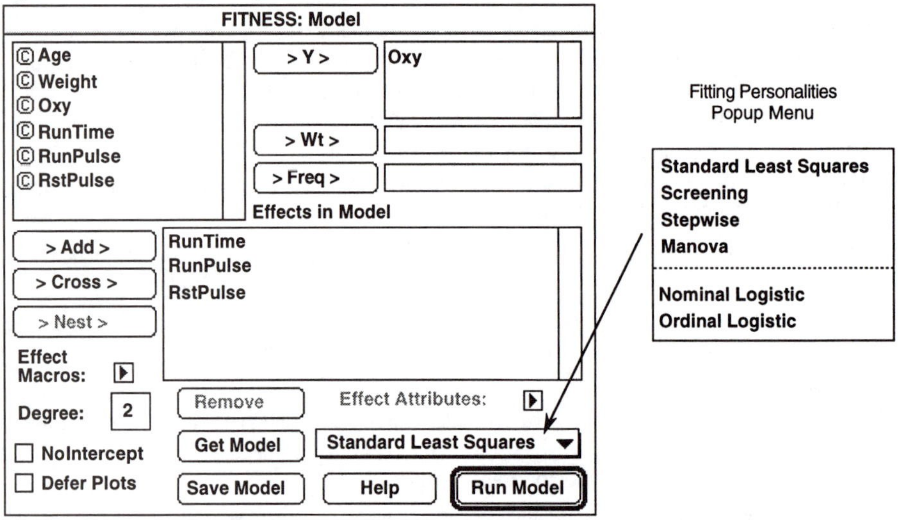

FITNESS: Model
Ⓒ Age
Ⓒ Weight
Ⓒ Oxy
Ⓒ RunTime
Ⓒ RunPulse
Ⓒ RstPulse
> Y > Oxy
> Wt >
> Freq >
Effects in Model
> Add >
> Cross >
> Nest >
RunTime
RunPulse
RstPulse
Effect Macros:
Degree: 2
NoIntercept
Defer Plots
Remove
Effect Attributes:
Get Model
Standard Least Squares
Save Model
Help
Run Model

Fitting Personalities Popup Menu
Standard Least Squares
Screening
Stepwise
Manova
Nominal Logistic
Ordinal Logistic

The following list briefly describes the different fitting techniques:

- **Standard Least Squares** gives a least squares fit for a single continuous response, accompanied by leverage plots and an analysis of variance table.
- **Screening** produces an exploratory screening analysis for single or multiple Y columns with continuous values.
- **Stepwise** gives a stepwise regression for a single continuous Y and all types of effects.
- **Manova** performs a multivariate model fit for multiple continuous Y variables and the effects you specify.
- **Nominal Logistic** for a single nominal response does a nominal regression by maximum likelihood.
- **Ordinal Logistic** for a single ordinal response does an ordinal cumulative logistic regression by maximum likelihood.

Analyze→Nonlinear Fit

Nonlinear Fit launches an interactive nonlinear fitting facility. You orchestrate the fitting process as a coordination of three important parts of JMP: the data table, the calculator, and the Nonlinear Fit platform:

- You define the model column and its prediction formula with the calculator. The formula is specified with parameters to be estimated. Use the **New Parameter** function in the calculator's **Parameters** list to define parameters and give them initial values.
- Launch the Nonlinear Fit platform with the response variable in the Y role and the model column with its fitting formula in the X role. If no Y column is given, then the X column formula is for residuals. If you have a loss column specified, then you may not need either an X or Y column specification.
- Interact with the platform through the control dialog, starting the iterations so that they can run in the background. When the fitting converges, you see a Solution Table and Graph. You can click **Confid. Limits** in the Nonlinear Fitting Control Panel at any time after you have estimates to generate likelihood confidence intervals on the parameter estimates.

Analyze→Correlation of Y's

Correlation of Y's explores how multiple variables relate to each other and how points fit that relationship. This platform helps you see correlations between two or more response (Y) variables, look for points that are outliers, and examine principal components to look for factors.

The Correlation of Y's platform appears showing a correlation matrix. Options show

- The inverse correlation matrix
- A partial correlation matrix
- Pairwise correlations with accompanying bar chart
- Nonparametric correlations
- A matrix of bivariate scatterplots with a plot for each pair of Y variables.

Correlation of Y's also shows a graphical outlier analysis that includes

- A Mahalanobis distance outlier plot
- A jackknifed multivariate distance outlier plot where the distance for each point is calculated excluding the point itself

There are options with these plots to save the distance scores.

You can also request principal components, standardized principal components, rotation of a specified number of components, and factor analysis information.

Analyze→Survival

The **Survival** command lets you analyze survival data two ways:

- Product-limit (Kaplan-Meier) life table survival computations with estimation of Weibull, log normal and exponential parameters.
- Regression analysis that tests the fit of an exponential, a Weibull, or a log normal distribution.

The Graph Menu

Graph menu commands produce windows that contain specialized graphs or plots with supporting tables and statistics.

The Graph platforms require role assignments. You can assign roles with the popup menu at the top of each spreadsheet column or with the **Assign Roles** command in the **Rows** menu. If roles were not previously assigned, the **Graph** platforms prompt you to assign them.

Note: Graphs produced by the **Bar/Pie Charts**, **Overlay Plot**, and **Control Charts** commands update automatically when you add rows to the current data table and then click the chart or graph to activate it.

Graph→Bar/Pie Charts

The **Bar/Pie Charts** command gives a chart for every numeric Y specified. The Y's are the statistics you want to chart. Initially, you see a vertical bar chart, but options include horizontal bar charts, line charts, step charts, needle charts, and pie charts.

Bar/Pie Charts assumes the data are summarized, giving a unique set of values for the X variables you specify. If multiple X variables have the same values, the chart facility assumes your data have not been summarized. It advises you to summarize them and displays the **Group/Summary** command dialog. When you complete the Group/Summary dialog and click OK on the dialog, the **Chart** command continues and displays the default bar chart.

You can specify as many as two X variables for grouping on the chart itself. The first X is the group variable, and the second X is the level (subgroup) variable. If you do not specify an X variable, then each row in the data table becomes a bar. The X variables do not have to be sorted, but each combination of the X's must yield a distinct category. Groups and levels display in the order that they occur in the data table.

Graph→Overlay Plots

The **Overlay Plot** command overlays a plot of a single numeric X column and all numeric Y variables. The axis can have either a linear or a log scale. Optionally, the plots for each Y can be shown separately with or without a common X axis.

By default, the values of the X variable are sorted in ascending order, and the points are plotted and connected in that order. You have the option of plotting the X values as they are encountered in the data table.

If the given X variable is not numeric, the **Overlay Plots** command calls **Bar/Pie Charts** described above, which displays an overlaid line chart.

Note: If you want scatterplots of two variables at a time with regression fitting options, use the **Fit Y by X** command instead of **Overlay Plots.**

Graph→Spinning Plot

Spinning Plot produces a three-dimensional spinable display of values from any three numeric columns in the data table. It also produces an approximation to higher dimensions through principal components, standardized principal components, rotated components, and biplots. Options let you save principal component scores, standardized scores, and rotated scores.

The Spinning Plot platform also gives factor-analysis-style rotations of the principal components to form orthogonal combinations that correspond to directions of variable clusters in the space. The method used is called a *varimax rotation,* and is the same method that is traditionally used in factor analysis.

Graph→Pareto Chart

The **Pareto chart** command gives charts that display counts or the relative frequency of problems in a quality-related process or operation. The defining characteristic of Pareto charts is that the bars are in descending order of values, which visually emphasizes the most important measures or frequencies.

When you select **Pareto Chart**, a variable selection dialog prompts you to assign variable roles. **Pareto Chart** uses a single Y variable, called a process variable, and gives

- A simple Pareto chart when you do not specify an X (classification) variable
- A one-way comparative Pareto chart when you specify a single X variable.
- A two-way comparative chart when there are two X variables

The Pareto chart facility does not distinguish between numeric and character variables, or between modeling types. All values are treated as discrete, and bars represent either counts or percentages.

Graph→Control Charts

The **Control Charts** command creates dynamic plots of sample subgroups as they are received and recorded. Control charts are a graphical analytic tool used for statistical quality improvement. Control charts can be broadly classified according to the type of data analyzed:

- Control charts for *variables* are used when the quality characteristic to be analyzed is measured on a continuous scale.
- Control charts for *attributes* are used when the quality characteristic is measured by counting the number of nonconformities (defects) in an item or by counting the number of nonconforming (defective) items in a sample.

The concepts underlying the control chart are that the natural variability in any process can be quantified with a set of control limits, and that variation exceeding these limits signals a special cause of variation. In industry, control charts are commonly used for studying the variation in output from a manufacturing process. They are typically used to distinguish variation due to *special* causes from variation due to *common* causes.

The Control Chart platform offers the following kinds of charts:

- Mean, range, and standard deviation attributes charts
- Individual measurement and moving range charts
- P, NP, C, and U variables charts
- UWMA and EWMA moving average charts

The Tools Menu

The **Tools** menu is a palette of special tools that determine the effect of mouse actions. The default tool is the arrow. The function of the arrow is territorial. For example,

- When the arrow is in a report window, it reveals text reports, accesses popup menus, displays point labels, and highlights histogram bars.
- When used in a spreadsheet, the arrow accesses popup menus at the top of each column, selects rows and columns, and selects text for editing.

Other tools cut or copy all or part of a report, access help windows, manipulate graphical displays, and highlight points of interest in a plot. These tools function in report windows, but not on the spreadsheet.

On the Macintosh you can *tear off* the **Tools** menu and drag it anywhere on your desktop. To tear off the **Tools** menu, click and drag it to the place you want it. Under Windows, the **Show Tools Palette** command in the **Window** menu displays a floating tools palette to drag and place wherever you want.

The hand tool (or grabber) is for direct manipulation or *grabbing* in plots and charts. In a text report, the hand tool behaves the same as the arrow tool. The hand behaves in graphs and plots as follows:

- On histograms you can use the grabber tool to change the number of bars in a histogram or o shift the boundaries of the bars on the axis.
- On spinning plots the grabber tool spins the plot. To spin a plot, grab the plot with the hand by holding down the mouse button, and then move the hand.
- In a scatterplot the grabber can drag the position of a column of scatterplots to a new position in the matrix.

The question mark icon accesses JMP Help. Select the help tool and then click graphs, plots, or tables to see help windows. Each help window persists as long as the cursor is within the window. You can SHIFT-click the question mark to bring up a help window, which persists until you close it. Some help windows have buttons to reveal further help details and persist automatically.

The brush tool is used for highlighting an area of points in plots. When you click, a rectangle appears. Move the rectangle over points to highlight them in the plot and in the active data table. SHIFT-click to extend the selection. OPTION-click (ALT-click under Windows) to change the size of the selection rectangle and also extend the selection. If the brush tool is not in a plot area, it behaves the same as the arrow tool.

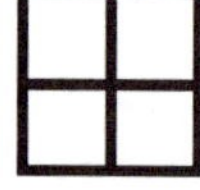

The crosshair is a movable set of axes used to measure points and distances in graphical displays. For example, use the crosshair on a fitted line or curve to identify the response value for any predicted value. The values where the crosshair intersects the vertical and horizontal axis appear automatically as you drag the crosshair within a plot.

The scissors are for selecting cut-and-paste territory from a report. Drag the scissors diagonally over any part of the report to define a rectangular area you want to copy to the clipboard. Hold down the OPTION key (ALT key under windows) and drag the scissors to select blocks of the report window. Use SHIFT-click and drag the scissors to select discontiguous sections of report areas.

The lasso tool lets you highlight an irregular area of points in plots. Drag the lasso around any set of points. When you release the lasso it automatically closes and highlights the points within the enclosed area. Use SHIFT-lasso to drag the lasso around discontiguous irregular areas of points.

The magnifying glass tool automatically zooms in on any area of a plot. When you click the magnifier, the point or area where you click becomes the center of a new view of the data. The scale of the new view is enlarged approximately 25%, giving you a closer look at interesting points or patterns. Use OPTION -click on the Macintosh or ALT-click under Windows at any time to restore the original plot. On a ternary plot you can drag the magnifier tool to zoom the triangular axes.

The annotate tool places a text box wherever you click in a JMP window. You can key in notes and remove them at a later time, draw lines to make a special point, or use the annotate tool to enhance a JMP graphical display. To remove an annotation box, drag it out of the plot frame.

The Window Menu

The **Window** menu helps you organize the windows produced during a JMP session. All open windows generated in a session are listed below the last line in the **Window** menu. When you click on a window name, that window becomes active.

The **Window** menu under Microsoft Windows is different than Macintosh **Window** menu. The first six commands (described next) are not in the Macintosh **Window** menu.

Window→Show Markers Palette

The **Show Markers Palette** opens a window that displays the markers palette given by the **Markers** command in the **Rows** menu. You can drag the Markers window to any convenient location on your desktop, and it remains open until you close it. This command serves the same function as *tearing off* the markers palette on the Macintosh.

Window→Show Colors Palette

The **Show Colors Palette** opens a window that displays the colors palette given by the **Colors** command in the **Rows** menu. You can drag the Colors window to any convenient location on your desktop, and it remains open until you close it. This command serves the same function as *tearing off* the colors palette on the Macintosh.

Window→Show Tools Palette

The Show Tools Palette opens a window that displays the **Tools** Menu and displays the JMP tools as a palette on your desktop. You can drag the Tools window to any convenient location on your desktop and it remains open until you close it. This command serves the same function as *tearing off* the **Tools** Menu on the Macintosh.

Window→Cascade

The **Cascade** command arranges open windows side by side so all of them are visible.

Window→Tile

The **Tile** command arranges open windows so that the title bar of each window is visible.

Window→Arrange Icons

The **Arrange Icons** command arranges all program-item icons for a selected group into rows. Or, if a group icon is selected, **Arrange Icons** arranges all group icons into rows.

Note: The remaining commands in the **Windows** menu are the same on the Macintosh and under Microsoft Windows.

Window→Redraw

The **Redraw** command redraws the active window. It is useful for cleaning up both spreadsheet views and graphical displays that have accumulated stray imperfections resulting from high-speed, dynamic handling of windows.

Window→Move to Back

The **Move To Back** command moves the active window behind all other windows generated by the current JMP session, leaving the next window in the sequence showing.

Window→Hide

The **Hide** command suppresses the display of the active window but does not close it. To reshow a hidden window, select its name from the list beneath the dotted line in the **Window** menu.

Window→New Data View

The **New data View** command displays a new spreadsheet view of an open data table. The new view is linked to the original view and all corresponding analysis windows. Changes made to a new view reflect on the original view when it is made active.

Window→Set Window Name

The **Set Window Name** command lets you change the name of any active JMP window. This is especially useful if you generate multiple untitled windows and need to distinguish among them during the JMP session.

Window→Close *type* Windows

The **Close *Type* Windows** command closes all windows of a given type, where the active window determines the ***Type***. For example, if a data table window is active, this command becomes **Close Data Windows**, and it closes all data tables when you select it.

The Help Menu (Microsoft Windows)

The Windows version of JMP has the standard Windows **Help** menu. Windows users will find the menu items similar to those in other applications. For details about the Windows Help menu in general, consult your Microsoft Windows User's Guide.

Help—>Contents

The **Contents** command displays the same help access window that the Macintosh gives with the **About JMP** command in its Apple menu. This is the topmost help window. This main JMP help window has buttons for further help. All the help documents are arranged hierarchically with buttons for accessing more detailed information.

Help—>Search for Help On

Search for Help displays a scrolling list of topics. You select a topic by typing in a topic name (or part of a topic name). After you select and show a topic, click the **Go To** button to go directly to help for that topic.

Help—>Statistical Guide

The **Statistical Guide** command displays a scrolling statistical index. When you click an analysis in the guide, directions on how to do the analysis appear. The directions tell you the JMP menu, command ,and options to use for the analysis or topic you selected.

Help—>How to Use Help

How to Use Help is the standard Windows help instructions found in most Windows applications.

Help—>About JMP

About JMP displays the JMP startup screen, which gives you license and version information.

Appendix B
Analyze and Graph Platform Commands and Options

Overview

When you have variables to examine and choose a **Graph** or **Analyze** command, a platform appears showing graphical and textual results. Most platforms have additional options you access in popup menus found on the platform. Every platform window has the following three icons found on the lower left border:

A check-mark popup menu lists options or commands that affect all graphs or reports in its specific platform. These commands can just affect the appearance of plots or can add optional features. Not all platforms have check menu items.

The Save (**$**) popup menu lists save commands for that specific platform. These commands usually create new columns in the current data table and fill them with values computed from the selected analysis. Occasionally Save commands create a new table in which to save values. Not all platforms have save commands.

The star (*) popup menu lists commands always available to all platforms. These include platform help, the option to reveal all or conceal all graphs and tables in a platform, the ability to add titles and footnotes annotation with font and justification specifics, and to turn on a print preview option.

In addition to the border menus, a plot or table can have popup menus specific to the analysis it represents. These menus are accessed with the following triangular popup menu icon.

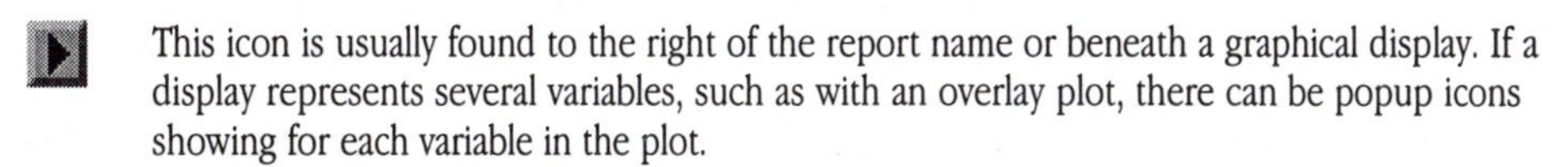

This icon is usually found to the right of the report name or beneath a graphical display. If a display represents several variables, such as with an overlay plot, there can be popup icons showing for each variable in the plot.

This Appendix lists all commands and options found in the Analyze and Graph platforms. It is organized for easy access; the commands are in the order they appear in the main menu, beginning at the top of the Analyze menu and finishing at the bottom of the Graph menu.

The menus for each platform are identified by their menu icon and are followed by a brief description of the respective list of options. If a platform has no additional options, its launch dialog and special features are discussed.

Appendix B Contents

Analyze→Distribution of Y

Platform Options for Analyze→Distribution of Y

All parts of the Distribution of Y report are optional. The popup menu accessed by the check-mark popup menu on the left side of the horizontal scroll bar lists the platform options. Each menu item is a toggle that turns the option on or off when selected.

Text Report

toggles the text reports for the response variables. Reports for nominal and ordinal values include a Frequencies table, which shows counts, proportions, and cumulative proportions. The Quantiles and Moments tables are text reports for continuous variables. They list selected quantiles and descriptive moments such as the mean, standard deviation, and 95% confidence intervals.

Histogram

toggles histograms for all response variables. Histogram of nominal and ordinal variables are bar charts with a bar for each value (level) of the variable. Histograms of continuous variables group the values into bars that illustrate the shape of the column's distribution. The hand tool from the **Tools** menu can change the number of bars or intervals.

Mosaic Plot

displays a mosaic bar chart for each nominal or ordinal response variable. A *mosaic chart* is a stacked bar chart in which each segment is proportional to its group's frequency count.

Outlier Box Plot

displays a box plot for each continuous variable. The ends of the box are the 25th and 75th quantiles, also called the quartiles. The line across the middle of the box identifies the median sample value. The lines on each end, sometimes called *whiskers*, extend from the ends of the box to the outermost data point that falls within the distance computed as 1.5 (interquartile range). The bracket at the edge of the box identifies the shortest half (the most dense 50%) of the observations.

Quantile Box Plot

displays a plot quantile plot for each continuous variable. The quantile box plot shows selected quantiles on the response axis. The median shows as a line across the middle of the box, and the 25th and 75th quantiles are the ends. The means diamond inside the box identifies the sample mean and the 95% confidence intervals.

Normal Quantile Plot

produces an additional graph for each continuous variable that is useful for visualizing the extent to which the variable is normally distributed. If a variable is normal, the normal quantile plot approximates a diagonal straight line. This kind of plot is also called a quantile-quantile plot or q-q plot. Optionally, points in the Normal Quantile Plot can be labeled using a label column and the **Label** command in the **Rows** menu.

Horizontal Layout

arranges text reports to the right of their corresponding graphical displays.

Normal Curve

superimposes a normal curve on the histogram of each continuous variable. The curve is constructed from the mean and standard deviation of the column.

Smooth Curve

fits a smooth curve to the histogram of each continuous variable using nonparametric density estimation. The smooth curve displays with a slider you can use to set the kernel standard deviation, which defines the range of Y values used to determine curve estimates.

Count Axis

adds an axis showing the frequency of column values represented by the histogram bars for nominal and ordinal responses.

Prob Axis

adds an axis showing the proportion of column values represented by histogram bars, for nominal and ordinal responses. This optioncan be used either alone or with the **Count Axis** option.

Uniform Axes

causes the axes for all numeric variables to be scaled the same minimum, maximum, and intervals, so that the distributions are easily compared.

Save Commands for the Analyze→Distribution of Y

You can save information computed from response variables using the Save ($) popup menu at the left of the horizontal scroll bar. Each command generates a new column in the current data table. All commands (except **Save Level Numbers**) apply only to continuous data. Note: You can select the Save ($) commands repeatedly. This enables you to save the same statistic multiple times under different circumstances, such as before and after combining histogram bars. If you use a Save ($) command multiple times and generate multiple columns of the same statistic, the column name for the statistic is numbered, *name*1, *name*2, and so forth, as needed to created unique column names. The Save ($)commands behave as described next:

Save Level Numbers

creates a new column, called Level *colname*, for each histogram in the report window. The level number of each observation corresponds to the histogram bar that contains the observation. The histogram bars are numbered from low to high beginning with 1.

Save Level Midpoints

creates a new column, called Midpoint *colname*, for each continuous histogram in the report window. The midpoint value for each observation is computed by adding half the level width to the lower level bound.

Save Ranks

creates a new column for each histogram in the report window. Each new column, called Ranked *colname,* contains a ranking for each of the corresponding column's values starting at 1. Duplicate response values are assigned consecutive ranks in order of their occurrence in the spreadsheet.

Save Ranks-averaged

creates a new column, called RankAvgd *colname*, for each histogram in the report window. If a value is unique, the averaged rank is the same as the rank. If a value occurs *k* times, the average rank is computed as the sum of the value's ranks divided by *k*.

Save Prob Scores

creates a new column, called Prob *colname,* for each response variable in the current data table. For N nonmissing scores, the probability score of a value is computed as the averaged rank of that value divided by (N + 1). This column is similar to the empirical cumulative distribution function.

Save Normal Quantiles

creates a new column, called N-Quantile *colname*, for each continuous histogram in the report window. The normal quantile values are computed from the cumulative probability distribution function for the normal distribution. The normal quantile values can be computed using the Normal Quantile (normQuant) JMP function on a column created by the **Save Ranks-Avg** command.

Save Standardized

creates a new column, called **Std** *colname*, for each continuous Y variable.

Individual-Plot Options for Analyze→Distribution of Y

The popup menu next to the histogram name for each continuous variable offers three commands that let you test the mean of the distribution, test the distribution for normality, and perform a capability analysis. The popup menu next to the histogram name for each nominal or ordinal variable tests each level to the hypothesized probability you enter.

Test Mean=value

prompts you to enter a hypothesized value for statistical comparison to a sample mean. You also have options to enter a standard deviation if know and to request the nonparametric Wilcoxon signed-rank statistic. After you click **OK**, the Test Mean=*value* table is appended to the bottom of the text reports and includes the one-tailed and two-tailed alternatives. The Wilcoxon signed-rank test uses average ranks for ties. The p values are exact for $n \le 20$ where n is the number of values not equal to the hypothesized value. For $n > 20$ a Student's t approximation (Iman and Conover) is used.

Test for Normality

tests that the distribution is normal. If $n \le 2000$, a Shapiro-Wilk W test is given. If $n > 2000$, a KSL (Kolmogorov-Smirnov-Lillifor) test is done. If the p value reported is less than .05 (or some other alpha), then you conclude that the distribution is not normal. If you conclude from these tests that the variable's distribution is not normal, the **Normal Quantile** command in the check-mark menu is useful to help assess the lack of normality in the distribution.

Set Spec Limits

gives a capability analysis for quality control applications. A dialog prompts you for **Lower Spec Limit**, **Upper Spec Limit**, and **Target**. These are optional fields, and only those fields you enter are part of the resulting Capability Analysis table. The platform appends the Capability Analysis table to the text report and shows the upper and lower specification limits drawn on the corresponding histogram.

Stem and Leaf

appends the conventional stem-and-leaf plot to the text report (Tukey 1977).

Test Probabilities

displays a dialog that lists the estimated probabilities for each level of the variable. You can enter any probability for each level and the report then shows a likelihood Ratio chi-square and a Pearson chi-square test that the estimated distribution and the hypothesized distribution you entered are the same.

Moments Report: Additional Statistics

The popup menu next to the Moments Report has the command **More Moments**, which expands the moments table to include a more comprehensive set of statistics, such as skewness, kurtosis, and coefficient of variation.

Analyze→Fit Y by X: Regression and Curve Fitting

Platform Options for Analyze→Fit Y by X: Regression and Curve Fitting

The **Fit Y by X** command analyzes pairs of variables, examining the distribution of a response variable Y as conditioned by the values of a factor X. When both X and Y are continuous, the platform begins with a scatterplot. You select commands and options that fit lines and curve, show them on the scatterlplot, and summarize the results with text reports. The popup menu beneath each plot frame gives the following commands:

Fit Mean
adds a horizontal line to the plot at the mean of the response variable.

Fit Line
adds a straight-line fit to the plot using least squares regression.

Fit Polynomial
fits a polynomial curve of the degree you select from the **Fit Polynomial** submenu. After you select the polynomial degree, the curve is fit to the data points using least squares regression.

Fit Spline
fits a spline with the flexibility you specify.

Fit Transformed
displays a dialog that allows you to specify the transformation you want for the X or Y (or both). Optional transformations include natural log, square root, square, reciprocal, and exponential

Density Ellipses
draws an ellipse that contains the specified mass of points as determined by the probability you choose from the **Density Ellipses** submenu. The density ellipsoid is computed from the bivariate normal distribution fit to the X and Y variables. The bivariate normal density is a function of the means and standard deviations of the X and Y variables, and the correlation between them. The **Other** selection lets you specify any probability greater than zero and less than or equal to one.

Paired t-Test
assumes the X and Y variables you specified are paired and performs a paired t test. A standard t test table summarizes the results, and a visual representation is shown on the scatterplot.

Grouping Variable
displays a dialog that prompts you to select a classification (grouping) variable. When a grouping variable is in effect, the Fit Y by X platform computes a separate analysis for each level of the grouping variable and overlays the regression curves or ellipses on the scatterplot.

Individual-Plot Fitting Options for Analyze→Fit Y by X: Regression and Curve Fitting

Each requested fit generates its own set of reports and has its own set of display and save options accessed by a popup menu beneath the scatterplot:

Line of Fit
displays or hides the line of fit.

Confid Curves Fit
displays or hides confidence curves for mean, line, and polynomial fits. This option is not available for the spline and density ellipse fits and is dimmed on those menus.

Confid Curves Indiv

displays the upper- and lower- 95% confidence limits for an individual predicted value. The confidence limits reflect variation in the error and in the parameter estimates.

Color

lets you choose from a palette of 65 colors to assign a color to each fit.

Save Predicteds

creates a new column in the current data table called Predicted *colname* with the model prediction formula as the data source. The formula computes the sample predicted values and computes values automatically for rows you add to the table.

Save Residuals

creates a new column in the current data table called Residuals *colname* that contains the residual values of the Y variable called *colname*. It is the difference between the actual (observed) value and the predicted value for that column. This option is not available for the mean fit or density ellipses and is dimmed on those menus.

Remove Fit

removes the fit from the graph and removes its text report.

Note: You can use the **Save Predicteds** and **Save Residuals** commands for each fit. If you use these commands multiple times or with a grouping variable, it is best to rename the resulting columns in the data table to reflect each fit.

Analyze→Fit Y by X: One-Way Analysis of Variance

The **Fit Y by X** command analyzes pairs of variables, examining the distribution of a response variable Y as conditioned by the values of a factor X. The plot for a continuous by nominal/ordinal analysis shows the vertical distribution of response points for each factor (X) value. The distinct values of X are sometimes called levels. The basic fitting approach is to compute the mean response within each group. If the group response variance is assumed to be the same across all groups, then a one-way analysis of variance tests whether the means of the groups are equal.

Analysis Options for Analyze→Fit Y by X: One-Way Analysis of Variance

The left popup menu, labeled **Analysis**, that shows beneath the plot frame gives the following options:

Quantiles

displays a table listing the minimum, 10%, 25%, median (50%), 75%, 90%, and maximum values of each group, and activates the **Quantile Boxes Display** option.

Means, Anova/t-test

displays the one-way analysis of variance report that has a summary table, a one-way analysis of variance, and a table that lists group frequencies, means, and standard errors computed with the pooled estimate of the error variance. If there are only two groups, a *t test* also shows. This option shows the **Means Diamonds** display.

Means, Std Dev, Std Err

displays a table listing the sample size, mean, standard deviation, and standard error of each group. This option displays **Means Dots**, **Error Bars**, and **Std Dev Lines**.

Compare Each Pair
displays a table with Student's *t* statistics for all combinations of group means and activates the **Comparison Circles** option in the Display popup menu beneath the scatterplot.

Compare All Pairs
displays a table showing Tukey-Kramer HSD (honestly significant difference) comparisons of group means and its corresponding comparison circles.

Compare with Best
displays a table showing Hsu's MCB (Multiple Comparison to the Best) comparisons of group means to the best (max or min) group and the appropriate comparison circles.

Compare with Control
displays a table showing Dunnett's comparisons of group means with a control group that you specify, with the appropriate comparison circles.

Nonpar-Wilcoxon
displays a Wilcoxon rank sum test if there are two groups and a Kruskal-Wallis nonparametric one-way analysis of variance if there are more than two groups.

Nonpar-Median
displays a two-sample median test or a k-sample Brown-Mood median test to compare group means.

Nonpar-VW
displays a van der Waerden test to compare group means.

UnEqual Variances
displays four tests for equality of group variances.

Set Alpha Level
lets you specify the alpha level for tests.

Display Options for Analyze→Fit Y by X: One-Way Analysis of Variance

The right popup menu, labeled Display, that shows beneath the plot frame gives these options:

Show Points
shows data points on the scatterplot.

Quantile Boxes
overlays quantile box plots on each response group that shows the quantiles listed in the Quantiles table.

Means Diamonds
draws a horizontal line through the mean of each group proportional to its X axis. The top and bottom points of the means diamond show the upper and lower 95% confidence points for each group.

Means Dots, Error Bars
identifies the mean of each group with a large marker and shows error bars one standard error above and below the mean.

Std Dev Lines
shows one standard deviation above and below the mean of each group of dotted lines.

Comparison Circles
show comparison circles computed for the multiple comparison method selected in the Analysis popup menu beneath the scatterplot.

Connect Means
connects the group means with a straight line.

X-Axis Proportional
makes spacing on X axis proportional to the sample size of each level.

Jitter
offsets points in the outlier box plot when they have the same value so that you can see more than one point.

Matching Lines
lets you specify a matching variable and draws lines between matching points.

Power Details for Analyze→Fit Y by X: One-Way Analysis of Variance

When you select the **Means /Anova ttest** from the **Analysis** popup menu, its One-way Anova table has an additional command in a popup next to the table name, called **Power Details.** This command, displays a dialog for you to specify what power calculations you want. JMP then gives calculations of statistical power and other details about a given hypothesis test.

Analyze→Fit Y by X: Contingency Tables

The **Fit Y by X** command analyzes pairs of variables, examining the distribution of a response variable **Y** as conditioned by the values of a factor **X**. When both the **X** and **Y** variables are nominal or ordinal, the Fit Y by X platform gives a two-way contingency table analysis for each pair of variables; a mosaic plot; Pearson and likelihood ratio chi-square tests; Fishers exact test if both variables have only two level; a kappa statistic when both variables have the same set of values; and optionally, a correspondence analysis.

Platform Options for Analyze→Fit Y by X: Contingency Tables

The popup menu icon at the top of the platform next to the variable name accesses the following options:

Mosaic Plot
graphically displays a two-way table. A *mosaic plot* is any plot divided into small rectangles such that the area of each rectangle is proportional to a frequency count of interest. There is a row for each response level and a column for each factor (sample) level. The borders of the table display the row, column, and grand totals.

The Crosstabs Table
initially appears as a two-way frequency table. There is a row for each response level and a column for each factor (sample) level. The borders of the table display the row, column and grand totals. The crosstabs table also has options, described next.

Tests
displays the likelihood ratio and Pearson chi-square statistics to test that the row and column variables are independent. If one variable considered as **Y** and the **X** is regarded as fixed, the chi-square statistics test that the distribution of the **Y** variable is the same across each **X** level. This is a test of marginal homogeneity. The Tests table is analogous to the Analysis of Variance table for continuous data. The negative log likelihood for categorical data plays the same role as sums of squares in continuous data.

If both variables have only two levels, Fisher's exact probabilities for the one-tailed tests and the two-tailed test are also shown. Also, the Kappa statistic (Agresti 1990) is shown in the Test table when the X and Y variables have the same set of values. Kappa measures the degree of agreement on a scale from zero to one. If two responses tend to agree, then most of the counts are on the diagonal.

Correspondence Analysis
is a graphical technique to show which rows or columns of a frequency table have similar patterns of counts. In the plot there is a point for each row and for each column. If two rows have similar row profiles, their points in the correspondence analysis plot are close together; likewise for columns. Although the distance between a row point and a column point has no meaning, the directions of columns and rows from the origin do have meaning, and the relationships will help interpret the plot. The technique is particularly useful for tables with large numbers of levels where deriving useful information from the table is more difficult.

Contingency Tables Report Options

The popup menu icon next to the Crosstabs table name accesses the following options, which let you decide what items to include in table cells:

Count is the cell frequency, margin total frequencies, and grand total (total sample size).

Total% is the percent of cell counts and margin totals to the grand total.

Row% is the percent of each cell count to its row total.

Col% is the percent of each cell count to its column total.

Expected is the expected frequency (E) of each cell under the assumption of independence. It is computed as the product of the corresponding row total and column total, divided by the grand total.

Deviation is the observed cell frequency (O) minus the expected cell frequency (E).

Cell ChiSq is the chi-square values computed for each cell as $(O - E)^2 / E$ where O is the observed cell frequency and E is the expected cell frequency.

Analyze→Fit Y by X: Logistic Regression

The **Fit Y by X** command analyzes pairs of variables, examining the distribution of a response variable Y as conditioned by the values of a factor X. When the Y column is nominal or ordinal and the X column is continuous gives a logistic regression analysis and displays a family of cumulative logistic probability curves. The is only one option available for this platform:

Inverse Prediction for Analyze→Fit Y by X: Simple Logistic Regression

To predict an exact X value for a given Y value is called *inverse prediction*. If your response has exactly two levels, a popup command shows beneath the plot that enables you to request inverse prediction.

Analyze→Fit Model: Dialog

The **Fit Model** command first displays the dialog that allows you to enter the model for your data. The Fit Model dialog is *nonmodal*, which means it persists until you explicitly close it. This is useful because often an analyst wants to change the model specification and fit several different models before closing the window.

The column selector list in the upper left of the dialog lists the columns in the current data table. When you click a column name, it highlights and responds to the action you choose with other buttons on the dialog.

Fit Model Dialog: Response Variable Role Assignment

To assign variables the **Y**, **Wt**, or **Freq** roles, select them and click the corresponding role button. If you want to remove preassigned variables from roles, highlight them and click **Remove.** The response role assignments are as follows:

- **Y** identifies one or more response (dependent) variables for the model.
- **Wt** is an optional role that identifies one column whose values supply weights for the response. Weighting affects the importance of each row in the model.
- **Freq** is an optional role that identifies one column whose values assign a frequency to each row for the analysis. The values of this variable determine how degrees of freedom are counted.

To assign a column the Label role, you must use the role assignment menu on the spreadsheet or the **Assign Roles** command in the **Cols** menu.

Fit Model Dialog: Effects Buttons

Suppose that a data table contains the variables X and Z with continuous values, and A, B, and C with nominal values. The following paragraphs describe the buttons that you can click in the Fit Model dialog to specify model effects.

> Add > To add a simple regressor (continuous column) or a main effect (nominal or ordinal column) as an X effect to any model, select the column from the column selector list and click **Add**. That column name appears in the **Effects in Model** list. As you add effects, be aware of the modeling type declared for that variable and the consequences that has in fitting the model:

- Variables with continuous modeling type become simple regressors.
- Variables with nominal modeling type are represented with *dummy* variables to fit separate coefficients for each level in the variable.
- Nominal and ordinal modeling types are handled alike, but with a slightly different coding.

If you mistakenly add an effect, select it in the **Effects in Model** list and click **Remove**.

> Cross > To create a compound effect by crossing two or more variables, first select the variables in the column selector list by SHIFT-clicking them. Then click **Cross**.

- Crossed nominal variables become interaction effects.
- Crossed continuous variables become multiplied regressors.
- Crossing a combination of nominal and continuous variables produces special effects suitable for testing homogeneity of slopes.

If you select both a column name in the selector list and an effect in the **Effects in Model** list, the **Cross** button crosses the selected column with the selected effect and adds this compound effect to the effects list.

> Nest > When levels of an effect B only occur within a single level of an effect A, then B is said to be *nested* within A and A is called the outside effect. To specify a nested effect:

1) Select the outside effects in the column selection list and click **Add** or **Cross**.
2) Select the outside effect in the **Effects in Model** list.
3) Select the nested variable in the column selection list and click **Nest**.

For example, suppose that you want to specify A*B[C]. Highlight both A and B in the column selection list. Then click **Cross** to see A*B in the Effects in Model list. Highlight the A*B term in the **Effects in Model** list and C in the column selection list. Click **Nest** to see A*B[C] as a model effect. Note: The nesting terms must be specified in order, from outer to inner. If B is nested within A, and C is nested within B, then the model is specified as: A, B[A], C[B,A].

Fit Model Dialog: Effect Macros

Commands from the **Effect Macros** popup menu automatically generate the effects for commonly used models. You can add or remove terms from an automatic model specification as needed.

Full Factorial

To look at many crossed factors, such as in a factorial design, use **Full Factorial**. It creates the set of effects corresponding to all crossings of all variables you select in the columns list. For example, if you select three variables A, B, and C, the **Factorial** selection generates A, B, A*B, C, A*C, B*C, A*B*C in the effect lists. To remove unwanted model interactions, highlight them and click **Remove**.

Factorial to a Degree

You can create a limited factorial by selecting **Factorial to a Degree** and entering the degree of interactions you want in the **Degree** box.

Factorial Sorted

The **Factorial Sorted** selection creates the same set of effects as **Full Factorial**, but lists them in order of degree. All main effects are listed first, followed by all two-way interactions, then all three-way interactions, and so forth.

Response Surface

In a response surface model, the object is to find the values of the terms that produce a maximum or minimum expected response. This is done by fitting a collection of terms in a quadratic model. The critical values for the surface are calculated from the parameter estimates and presented with a report on the shape of the surface.

To specify a response surface effect, select two or more variables from the column list. When you select **Response Surface** from the **Effect Macros** menu, response surface expressions appear in the **Effects in Model** list.

Mixture Response Surface

The mixture design omits the squared terms and the intercept and attaches the **{RS}** effect attribute to the main effects.

Polynomial to Degree

A polynomial effect is a series of terms that are powers of one variable. To specify a polynomial effect:

1) Click a variable in the column list.
2) Enter the degree of the polynomial in the **Degree** box
3) Use the **Polynomial to Degree** command in the **Effects Macros** menu

Fit Model Dialog: Effect Attributes Popup Selections

The **Effect Attributes** popup menu has two attributes you can assign to an effect in a model:

Random Effect

If you have multiple error terms or random effects, as with a split-plot or repeated measures design, you can highlight them in the effects list and select **Random Effect** from the **Effect Attributes** menu. This causes JMP to attempt construction of tests that have the right properties.

Response Surface

If you have a response surface model, you can use this attribute to identify which factors participate in the response surface. This attribute is automatically assigned if you use the **Response Surface** effects macro. You need only identify the main effects, not the crossed terms to obtain the additional analysis done for response surface factors.

Fit Model Dialog: Fitting Personalities

The Model Specification dialog in JMP serves many different fitting methods. Rather than have a separate dialog for each method, there is one dialog with a choice of *fitting personality* for each method. The following list briefly describes each type of model. Fitting personalities are also summarized in Appendix A, "The JMP Main Menu." Options for the fitting platforms are listed in this appendix.

Standard Least Squares

JMP models continuous responses in the usual way through fitting a linear model by least squares. Features include leverage plots, least squares means, contrasts, and output formulas. The standard least squares personality allows only one continuous Y and is the default for this situation.

Screening Models

The screening personality of the fitting platform is designed to analyze experimental data where there are many effects but few observations. Traditional statistical approaches focus on the residual error. However, because in near-saturated designs there is little or no room for estimating residuals, the focus is on the prediction equation and the effect sizes. Of the many effects that a screening design can fit, you expect a few important terms to stand out in comparison to the others. Also, because the goal of the experiment is usually to optimize some response rather than show statistical significance, the factor combinations that optimize the predicted response are of overriding interest. The screening personality allows multiple continuous Y variables.

Stepwise Regression

Stepwise regression is an approach to selecting a subset of effects for a regression model. The stepwise feature computes estimates that are the same as those of other least squares platforms, but it facilitates searching and selecting among many models. The stepwise personality allows only one continuous Y variable.

Nominal Logistic Models

If the response is nominal, the Fit Model platform fits a linear model to a multilevel logistic response function using maximum likelihood.

Ordinal Logistic Models

If the response variable has an ordinal modeling type, the Fit Model platform fits the cumulative response probabilities to the logistic distribution function of a linear model using maximum likelihood.

Fit Model Dialog: Saving and Retrieving Models

To save and later retrieve a model, use two buttons labeled **Get Model** and **Save Model** on the lower part of the Fit Model dialog:

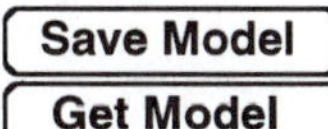

Save Model saves the model showing in the Fit Model dialog. **Get Model** retrieves and displays a model previously saved from a Fit Model dialog. The model is saved in an ordinary text file, which can be edited with any text editor.

Fit Model Dialog: Other Dialog Features

Two check boxes at the lower left of the dialog give you additional model options:

No Intercept

omits the intercept parameter from the model.

Defer Plots

initially suppresses all plots and displays only tables for the standard least squares personality. This speeds up processing and leaves you the option of requesting plots later.

Analyze→Fit Model: Standard Least Squares

JMP models continuous responses in the usual way through fitting a linear model by least squares. Features include leverage plots, least squares means, contrasts, and output formulas. The **Standard Least Squares** selection from the fitting personality popup menu on the Fit Model dialog allows only one continuous Y variable (response).

Platform Options for Analyze→Fit Model: Standard Least Squares

There are many options with the Fit Model report that let you tailor the display and request additional tables. The following options are listed on the check-mark popup icon on the lower left of the window border:

Plot Y by Predicted

displays the observed values by the predicted values of Y. This is the leverage plot for the whole model.

Plot Residuals

displays the residual values by the predicted values of **Y**.

Plot Effect Leverage

displays the leverage plots for each effect in the model, showing the point-by-point composition of the test for that effect.

Show Points

alternately hides or displays the points on all the leverage plots.

Power Details

You complete the dialog to see the details for LSV (least significant value), LSN (least significant number), and power for all parameter estimates. The dialog lets you specify a range for which details are to be computed, significance level, standard error of the residual, raw effect size, and sample size.

These are the LSV (least significant value), LSN (least significant number), and adjusted power for an alpha of .05 for each parameter in the linear model.

Custom Test

enables you to test a custom hypothesis that is not already tested. This option gives you a dialog to construct tests in terms of all the parameters.

Seq SS (Type I)

(Sequential [Type I] sums of squares) shows the reduction in residual sum of squares as each effect is entered into the fit. They are model-order dependent. Each effect is adjusted only for the preceding effects in the model. Therc a number of models for which the Type I hypotheses are considered appropriate.

Inverse Prediction

displays a dialog that lets you ask for a specific value of one independent (**X**) variable, given a specific value of a dependent variable. The values of other independent effects are held constant at their mean value for the inverse prediction computation. The inverse prediction computation (also called calibration) includes confidence limits.

Durbin-Watson Test

displays the Durbin-Watson statistic to test whether or not the errors have first-order autocorrelation. The autocorrelation of the residuals is also shown. This Durbin-Watson table is only appropriate for time-series data when you suspect that the errors are correlated across time.

Correlation of Estimates

displays an additional table showing the correlations of the model parameter estimates.

Save Commands for Analyze→Fit Model: Standard Least Squares

When you click the save ($) icon on the lower left of the window border, a popup menu offers the following choices. Save commands generate one or more new columns in the current data table. The new columns are given names that identify the save command that was selected and the response variable in the model:

Save Predicted Values

creates a new column called **Predicted** *response variable* containing the predicted values computed by the specified model. Use the **Save Prediction Formula** option described later in this list to always see predicted values for all rows.

Save Residuals

creates a new column called Residual *response variable* containing the residuals, which are the observed response values–predicted values.

Save Prediction Confidence

creates two new columns called Lower95% Pred *response variable* and Upper95% Pred *response variable*. The new columns contain the lower and upper 95% confidence limits for the line of fit.

Save Indiv Confidence

creates two new columns called Lower95% Indiv *response variable* and Upper95% Indiv *response variable*. The new columns contain lower and upper 95% confidence limits for individual response values.

Save Studentized Residual

creates a new column called Studentized Resid *response variable*. The new column values are the residuals divided by their standard error.

Save h[i]

creates a new column called h *response variable*. The new column values are the diagonal values of the matrix, $X(X'X)^{-1}X'$, sometimes called *hat* values.

Save Std Error of Predicted

creates a new column, called StdErr Pred *response variable,* containing the standard errors of the predicted values.

Save Std Error of Residual

creates a new column called, StdErrResid *response variable,* containing the standard errors of the residual values.

Save Std Error of Individual

creates a new column, called StdErr Indiv *response variable,* containing the standard errors of the individual predicted values.

Save Effect Leverage Pairs

creates a set of new columns that contain the values for each leverage plot. The new columns consist of an X and Y column for each effect in the model.

Save Cook's D Influence

saves the Cook's D influence statistic. Influential observations are those that, according to various criteria, appear to have a large influence on the parameter estimates.

Save Prediction Formula

creates a new column, called Pred Formula *response variable*, containing the predicted values computed by the specified model. It differs from the **Save Predicted Values** column in that the prediction formula is saved with the new column. This is useful for predicting values in new rows or for obtaining a picture of the fitted model.

The prediction formula requires considerable space for large models. If you do not need the formula with the column of predicted values, use the **Save Predicted Values** option.

Save StdErr Pred Formula

creates a new column, called Pred SE *response variable*, containing the standard error of the predicted values. It is the same as the **Save Std Error of Predicted** option, but saves the formula with the column. This option only works if all effects are continuous and can produce a very large formula.

Parameter Estimates Table Options

The popup menu icon next to Parameter Estimates table name has options that allows you to show or hide columns in the table. The information (options) for each parameter include **Std Error, T-test, Prob >ltl, 95% Confid. Limits, Stdized Beta,** and **VIF.**

Effect Details Options

Each effect has a popup menu icon next to its name. The menu items let you request more information about that effect. These commands are dimmed if they are inappropriate for a particular effect.

Add Contrast
displays a dialog that allows you to specify the contrasts you want with respect to that effect. This command is enabled only for pure classification effects.

Plot Effect
plots the profile of the least squares means for nominal and ordinal main effects main effects and two-way interactions. The screening fit facility also offers interaction plots in a different format.

Power Details
displays the Power calculation dialog for the specific effect (as opposed the whole model option described previously). You complete the dialog to see the details for LSV (least significant value), LSN (least significant number), and power for that effect. The dialog lets you specify a range for which are details to be computed. Power calculations include significance level, standard error of the residual, raw effect size, and sample size.

Power Details Plot Option

The popup command located next to the Power table gives you the command **Plot Power by N**, which plots the Power by Number columns from the Power Details table.

Analyze→Fit Model: Screening Models

The screening personality of the fitting platform is designed to analyze experimental data where there are many effects but few observations. Traditional statistical approaches focus on the residual error. However, in near-saturated designs, the focus is on the prediction equation and the effect sizes because there is little or no room for estimating residuals. Of the many effects that a screening design can fit, you expect a few important terms to stand out in comparison with the others. Also, because the goal of the experiment is usually to optimize some response rather than to show statistical significance, the factor combinations that optimize the predicted response are of overriding interest.

Platform Options for Analyze→Fit Model: Screening Models

The check-mark popup menu lists the four kinds of screening analyses:

Profiler
displays the Prediction Profile plot, which shows the predicted response as one variable changes while the others are held constant. The Prediction Profiler displays *prediction traces* for each X variable. A prediction trace is the predicted response as one variable is changed while the others are held constant at the current values. The Prediction Profiler recomputes the traces as you drag

a vertical dotted line that varies the value of an X variable. The Prediction Profiler is a way of changing one variable at a time and looking at the effect on the predicted response.

Cube Plots

shows predicted values in a cube layout. A *cube plot* is a set of predicted values for the extremes of the factor ranges, laid out on the vertices of one or more cubes. If a factor is nominal, the vertices are the first and last level. To change the layout so that terms map to different cube coordinates, click one of the term names in the first cube and drag it to the coordinate you want.

Effect Screening

screens parameter estimates for evidence of an active contribution to the fit. In experimentation, it is often the effect size rather than the statistical significance that is of interest. Often there are very few degrees of freedom left after model fitting to estimate an error variance. A common approach is to assume that many of the effects are zero and that their estimates reflect random noise. Then make inferences based on the distribution of effect sizes (parameter estimates).

The **Effect Screening** display shows a Normal Plot. The parameter estimates show on the vertical axis and the normal quantiles on the horizontal axis. If all effects are due to random noise, they follow a straight line with slope σ, the standard error. The line with slope equal to the Lenth's PSE estimate is shown in blue. If there are degrees of freedom for error, then a line with slope equal to the root mean squared error (RMSE) shows in red. Lenth's PSE is computed using the normalized estimates and disregards the intercept. The Normal plot helps you pick out effects that deviate from the normal lines. Estimates that deviate substantially are labeled.

Interaction Plots

shows a matrix of interaction profile plots for the two-factor effect. In an interaction plot, evidence of interaction shows as nonparallel lines.

Prediction Profiler Report Options

The Prediction Profiler, described above, has a popup command called **Desirability Functions**. This command appends a new row to the bottom of the profiler plot matrix, dedicated to measuring desirability. The row has a plot for each factor showing its *desirability trace*. It also adds a column that has a three-point desirability function for each Y variable. The overall desirability measure shows at the left of the row of desirability traces. To use a variable's desirability function, drag each of the three function points vertically to represent a response value and horizontally to choose the desirability of that response. As you drag a desirability function point, the changing response value and the desirability value (between 0 and 1) show in the cell labeled **Desirability** at the lower right of the plots. The dotted line is the response for the current factor settings. The overall desirability shows to the left of the row of desirability traces.

Analyze→Fit Model: Stepwise Regression Control Panel

Stepwise regression is an approach to selecting a subset of effects for a regression model. The **Stepwise** feature computes estimates that are the same as those of other least squares platforms, but it facilitates searching and selecting among many models. The Stepwise selection allows only one continuous Y. It is used when there is little theory to guide the selection of terms for a model, and the modeler, in desperation, wants to use whatever seems to provide a good fit.

The approach is somewhat controversial. The significance levels on the statistics for selected models violate the standard statistical assumptions because the model has been selected, rather than tested within a fixed model. On the positive side, the approach has been of practical use for 30 years in helping reduce models to predict many kinds of responses. The book, *Subset Selection in Regression*, by A. J. Miller brings statistical sense to model selection statistics.

Stepwise lets you run experimental models with a control panel. You can change model characteristics by entering new values into the control panel. Then rerun the model as often as you want. You use the control panel as follows:

Prob to Enter is the significance probability that must be attributed to a regressor term for it to be considered as a forward step. Click the field to enter a value.

Prob to Leave is the significance probability that must be attributed to a regressor term in order for it to be considered as a backward step. Click the field to enter a value.

Direction is the popup menu s with selections Forward, Backward, and Mixed.

Forward brings in the regressor that most improves the fit, given that term is significant at the level specified by Prob to Enter.

Backward removes the regressor that affects the fit the least, given that term is not significant at the level specified in Prob to Leave.

Mixed alternates forward and backward steps. It includes the most significant term that satisfies Prob to Enter and removes the least significant term satisfying Prob to Leave. It continues removing terms until the remaining terms are significant, and then changes to the forward direction.

Buttons on the controls panel let you control the stepwise processing:

Go starts the selection process. The process continues to run in the background until the model is finished.

Stop stops the background selection process.

Enter All enters all unlocked terms into the model.

Remove All removes all terms from the model.

Make Model forms a model for the Fit Model Dialog from model currently showing in the Current Estimates table. In cases where there are nominal or ordinal terms, Make Model can create new data table columns to contain terms that are needed for the model.

Analyze→Fit Model: Categorical Models (Nominal or Ordinal Responses)

If the model response is nominal, the Fit Model platform fits a linear model to a multilevel logistic response function using maximum likelihood. Likelihood Ratio test statistics are computed for the whole model. Lack of Fit tests and Wald test statistics are computed for each effect in the model. Options give Likelihood ratio tests for effects, and confidence limits and odds ratios for the maximum likelihood parameter estimates.

If the response variable is ordinal, the platform fits the cumulative response probabilities to the logistic distribution function of a linear model using maximum likelihood. Likelihood Ratio test statistics are provided for the whole model and lack of fit.

Platform Options for Analyze→Fit Model: Nominal or Ordinal Responses

The check-mark menu on the window border gives you the four additional commands :

The Likelihood Ratio Tests

produces a table that shows the Likelihood Ratio chi-square tests, which are calculated as twice the difference of the log likelihoods between the full model and the model constrained by the hypothesis to be tested (the model without the effect). These tests can take time to do because each test requires a separate set of iterations.

Confidence Intervals for Effects

requests profile likelihood confidence intervals for the model parameters. When you select **Confidence Intervals** a dialog prompts you for α to compute the $1-\alpha$ confidence intervals, or you can use the default of $\alpha = .05$.

Odds Ratios

extends the Parameter Estimates table to show the Odds Ratios. If you had previously selected the **Confidence Intervals** command, the Parameter Estimates table also shows confidence limits for the odds ratios. The odds ratio is calculated as

$$\exp((x_i\max - x_i\min) \bullet b$$

where $x_i$max and $x_i$min are the maximum and minimum values of the *i*th column that corresponds to the parameter estimate, b. This ratio measures how the fitted probability is multiplied as the regressor changes from its minimum to its maximum (Hosmer and Lemeshow 1989).

Inverse Prediction

finds the X value that results from a specified probability and gives a fiducial confidence interval for this prediction.

Save Commands for Analyze→Fit Model: Nominal or Ordinal Responses

If you have ordinal or nominal response models, the Save ($) menu offers the **Save Prob Formulas** command. If the response is numeric and is assigned the ordinal modeling type, the **Save Quantiles** and **Save Expected Values** commands are also available.

Save Prob Formulas

creates columns in the current data table that save formulas for linear combinations of the response levels, prediction formulas for the response levels, and a prediction formula giving the most likely response.

For a nominal response model with *r* levels, JMP creates

- Columns called Lin[*j*] that contain a linear combination of the regressors for response levels $j = 1,\ldots, r-1$.
- Acolumn called Prob[*r*], with a formula for the fit to the last level, *r*.
- Columns called Prob[*j*] for $j < r$ with a formula for the fit to level *j*.

- A column called **MostLikely** *responsename* that picks the most likely level of each row based on the computed probabilities.

For an ordinal response model with *r* levels, JMP creates

- aAcolumn called **Linear** that contains the formula for a linear combination of the regressors without an intercept term.
- Columns called **Cum**[*j*], each with a formula for the cumulative probability that the response is less than or equal to level *j*, for levels $j = 1,\ldots, r - 1$. There is no **Cum**[*r*] that is 1 for all rows.
- Columns called **Prob**[*j*], for $1 < j < r$, each with the formula for the probability that the response is level *j*. **Prob**[*j*] is the difference between **Cum**[*j*] and **Cum**[*j* - 1]. **Prob**[1] is **Cum**[1], and **Prob**[*r*] is 1 - **Cum**[*r* - 1].
- aAcolumn called **MostLikely** *responsename* that picks the most likely level of each row based on the computed probabilities.

Save Quantiles

creates columns in the current data table named **OrdQ.05**, **OrdQ.50**, and **OrdQ.95** that fit the quantiles for these three probabilities.

Save Expected Values

creates a column in the current data table called **Ord Expected** that contains the expected values of the response variable. The column also has the formula that is the linear combination of the response values with the fitted response probabilities for each row.

Analyze→Fit Model: Summary of Tables from General Fitting Platform

Table B.1 is an alphabetical list of most of the tables produced by the Fit Model platform.

Table B.1 Summary of Tables Given by the General Fitting Platform

Table Name	Fitting Personality	Model Effects
Analysis of Loglikelihood	nominal/ordinal	standard
Analysis of Variance	standard LS/screening	standard
Contour Plot Specification	standard LS/screening	response surface
Correlation of Estimates	standard LS, screening	standard
Current Estimates	stepwise	standard
Custom Test	standard LS	standard
Durbin-Watson	standard LS	standard
Effect Likelihood-Ratio Tests	nominal/ordinal	standard
Effect Test *all effects in single table*	standard LS/screening	standard
Effect Test *individual tables*	standard LS	standard
Eigenstructure	standard LS/screening	response surface
Expected Mean Squares	standard LS	random effect
Inverse Prediction	standard LS/nominal/ordinal	standard
Iteration History	ordinal/nominal	standard
Lack of Fit	standard LS	*for replicated values*

continued next page

Table B.1 (continued) Summary of Tables Given by the General Fitting Platform

Least Squares Means	standard LS	standard
Odds Ratios	Ordina/Nominal	standard
Parameter Estimates	all	standard
Partial Correlation	Manova	standard
Power Details	standard LS	standard
Response Surface	standard LS/screening	response surface
Risk Ratios	proportional hazard	standard
Sequential (Type I) Tests	standard LS	standard
Singularity Details	standard LS	linear combinations
Solution	standard LS/screening	response surface
Step History	stepwise	standard
Summary of Fit	all	standard
Test Denominator Synthesis	standard LS	random effects
Test wrt Random Effects	standard LS	random effects
Variance Components Estimates	standard LS	random effects

Analyze→Fit Nonlinear

The Nonlinear Fit platform begins with an interactive panel for controlling the iteration process. The Nonlinear Fitting Control Panel has

- Buttons to start, stop, and step through the fitting process, and to reset parameter values.
- Fitting options to specify loss functions and computational methods.
- A *processing messages* area, which shows the progress of the fitting processing
- A list of current and limit convergence criteria and step counts, current parameter estimates, and error sum of squares.
- Options to specify the alpha level for confidence intervals and delta for numerical derivatives.

Save Commands for Analyze→Fit Nonlinear

The only optional command after the fitting is complete lets you save the sum of squares error in a new data table to use for plotting. The **Save SSE Grid Table** command in the Save ($) border menu opens the Specify Grid for Output panel, which lets you enter the maximum and minimum for each parameter and the number of points you want in the grid.

Analyze→Correlation of Y's

Platform Options for Analyze→Correlation of Y's

By default, the **Correlation of Y's** platform displays a standard correlation matrix. The checkmark popup menu lists additional correlations options and other techniques that allow you to look at multiple variables:

Correlation-Multivar

gives a matrix of correlation coefficients, which summarizes the strength of the linear relationships between each pair of response (Y) variables. This correlation matrix only uses the observations that have nonmissing values for all variables in the analysis. The **Correlations-Pairwise** option described later computes correlations using all values for a given pair of variables. These numbers are the Pearson product-moment correlations.

Correlations-Inverse

provides useful multivariate information. The diagonal elements of the inverse correlation matrix are a function of how closely the variable is a linear function of the other variables. In the inverse correlation, the diagonal is $1/(1-R^2)$ for the fit of that variable by all the other variables. If the multiple correlation is zero, the diagonal inverse element is 1. If the multiple correlation is 1, then the inverse element becomes infinite and is reported missing.

Correlation-Partial

shows the partial correlations of each pair of variables after adjusting for all the other variables. It happens that this is the negative of the inverse correlation matrix scaled to unit diagonal.

Correlations-Pairwise

lists the Pearson product-moment correlations for each pair of Y variables, using all available values. The count values differ if any pair has a missing value for either variable. These are the same correlations you get if you use the Fit Y by X platform for each pair one at a time and request the **Density Ellipses** option. The Pairwise Correlations report also shows significance probabilities and compares the correlations with a bar chart.

Correlations-Nonpar

displays the Nonparametric Measures of Association table. You can request Spearman's Rho, Kendall's tau-b, and the Hoeffding D statistics for each pair of variables. The Nonparametric Measures of Association report also shows significance probabilities for all measures and compares them with a bar charts.

Scatterplot Matrix

shows scatterplot for each pair of response variables arranged in a matrix. The cells of the scatterplot matrix are size-linked so that the stretch box from any cell resizes all the scatterplot cells. A 95% bivariate normal density ellipse is imposed on each scatterplot. If the variables are bivariate normally distributed, this ellipse encloses approximately 95% of the points (when the number of points is large). The correlation of the variables is seen by the collapsing of the ellipse along the diagonal axis. If the ellipse is fairly round and is not diagonally oriented, the variables are uncorrelated.

Outlier Analysis

displays a plot that shows the *Mahalanobis* distance of each point from the multivariate mean (centroid). The standard Mahalanobis distance depends on estimates of the mean, standard deviation, and correlation for the data. The distance is plotted for each observation number. The extreme multivariate outliers are identified by highlighting the points with the largest distance values. The outlier analysis report has additional options describe later.

Prin comp/Factor

displays a report of linear combinations of the original variables. The linear combinations are computed such that the first principal component has maximum variation, the second principal component has the next most variation subject to being orthogonal to the first, and so on. Principal components analysis is also implemented on the Spinning Plot platform, which is the preferable approach unless you have too many variables to show graphically.

The principal components text report summarizes the variation of the specified Y variables with principal components. The principal components are derived from an eigenvalue decomposition of the correlation matrix of the variables. The details in the report show how the principal components absorb the variation in the data and in which direction the components lie in terms of the variables. The principal component points are derived from the eigenvector linear combination of the standardized variables.

Prin comp/Covar

performs the same principal components and factor analysis described above, but assumes the Y columns form a covariance matrix.

Nonparametric Correlations Report Options

The popup menu next to the Nonparametric Measures of Association table lists the three available nonparametric measures:

Spearman's Rho

shows the table of correlation coefficients computed on the ranks of the data values instead of on the values themselves.

Kendall's tau b

displays coefficients based on the number of concordant and discordant pairs of observations.

Hoeffding D

uses a scale that ranges from –.5 to 1, with only large positive values indicating dependence.

Scatterplot Matrix Options

The popup menu next to the Scatterplot matrix name provides these two commands:

Density Ellipses

shows or hides 95% ellipses on the scatterplots.

Color

lets you select from a palette of colors to change the color of the density ellipses if you are working on a color monitor.

Outlier Analysis Report Options

The popup menu next to the Outliers analysis name provides these two commands:

Outlier Distances

shows the *Mahalanobis* distance of each point from the multivariate mean (centroid). The distance is plotted for each observation number. The standard Mahalanobis distance depends on estimates of the mean, standard deviation, and correlation for the data.

Jackknifed Distances

calculates an alternate distance using a *jackknife* technique. The distance for each observation is calculated with estimates of the mean, standard deviation, and correlation matrix that do not include the observation itself.

Save Outlier Distance

saves the Mahalanobis distance computations as a new column in the current data table.

Save Jackknifed Distance

saves the jackknifed distance computations as a new column in the current data table.

Principal Components Report Options

The popup menu next to the Prin. Components/Factor Analysis table name gives you options to see the results of rotating principal components and to save principal component scores.

Rotate/Factor Analysis

gives factor-analysis-style rotations of the principal components. Selecting this command causes a prompting dialog requesting the number of components to include, whether to use the covariance matrix or the correlation matrix in calculations, and whether to standardize the data or not.

Rotations are used to better align the directions of the factors with the original variables so that the factors may be more interpretable. You hope for clusters of variables that are highly correlated to define the rotated factors. The rotation's success at clustering the interpretability is highly dependent on the number of factors you choose to rotate.

The rotation method is called *varimax* rotation and is the same as that traditionally used in factor analysis. The rotated components correspond to what Kaiser (1963) calls the *Little Jiffy* approach to factor analysis. The varimax method tries to make elements of this matrix go toward 1 or 0 to show the clustering of variables.

Save Prin Comp
Save Rotated Comp

save the scores for all the principal components or rotated components you request as new columns in the current data table.

Analyze→Survival

Platform Options for Analyze→Survival

The check-mark popup menu accesses options for the Survival platform. All of the platform options alternately hide or display information. The following list summarizes the platform options:

Show Points

shows the sample points at each step of the survival plot. Failures are shown at the bottom of the steps, and censorings are indicated by points above the steps.

Show Combined

displays the survival curve for the combined groups in a thick gray line in the Survival Plot.

Survival Plot

displays overlaid survival plots for each group.

Exponential Plot

plots **-log(Surv)** by time for each group where **Surv** is the product-limit estimate of the survival distribution. Lines that are approximately linear empirically indicate the appropriateness of using an exponential model for further analysis.

Weibull Plot

plots **log(-log(Surv)** by the log of time for each group, where Surv is the product-limit estimate of the survival distribution. A Weibull plot that has approximately parallel and straight lines indicates a Weibull survival distribution model is appropriate to use for further analysis.

Lognormal Plot

plots the inverse cumulative normal distribution of 1 minus the survival (denoted **Probit(1-Surv)**) by the log of time for each group, where Surv is the product-limit estimate of the survival distribution. A log normal plot that has approximately parallel and straight lines indicates a log normal distribution is appropriate to use for further analysis.

Competing Causes

prompts for a column in the data table that contains names of causes of failure. Then for each cause the estimation of the Weibull model is performed using that cause to indicate a failure event, and other causes to indicate censoring. The fitted distribution is shown by a dashed line in the Survival Plot.

Save Commands for Analyze→Survival

The **Save Estimates** command in the Save ($) menu saves the information in the survival estimates along with multiple columns of information about the survival analysis. There is also a **Save Estimates** command specific for the competing risks analysis, which appears only after requesting **Competing Causes** from the check-mark menu.

Survival Plots Options

When you select the **Exponential Plot**, **Weibull Plot**, or **Lognormal Plot** from the platform check menu, a popup command next to the plot name estimates the distribution parameter and appends a table of estimates to the report.

Competing Risks Analysis Options

When you select the Competing Causes, a popup menu icon shows next to the competing causes report name for each group. These options let you omit causes from the competing risk analysis and save information about the analysis.

Omit Causes

prompts you for one or more cause values to omit and again calculates competing risks. The results are overlaid on the survival plot.

Save Estimates

creates a new JMP table that lists the causes of failure and the alpha and beta Weibull parameter estimates for each cause for each group.

Save cause-by-cause coord

adds a new column to the current table called **log(-log(Surv))**. This information can produce useful plots. For example, use **Tables→Group/Summary** with the **By-Groups** in effect. Then use the **Graph→Overlay Plots** to see plots of **log(-log(Surv))** by **log(time)**.

Graph→Bar/Pie Charts

Platform Options for Graph→Bar/Pie Charts

The check-mark popup menu at the lower left of the display window accesses options that let you tailor the appearance of the Bar/Pie charts. These *platform* options affect every plot in the report window.

Vertical
changes a horizontal bar chart or a pie chart to a vertical bar chart.

Horizontal
changes a vertical bar chart or a pie chart to a horizontal bar chart.

Pie
changes the chart type to pie chart.

Bar
redraws the default bar chart.

Line
replaces a bar chart with a line chart connecting the plotted points. You can also choose the **Mark Points** option to show or hide the points, and **Connect Line** to suppress the connecting lines and show points only.

Needle
replaces each bar with a line drawn from the axis to the plotted valuc for all charts in the window. The **Needle Chart** option is also available for individual charts.

Overlay
displays a single overlaid chart when you have more than one Y variable. You can assign any chart type to the individual charts and overlay them.

Stack
applies only when you have a second grouping variable (called a subgroup variable). **Stack** combines the bars of the grouping variable and creates a single divided bar for each level of the subgroup variable. The bar segments are proportional to the levels of the grouping variable. If there is more than one Y variable you can overlay the stacked charts.

Thin Needles
uses .25 point line when drawing a needle chart.

Thick Lines
uses a 2 point line to connect points.

Large Markers
uses larger than usual markers for points.

Use Pattern
changes the solid fills of bars to patterned fills.

Vertical Legend
displays the legend for overlaid plots across the bottom of the chart instead of listed beneath the chart.

Hide Options Buttons
hides the single-chart popup menu buttons beneath the plots. This is useful to print plots when you don't want the option buttons to show. The single-chart options are still available by OPTION-clicking (ALT-click under Windows) in the plot area.

Single-Chart Options

Popup menu icons labeled with the name of each Y statistic you requested show beneath the plot frame. This menu also appears if you OPTION-click (ALT-click under Windows) in the chart's display area. You can modify an individual charts with these *single-chart* selections:

Bar Chart
displays a bar for each level of the chart variables. The default chart is a bar chart.

Line Chart
replaces a bar chart with a line chart, which connects each point with a straight line. You can also choose the **Mark Points** option to show or hide the points. **Line Chart** is also available as a whole-window option, which applies to all charts at once.

Needle Chart
replaces each bar with a line drawn from the axis to the plotted value. **Needle Chart** is also available as a whole window option, which applies to all charts at once.

Show Points
toggles the point markers on a line or needle chart off and on.

Connect Line
toggles the line connecting points off, leaving a point chart.

Overlay Color
assigns color to a chart to identify it when overlaid with other charts.

Overlay Marker
assigns plot points a marker to identify them in overlaid charts.

Overlay Pattern
assigns bars a fill pattern to identify them in overlaid charts.

Label Category Axis
toggles the labels across the X-axis (category labels) off and on. Note: The overlay patterns show on bars only when the **Overlay** and **Use Patterns** options are in effect.

Individual-Level Options

Popup menu icons labeled with the name of the X grouping variable you used show beneath the plot frame. To use these *individual-chart options*, click one or more bars, points, or pie slices to select chart levels and their corresponding rows in the active data table. The n use the popup menu to specify a **Color, Marker,** or **Pattern** for the selected groups. Note: The **Use Pattern** option for the whole window must be in effect for the **Level Pattern** individual-level option to be active.

Graph→Overlay Plots

Platform Options: Graph→Overlay Plots

The check-mark popup menu at the lower left of the display window accesses options that let you tailor the appearance of the overlay plots. These *platform options* affect every plot in the report window.

Overlay
overlays plots for all columns assigned the Y role. Plots initially appear overlaid with the **Connect Points** option in effect. When you turn the **Overlay** option off, the plots show separately.

Needle

joins the overlaid Y values vertically at each X value. This option can be used with or without the **Connect Points** option in effect for the individual Y variables.

Range

connects the lowest and highest points at each X value with a line with bars at each end. Note: The **Needle** and **Range** options are mutually exclusive.

Separate X Axis

lets you print the X axis scale values only once on the last plot in the window. When **Separate X Axis** is selected, the X axes for other Y variables show tick marks, but show no scale values.

Rescale Axis

rescales the axes when points are added to the current data table and automatically update the plot.

Single-Plot Options

Popup menu icons labeled with the name of each Y variable show beneath the plot frame. You can modify an individual plot with these *single-plot* selections:

Show Points

alternately shows or hides points.

Connect

alternately connects the points with lines. You can use **Connect** without showing points.

Step

joins the position of the points with a discrete step by drawing a straight horizontal line from each point to the X value of the following point, and then a straight vertical line to that point. You can use **Step** without showing points.

Needle

draws a straight vertical line from each point to the X axis. Note: The **Connect**, **Step**, and **Needle** options are mutually exclusive.

Overlay Markers

assigns markers to plotted points using the standard JMP marker palette.

Connect Color

displays the standard JMP color palette for assigning colors to lines that connect points.

Graph→Spinning Plot

Platform Options for Spinning for Graph→Overlay Plots

The check-mark popup menu at the lower left of the display window accesses options that let you request principal component analysis.

Add Variables

brings up a variable selection dialog to add more variables to the spin components list.

Principal Components

calculates principal components on the set of variables in the components list and adds the principal components to the list. You can do this repeatedly with different current data values and variables.

Std. Prin. Components

calculates principal components the same as the **Principal Components** command, but scales the principal component scores to have unit variance instead of canonical variance. Use this option if you want GH' rather than JK' biplots. The interpoint distance shown by GH' biplots is less meaningful, but the angles of the biplot rays measure correlations better.

Rotated Components

prompts you to specify the number of factors you want to rotate and then computes that many rotated component scores.

Biplot Rays

alternately shows or hides rays on the plot corresponding to the principal components on the variables, or the variables on the principal components.

Remove Prin. Comp.

removes the principal components and their report from the session.

Save Commands for Spinning for Graph→Overlay Plots

The Save ($) menu offers these two commands:

Save Prin Comp Scores

saves the number of current principal component scores you request as columns in the current data table. These columns include their formulas and are locked. For *n* variables in the components list, *n* principal component columns are created and named Prin Comp 1, Prin Comp 2, ... Prin Comp *n*.

Save Rotated Scores

saves the rotated component scores as columns in the current data table. These columns include their formulas and are locked. If you requested *n* rotated components then *n* rotated component columns are created and named Rot Comp 1, Rot Comp 2, ... Rot Comp *n*.

Options to Tailor the Spinning Plot Appearance

The popup menu beside the spin control panel menu provides a variety of display options. All the display options except scale down, scale up, and home in the control panel are toggles that turn the feature on or off. The feature is in effect if there is a check-mark next to it.

White Background

alternates the background of the Spin platform between black and white. This option is independent of the background option in the Preferences window. Also, printing or cut-and-paste operations always use a white spin background, regardless of this option.

Axis Lines

shows or suppresses the X, Y, and Z axes in the spinning plot.

Box

draws a reference cube around the point cloud.

Rays

draws a line from each point to the origin of the plot.

Color

specifies that the points show in the current row-state color assigned to rows in the data table instead of black or white only.

Depth Cue

sets on or off the display of points in two depths (near points bright, distant points dim).

Markers vs. Dots
specifies that the points show as the current row-state markers assigned to rows in the data table instead of as uniform large dots.

Small Dots
displays all the points as small dots and speeds up the spinning a bit. This option takes precedence over the **Markers vs. Dots** option.

Background Coast
specifies whether coasting spins (continuous spin when you SHIFT-click with either spin icons or the hand tool) are dedicated tasks or background tasks.

Fast Mode
helps the spin rotate smoothly if you have thousands of points. However, this option does not support colors, markers, or other row states. It is most effective if the size and scale of the plot are left in their initial states. This option is only available in 8-bit video (256 color) mode.

Save as Movie (Macintosh only)
builds a stand-alone movie file on disk that can be viewed with any application that supports QuickTime movies, such as Apple's Simple Movie Player application. To record a movie select **Save as Movie** and give the new movie file a name and disk location. From that point on, any movement of the spinning plot generates a movie frame. Use the Spin Control to rotate the plot about any axis or zoom in and out. Other display options such as **Box** and **Rays** show in movies. Note: To stop recording, uncheck the **Save as Movie** option.

You record spin movies by moving the plot with either the hand tool or the spin controls. Anytime the spinning plot is in motion, frames are being recorded. No frames are generated while the spin plot is sitting idle.

Copy as Movie (Macintosh only)
builds a movie file on disk, and places movie information on the clipboard. This enables you to paste a movie into another application that supports QuickTime movies. To **Copy as Movie**, follow the same steps described for **Save as Movie**.

Graph→Pareto Charts

Platform Options for Graph→Pareto Charts

The check-mark menu at the lower left of the display window accesses options that let you tailor the appearance of the Pareto charts.

% Scale
toggles between count and percent display.

Pie Chart
changes bar chart to pie chart.

Horizontal Layout
arranges single X charts in either a row or a column.

Cum % Curve
adds cumulative percent curve above bars.

Curve Color
shows palette of colors to color cumulative curve.

Cum % Axis
shows cumulative percent axis on vertical right axis.

Cum % Points
shows points on the cumulative line at the midpoints of the bars.

Reorder Horizontal and **Reorder Vertical**
work together to reorder the rows and columns of the matrix of Pareto charts by selected values of the X variables, and to redefine to the key cell.

Category Legend
toggles between labeled bars and category legend.

N Legend
toggles the N= legend on and off.

Individual-Bar Options

When you select bars, you can access a menu of options by clicking the popup menu icon beneath the Pareto charts report, or by OPTION-clicking (ALT-clicking under Windows) anywhere in the chart area. These options apply to bars within charts instead of to the charts as a whole. You can highlight a bar by clicking on it, and you can extend the selection in standard fashion by SHIFT-clicking adjoining bars. Use COMMAND-click to select multiple bars that are not contiguous.

Other control charts can use all of the following options:

Combine
combines selected (highlighted) bars.

Separate
separates selected bars into their original component bars.

Move to First
moves one or more highlighted bars to the right (last) position.

Move to Last
moves one or more highlighted bars to the left (first) position.

Color
shows the color palette to color one or more highlighted bars

Pattern
shows the pattern palette to place a pattern on highlighted bars.

Label
displays the bar value at the top of all highlighted bars.

Graph→Control Charts

The **Control Charts** command in the **Graph** menu offers a variety of ways to analyze and monitor process data with control charts. It displays control charts that update dynamically as samples are received. Control charts are a graphical and analytical tool for deciding whether a process is in a state of statistical quality control. Shewhart control charts are broadly classified into control charts for variables and control charts for attributes. Moving average charts are special kinds of control charts for variables.

Types of Charts given by Graph→Control Charts

You select the type of chart you want from the popup menu on the Control Chart dialog:

Mean (mean)
displays subgroup means (averages).

R chart (Range)
displays subgroup ranges (maximum – minimum).

S chart (standard deviation)
displays subgroup standard deviations.

IM chart (individual measurement)
displays individual measurements.

MR chart (moving range)
displays moving ranges of two or more successive measurements. Moving ranges are computed for the number of consecutive measurements you enter.

UWMA chart (uniformly weighted moving average)
displays the average of the *w* most recent subgroup means, including the present subgroup mean. When you obtain a new subgroup sample, the next moving average is computed by dropping the oldest of the previous *w* subgroup means and including the newest subgroup mean.

EWMA chart (exponentially weighted moving average also referred to as a Geometric Moving Average [GMA) chart] displays the weighted average of all the previous subgroup means, including the mean of the present subgroup sample. The weights decrease exponentially going backward in time.

P chart (proportion of nonconforming or defective items in a subgroup sample)
displays the proportion of nonconforming (defective) items in a subgroup sample.

NP chart (number of nonconforming or defective items in a subgroup sample)
displays the number of nonconforming (defective) items in a subgroup sample.

C chart (proportion of nonconformities or defects subgroup sample)
displays the number of nonconformities (defects) in a subgroup sample that usually consists of one *inspection* unit.

U chart (number of nonconforming or defects per unit)
displays the number of nonconformities (defects) per unit in a subgroup sample with an arbitrary number of inspection units.

Platform-Options for Graph→Control Charts

The check-mark popup menu on the window border lists options that affect the whole report window. The following three options show for all control charts:

List of Charts
shows or hides all charts in a window.

Hide Options Buttons
shows or hides the popup menu icon beneath each plot. For example, if you want to use control charts for presentation purposes, you can hide the options buttons.

Connect thru Missing

connects points when some samples have missing values.

The control limits or limit parameter values must be in a JMP data table, referred to as the *Limits Table.* When you specify the **Control Charts** command, you can retrieve the Limits Table with the **Retrieve** button in the Control Charts dialog. The easiest way to create a Limits Table is to save results computed by the Control Charts platform. The **Save Limits** command in the save ($) menu creates the Limits Table automatically from the sample values. The type of data saved in the table varies according to the type of control charts in the analysis window.

Individual-Chart Options

The popup menu of chart options available to each chart individually appears when you click the icon beneath the chart or OPTION-click (or ALT-click) in the chart space. Other control charts can use all of the following options:

Box Chart

superimposes a boxplot on the subgroup means plotted in a Mean chart. The boxplot shows the subgroup maximum, minimum, 75th percentile, 25th percentile, and median. Markers for subgroup means show unless you deselect the **Show Points** option. The control limits displayed apply only to the subgroup mean.

Needle

connects plotted points to the center line with a vertical line segment.

Show Points

toggles between showing and not showing the points representing summary statistics. Initially, the points show. You can use this option to suppress the markers denoting subgroup means when the **Box Chart** display option is in effect.

Connect Points

toggles between connecting and not connecting the points.

Connect Color

enables you to choose the color of the line segments used to connect points. The color palette has the same choices as the color palette in the **Rows** menu.

Markers

lets you specify a common marker for all points on the plot. The marker palette has the same choices as the marker palette in the **Rows** menu. By default, the marker for all points is an **X**. If you select the first small dot at the upper left of the marker palette, JMP uses active row state markers from the first row of each sample in the data table.

Center Line

initially displays the center line in green. Deselecting **Center Line** removes the center line and its legend from the chart.

Control Limits

toggles between showing and not showing the chart control limits.

Lines Legend

toggles between showing and not showing the chart control line legend.

Tests

shows a submenu that enables you to choose which tests to mark on the chart when the test is positive. Tests apply only for charts whose limits are 3σ limits and whose control limits do not vary with subgroup samples. Tests 1 to 4 can be applied to attributes charts. Tests 1 to 8 can be applied to Mean charts and Individual Measurement charts. If tests do not apply to a chart, the **Tests** option is dimmed. These special tests are also referred to as the Western Electric rules.

Show Zones

toggles between showing and not showing the *zone lines* with the tests for special causes.

Appendix C
Calculator Functions

Overview

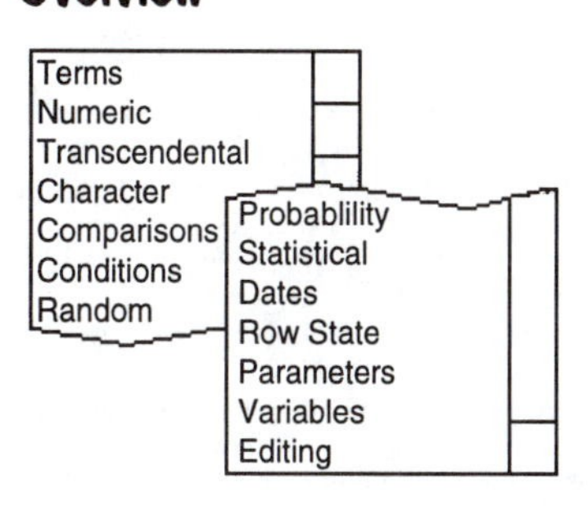

The *function browser* groups the calculator functions by the topics shown at left. To enter a function, highlight an expression and click any item in the function browser topics or use the function's keyboard shortcut. See the last section in this Appendix for a list of keyboard equivalents.

This appendix begins with an alphabetical list of the functions supported by the JMP calculator. Next there is a section for each of the following function groups with a brief description of its members:

- **Terms** are commonly used constants or functions such as e, π, row number, and others.
- **Numeric** functions include, **Absolute Value**, **Maximum**, **Minimum**, and others.
- **Transcendental** are standard functions such as logs and trigonometric functions.
- **Character** functions operate on character arguments such as the Substring and length.
- **Comparisons** are the standard logical comparisons such as less than, less than or equal to, not equal to, and so forth.
- **Conditions** are the logical functions **Not**, **And**, and **Or**. They also include programming-like functions such as **Assign**, **If/Otherwise**, **Match**, and **Choose**.
- **Random** functions generate uniform or normal random numbers. There is also a function to randomize the order of table rows.
- **Probability** functions compute probabilities and quantiles for the normal, Student's t, chi-square, and F distributions.
- **Statistical** lists a variety of functions that calculate standard statistical quantities such as the mean or standard deviation.
- **Dates** require arguments with the *date* data type, which is interpreted as the number of seconds since January 1, 1904. Date functions return values such as day or month of the year, compute dates, and can find intervals.
- **Row State** functions assign or detect row state status of color, marker, label, hidden, excluded, or selected.
- **Parameters** are named constants that you create and can use in any formula.
- **Variables** are named, temporary variables that you assign an expression and use in other expressions for a given column.
- **Editing** functions let you alter the size of the font used in the formula display, replace a formula with its derivative, or discard any changes made to a formula.

Appendix C Contents

Alphabetic List of Calculator Functions

This section lists the calculator function names spelled as they are found in the Calculator, followed by their respective function group in parentheses. In the sections that follow, the functions are referenced and described by function group. To look up a specific function, refer to the Table of Contents for the page number of its function group.

# of non-missing (Statistical)
a day (Dates)
a minute (Dates)
a week (Dates)
a year (Dates)
an hour (Dates)
Absolute Value (Numeric)
And (Conditions)
Assignments (Conditions)
Binomial (Random)
Cauchy (Random)
Ceiling (Numeric)
Chi-square Distribution (Probability)
Chi-square Quantile (Probability)
Char to Num (Transcendental)
Character to date (Dates)
Chi-Square Distribution (Probability)
Chi-Square Quantile (Probability)
Choose (Conditions)
Color (Row State)
Color of (Row State)
Common Log (Transcendental)
Concatenate (Transcendental)
Contains (Transcendental)
Convert to Uppercase (Transcendental)
Convert (Transcendental)
Cosine (Transcendental)
Count (Numeric)
Date (Dates)
Date to abbrev. character (Dates)
Date to long character (Dates)
Date to short character (Dates)
Day (Dates)
Day of week (Dates)
Day of year (Dates)
a Day (Dates)
e (Terms)
equal to (Comparisons)
Exclude (Row State)
Excluded (Row State)
Exponential (Random)
F Distribution (Probability)
F Quantile (Probability)
Floor (Numeric)
Gamma (Random)
Gamma (Transcendental)
Geometric (Random)
greater than (Comparisons)
greater than or equal to (Comparisons)
Hide (Row State)
Hidden (Row State)
an Hour (Dates)
Hue (Row State)
Hyperbolic Cosine (Transcendental)
Hyperbolic Sine (Transcendental)
Hyperbolic Tangent (Transcendental)
i (row #) (Terms)
If, otherwise (Conditions)
Incomplete Gamma (Transcendental)
Inverse Cosine (Transcendental)

Inverse Hyperbolic Cosine (Transcendental)
Inverse Hyperbolic Sine (Transcendental)
Inverse Hyperbolic Tangent (Transcendental)
Inverse Sine (Transcendental)
Inverse Tangent (Transcendental)
item (Transcendental)
Label (Row State)
Labeled (Row State)
Larger Font (Editing)
Length (Transcendental)
less than (Comparisons)
less than or equal to (Comparisons)
Log Gamma (Transcendental)
Marker (Row State)
Marker of (Row State)
Match (Conditions)
Mean (Statistical)
Minimum (Numeric)
a Minute (Dates)
Modulo (Numeric)
Month (Dates)
Munger (Character)
n (# of rows) (Terms)
Natural Log (Transcendental)
Negative Binomial (Random)
New Combination (Row State)
New Parameter (Parameters)
New Variable (Variables)
Normal Distribution (Probability)
Normal Quantile (Probability)
Normal (Random)
Not (Conditions)
not equal to (Comparisons)
Num to Char (Transcendental)
number of non-missing (Statistical)
Of (Statistical)
Or (Conditions)
Parameter or New Parameter (Parameters)
pi or π (Terms)
Poisson (Random)
Product (Statistical)
Quantile (Statistical)
Random Number Seed (Random)
Revert (Editing)
Round (Numeric)
Select (Row State)
Selected (Row State)
Shade (Row State)
Shuffle (Random)
Sine (Transcendental)
Smaller Font (Editing)
Std. Deviation (Statistical)
Student's t Distribution (Probability)
Student's t Quantile (Probability)
Subscript (Terms)
substring (Character)
Sum (Statistical)
Take Derivative (Editing)
Tangent (Transcendental)
This row (Row State)
Today (Dates)
Triangular (Random)
Trim (Transcendental)
Uniform (Random)
Variable or New Variable (Variables)
a Week (Dates)
Week of year (Dates)
Word (Transcendental)
Year (Dates)
a Year (Dates)

Terms Functions

The **Terms** selector list consists of terms commonly used in formulas. To select from this list, highlight an expression and click on one of the following terms term:

π (written as pi under Windows)
: is the numeric constant pi, 3.14159265....

e
: is the numeric base of natural logarithms, 2.7182818....

i (row #)
: is the current row number when an expression is evaluated for that row. You can incorporate **i** in any expression, including those used as column name subscripts. The default subscript of a column name is **i** unless otherwise specified.

n (# of rows)
: is the total number of rows in the active data table.

Subscript
: enables you to use a column's value from a row other than the current row. Highlight a column name in the formula display and click **Subscript** to display the column's default subscript **i**. The subscript **i** shows highlighted and can be changed to any numeric expression. To remove a subscript from a variable in a formula, select the subscript and delete it. Then delete its missing box.

Numeric Functions

Numeric functions require numeric values as arguments and calculate numeric values as results. If you have numbers stored as characters and want to use them in a numeric function, first convert them to numbers with the **num-to-char** function found in the list of character functions. JMP supports the following numeric functions:

Floor
: returns the largest integer less than or equal to its argument.

Ceiling
: returns the smallest integer greater than or equal to its argument.

Round
: rounds the first argument to the number of decimal places given by the second argument.

Modulo
: returns the remainder when the second argument is divided into the first.

Absolute Value
: returns a positive number of the same magnitude as the value of its argument.

Maximum
: takes the maximum of its numeric or character arguments. **Maximum** ignores missing values. Use the insert and delete buttons in the calculator panel to add blank **Maximum** function arguments or remove unwanted arguments. The range of values can be computed as **Maximum–Minimum**.

Minimum
: takes the minimum of its numeric or character arguments. **Minimum** ignores missing values.

Count

creates a list of values beginning with the **from** value and ending with the **to** value. The number of steps specifies the number of values in the list between and including the **from** and **to** values. Each value determined by the first three arguments of the **count** function occurs consecutively the number of times you specify with the **time** argument. When the **to** value is reached, **count** starts over at the **from** value.

Transcendental Functions

The calculator supports the following logarithmic functions for any base, selected trigonometric functions and gamma functions.

Natural Log

calculates the natural logarithm (base e) of its argument.

Common Log

calculates a common logarithm (base 10). To change the default log base from 10 to another value, highlight the base value , type a value into constant entry box and press either the **Constant** button on the calculator or the ENTER key. Both the log base and the log argument can be numeric expressions.

Sine and **Cosine**

calculate the sine and cosine of their respective arguments given in radians.

Tangent

calculates the tangent of an argument given in radians.

Inverse Sine

Inverse Cosine

return the inverse sine (arcsine) and inverse cosine (arccosine) of their respective arguments. The returned value is measured in radians.

Inverse Tangent

returns the inverse tangent (arctangent) of its argument measured in radians.

Hyperbolic Sine

Hyperbolic Cosine

return the hyperbolic sine and hyperbolic cosine of their respective arguments.

Hyperbolic Tangent

returns the hyperbolic tangent of its argument.

Inverse Hyperbolic Sine

Inverse Hyperbolic Cosine

return the inverse hyperbolic sine and inverse hyperbolic cosine of their respective arguments.

Inverse Hyperbolic Tangent

returns the inverse hyperbolic tangent of its argument.

Gamma denoted by Γ(i), is defined by

$$\Gamma(i) = \int_0^\infty x^{i-1} e^{-x} dx$$

In JMP, this formula computes gamma for each row using i, which is the current row number. It is used in the formula for the gamma distribution and other probability distributions.

Incomplete Gamma

is the same as **Gamma** except the integral is incomplete and it has a second parameter called shape in the JMP formula. The incomplete gamma is written

$$\text{igamma}(a,i) = \frac{1}{\Gamma(a)} \int^{i} x^{a-1}$$

Log Gamma

is the natural log of the result of the gamma function evaluation. You get the same result using the natural log function in the calculator with the **Gamma** function. However, the **Log Gamma** function computes more efficiently than do the ln and the Ln **Gamma** functions together.

Squash

is an efficient computation of the function

$$\frac{1}{1+e^{\cdot}}$$

where x is any column, expression, or temporary variable.

Character Functions

Character functions accept character arguments or return character strings and can convert the data type of a value from numeric to character, or from character to numeric. The following list is in the order you find them listed in the **Character** function group in the calculator.

Munger munger(☐, starting at 1, find "find", replace with "replace")

computes new character strings from existing strings by inserting or deleting characters. It can also produce substrings, calculate indexes, and perform other tasks depending on how you specify its arguments. **Munger's** four arguments are

Search String is a character expression. **Munger** applies the other three arguments to this string to compute a result.

Offset or **Starting Position** is a numeric expression. If the offset is greater than the search string's length, **Munger** uses the string's length as the offset.

Find Value is a character or numeric expression. Use a character string as search criterion, or use a positive integer to represent the number of consecutive characters to find. If you specify a negative integer as the **Find Value, Munger** searches to the end of the string.

Replace String can be a string or a missing term. If it is a string, **Munger** replaces the search criterion with the **Replace String** to form the result. If it is a missing term, **Munger** calculates either a substring or an index depending on whether the **Find Value** is an integer or a string, respectively.

|| (concatenate character strings)

produces a string with the function's second character argument appended to the first.

Substring substring(☐, starting at 1, of length ☐)

function produces extracts the characters that are the portion of the first argument beginning at the position given by the second argument and ending with the value in the third argument. The first argument can be either a character column or a literal value. The starting argument and the length argument can be numbers of expressions that evaluate to numbers.

Contains □ contains □

returns the numeric position within the first argument of the second argument if it exists, and a zero otherwise.

Trim

produces a new character string from its argument, removing any trailing blanks.

Word and **Item** word(1, of □, delimited by " ") and item(1, of □, delimited by ",")

both extract a word from a character string. The delimiter you enter as the last argument defines what a word is, and the first argument tells the position of the word you want. For example, to extract the last name in the following examples, use a blank as the delimiting character and ask for the second word.

The **item** function is different than the **Word** function because of the way they treat word delimiters. If a delimiter is found multiple times or you enter a delimiter with multiple characters, the **word** function treats them as a single delimiter. The **item** function uses each delimiter to define a new word position.

Length

function calculates the length of its argument.

Convert to Lowercase

converts any upper case character found in its argument to the equivalent lowercase character.

Convert to Uppercase

converts any lowercase character found in its argument to the equivalent lower case character.

Num to Char

produces a character string that corresponds to the digits in its numeric argument.

Char to Num

produces a numeric value that corresponds to its character string argument when the character string consists of numbers only. All other character data produce a missing value.

Comparisons

Comparisons are functions that compare the ranks of their arguments. Each comparison relationship evaluates as *true* or *false* based on numeric magnitudes or character rankings. **Comparisons** are most useful when you include them in conditional expressions, but they can also stand alone as numeric expressions. A true relationship evaluates as 1, and a false relationship evaluates as 0.

The relational symbols are

- < (less than)
- > (greater than)
- ≥ (greater than or equal to)
- = (equal to)
- ≤ (less than or equal to)
- ≠ (not equal to)

A relational symbol's arguments can be any two expressions. However, both arguments in a **Comparison** function must be of the same data type. Each data type accepts a missing term. A missing term is considered less than all other values.

The calculator uses the International Utilities package when comparing character strings. This package contains different rankings for each international character set and takes diacritical marks into consideration.

Conditions

JMP offers four types of conditional expressions (called *conditionals* for short) in the **Conditions** list. They are **Assignments, If, Match**, and **Choose**. These expressions let you build a sequence of clauses paired with *result expressions*. Constructing a sequence of clauses is the way you *conditionally* assign values to cells in a calculated column. The **Conditions** list contains the following functions and the standard logical operators, **Not, And,** and **Or**:

□⇐□
results □

The **Assignments** function is used to assign expressions to temporary variables, which are then used in a complex equation. This technique can greatly simplify building an equation.

{□, if □
□, otherwise

When you highlight an expression and click **If**, the calculator creates a new conditional expression with one **If** clause and one **otherwise**. You can add clauses if needed.

{match □:
□, when □
□, otherwise

When you highlight an expression and click **Match**, the calculator creates a new conditional expression with one **when** clause and one **otherwise** clause. The highlighted expression becomes the **Match** argument. To enter a conditional expression, fill the four missing terms with expressions. Use insert [^] and delete [delete] to add a new **when** clause or to remove clauses.

{choose □:
□, choice 1
□, otherwise

Choose is a special case of **Match** in which the arguments of **when** clauses are a sequence of integers starting at 1. **Choice** clauses replace **when** clauses in the **Choose** conditional. When you highlight an expression and click **Choose**, the calculator creates a new conditional expression with one **choice** clause and one **otherwise** clause. The highlighted expression becomes the **Choose** argument.

Not, **And**, and **Or**

When its argument is false, the **Not** function evaluates as 1. Otherwise, the **Not** function evaluates as 0. **And** evaluates as 1 when both of its arguments are true. Otherwise it evaluates as 0 . **Or** evaluates as 1 when either of its arguments is true. If both of its arguments are false, then the **Or** expression evaluates as 0.

Random Number Functions

Random number functions generate real numbers by effectively "rolling the dice" within the constraints of the specified distribution. The random number functions in JMP appear in formulas preceded by a "?" to indicate randomness. Each time you click **Evaluate** in the calculator window, these functions produce a new set of random numbers. JMP supports the following random number functions. The functions are listed in the order you find them in their calculator function group.

Uniform

?uniform generates random numbers uniformly between 0 and 1. This means that any number between 0 and 1 is as likely to be generated as any other. The result is an approximately even distribution. You can shift the distribution and change its range with constants. For example , **5 + ?uniform •20** generates uniform random numbers between 5 and 25.

Normal

?normal generates random numbers that approximate a normal distribution with a mean of 0 and standard deviation of 1. The normal distribution is bell shaped and symmetrical. You can modify the Normal function with constants to specify a normal distribution that has a different mean and standard deviation. For example, **5+?normal•2** generates a normal distribution with a mean of 5 and standard deviation of 2.

Exponential

?exponential generates a single parameter exponential distribution for the parameter lambda=1. You can modify the Exponential function to use a different lambda. For example, **.1•?exponential$^{-.1}$** generates an exponential distribution for lambda=.1. The exponential distribution is often used to model simple failure time data, where lambda is the failure rate.

Cauchy

?cauchy generates a Cauchy distribution with location parameter 0 and scale parameter 1. The Cauchy distribution is bell shaped and symmetric but has heavier tails than the normal distribution. A cauchy variate with location parameter alpha and scale parameter beta can be generated with the formula **alpha+beta•?cauchy**.

Gamma

?gamma(alpha ☐) gives a gamma distribution for the parameter, alpha, you enter as the function argument. The gamma distribution describes the time until the kth occurrence of an event. The gamma distribution can also have a scale parameter, beta. A gamma variate with shape parameter alpha and scale beta can be generated with the formula **beta •?gamma(alpha)**. If 2•alpha is an integer, a chi-square variate with 2•alpha degrees of freedom is generated with the formula **2•?gamma(alpha)**.

Triangular

?triangular(mid ☐) generates a triangular distribution of numbers between 0 and 1, with the midpoint you enter as the function argument. You can add a constant to the function to shift the distribution, and multiply to change its span.

Shuffle

?shuffle selects a row number at random from the current data table. Each row number is selected only once. When Shuffle is used as a subscript, it returns a value selected at random from the column that serves as its argument. Each value from the original column is assigned only once as **Shuffle**'s result.

Poisson

?poisson(lambda □) generates a Poisson variate based on the value of the parameter, lambda, you enter as the function argument. Lambda is often a rate of events occurring per unit time or unit of area. Lambda is both the mean and the variance of the Poisson distribution.

Binomial

?binomial(□ , probability □) returns random numbers from a binomial distribution with parameters you enter as function arguments. The first argument is n, the number of trials in a binomial experiment. The second argument is p, the probability that the event of interest occurs. When n is 1, the binomial function generates a distribution of Bernoulli trials. For example, n=1 and p=.5, gives the distribution of tossing a fair coin. The mean of the binomial distribution is np, and variance is $np(1-p)$.

Geometric

?geometric(probability □) returns random numbers from the geometric distribution with the parameter you enter as the function argument. The parameter, p, is the probability that a specific event occurs at any one trial. The number of trials until a specific event occurs for the first time is described by the geometric distribution. The mean of the geometric distribution is $1/p$, and the variance is $(1-p)/p^2$.

Negative Binomial

?negBinomial(□ , probability □) generates a negative binomial distribution for the parameters you enter as function arguments. The first parameter is the number of successes of interest (r) and the second argument is the probability of success (p). The random variable of interest is the number of failures that precede the rth success. In contrast to the binomial variate where the number of trials is fixed and the number of successes is variable, the negative binomial variate is for a fixed number of successes and a random number of trials. The mean of the negative binomial distribution is $(r(1-p))/p$ and the variance is $(r(1-p))/p^2$.

Random Number Seed

This function lets you start a random number sequence with a seed you specify. To use the Random Number Seed function, assign it a value using the Assignment function found in the Conditions functions, and use the random number function you want as the results clause of the Assignment function. This example uses the number 1234567 as the seed to generate a sequence of uniform random numbers.

$$\text{seed} \Leftarrow \begin{cases} 1234567, & \text{if } i=1 \\ \text{seed}, & \text{otherwise} \end{cases}$$

results ?uniform

Probability Functions

Probability functions calculate quantiles and probabilities for normal, chi–square, Student's t, and F distributions.

Normal Distribution accepts a quantile argument. It returns the probability that an observation from the standard normal distribution is less than or equal to the specified quantile.

Normal Quantile accepts a probability argument p. It returns the pth quantile from the standard normal distribution.

Chi–Square Distribution
accepts three arguments: a quantile, a degrees-of-freedom, and a noncentrality parameter. It returns the probability that an observation from the chi-square distribution with the specified noncentrality parameter and degrees of freedom is less than or equal to the given quantile. The **Chi–square Distribution** function accepts integer and noninteger degrees of freedom. It is centered at 0 by default. The **Chi–Square Quantile** function is the inverse of the **Chi–Square Distribution** function.

Chi–Square Quantile
accepts three arguments: a probability p, a degrees of freedom, and a noncentrality parameter. It returns the pth quantile from the chi–square distribution with the specified noncentrality parameter and degrees of freedom. The **Chi–Square Quantile** function accepts integer and noninteger degrees of freedom. It is centered at 0 by default. The **Chi–Square Distribution** function is the inverse of the **Chi–Square Quantile** function.

Student's t Distribution
accepts three arguments: a quantile, a degrees of freedom, and a noncentrality parameter. It returns the probability that an observation from the Student's t distribution with the specified noncentrality parameter and degrees of freedom is less than or equal to the given quantile. The **Student's t Distribution** function accepts integer and noninteger degrees of freedom. It is centered at 0 by default. The **Student's t Quantile** function is the inverse of the **Student's t Distribution** function.

Student's t Quantile
accepts three arguments: a probability p, a degrees of freedom, and a noncentrality parameter. It returns the pth quantile from the Student's t distribution with the specified noncentrality parameter and degrees of freedom. The **Student's t Quantile** function accepts integer and noninteger degrees of freedom. It is centered at 0 by default. The **Student's t Distribution** function is the inverse of the **Student's t Quantile** function.

F Distribution
accepts four arguments: a quantile, a numerator and denominator degrees of freedom, and a noncentrality parameter. It returns the probability that an observation from the F distribution with the specified noncentrality parameter and degrees of freedom is less than or equal to the given quantile. The **F Distribution** function accepts integer and noninteger degrees of freedom. By default it is centered at 0 and has 1 numerator degree of freedom. The **F Quantile** function is the inverse of the **F Distribution** function.

F Quantile
accepts four arguments: a probability p, a numerator and denominator degrees of freedom, and a noncentrality parameter. It returns the pth quantile from the F distribution with the specified noncentrality parameter and degrees of freedom.

Statistical Functions

The calculator evaluates statistical functions differently from other functions. Most functions evaluate data for the current row only. However, all statistical functions require a set of values upon which to operate.

Note: The **Sum** and **Product** functions always evaluate for an explicit range of column values. All other statistical functions *always* evaluate for $i = 1$ to n values *on every row*.

Except for the **# of non–missing** function, statistical functions apply only to numeric data. The calculator excludes missing numeric values from its statistical calculations.

# of non–missing

counts the number of *non*–missing values in the column you specify. A *missing numeric value* occurs when a cell has no assigned value or as the result of an invalid operation (such as division by zero). Missing character values are null character strings. In formulas for row state columns, an excluded row state characteristic is treated as a missing value. The calculator interprets other missing values according to their data types.

Quantile

computes the value at which a specific percentage of the values is less than or equal to that value. For example, the value calculated as the 50% quantile, also called the *median*, is greater than or equal to 50% of the data. Half of the data values are less than the 50*th* quantile.

The **Quantile** function's subscript, p, represents the quantile percentage divided by 100. The 25% quantile, also called the lower quartile, corresponds to $p = .25$, and the 75% quantile, called the upper quartile, corresponds to $p = .75$. The default value of p is 0.5 (the median). Note that $p = 0$ gives the minimum of a column and $p = 1$ returns the maximum.

The calculator computes a quantile for a column of N nonmissing values by arranging the values in ascending order. The subscripts of the sorted column values, $y_1, y_2, \ldots, y_N$, represent the ranks in ascending order. The pth quantile value is calculated using the formula $I = p \bullet (N+1)$ where p is the quantile and **N** is the total number of nonmissing values. If I is an integer, then the quantile value is $y_p = y_i$. If I is not an integer, then the value is interpolated by assigning the integer part of the result to i and the fractional part to f, and by applying the formula

$$q_p = |1 - f)y_i + (f)y_{i+1}$$

Mean

calculates the mean (or arithmetic average) of the numeric values identified by its argument. The mean of a variable, y, is denoted by $\overline{y}$ and computed according the formula shown to the right.

$$\overline{y} = \frac{\sum_{i=1}^{n} y_i}{N}$$

Std. Deviation

measures the spread around the mean of the distribution identified by its argument. The standard deviation, s, of a variable, y, is computed by the formula shown here.

$$s = \sqrt{s^2}, \text{ where } s^2 = \frac{\sum_{i=1}^{N} |y - \overline{y})^2}{N-1}.$$

Sum

uses the summation notation shown above. To calculate a sum, replace the missing term with an expression containing the index variable j. **Sum** repeatedly evaluates the expression for j = 1, j = 2, through j = n and then adds the nonmissing results together to determine the final result. The

Sum function uses the summation notation shown above. To calculate a sum, replace the missing term with an expression containing the index variable j. **Sum** repeatedly evaluates the expression for j = 1, j = 2, through j = n and then adds the nonmissing results together to determine the final result. You can replace **n**, the number of rows in the active spreadsheet, and the index constant, 1, with any expression appropriate for your formula.

Product

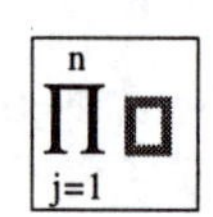

uses the notation shown to the right. To calculate a product, replace the missing term with an expression containing the index variable j. **Product** repeatedly evaluates the expression for j = 1, j = 2, through j = n and multiplies the nonmissing results together to determine the final result. You can replace **n**, the number of rows in the active spreadsheet and the index constant, 1, with any expression appropriate for your formula.

Index

works with **Sum** and **Product**, usually as a column subscript (see **Terms**). The calculator matches a new index to the first summation or product index variable it finds while searching outward from the highlighted expression. Each time you choose **Index** while one expression remains highlighted, the calculator advances it to the next index out.

Sometimes, it is useful to designate an expression containing a summation or product as the argument of a summation or product function. This is called *nesting*. The calculator automatically assigns a unique index variable for each Σ and Π, assigning *j* as the left-most index and reassigning the existing indexes as necessary. All corresponding index variables within the formula also adjust as necessary.

Note: When columns are arguments of a summation or a product function they are automatically assigned the index of the innermost Σ or Π. The **Index** function can be applied to the column subscript to choose an index variable at another nesting level as described above.

Of

is usually used with any of the statistical functions, which enables you to compute statistics across columns. For each row,

std of(*hist0*, *hist1*, *hist3*, *hist5*)

computes the standard deviation of the variables in the argument list.

To create a list of arguments for the **Of** operator, use either the COMMA key or the insert clause button on the calculator panel. Use the DELETE key or the delete clause button to remove unwanted arguments.

Date & Time Functions

JMP stores dates and times in numeric columns, using the Macintosh standard of the *number of seconds since January 1, 1904*. When a column has date values, you can assign a date format to that column using the **Date & Time** format popup menu in the Column Info dialog so that they display in a familiar form. The calculator supports JMP dates with the following functions:

a Minute, an Hour, a Day, a Week, a Year

converts from the units of the function name to the equivalent number of seconds. The argument must be a number or numeric expression.

Today
returns the number of seconds between January 1, 1904, and the current date.

Date
accepts numeric expressions for day, month, and year and returns the associated JMP date.

Day, Month, Year
return the day of the month, the month as a number from 1 to 12, and a four-digit year, respectively. The argument for the **Day**, **Month**, and **Year** functions is interpreted as a JMP date.

Day of week, Day of year, Week of year
return the numeric values that represent their arguments. The argument for **Day of week, Day of year**, and **Week of year** functions is a JMP date. **Day of week** returns a number from 1 to 7 where 1 represents Sunday, **Day of year** returns the number of days from the beginning of the year, and **Week of year** returns a number from 1 to 52.

Date to short character, Date to long character, Date to abbrev. character
create character strings that are the formatted representation of the argument. The argument for these functions is a column that contains a formatted JMP date. The converted character string is no longer recognized as a JMP date.

Character to Date
returns the appropriate JMP date value for its argument, which is a date character string, such as "Wednesday, March 20, 1991," or any valid string recognized by your machine as a date.

Row State Functions

Formulas process row state data just as they process character and numeric data. There are six row state conditions called **Select**, **Hide**, **Exclude**, **Label**, **Color**, and **Marker**:

- **Select** rows to identify data applicable to JMP commands.
- **Hide** rows to hide points in report displays.
- **Exclude** rows to eliminate data from analysis calculations and displays.
- **Label** rows to identify points in plots.
- Use one or more of the 64 JMP colors to distinguish points in report displays.
- Use one or more of the 8 JMP markers to distinguish points in report displays.

A row can be assigned any combination of row states. A list of multiple row states is called a *row state combination*. Row state functions are described as follows:

New Combination
generates a row state combination with two arguments. The currently selected expression becomes the first argument when you choose **New Combination**. Replace each argument with an expression that evaluates to a row state.

Use the insert and delete buttons in the calculator panel to add more arguments or remove unwanted arguments.

row
denoted *row* in formulas, returns the active row state condition of the current row as true or false.

You can use this function to conveniently write conditional clauses that depend on the status of the current row.

Hue

function returns the color from the JMP Hue Map that corresponds to its integer argument. JMP hues are numbered 0 through 11, but larger integers are treated as *modulo 12.* The **Hue** function does not map to black, gray, or white. Hues of 0 and 11 map to red and magenta. The formula

$$\text{hue}\left(\frac{z-\text{quantile}_0\, z}{\text{quantile}_1\, z-\text{quantile}_0\, z}\bullet 12\right)$$

assigns row state colors in a chromatic spread based on the value of z. If you hold down the OPTION key (ALT key under Windows) and choose **Hue**, the following formula appears:

$$\text{hue}\left(\begin{array}{|l} \boldsymbol{min1}\Leftarrow\text{quantile}_0\, z \\ \boldsymbol{max1}\Leftarrow\text{quantile}_1\, z \\ \text{results}\ \dfrac{z-\boldsymbol{min1}}{\boldsymbol{max1}-\boldsymbol{min1}} \end{array}\bullet 12\right)$$

This formula assigns row state colors in the same way as the first formula but creates special variables *minl* and *maxl* to form a simplified results formula. Special variables are discussed later in this Appendix.

Shade

assigns five shade levels to a color or hue. A shade of –2 is darkest and shade of + 2 is lightest. A shade of zero is a pure color. If you hold down the OPTION key and choose **Shade**, the following formula appears:

$$\text{shade}\left(\begin{array}{|l} \boldsymbol{min2}\Leftarrow\text{quantile}_0\, z \\ \boldsymbol{max2}\Leftarrow\text{quantile}_1\, z \\ \text{results}\ \dfrac{z-\boldsymbol{min2}}{\boldsymbol{max2}-\boldsymbol{min2}} \end{array}\bullet 5-2\right)$$

This formula assigns shade values based on the value of z. However, to assign all shades of all the colors in the colors palette, you need to use the **Hue** and **Shade** assignments together as follows:

$$\ll \text{hue}\left(\begin{array}{|l} \boldsymbol{min2}\Leftarrow\text{quantile}_0\, z \\ \boldsymbol{max2}\Leftarrow\text{quantile}_1\, z \\ \text{results}\ \dfrac{z-\boldsymbol{min2}}{\boldsymbol{max2}-\boldsymbol{min2}} \end{array}\bullet 12\right),\ \text{shade}\left(\begin{array}{|l} \boldsymbol{min1}\Leftarrow\text{quantile}_0\, z \\ \boldsymbol{max1}\Leftarrow\text{quantile}_1\, z \\ \text{results}\ \dfrac{z-\boldsymbol{min1}}{\boldsymbol{max1}-\boldsymbol{min1}} \end{array}\bullet 5-2\right)\gg$$

The formula above uses the **Combination** function described at the beginning of this section. The first argument in the **Combination** function is OPTION-Hue (or ALT-**Hue**) for the variable x, and the second argument is OPTION-Shade (or ALT-**Shade**) for y.

Color

returns the color from the JMP Color Map that corresponds to its integer argument. JMP colors are numbered 0 through 12, but larger integers map to color indices are treated as *modulo12.* Zero maps to black.

Marker

returns markers from the JMP Marker Map that correspond to its integer argument. Markers are numbered 0 through 7, but larger integers that map to marker indices are treated as *modulo 8.*

Exclude, Label, Hide, and Select

accept expressions that evaluate as either 1 or 0 (true or false) as an argument. Enter 1 to activate one of these row states or 0 to deactivate a row state.

Color of

accepts a row state argument and returns a number from the JMP Color Map that corresponds to the color argument. Use the row state popup menu on the calculator panel to select a color argument for the **Color of** function.

Marker of

accepts a row state argument and returns a number from the JMP Marker Map that corresponds to the marker argument.

Excluded, Labeled, Hidden, and Selected

accept any row state expression and return a 1 if the row state is active or 0 if the row state is inactive. These characteristics are inactive by default.

Parameters

Parameters are named constants created in the calculator that can be used in any formula. Numeric parameters are most useful in formulas created for nonlinear fitting.

Clicking the **New Parameter** item at the top of the list brings up a dialog you use to assign a name, value, and data type to the new parameter. You can create as many parameters as you need. After a parameter is created, it shows at the bottom of the function browser parameter list.

You assign a data type to a column (numeric, character, or row state) with the **Data Type** popup menu in parameter's dialog. Click **Done** when the dialog is complete or **Cancel** to exit the dialog without creating a new parameter. To change parameter settings or remove a parameter, OPTION–click (ALT–click under Windows) the parameter name. This displays the dialog again, with the **Remove** button activated.

Parameters are added to formulas in much the same way that variables are added. To insert parameters into your formula, click the parameter name in the function browser list. Parameters are easy to recognize in formulas because they are displayed in **bold** type. For example, in the formula

$$\mathrm{b0} \bullet \left(1 - e^{-\mathrm{b1} \bullet x}\right)$$

b0 and b1 are user-defined parameters.

Parameter data types must be valid in the context of the expressions where they are used. If a parameter data type is invalid, an error message appears. When a parameter is used in a model for the nonlinear platform, the initial value or starting value should be given as the parameter value. After completing a nonlinear fit or after using the **Reset** button in the nonlinear control panel, the parameter's value is the most recent value computed by the nonlinear platform.

When you paste a formula with parameters into a column, the parameters are automatically created for that column unless it has existing parameters with the same names.

Variables

The Variables function lets you define temporary variables to use in expressions. Temporary variables exist only for the column in which they are defined.

Clicking **New Variable** brings up a dialog you use the dialog to assign a name and data type (Numeric, Character, or Row State) to the new variable. For example, if **Model** is a data table column and **Sigma** is a parameter, the formula for a Weibull loss function is

$$\begin{cases} \dfrac{Model}{\textbf{sigma}} - e^{\left(\frac{Model}{\textbf{sigma}}\right)} - \ln \textbf{sigma}, & \text{if } Censor \neq 0 \\ -e^{\left(\frac{Model}{\textbf{sigma}}\right)}, & \text{otherwise} \end{cases}$$

This complicated formula can be simplified. Choose **Assignments** from the **Conditions** function list and use the insert button on the calculator panel to create the assignment structure shown on the left in the example below. Then create two temporary variables, *t1* and *t2*, and use them as shown in second assignment function. The results expression is constructed by substituting *t1* and *t2* into the Weibull function, which gives the results expression shown on the right in the example below.

$$\begin{array}{l} \square \Leftarrow \square \\ \square \Leftarrow \square \\ \text{results } \square \end{array} \qquad \begin{array}{l} t1 \Leftarrow \dfrac{Model}{\textbf{sigma}} \\ t2 \Leftarrow e^{t1} \\ \text{results - } \begin{cases} t1 - t2 - \ln \textbf{sigma}, & \text{if } Censor \neq 0 \\ -t2, & \text{otherwise} \end{cases} \end{array}$$

Editing Functions

Editing features can help you build formulas with ease. Each feature is described as follows:

Revert
: discards all changes made to the formula after the data table was last saved.

Take Derivative
: takes the derivative of a formula with respect to any highlighted column or parameter within the formula. The derivative replaces the original formula in the formula display.

To take a derivative, first create a formula. Next, select any occurrence of a column or parameter within the formula. Click **Take Derivative** to replace the original formula by its derivative with respect to the selected term. For example, the derivative of

$$\text{b0} \bullet \left(1 - e^{-\text{b1} \bullet x}\right)$$

with respect to b1 shows as

$$\text{b0} \bullet \text{x} \bullet e^{-\text{b1} \bullet x}$$

Use the Undo command to revert to the original formula.

Smaller Font
decreases the type sizes in the formula display. Using **Smaller Font**, you can view a large formula without scrolling. You can also use **Smaller Font** to adjust a formula's size before copying it.

Larger Font
increases the type sizes in the formula display. Using **Larger Font**, you can get a better view of the formula. You can also use **Larger Font** to adjust a formula's size before copying it.

Note: The initial type style and size of characters in the formula display depend on the selection of fonts available in your machine's operating system. On the Macintosh the calculator first searches for the Times font and then the New York font. If neither font is available, formulas display in your system's default application font (usually Geneva). The calculator automatically chooses the largest type size available that is smaller than 24 point.

Appendix D
What's in JMP that's not in JMP IN?

JMP IN is a product exclusively for students. It is derived from the product JMP by omitting a number of features, as described in this appendix.

Restricted Use

Despite the differences with JMP, JMP IN is a big product that is worth hundreds of dollars in the commercial marketplace, but is sold at a fraction of that with the understanding that it is only for student use. We depend on student stores and booksellers to restrict the sales of JMP IN, to not sell it for commercial use. Please respect the restriction so that we can continue to serve students at a low price. If you want to use JMP professionally, please get the professional version of JMP.

Documentation

JMP has more documentation than JMP IN. This book, sold with JMP IN, is customized for students learning statistics. It omits details, especially for features we consider unlikely to be used by students taking statistics courses; this book focuses on statistical concepts.

Technical Support

The professional JMP product comes with technical support from the developers, SAS Institute. The student version, JMP IN, has more limited technical support provided from the distributor, Duxbury.

MacOS Interface Features

JMP IN does not support the following JMP system interface features on the Mac:

- **Apple Events** (other than the System 7 basic Core Suite): On the Macintosh, JMP can be scripted through Apple Events. This is how you can program or script JMP to do repetitive operations.
- **Quicktime**: On the Mac JMP can make movies of spinning plots and response surface contours.

- **Instrument**: On the Mac JMP can be wired to an instrument through the Comm Toolbox, so that, for example, you can hook up a digital instrument. JMP reads the data as it is collected and produces a real-time Shewhart control chart.
- **Clear Access:** On the Mac, JMP can interface with a product called Clear Access® to get data from other data bases.

Analysis Platforms

Fit Model: Standard Least Squares JMP IN does not do the exact p value for the Durbin-Watson statistic.

Fit Model Screening. JMP IN does not support Pareto effect charts, the Box-Meyer Bayes plots, or Box-Cox transformation plots.

Fit Model: JMP IN does not have this Log Variance analysis, which was an experimental feature in JMP to provide for maximum likelihood variance modeling through log-linear models.

Fit Model: Proportional Hazards, Parametric Survival, D-Optimal are removed .

Fit Model: The Clustering platform is not included with JMP IN. In JMP, it does both hierarchical and k-means clustering.

Fit Model: JMP IN supports the Kaplan Meier platform or the Cox Proportional Hazards Model. The parametric survival model can be done using the nonlinear platform. JMP IN does not support several other survival models that JMP does.

Graph Platforms

Ternary Plots: Removed from JMP IN.

Contour Plots: Removed from JMP IN.

Control Charts: JMP IN does not support the following three features: CUSUM control charts, periodograms, and variability charts.

Design of Experiments

JMP IN does not support :

- Cotter Designs.
- mixture designs.
- mixed level designs,
- 3-level fractional factorials.
- D-Optimal design search

References

Agresti, A. (1984), *Analysis of Ordinal Categorical Data*, New York: John Wiley and Sons, Inc.

Agresti, A. (1990), *Categorical Data Analysis*, New York: John Wiley and Sons, Inc.

Anderson, T.W. (1971), *The Statistical Analysis of Time Series*, New York: John Wiley and Sons.

Anscombe, F.J. (1973), American Statistician 27, 17–21.

Belsley, D.A., Kuh, E., and Welsch, R.E. (1980), *Regression Diagnostics*, New York: John Wiley and Sons.

Box G.E.P. and Jenkins (1976), *Time Series Analysis:Forecasting and Control*, San Francisco:Holden-Day.

Box, G.E.P., Hunter,W.G., and Hunter, J.S. (1978), *Statistics for Experimenters*, New York: John Wiley and Sons, Inc.

Cochran, W.G. and Cox,G.M. (1957) Experimental Designs, 2nd edition, New York: John Wiley and Sons.

Daniel, C. (1959), "Use of Half–normal Plots in Interpreting Factorial Two–level Experiments," *Technometrics*, 1, 311–314.

Eppright, E.S., Fox, H.M., Fryer, B.A., Lamkin, G.H., Vivian, V.M., and Fuller, E.S. (1972), "Nutrition of Infants and Preschool Children in the North Central Region of the United States of America," *World Review of Nutrition and Dietetics*, 14.

Eubank, R.L. (1988), *Spline Smoothing and Nonparametric Regression*, New York: Marcel Dekker, Inc.

Gabriel, K.R. (1982), "Biplot," *Encyclopedia of Statistical Sciences*, Volume 1, eds. N.L.Johnson and S. Kotz, New York: John Wiley and Sons, Inc., 263–271.

Hajek, J. (1969), *A Course in Nonparametric Statistics*, San Francisco: Holden-Day.

Hosmer, D.W. and Lemeshow, S. (1989), *Applied Logistic Regression*, New York: John Wiley and Sons.

Iman, R.L. (1995), *A Data-Based Approach to Statistics*, Belmont, CA:Duxbury Press.

Iman, R.L. and Conover, W.J. (1979), "The Use of Rank Transform in Regression," *Technometrics*, 21, 499-509.

John, P.M. (1971), *Statistical Design and Analysis of Experiments*, New York, Macmillan.

Kaiser, H.F. (1958), "The varimax criterion for analytic rotation in factor analysis" *Psychometrika*, 23, 187–200.

Kemp, A.W. and Kemp, C.D. (1991) Weldon's dice data revisited. The American Statistician, 45 216-222.

Koehler, G. and Dunn, J.D. (1988), "The Relatiohnship Between Chemical Structure and the Logarithm of the Partition," Quantitative Structure Activity Relationships, 7.

Lenth, R.V. (1989), "Quick and Easy Analysis of Unreplicated Fractional Factorials," *Technometrics*, 31, 469–473.

Linnerud (see Rawlings (1988))

Miller, A.J. (1990), *Subset Selection in Regression*, New York: Chapman and Hall.

Moore, D.S. and McCabe, G. P. (1989), *Introduction to the Practice of Statistics*, New York and London: W. H. Freeman and Company.

Myers and McCaulley pp46.48 Briggs-Meyers test reference

Nelson, L. (1984), "The Shewhart Control Chart – Tests for Special Causes," *Journal of Quality Technology*, 15, 237–239.

Nelson, L. (1985), "Interpreting Shewhart X Control Charts," *Journal of Quality Technology*, 17, 114–116.

Rawlings, J.O. (1988), *Applied Regression Analysis: A Research Tool*, Pacific Grove CA: Wadsworth and Books/Cole.

Sall, J.P. (1990), "Leverage Plots for General Linear Hypotheses," *American Statistician*, 44, (4), 303–315.

SAS Institute (1986), *SAS/QC User's Guide, Version 5 Edition*, SAS Institute Inc., Cary, NC.

SAS Institute (1987), *SAS/STAT Guide for Personal Computers, Version 6 Edition*, Cary NC: SAS Institute Inc.

SAS Institute (1988), *SAS/ETS User's Guide, Version 6 Edition*, Cary NC: SAS Institute Inc.

SAS Institute (1989), "SAS/ Technical Report P–188: SAS/QC Software Examples, Version 6 Edition," SAS Institute Inc., Cary, NC.

SAS Institute Inc. (1989), *SAS/STAT User's Guide, Version 6, Fourth Edition, Volume 2*, Cary, NC: SAS Institute Inc., 1165-1168.

Snedecor, G.W. and Cochran, W.G. (1967), *Statistical Methods*, Ames Iowa: Iowa State University Press.

Simpson, E.H., 1951, The interpretation of interaction in contingency tables, *JRSS* B13: 238-241.

Stigler, S.M. (1986), *The History of Statistics*, Cambridge: Belknap Press of Harvard Press

Neter, J. and Wasserman, W. (1974), Applied Linear Statistical Models, Homewood, IL: Richard D Irwin, Inc.

Theil and Fiebig, 1984, *Exploiting Continuity*, Cambridge Mass: Ballinger Publishing Co.

Tukey, J. (1953), "A problem of multple comparisons," Dittoed manuscript of 396 pages, Princeton University.

Tversky and Gilovich, 1989, "The Cold Facts About the Hot Hand in Basketball," *CHANCE*, 2, 16-21.

Wardrop, Robert 1995, "Simpson's Paradox and the Hot Hand in Basketball", *American Statistician*, Feb 49:1, 24-28.

Yule, G.U., 1903, "Notes on the theory of association of attrbiutes in statistics," *Biometrika* 2: 121-134.

Index

A

B

C

D

E

F

G

H

I

J

K

L

M

N

S

T

U

V

W

X

Z

Technology License Notices

JMP software contains portions of the file translation library of MacLinkPlus, a product of DataViz Inc., 55 Corporate Drive, Trumbull, CT 06611, (203) 268-0030.

JMP for the Power Macintosh was compiled and built using the CodeWarrior C compiler from MetroWorks Inc.